Lecture Notes in Bioinformatics 3240

Subseries of Lecture Notes in Computer Science

T0181285

Inge Jonassen   Junhyong Kim (Eds.)

Algorithms in Bioinformatics

4th International Workshop, WABI 2004
Bergen, Norway, September 17-21, 2004
Proceedings

 Springer

Volume Editors

Inge Jonassen
University of Bergen
Department of Informatics and Computational Biology Unit, HIB
5020 Bergen, Norway
E-mail: Inge.Jonassen@ii.uib.no

Junhyong Kim
University of Pennsylvania
Goddard Laboratories, Biology Dept., School of Arts and Sciences
415 S University Avenue, Philadelphia, PA 19104, USA
E-mail: junhyong@sas.upenn.edu

Library of Congress Control Number: 2004111461

CR Subject Classification (1998): F.1, F.2.2, E.1, G.1, G.2, G.3, J.3

ISSN 0302-9743
ISBN 3-540-23018-1 Springer Berlin Heidelberg New York

Springer is a part of Springer Science+Business Media

springeronline.com

© Springer-Verlag Berlin Heidelberg 2004
Printed in Germany

Typesetting: Camera-ready by author, data conversion by Olgun Computergrafik
Printed on acid-free paper   SPIN: 11319306   06/3142   5 4 3 2 1 0

Preface

It gives us great pleasure to present the proceedings of the 4th Workshop on Algorithms in Bioinformatics (WABI 2004) which took place in Bergen, Norway, September 17–21, 2004. The WABI 2004 workshop was part of a five-conference meeting, which in addition to WABI, included ESA, WAOA, IWPEC, and ATMOS, hosted by the University of Bergen, Norway. See http://www.ii.uib.no/algo2004/ for more details.

The Workshop on Algorithms in Bioinformatics covers research on all aspects of algorithmic work in bioinformatics. The emphasis is on discrete algorithms that address important problems in molecular biology. These are founded on sound models, are computationally efficient, and have been implemented and tested in simulations and on real datasets. The goal is to present recent research results, including signficant work in progress, and to identify and explore directions of future research.

Original research papers (including significant work in progress) or state-of-the-art surveys were solicited on all aspects of algorithms in bioinformatics, including, but not limited to: exact and approximate algorithms for genomics, genetics, sequence analysis, gene and signal recognition, alignment, molecular evolution, phylogenetics, structure determination or prediction, gene expression and gene networks, proteomics, functional genomics, and drug design.

We received 117 submissions in response to our call for papers, and were able to accept 39 of these. In addition, WABI hosted one invited distinguished lecture, given to the entire ALGO 2004 conference, by Dr. Marie France Sagot of the INRIA Rhône-Alpes laboratories in France.

We would like to sincerely thank all the authors of sumbitted papers, and the participants of the workshop. We also thank the program committee for their hard work in reviewing and selecting the papers for the workshop. We were fortunate to have on the program committee the following distinguished group of researchers:

Pankaj Kumar Agarwal (Duke University)
Amihood Amir (Bar-Ilan University)
Alberto Apostolico (Purdue University)
Gary Benson (MSSN, New York)
Alvis Brazma (EMBL-EBI, UK)
Olivier Gascuel (LIRMM, Montpelier)
Raffaele Giancarlo (University of Palermo)
David Gilbert (University of Glasgow)
Jan Gorodkin (KVL, Denmark)
Roderic Guigo (University of Pompeu Fabra)
Jacques van Helden (Université Libre de Bruxelles)
Daniel Huson (University of Tubingen)
Gregory Kucherov (Loria, France)
Nadia El-Mabrouk (University of Montreal)
Inge Jonassen (University of Bergen)

Junhyong Kim (University of Pennsylvania)
Jens Lagergren (KTH, Sweden)
Gad M. Landau (University of Haifa)
Thierry Lecroq (Université de Rouen)
Bernard Moret (University of New Mexico)
Vincent Moulton (University of Uppsala)
Roderic D.M. Page (University of Glasgow)
David Sankoff (University of Ottawa)
Joao Carlos Setubal (Virginia Polytechnic Institute)
Jens Stoye (University of Bielefeld)
Esko Ukkonen (University of Helsinki)
Lisa Vawter (Aventis Inc., USA)
Jaak Vilo (Egeen Inc., Estonia)
Alfonso Valencia (CNB-CSIC, Spain)
Martin Vingron (Max Planck Inst., Berlin)
Tandy Warnow (University of Texas)
Peter S. White (University of Pennsylvania)
Louxin Zhang (National University of Singapore)

We would also like to thank the co-reviewers who assisted the program committee members in their work: Ali Al-Shahib, Lars Arvestad, Vineet Bafna, Vikas Bansal, Ugo Bastolla, Ann-Charlotte Berglund, Allister Bernard, Laurent Brehelin, Dave Bryant, Trond Hellem Bø, Sergio Carvalho Jr., Robert Castel, Sergi Castellano, Benny Chor, Richard Christen, Matteo Comin, Richard Coulson, Eivind Coward, Miklós Csürös, Tobias Dezulian, Bjarte Dysvik, Ingvar Eidhammer, Isaac Elias, Eduardo Eyras, Pierre Flener, Kristian Flikka, Eva Freyhult, Ganesh Ganapathy, Clemens Groepl, Stefan Grunewald, Yann Guermeur, Michael Hallett, Sylvie Hamel, Chao He, Danny Hermelin, Pawel Herzyk, Matthias Höchsmann, Katharina Huber, Torgeir Hvidsten, Johan Kåhrström, Hans-Michael Kaltenbach, Michael Kaufmann, Carmel Kent, Mikko Koivisto, Tsvi Kopelowitz, Arnaud Lefebvre, Alice Lesser, Zsuzsanna Liptak, Olivier Martin, Gregor Obernosterer, Sebastian Oehm, Kimmo Palin, Kjell Petersen, Cinzia Pizzi, Mathieux Raffinot, Sven Rahmann, Pasi Rastas, Knut Reinert, Eric Rivals, Kim Roland Rasmussen, Mikhail Roytberg, Gabriella Rustici, Anastasia Samsonova, Erik Sandelin, Stefanie Scheid, Alexander Schliep, Beng Sennblad, Rileen Sinha, Örjan Svensson, Jinsong Tan, Gilleain Torrance, Aurora Torrente, Dekel Tsur, Juris Viksna, Alexey Vitreschak, Li-San Wang, Oren Weimann, R. Scott Winters, Peng Yin, Tomasz Zemojtel, Zhonglin Zhou, Michal Ziv-Ukelson.

We also would like to thank the WABI steering committee, Gary Benson, Olivier Gascuel, Raffaele Giancarlo, Roderic Guigo, Dan Gusfield, Bernard Moret, and Roderic Page, for inviting us to co-chair this program committee, and for their help in carrying out that task.

Finally we are grateful to Ole Arntzen at the Dept. of Informatics, University of Bergen, who helped with technical issues.

July 2004   Inge Jonassen and Junhyong Kim
                                                 WABI 2004 Program Co-chairs

Table of Contents

Reversing Gene Erosion – Reconstructing Ancestral Bacterial Genomes from Gene-Content and Order Data

Joel V. Earnest-DeYoung[1], Emmanuelle Lerat[2], and Bernard M.E. Moret[1]

[1] Dept. of Computer Science, Univ. of New Mexico, Albuquerque, NM 87131, USA
{joeled,moret}@cs.unm.edu
[2] Dept. of Ecology and Evolutionary Biology, Univ. of Arizona, Tucson, AZ 85721, USA
lerat@email.arizona.edu

Abstract. In the last few years, it has become routine to use gene-order data to reconstruct phylogenies, both in terms of edge distances (parsimonious sequences of operations that transform one end point of the edge into the other) and in terms of genomes at internal nodes, on small, duplication-free genomes. Current gene-order methods break down, though, when the genomes contain more than a few hundred genes, possess high copy numbers of duplicated genes, or create edge lengths in the tree of over one hundred operations. We have constructed a series of heuristics that allow us to overcome these obstacles and reconstruct edges distances and genomes at internal nodes for groups of larger, more complex genomes. We present results from the analysis of a group of thirteen modern γ-proteobacteria, as well as from simulated datasets.

1 Introduction

Although phylogeny, the evolutionary relationships between related species or taxa, is a fundamental building block in much of biology, it has been surprisingly difficult to automate the process of inferring these evolutionary relationships from modern data (usually molecular sequence data). These relationships include both the evolutionary distances within a group of species and the genetic form of their common ancestors.

In the last decade, a new form of molecular data has become available: gene-content and gene-order data; these new data have proved useful in shedding light on these relationships [1–4]. The order and the orientation in which genes lie on a chromosome change very slowly, in evolutionary terms, and thus together provide a rich source of information for reconstructing phylogenies. Until recently, however, algorithms using such data required that all genomes have identical gene content with no duplications, restricting applications to very simple genomes (such as organelles) or forcing researchers to reduce their data by equalizing the gene content (deleting all genes not present in every genome and all "copies" of each gene, e.g., using the *exemplar* strategy [5]). The former was frustrating to biologists wanting to study more complex organisms, while the latter resulted in data loss and consequent loss of accuracy in reconstruction [6].

Our group recently developed a method to compute the distance between two nearly arbitrary genomes [7, 8] and another to reconstruct phylogenies based on gene-content and gene-order in the presence of mildly unequal gene content [6]. In this paper, we bring together these methods in a framework that enables us to reconstruct the genomes

I. Jonassen and J. Kim (Eds.): WABI 2004, LNBI 3240, pp. 1–13, 2004.
© Springer-Verlag Berlin Heidelberg 2004

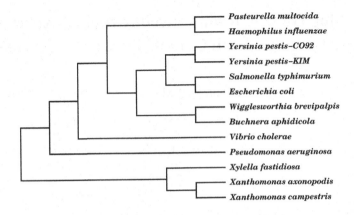

Fig. 1. The 13 gamma-proteobacteria and their reference phylogeny [9].

of the common ancestors of the 13 modern bacteria shown in Fig. 1 (from [9]). Gamma-proteobacteria are an ancient group of bacteria, at least 500 million years old [10]; the group includes endosymbiotic, commensal, and pathogenic species, with many species playing an important medical or economic role. The evolutionary history of the group is quite complex, including high levels of horizontal gene transfer [11–13] and, in the case of *B. aphidicola* and *W. brevipalpis*, massive levels of gene loss. These factors make a phylogenetic analysis of this group both interesting and challenging.

The rest of this paper is organized as follows. Section 2 presents the problem. Section 3 summarizes prior work on phylogenetic reconstruction from gene-content and gene-order data. Section 4 presents our framework for tackling the problem of ancestral genome reconstruction given a reference phylogeny; it is itself divided into three subsections, one on each of our three main tools: median-finding, content determination, and gene clustering. Section 5 discusses our approach to the testing of our framework: given that we have only one dataset and that ancestral genomes for that dataset are entirely unknown, our testing was of necessity based on simulations. Section 6 presents the results of this testing.

2   The Problem

We phrase the reconstruction problem in terms of a parsimony criterion:

> Given the gene orders of a group of genomes and given a rooted tree with these genomes at the leaves, find gene orders for the internal nodes of the tree that minimize the sum of all edge lengths in the tree.

The length of an edge is defined in terms of the number of evolutionary events (permissible operations) needed to transform the genome at one end of the edge into the genome at the other end. The permissible operations in our case are inversions, insertions (and duplications), and deletions; all operations are given the same cost in computing edge lengths. Restricting rearrangements to inversions follows from findings that the inversion phylogeny is robust even when other rearrangements, such as transpositions, are used in creating the data [14]. Our assignment of unit costs to all operations simply reflects insufficient biological knowledge about the relative frequency of these operations.

In our setting, one insertion may add an arbitrary number of genes to a single location, and one deletion may remove a contiguous run of genes from a single location, a convention consistent with biological reality. Gene duplications are treated as specialized insertions that only insert repeats. Finally, on each edge a gene can either be inserted or deleted, but not both; the same holds for multiple copies of the same gene. Allowing deletion and insertion of the same genes on the same edge would lead to biologically ridiculous results such as deleting the entire source genome and then inserting the entire target genome in just two operations.

Finding internal labels that minimize edge distances over the tree has been addressed by our group – this is the main optimization performed by our software suite GRAPPA [15]. However, even the most recent version of GRAPPA [6] is limited to relatively small genomes (typically of organellar size, with fewer than 200 genes), with modestly unequal content and just a few duplications. In stark contrast, the bacterial genomes in our dataset contain 3,430 different genes and range in size from 540 to 2,987 genes, with seven containing over 2,300 genes; moreover, these genomes contain a large number of duplications, ranging from 3% to 30% of the genome. Thus, in our model, most pairwise genomic distances are very large: a simple pairwise comparison along the tree of Fig. 1 indicates that some edges of the tree must represent at least 300 events. Such lengths are at least an order of magnitude larger than GRAPPA can handle. The large genome size, vastly unequal gene content, large number of duplications, and large edge lengths all combine to make this dataset orders of magnitude more difficult to analyze than previously analyzed genome sets.

3 Prior Work

A thorough recent review of the current work in phylogenetic reconstruction based on gene content and gene order appears in [16]; we review only the relevant points here.

The GRAPPA software package [17] computes internal labels in two phases. First, it initializes internal labels of the tree by some method. Then it iteratively refines labels until convergence: each newly labeled (or relabeled) node is pushed on a queue and, while the queue is not empty, the node at the head of the queue is removed, a new label computed for it (by computing the median of its three neighbors), and, if the new label reduces the total distance to the three neighbors, the existing label is replaced with the improved label and the three neighbors are placed on the queue. Thus GRAPPA relies on the computation of the *median* of three genomes, that is, a fourth genome which minimizes the sum of the number of operations needed to convert it into each of the three given genomes. GRAPPA finds optimal inversion medians with an algorithm that runs in worst-case exponential time, but finishes quickly when the edge lengths are small (10 to 40 operations per edge) [6, 18]. GRAPPA treats groups of genes that occur in the same order and orientation in all genomes as a single genetic unit; this *condensation* step reduces computational costs and does not affect the final result [19].

Our group developed a method to find the distance between two genomes with arbitrary gene content [7, 8]; this method relies on a *duplication-renaming* heuristic that matches multiple copies of genes between genomes and renames each pair and each unmatched copy to a new, unused gene number. Thus arbitrary genomes are converted into duplication-free genomes. We proved that, given two genomes with unequal gene content and no duplications, any optimal sorting sequence can be rearranged to contain first all insertions, then all inversions, and finally all deletions – a type of *normal form* for edit sequences [7]. (Deletions here are genes unique to the source genome, while insertions are genes found only in the target genome.) Using the genomes produced by the duplication-renaming method, an optimal inversion sequence can be calculated in time quadratic in the size of the consensus genomes [20, 21]. The number of deletions is calculated by counting the number of Hannenhalli-Pevzner cycles that contain deletions, as described in [22]. Finally, the number of insertions is estimated by calculating all possible positions in the source genome to which the inversion sequence could move insertions, then choosing the final position for each insertion that minimizes the number of groups of inserted genes.

In some genomes, especially bacterial ones, genes with similar function are often located together on one strand of a chromosome; these functional units are called *operons*. In bacteria, at least, while the order of genes in an operon may change, the gene content of the operon is much less likely to do so [23]. In gene-order data, an operon appears as a cluster of gene numbers with the same sign, with content, but not order, preserved across genomes. Heber and Stoye developed a linear-time *cluster-finding* algorithm to identify these operon-like clusters within equal-content genomes [24].

McLysaght *et al.* [4] reconstructed ancestral genomes for a group of poxviruses; she determined gene content by assuming that the phylogenetic tree contained a single point of origin for each gene family in the modern genomes. Each point of origin was assigned to that internal node which minimized the number of loss events necessary to achieve the gene content of the leaf genomes.

4   Designing an Algorithmic Framework

To address the problem of reconstructing ancestral genomes at the level of complexity of gamma-proteobacteria, we use condensation of gene clusters in order to reduce the size of the genomes, describe a procedure similar to that of McLysaght *et al.* to determine the gene content of every internal node, and present a new heuristic to compute the median of three very different genomes.

4.1   Medians

We use the queue-based tree-labeling heuristic described in Section 3. Since leaves contain the only labels known to be correct, we update the nodes in order of their distance from the leaves, as shown in Fig. 2. The heart of the top-level heuristic is the median computation. Exact median-finding algorithms are limited to small genomes, small edge lengths in the tree, and few changes in content – and none of these properties holds in our problem. We therefore pursue a simple heuristic inspired by geometry. The median of a triangle in the plane can be found by drawing a line from one vertex to the middle of the opposite segment, then moving two thirds of the way along this line. By analogy, we generate a sorting sequence from one genome to another (an edge of the triangle),

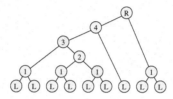

Fig. 2. Internal nodes ordered by their distance from the leaves. Nodes with lower indices will be labeled first; no label is generated for the root.

then choose a genome halfway along this sorting sequence and generate a new sorting sequence from it to the third genome, stopping one-third along the way.

We extend the method of Marron *et al.* [7] to enumerate all possible positions, orientations, and orderings of genes after each operation. Deleted genes at the endpoint of an inversion are moved to the other endpoint if doing so avoids "trapping" the deleted genes between two consensus genes that are adjacent in the target genome. Inserted genes are moved so as to remain adjacent to one of the two consensus genes between which they lie in the target genome. We can thus generate the genomes produced by "running" a portion of the sorting sequence, then use these intermediate genomes for the median heuristic just described, all in polynomial time.

This handling of inserted genes leads to an *overestimate* of the edit distance, which Marron *et al.* showed at most doubles the number of operations [7]. Their original method calculates all possible positions in the source genome to which the inversion sequence could move insertions and chooses the final position (for each insertion) to minimize the number of groups of inserted genes; it may *underestimate* the edit distance because the grouping of inserted genes may require an inversion to join inserted genes and simultaneously split deleted genes, which is not possible. We compared pairwise distances produced by our method and by theirs to get an upper bound on the overestimation: average and maximum differences between the overestimate and underestimate were 11.3% and 24.1%, respectively.

4.2   Gene Content

We predetermine the gene content of every internal node before computing any median: once the gene content of an internal node is assigned, it remains unchanged. Since the tree is rooted, we know the direction of time flow on each tree edge; we also assume that deletions are far more likely than insertions, The number of copies of each gene g is decided independently of all other genes; at internal node i, it is set to the maximum number of copies of g found in any of the leaves in i's subtree if: (i) there are leaves both inside and outside i's subtree that contain g; or (ii) there are leaves containing g in each half of i's subtree. Otherwise the number of copies of gene g in node i is set to zero.

This value can be calculated in $O(NG)$ time, where N is the number of nodes in the tree and G is the number of distinct genes in all the leaves, as follows. For each node in the tree, we determine the maximum number of copies of each gene from among the leaves of that node's subtree, using a single depth-first traversal. We use a second depth-first traversal to set the actual number of copies of each gene at each internal node. If

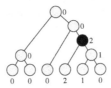

Fig. 3. The number of copies of a gene in internal nodes.

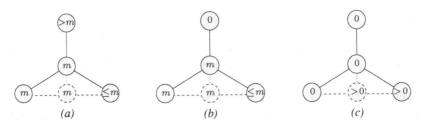

Fig. 4. Cases where the median and its neighbors have different numbers of copies of a gene. Solid lines are tree edges; dashed and dotted lines are fractions ($\frac{1}{2}$ and $\frac{1}{3}$, resp.) of sorting sequences.

either of the root's children has a subtree maximum of zero, then we set the root's actual number to zero as well. For each internal node other than the root, if its parent's actual number of copies is zero and at least one of its two children's subtree maximum is zero, then we set the number of copies for the gene to zero; otherwise we set the number of copies to the node's subtree maximum for the gene.

Internal nodes will thus possess at least as many copies of a gene as the majority consensus of their neighbors' gene contents. An internal node will always possess a copy of a gene if two or more of its neighbors do. (We consider the two children of the root to be neighbors.) Moreover, if a node is the nearest common ancestor of all genomes possessing the gene, it may have more copies of the gene than its parent and one of its children, as in the case of the black node in Fig. 3. The gene content of intermediate genomes along sorting sequences is a union of the gene contents of the starting genomes, because the sorting sequence of operations that we use always involves first insertions, then inversions, and finally deletions. Therefore, when calculating medians from sorting sequences, we face three cases in which the number of copies of a gene differ between the intermediate genome, the median genome, and the median's parent – see Fig. 4. In Fig. 4a, the intermediate genome has the same number of copies as the median, but fewer than the parent, as with the black node's right child in Fig. 3. Each copy in the parent that is not matched by the duplication-renaming algorithm will be excluded from the median genome. The case of Fig. 4b only arises when the median genome is the nearest common ancestor of all genomes containing the gene in question, as with the black node in Fig. 3. Genomes along the intermediate sequence have the same number of copies as the median, while the parent of the median contains no copy at all. Finally, the case of Fig. 4c can only arise when the right child of the median is the nearest common ancestor of all genomes containing the gene, as with the parent of the black node in Fig. 3.

Biologically, this process of finding which duplicates to include in the median corresponds to matching orthologous duplicates of each gene between genomes and to discarding unmatched paralogous duplicates. Since the original nucleotide sequences are abstracted away before the analysis begins, this ortholog matching is decided entirely on the basis of which other genes are located next to the different homologs. Fortunately, orthologs and paralogs that can be distinguished by a nucleotide-based analysis are assigned different gene numbers before our analysis begins. Therefore, our method represents a reasonable way to integrate both nucleotide and gene-order data in differentiating orthologous and paralogous homologs of genes.

4.3 Cluster Condensation

To extract information from larger and more complex biological datasets, we need fast algorithms with fast implementations; faster processing also enables a more thorough analysis and thus produces results of higher quality. The key factor here is the size of the genomes – their number is a much smaller issue. We thus developed a technique to identify and condense gene clusters in order to reduce the size of the genomes. Our approach generalizes that used in genomes with equal content [24]; in contrast, GRAPPA only condenses identical subsequences of genes, because it aims to preserve the identity of edit sequences. Our method allows the condensation of clusters based only on content (not order, at least as long as genes stay on the same strand) and also handles the difficult cases that arise out of unequal gene content (such as an insertion within a cluster).

To identify clusters, we first use the duplication-renaming technique of Marron et al. to create duplication-free genomes. After renaming, we remove any genes not present in all of the genomes under examination. This step creates a group of genomes with equal gene content. We then use the cluster-finding algorithm of Heber and Stoye [24] to find equivalent clusters of genes within the equal-content genomes. Once clusters are identified, each one is condensed out of the original genomes and replaced with a single marker (as if it were a single gene). In a set of genomes with unequal gene content, there can be genes inside a cluster that are not present in the corresponding equal-content genomes. We deal with these genes in one of two ways. If every occurrence of that gene is located inside the cluster in each of the genomes that possesses the gene, then the gene is condensed along with the rest of the cluster. Otherwise, the extra gene is moved to one side of the cluster and the cluster condensed. When a median genome is computed, a median for each cluster is also computed and each cluster's marker in the median genome is replaced with the cluster's median. At this point, if any extra genes moved to the side of the cluster are still beside it, they are moved back inside the cluster to a position similar to their original one.

4.4 Putting It All Together

Ancestral genome reconstructions are performed using these three main components. Initialization of the internal nodes of the tree is done from the leaves up by taking either the midpoint or one of the two endpoints (along the inversion portion of an edit sequence) of an internal node's two children and discarding any genes not allowed by the median gene content. This method accounts for all three of the cases in Fig. 4 and produces labels with the desired gene content. New medians are computed locally node by node in a postorder traversal of the tree, so as to propagate information from the leaves towards the root. Whenever a median is found that reduces the local score at a node, it immediately replaces the previous label at that node; that node and all its neighbors are then marked for further update.

5 Testing

We used our label reconstruction method on the bacterial dataset as well as on simulated datasets. With simulated datasets, we know the true labels for the internal nodes as well as the exact evolutionary events along each edge, so that we can test the accuracy of the reconstruction. The goal of our experiments was to generate datasets roughly comparable to our biological dataset so that our experimental results would enable us to predict a range of accuracy for the results on the biological dataset.

The simulated data were created using the tree of Fig. 1; edge lengths were assigned to the tree based on our best estimate of the edge lengths for the bacterial genomes. To keep the data consistent, edge lengths were interpreted as the number of operations per gene rather than as an absolute number, allowing us to use the same relative value for genomes of different sizes. The tree was labeled by first constructing a root genome. The number of genes g and the total size n of the root genome were set as variable user parameters. One of each gene from 1 to g was added to the root genome, after which $n - g$ additional genes were chosen uniformly at random in the range $[1, g]$ and added to the root genome. The root genome was then randomly permuted and each gene assigned a random sign. The other nodes were then labeled from the root by evolving the genomes down the tree according to the prescribed number of operations. The allowed operations were insertions, deletions, and inversions. Although the total number of operations was fixed, the proportion of each of the three types of operations was left as a variable parameter by setting the ratio of inversions to insertions to deletions. This mix of operations was used over all edges of the tree.

The characteristics of each type of operation were determined separately. The length of each inversion was chosen uniformly at random between 1 and half the size of the genome, with a start point chosen uniformly at random from the beginning of the genome to the size of the genome minus the length of the inversion. The average insertion length was set via a user parameter as a portion of the size of the root genome and was used unchanged over the entire tree, while the actual length of each insertion was drawn from a Poisson distribution with this expectation and its location was chosen uniformly at random from the beginning to the end of the genome. In moving from the root to the leaves, it was assumed that a particular gene could only be inserted along one edge of the tree – multiple insertions of the same gene, even along separate paths, were not allowed. The average deletion length was chosen as a user-specified portion of the genome from which genes were being deleted, thus varying from edge to edge as well as along each edge with each successive deletion, while the actual size of each deletion was drawn from a Poisson distribution with this expectation and with a start location chosen uniformly at random from the beginning of the genome to the size of the genome minus the length of the deletion. With the constant expected insertion length, genomes grow linearly in the absence of deletions, while, with a proportional expected deletion length, genomes shrink exponentially in the absence of insertions. When both insertions and deletions are used, genomes farther from the root tend towards a stable size. Along each edge, the prescribed number of insertions are performed first, then inversions, and finally deletions. Once all nodes have been been assigned genomes, the resulting leaf genomes are fed into our reconstruction procedure. The results of the reconstruction, in terms of gene content and gene order at each internal node, are compared with the "true" tree, i.e., that generated in the simulation.

We constructed trees using five different models: an "inversion-only" model, a "no-deletions" model with a 6:1 inversion-to-insertion ratio, a "no-insertions" model with a 6:1 inversion-to-deletion ratio, a "low-insertion/deletion" model with a 40:4:1 ratio of inversions to deletions to insertions, and a "high-insertion/deletion" model with a 30:10:3 ratio of inversions to deletions to insertions. The average insertion length was set to 2% of the root genome and the average deletion length to 3% of the local genome.

In order to test the efficacy of cluster condensation, we tested the technique on triples among the bacterial genomes that lie close to each other on the tree in Fig. 1. Triples were chosen by selecting internal nodes, then, for each of the three edges leading out from the internal node, by choosing a nearby leaf reachable by following the edge. For each set of three genomes, we measured the sum of the lengths of all clusters that were found.

6   Results

Our discussion and summaries of results refer to Fig. 5. Reconstruction of ancestral genomes for the bacterial genomes takes around 24 hours on a typical desktop computer. The midpoint-initialization proved quite strong: the only genomes to be updated in the subsequent local improvement procedure were the two children of the root (nodes 1 and 6 in Fig. 5), the only neighboring genomes in which one neighbor was not used to create the other. When we used endpoint-initialization, three internal nodes were up-

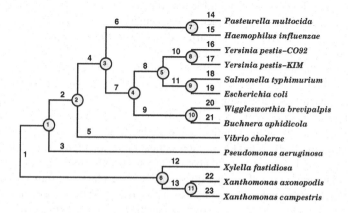

Fig. 5. The bacterial tree with numbered edges and internal nodes.

Table 1. Error Percentage in Tree Scores.

	Avg error	Min error	Max error
Inversion only	63.2%	57.3%	67.4%
No deletions	62.6%	54.8%	70.7%
No insertions	45.2%	37.6%	54.3%
Low insertion/deletion	56.4%	46.7%	64.8%
High insertion/deletion	34.9%	25.1%	46.4%

dated (nodes 3, 4, and 6 in Fig. 5) and the score of the entire tree was lowered by 2.8%. This finding may indicate that the initialization is very good, but it may also reflect the large numbers of local optima in the search space – a similar finding was reported for the simpler GRAPPA [19]. It should be noted that, when calculating medians, only four different midpoints in the child-to-child sorting sequence are used; from each of these midpoints, only three midpoints in the sorting sequence from the intermediate genome to the parent are tested. Thus we only perform a very shallow search and could easily miss a better solution. Interestingly, though, when we did a slightly more thorough search with ten midpoints from child to child and four midpoints from intermediate to parent, using endpoint-initialization, the tree score was slightly worse than in the shallower analysis, although the search, which took about 3.5 times longer, updated the same three internal nodes. Of course, this larger search remains very shallow; going beyond it will require a much more efficient implementation of the duplication-renaming heuristic of Marron *et al.* [7] – in our current version, it uses up over 90% of the computing time.

We simulated 100 labelings of the tree with a root genome size of 200 genes for each of the five previously described scenarios. Endpoint-initialization was used in all scenarios. The leaf genomes produced in our simulations ranged in size from 70 genes to 400 genes. We compared the predicted gene content of the internal nodes with the actual gene content. As expected (due to our restriction on generation), the predicted gene content always matched, except when a gene copy that was present at an internal node was lost in all leaves. Failure to detect this kind of missing gene is unavoidable in an analysis since the deletion from all leaves means that no historical record is left to attest the presence of that gene in ancestral genomes. When we compared the number of operations over all edges in reconstructed trees versus the original simulated tree, the score for the tree was fairly inaccurate, consistently overestimating the true score, as illustrated in Table 1. The rather tight distribution for the tree indicates that the error is not a random process, but a result of some aspect of our reconstruction method, one that may lend itself to reverse mapping.

We compared edge lengths in the reconstructed trees with those in the true trees by calculating the ratio of the lengths for each edge (Fig. 6). A perfect reconstruction would give a ratio of 1.0; as the figure shows, most ratios are higher, with edges further from leaves having larger ratios (and also larger variances). About half of the 23 edges are within a factor of two and another quarter are within a factor of four.

We calculated the number of operations needed to convert the reconstructed genome labels at internal nodes into the corresponding labels from the true tree. We normalized these values by dividing them by the size of the true tree genome, thus a perfect recon-

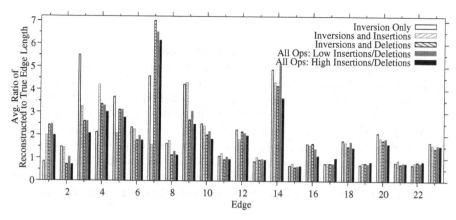

Fig. 6. Average ratio of reconstructed to true edge lengths for each edge, numbered as in Fig. 5.

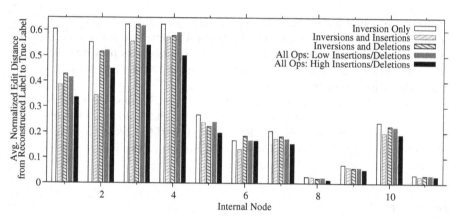

Fig. 7. Average normalized edit distance from reconstructed to true labels for each internal node of the tree, numbered as in Fig. 5.

struction would give edit distances of zero. Here again, internal nodes closer to leaves are closer to the true ancestral gene orders.

Finally, we tested cluster condensation on triples of closely-related bacterial genomes. The number of genes that fell into clusters, and thus the number of genes that could be condensed away, is a lower bound on the clustering potential in the actual tree. Condensation would remove the same number of genes from each genome, so the maximum possible condensation is determined by the smallest of the three genomes considered. In the cases we examined, it was possible to condense away on average 21% (ranging from 13% to 31%) of the size of the smallest genome. Computationally, however, the cluster condensation is heavily dependent on the duplication-renaming heuristic, the slowest of the various algorithmic components; thus, the benefits of working with smaller genomes will not be apparent until the time required to condense the genomes can be substantially reduced.

7   Conclusions

We have successfully produced a framework under which we are able to compute ancestral gene orders for modern bacteria. The number of operations over the tree is somewhat inaccurate in absolute terms, but rather accurate in relative terms – the error is a systematic bias towards overestimation. Accuracy is, unsurprisingly, greater for internal nodes and edges closer to the leaves (the modern data). We also have shown that, under certain simplifying assumptions, we are able to recover consistently the gene content of the ancestral genomes of simulated genomes. The size and complexity of the genomes mean that only a very shallow search of the space of possible ancestral genomes is possible: our results are undoubtedly heavily impacted by that problem, but we have pushed the size boundary for phylogenetic analysis with gene orders by an order of magnitude.

Acknowledgments

We thank our colleagues at the U. of Arizona for data, analysis, and advice: Nancy Moran (E. Lerat's postdoctoral advisor) and Howard Ochman and his postdoctoral student Vincent Daubin. We also thank Jens Stoye (U. Tübingen) for providing the source code to the cluster-finding program and Mark Marron and Krister Swenson (U. of New Mexico) for many useful discussions.

Research on this topic at the University of New Mexico is supported by the National Science Foundation under grants EF 03-31654, IIS 01-13095, IIS 01-21377, and DEB 01-20709 and by the NIH under grant 2R01GM056120-05A1 (through a subcontract to the U. of Arizona); research at the University of Arizona on this topic is supported by the NIH under grant 2R01GM056120-05A1.

References

1. Cosner, M., Jansen, R., Moret, B., Raubeson, L., Wang, L.S., Warnow, T., Wyman, S.: An empirical comparison of phylogenetic methods on chloroplast gene order data in Campanulaceae. In Sankoff, D., Nadeau, J., eds.: Comparative Genomics: Empirical and Analytical Approaches to Gene Order Dynamics, Map Alignment, and the Evolution of Gene Families. Kluwer Acad. Publ. (2000) 99–121
2. Waterston, R., et al.: Initial sequencing and comparative analysis of the mouse genome. Nature **420** (2002) 520–562
3. Hannenhalli, S., Chappey, C., Koonin, E., Pevzner, P.: Genome sequence comparison and scenarios for gene rearrangements: A test case. Genomics **30** (1995) 299–311
4. McLysaght, A., Baldi, P., Gaut, B.: Extensive gene gain associated with adaptive evolution of poxviruses. Proc. Nat'l Acad. Sci. USA **100** (2003) 15655–15660
5. Sankoff, D.: Genome rearrangement with gene families. Bioinformatics **15** (1999) 990–917
6. Tang, J., Moret, B., Cui, L., dePamphilis, C.: Phylogenetic reconstruction from arbitrary gene-order data. In: Proc. 4th Int'l IEEE Conf. on Bioengineering and Bioinformatics BIBE'04, IEEE Press (2004)
7. Marron, M., Swenson, K., Moret, B.: Genomic distances under deletions and insertions. In: Proc. 9th Int'l Conf. Computing and Combinatorics (COCOON'03). Volume 2697 of Lecture Notes in Computer Science., Springer Verlag (2003) 537–547

8. Swenson, K., Marron, M., Earnest-DeYoung, J., Moret, B.: Approximating the true evolutionary distance between two genomes. Technical Report TR-CS-2004-15, Univ. of New Mexico (2004)

9. Lerat, E., Daubin, V., Moran, N.: From gene trees to organismal phylogeny in prokaryotes: The case of the γ-proteobacteria. PLoS Biology **1** (2003) 101–109

10. Clark, M., Moran, N., Baumann, P.: Sequence evolution in bacterial endosymbionts having extreme base composition. Mol. Biol. Evol. **16** (1999) 1586–1598

11. Lawrence, J., Ochman, H.: Amelioration of bacterial genomes: Rates of change and exchange. J. Mol. Evol. **44** (1997) 383–397

12. Parkhill, J., et al.: Complete genome sequence of a multiple drug resistant *Salmonella enterica* serovar Typhi CT18. Nature **413** (2001) 848–852

13. Stover, C., et al.: Complete genome sequence of *Pseudomonas aeruginosa* PAO1, an opportunistic pathogen. Nature **406** (2000) 959–964

14. Moret, B., Tang, J., Wang, L.S., Warnow, T.: Steps toward accurate reconstructions of phylogenies from gene-order data. J. Comput. Syst. Sci. **65** (2002) 508–525

15. Bader, D., Moret, B., Warnow, T., Wyman, S., Yan, M.: (GRAPPA (Genome Rearrangements Analysis under Parsimony and other Phylogenetic Algorithms)) www.cs.unm.edu/~moret/GRAPPA/

16. Moret, B., Tang, J., Warnow, T.: Reconstructing phylogenies from gene-content and gene-order data. In Gascuel, O., ed.: Mathematics of Evolution and Phylogeny. Oxford Univ. Press (2004)

17. Tang, J., Moret, B.: Scaling up accurate phylogenetic reconstruction from gene-order data. In: Proc. 11th Int'l Conf. on Intelligent Systems for Molecular Biology ISMB'03. Volume 19 (Suppl. 1) of Bioinformatics. (2003) i305–i312

18. Moret, B., Siepel, A., Tang, J., Liu, T.: Inversion medians outperform breakpoint medians in phylogeny reconstruction from gene-order data. In Guigó, R., Gusfield, D., eds.: Proc. 2nd Int'l Workshop on Algorithms in Bioinformatics WABI'02. Volume 2452 of Lecture Notes in Computer Science., Springer Verlag (2002) 521–536

19. Moret, B., Wyman, S., Bader, D., Warnow, T., Yan, M.: A new implementation and detailed study of breakpoint analysis. In: Proc. 6th Pacific Symp. on Biocomputing (PSB'01), World Scientific Pub. (2001) 583–594

20. Bergeron, A.: A very elementary presentation of the Hannenhalli-Pevzner theory. In: Proc. 12th Ann. Symp. Combin. Pattern Matching (CPM'01). Volume 2089 of Lecture Notes in Computer Science., Springer Verlag (2001) 106–117

21. Bergeron, A., Stoye, J.: On the similarity of sets of permutations and its applications to genome comparison. In: Proc. 9th Int'l Conf. Computing and Combinatorics (COCOON'03). Volume 2697 of Lecture Notes in Computer Science., Springer Verlag (2003) 68–79

22. El-Mabrouk, N.: Genome rearrangement by reversals and insertions/deletions of contiguous segments. In: Proc. 11th Ann. Symp. Combin. Pattern Matching (CPM'00). Volume 1848 of Lecture Notes in Computer Science., Springer Verlag (2000) 222–234

23. Overbeek, R., Fonstein, M., D'Souza, M., Pusch, G., Maltsev, N.: The use of gene clusters to infer functional coupling. Proc. Nat'l Acad. Sci. USA **96** (1999) 2896–2901

24. Heber, S., Stoye, J.: Algorithms for finding gene clusters. In: Proc. 1st Int'l Workshop on Algorithms in Bioinformatics WABI'01. Volume 2149 of Lecture Notes in Computer Science., Springer Verlag (2001) 252–263

Reconstructing Ancestral Gene Orders Using Conserved Intervals

Anne Bergeron[1], Mathieu Blanchette[2], Annie Chateau[1], and Cedric Chauve[1]

[1] LaCIM, Université du Québec à Montreal, Canada
[2] McGill Center for Bioinformatics, Canada

Abstract. Conserved intervals were recently introduced as a measure of similarity between genomes whose genes have been shuffled during evolution by genomic rearrangements. Phylogenetic reconstruction based on such similarity measures raises many biological, formal and algorithmic questions, in particular the labelling of internal nodes with putative ancestral gene orders, and the selection of a good tree topology. In this paper, we investigate the properties of sets of permutations associated to conserved intervals as a representation of putative ancestral gene orders for a given tree topology. We define set-theoretic operations on sets of conserved intervals, together with the associated algorithms, and we apply these techniques, in a manner similar to the Fitch-Hartigan algorithm for parsimony, to a subset of chloroplast genes of 13 species.

1 Introduction

The information contained in the order in which genes occur on the genomes of different species has proved very useful for inferring phylogenetic relationships (see [18] for a review). Together with phylogenetic information, ancestral gene order reconstructions give some clues about the conservation of the functional organisation of genomes, towards a global knowledge of life evolution. With a few exceptions [16], phylogeny reconstruction techniques using gene order data rely on the definition of an evolutionary distance between two gene orders. These distances are usually computed as the minimal number of rearrangement operations needed to transform one genome into another, for a *fixed* set of rearrangement operations. Since most choices lead quickly to hard problems, the set of operations is usually restricted to reversals, translocations, fusions or fissions, in which case a linear-time algorithm exists ([1, 13, 14] and [3] for a review). However, this choice of rearrangement operations is more dictated by algorithm necessity than by biological reality, as rearrangements such as transpositions and inverted transpositions could be quite common in some genomes (see [6] for heuristics dealing with these types of rearrangements).

A family of phylogenetic approaches dubbed "distance-based" methods only rely on the ability to compute pair-wise evolutionary distances between extant species, which are then fed into an algorithm such as neighbor-joining (see [11] for a review) to infer a tree topology and branch lengths for the species considered. While these approaches have proved very useful for phylogenetic inference

I. Jonassen and J. Kim (Eds.): WABI 2004, LNBI 3240, pp. 14–25, 2004.

[22], they provide information neither about the putative ancestral gene orders nor about the evolutionary process that led to the extant species. In contrast, parsimony-based approaches attempt to identify the rearrangement scenario (including tree topology and gene orders at the internal nodes) that minimizes the number of evolutionary events required. This formulation usually leads to much more difficult computational problems [9], although good heuristics have been developed for breakpoint [5, 19, 21] and reversal [8, 17, 23] distances. It provides a candidate explanation, in terms of ancestral gene orders and rearrangements applied on them, for the modern gene orders. However, these methods only provide us with one (or a small number of) possible hypothesis about ancestral gene orders, with no information about alternate optimal or near-optimal solutions.

In this study, we develop the mathematical tools and algorithms required to describe and infer a set of likely ancestral gene orders at each internal node of a phylogenetic tree with a *given* topology. We use the notion of *conserved intervals*, introduced in [4], as a measure of similarity for sets of permutations representing genomes with equal gene contents. In short, a conserved interval is a generalization of the notion of gene adjacency, corresponding to a constraint on the ordering of the genes. This type of representation has several properties that make it is particularly useful in the study of gene order: (i) it is a compact representation of a rich set of gene orders (e.g. putative ancestral gene orders), (ii) it provides computationally tractable operations on these sets (some originally described in [4], others reported here), (iii) it is intimately related to the reversal distance computation [3], although it behaves well even in the presence of other types of intra-chromosomal rearrangements like transpositions and inverted transposition, and (iv) it is particularly effective at dealing with short rearrangement events, which seem to be the most common in mitochondrial and chloroplastic genomes [20].

In Section 2, we introduce the notion of conserved intervals and illustrate it using a small example. Section 3 reviews the main definitions and properties associated to conserved intervals, and Section 4 gives new fundamental results on the operations on sets of conserved intervals, together with the associated algorithms, in Section 5. In particular, we show how an algorithm, conceptually similar to the Fitch-Hartigan algorithm ([12, 15]) for character-based parsimony, can be build upon the defined set of operations. The output of this algorithm is a hypothesis regarding ancestral gene orders, in the form of a set of conserved intervals at each node of the tree. The results obtained on chloroplastic genomes, reported in Section 6, indicate that the algorithm seems effective at capturing specifically a set of likely ancestral gene orders.

2 Looking for Ancestors

We assume that gene orders are represented by signed permutations where each element corresponds to a different gene and its sign represents the gene orientation.

Definition 1 (Conserved interval [4]). *A conserved interval of a set of signed permutations is an interval* $[a, b]$ *such that a precedes b, or* $-b$ *precedes* $-a$, *in* *each permutation, and the set of elements, without signs, between a and b is the same in each permutation.*

Consider the following genomes P and Q represented by signed permutations on the set $\{1, 2, 3, 4, 5, 6\}$: $P = (1\ 2\ 3\ 4\ 5\ 6)$ and $Q = (1\ -2\ 3\ -5\ 4\ 6)$. The conserved intervals of P and Q are $I(\{P, Q\}) = \{[1, 3], [3, 6], [1, 6]\}$. A practical representation of conserved intervals is to choose one signed permutation of the set, box its elements, and join the extremities of conserved intervals that are not the union of smaller ones with larger boxes. For example, the conserved intervals of P and Q can be represented as:

$$I_1 = \boxed{1}\,\boxed{2}\,3\,\boxed{4}\,\boxed{5}\,\boxed{6}$$

We associate, to such a representation, the set of all signed permutations that share the same conserved intervals. Graphically, this set can be obtained by reversals and transpositions that do not "break" any box. The set $Perm(I_1)$ of signed permutations sharing the conserved intervals I_1 would thus contains 16 permutations, obtained by reversing elements 2, 4 or 5, or by transposing elements 4 and 5. (Note that the extreme points 1 and 6 are considered fixed.)

Suppose now that two other signed permutations are added in the set of genomes under study: $R = (1\ -2\ -3\ 5\ 4\ 6)$ and $S = (1\ -3\ 2\ 5\ -4\ 6)$. Their conserved intervals are represented as:

$$I_2 = \boxed{1}\,\boxed{-2}\,\boxed{-3}\,5\,\boxed{4}\,\boxed{6}$$

and the set of associated permutations contains also 16 permutations.

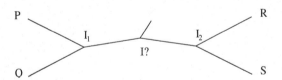

Fig. 1. A Tree Topology for the Permutations $\{P, Q, R, S\}$.

If we are given the tree topology of Fig. 1, it would seem natural to label the parent of $\{P, Q\}$ with I_1, and the parent of $\{R, S\}$ with I_2. Indeed, under most reasonable rearrangement scenarios, the ancestors are respectively in $Perm(I_1)$ and $Perm(I_2)$ [4]. What should be the label I of the ancestral node? Computing the conserved intervals of the permutations $\{P, Q, R, S\}$ yields the trivial interval $[1, 6]$. However, in this example, the two sets of signed permutations associated to I_1 and I_2 have a non-empty intersection consisting of the four permutations: $(1\ 2\ 3\ 5\ 4\ 6)$, $(1\ -2\ 3\ 5\ 4\ 6)$, $(1\ 2\ 3\ 5\ -4\ 6)$, and $(1\ -2\ 3\ 5\ -4\ 6)$.

Thus, an interesting label for the ancestral node could be the set of conserved intervals of these four permutations:

$$I = \boxed{1}\,\boxed{2}\,3\,5\,\boxed{4}\,\boxed{6}$$

Note that this set contains all conserved intervals of both sets I_1 and I_2, together with the adjacency $[3, 5]$. The distinctive characteristic of the two subgroups of the tree of Fig. 1 is the alternate ways in which the adjacency $[3, 5]$ is broken.

When the intersection of the two sets is empty, we will show, in Sections 4 and 5, that it is possible to keep a subset of each set of signed permutations, and then compute conserved intervals of the union of these sets.

3   Basic Properties of Conserved Intervals

Let G be a set of signed permutations, we will denote by $I(G)$ the set of conserved intervals of G. Sets of conserved intervals are highly structured, which was not readily apparent with the simple examples of Section 2. For example, consider the following set G of two signed permutations: $P =$ (1 2 3 4 5 6 7 8 9) and $Q =$ (1 -6 -5 3 4 -2 -8 -7 9). Then the set $I(G)$ is represented by the following diagram, based on the permutation P.

$$I(G) = \boxed{1}\;\boxed{2}\;\boxed{3}\;\boxed{4}\;\boxed{5}\;\boxed{6}\;\boxed{7}\;\boxed{8}\;\boxed{9}$$

A conserved interval that is not the union of smaller conserved intervals is called *irreducible*. For example, in $I(G)$, all intervals are irreducible except the interval $[2, 6]$. Irreducible intervals share at most one endpoint, as made precise by the following proposition:

Proposition 1 ([4]). *Let G be a set of signed permutations. Let $[a, b]$ and $[c, d]$ be two irreducible intervals of $I(G)$. Then $[a, b]$ and $[c, d]$ are either disjoint, nested with different endpoints, or overlapping on one element.*

Successive irreducible intervals that overlap on one element form *chains*. Chains are denoted by the successive common elements of the overlapping intervals, such as the chain $[2, 5, 6]$ in $I(G)$. It is easy to see that each conserved interval of a set of signed permutations is either irreducible, or is a chain.

Because the set of conserved intervals of a given set of signed permutations has some structural properties, a collection C of intervals is not necessarily the set of conserved intervals of a set of signed permutations. However, it is always possible to construct the smallest collection that contains C, and that is the set of conserved intervals of a set of signed permutations.

Definition 2 (Closure of a set of intervals). *Let U be a set of intervals of a signed permutation $P = (p_1, \ldots, p_n)$. The closure of U, denoted by U^*, is the smallest set of intervals containing U and such that, for any pair $([p_i, p_j], [p_k, p_l])$ of intervals in U^*, such that $i \leq k \leq j \leq l$, then $[p_i, p_k]$, $[p_k, p_j]$, $[p_j, p_l]$ and $[p_i, p_l]$ are in U^*, provided that they have more than one element.*

Consider the set of intervals $U = \{[1,3], [3,6], [1,5], [5,6]\}$ of the identity permutation. Its closure is given by: $U^* = \{[1,3], [1,5], [1,6], [3,5], [3,6], [5,6]\}$.

Given a set of intervals I, the maximal set of signed permutations that have all the conserved intervals of I is denoted by $Perm(I)$. Again, not all sets of signed permutations can be constructed in this way.

Definition 3 (Saturated set of permutations). *A set of signed permutations G is* saturated *if G is the set of signed permutations that have all the conserved intervals of $I(G)$, that is to say $G = Perm(I(G))$.*

For example, the set $\{(1 \ -2 \ 3 \ 5 \ 4 \ 6), (1 \ 2 \ 3 \ 5 \ -4 \ 6)\}$ is not saturated, because both permutations $(1 \ 2 \ 3 \ 5 \ 4 \ 6)$, and $(1 \ -2 \ 3 \ 5 \ -4 \ 6)$ share the same conserved intervals. These four permutations form a saturated set since they are the only ones that have the conserved intervals:

$$I = \boxed{1}\ \boxed{2}\ \boxed{3\ 5}\ \boxed{4}\ \boxed{6}$$

4   Operations on Sets of Conserved Intervals

We now turn to the problem of computing the conserved intervals of unions and intersections of sets of signed permutations. The first result is the basis of a linear-time algorithm to compute the conserved intervals of the union of two sets of signed permutations.

Theorem 1 (Conserved intervals of a union [4]). *Let G_1 and G_2 be two sets of signed permutations on the set $\{1, \ldots n\}$, then $I(G_1 \cup G_2) = I(G_1) \cap I(G_2)$.*

However, there is no such simple characterization of the conserved intervals of an intersection of arbitrary sets of signed permutations. In order to have a dual property, we must assume that the sets G_1 and G_2 are saturated, but this will be the case in the algorithms we describe in Section 5.

Theorem 2 (Conserved intervals of an intersection). *Let G_1 and G_2 be two saturated sets of signed permutations on the set $\{1, \ldots n\}$. If $G_1 \cap G_2 \neq \emptyset$. Then $I(G_1 \cap G_2) = (I(G_1) \cup I(G_2))^*$.*

Note that the right hand side of the above equation is well defined, since the intersection of G_1 and G_2 is not empty, thus all intervals of $I(G_1)$ and of $I(G_2)$ can be represented using the same permutation.

Testing whether $G_1 \cap G_2$ is empty is not elementary, and is at the heart of the algorithmic complexity of constructing intersections. The next definition introduces the basic concept of *filtering* a set of signed permutations with an interval.

Definition 4 (Filtering sets of permutations). *Let $[a, b]$ be an interval of a signed permutation P, and G a saturated set of signed permutations. The* filtered *set $G_{[a,b]}$ is the subset of all signed permutations of G that have the conserved interval $[a, b]$. The set of conserved intervals of $G_{[a,b]}$ is denoted by $I(G)_{[a,b]}$.*

For example, consider the following set of conserved intervals, and the corresponding saturated set G.

$$I = \boxed{1}\ \boxed{2\ 3}\ \boxed{4}\ \boxed{5}\ \boxed{6}\ \boxed{7}\ \boxed{8}.$$

Let $P = (1\ 3\ 2\ 4\ 5\ -7\ 6\ 8)$. Filtering G with the interval $[4, -7]$ of P yields the following set of conserved intervals:

$$I(G)_{[4,-7]} = \boxed{1}\ \boxed{\boxed{2}\ \boxed{3}}\ \boxed{\boxed{4}\ \boxed{5}\ \boxed{-7}}\ \boxed{6}\ \boxed{8}.$$

However, filtering G with the interval $[1, 3]$ of P would yield the empty set, since no permutation of G has the conserved interval $[1, 3]$.

Proposition 2. *Let G be a saturated set of signed permutations. Let $[a, b]$ and $[c, d]$ be two intervals. Then $G_{[a,b]}$ is saturated, and $(G_{[a,b]})_{[c,d]} = (G_{[c,d]})_{[a,b]}$.*

Theorem 3. *Let G_1 and G_2 be two saturated sets of signed permutations. Let $J_1, \ldots, J_k$ be the set of irreducible conserved intervals of G_1, then $G_1 \cap G_2 = \emptyset$ if and only if $(I(G_2))_{J_1,\ldots,J_k} = \emptyset$. Moreover, if $G_1 \cap G_2 \neq \emptyset$, then we have $I(G_1 \cap G_2) = (I(G_2))_{J_1,\ldots,J_k}$.*

Together with Proposition 2, this theorem yields an algorithm to compute the intersection of two saturated sets of signed permutations using successive filtering. Indeed, if there is a step in which filtering produces an empty result, then the intersection is empty.

However, even when the intersection is empty, there might still exists a non-empty subset of, say G_1, that shares conserved intervals with G_2. Such conserved intervals are likely to have been shared by a common ancestor. Some care must be taken in order to properly define these collections. Indeed sets of intervals can be *conflicting*:

Definition 5. *A set S of conserved intervals is* conflicting *with respect to a saturated set G of signed permutations if $G_S = \emptyset$ and $\forall I \in S, G_{S \setminus \{I\}} \neq \emptyset$.*

In Section 6 we will see that, when G_1 and G_2 are filtered with collections of conserved intervals in which conflicting subsets are removed, we can obtain ancestral gene orders that are extremely well-defined.

Conjecture 1. Let G_1 be a saturated set of signed permutations, and S be the set of irreducible intervals of G_1, then it is possible to identify, in polynomial time, all conflicting subsets of S with respect to a saturated set of signed permutations G_2.

5   Algorithms

We discuss now the two main algorithmic issues raised in the previous sections: filtering and ancestors labelling. We first describe how to represent a set of conserved intervals as a *PQ-tree*, then we outline a linear time filtering algorithm. Finally, we describe an ancestor labelling algorithm based on the principle of the Fitch-Hartigan parsimony algorithm.

Conserved Intervals and PQ-Trees. A *PQ*-tree is a data structure used to represent in a compact way a set of permutations [7]. Here we adapt this data structure to represent sets of conserved intervals. This idea was briefly introduced in [4].

We define a variant of *PQ-trees* as *ordered trees* with three types of nodes: n *leaves* that are labelled with signed elements of $\{1, \ldots, n\}$, and internal nodes that are either *round* or *square*. The root is always a square node, and all the children of a round (resp. square) node are square (resp. round) nodes or leaves. Moreover, among the children of a square node, the first and last are leaves and there cannot be two consecutive round nodes. It follows that the total number of nodes of a *PQ*-tree is linear in n. The relationship between PQ-trees and conserved intervals is as follows: a round node represents the free elements and conserved intervals that are inside a box of the box representation, and a square node represents a maximal chains of intervals. The children of a square node are either round nodes, or the endpoints of the irreducible intervals of the maximal chain it represents. See Fig. 2.

Fig. 2. Two representations of the same set of conserved intervals.

Enumerating, during a depth-first traversal, the leaves of a *PQ*-tree representing a set I of conserved intervals, gives a permutation in $Perm(I)$. Considering *PQ*-trees as ordered trees implies that there are as many different *PQ*-trees representing I as there are permutations in $Perm(I)$, and that each of these different trees, when traversed as described above, gives a different permutation of $Perm(I)$. Indeed, performing one of the following transformations on a *PQ*-tree T does not change the set I of conserved intervals it represents, but implies that the new ordered tree obtained represents a different permutation of $Perm(I)$: changing the sign of a leaf incident to a round vertex, reordering the children of a round vertex, reversing the order of the children of a square vertex (except the root), and changing in the same time the signs of all leaves present among these children. It is straightforward to design simple data structures to implement *PQ*-trees that allow to perform transformations of a node of a *PQ*-tree in constant time.

Filtering a Set of Conserved Intervals. We now outline the basic steps in the construction of the filtering algorithm[1].

Let T be a PQ-tree representing a set G of unsigned permutations on the set $\{1, \ldots n\}$ and $S \subseteq \{1, \ldots, n\}$ a given set of elements. The *reduction* of T over S yields the PQ-tree that represents the set G_S of all permutations of G in which

---

[1] The full technical details of the implementation of this algorithm will appear in [2].

the elements of S appear consecutively. A reduction can be computed in linear time with respect to n [7].

A conserved interval $[a, b]$ with inner elements $x_1, \ldots, x_k$ yields three sets that should appear consecutively in all permutations: $\{x_1, \ldots, x_k\}$, $\{a, x_1, \ldots, x_k\}$ and $\{x_1, \ldots, x_k, b\}$. Moreover, signed permutations on n elements can be coded by unsigned permutations on $2n$ elements by replacing $+i$ by $2i - 1, 2i$, and $-i$ by $2i, 2i - 1$. Thus, filtering a set G of signed permutations amounts to perform three reductions (five in the unsigned case) on the PQ-tree representing the unsigned versions of permutations in G. We have:

Proposition 3. *Let $[a, b]$ be an interval of a permutation P on n elements, and G a saturated set of signed permutations, then the set of conserved intervals of $G_{[a,b]}$ can be computed in $O(n)$ time and space.*

Ancestors Labelling. We now describe an algorithm for inferring putative ancestral genes orders for a phylogenetic tree with a given topology and with gene orders at the leaves. The algorithm is similar in spirit to the Fitch-Hartigan algorithm ([12], [15]) for character-based parsimony, and consists of two labelling phases: a bottom-up labelling and a top-down refinement of this labelling.

Bottom-Up Labelling. In a first phase, during a bottom-up traversal of the tree, each ancestral node is labelled with a set of conserved intervals and the associated saturated set of signed permutations. Let x be a node with children y and z, and assume that y and z are already labelled by saturated sets of signed permutations G_y and G_z, with sets of conserved intervals I_y and I_z. Intuitively, we choose the label I_x that has as many intervals in common with I_y and I_z as possible. If $G_y \cap G_z \neq \emptyset$, then we set $I_x = I(G_y \cap G_z)$. If $G_y \cap G_z = \emptyset$, then I_y and I_z contain some conflicting intervals that need to be removed. We first identify S_y the subset of I_y that contains intervals that do not belong to conflicting subsets with respect to G_z. We obtain S_z similarly, with respect to G_y. Finally, we set $I_x = I((G_y)_{S_z} \cup (G_z)_{S_y})$ and $G_x = Perm(I_x)$. The algorithm proceeds up the tree until a label for the root is obtained.

Top-Down Refinement. While the root of the tree was assigned a label I_{root} based on all the leaves of the tree, this is the not the case for the other internal nodes, which were so far inferred based only on the leaves of the subtree of which they are the root. To let the information about all leaves be used to establish ancestral genomes, we proceed to a second phase, again similar to the second phase of the Fitch-Hartigan algorithm, where the conserved interval I_x of node x are used to refine the conserved intervals of the children of x. For any child y of node x, we first compute S_{xy}, the subset of I_x that contains intervals that do not belong to conflicting subsets with respect to G_y. We then refine I_y as $I_y = (I_y \cup S_{xy})^*$, and obtain $G_y = Perm(I_y)$.

By Theorem 2 and assuming that Conjecture 1 holds, the running time of the whole labelling procedure is polynomial in the number of genes and the number of leaves of the tree.

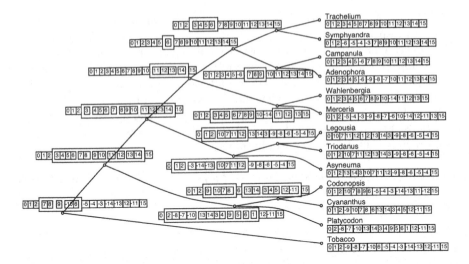

Fig. 3. Reconstructed conserved intervals for internal nodes of the campanulaceae phylogeny, after bottom-up labelling.

6 Chloroplast Genomes

To assess the specificity of the ancestral gene order reconstruction method described above, we tested our algorithm on a subset of gene segments of the chloroplast genomes of 13 species of plants previously studied by Cosner et al.[10].

Based on a phylogenetic tree previously reported for these species, the two phases of the ancestral gene order reconstruction were performed, and the inferred sets of conserved intervals are illustrated in Fig. 4. For example, in the first phase of the algorithm, when building the set of conserved intervals for the ancestor of Legousia and Triodanus: since both sets contain single permutations, the label of the ancestor is $I(G_{Leg} \cup G_{Tri})$. This yields the following representation of the conserved intervals:

$$A_{Leg,Tri} = \boxed{0 \; \boxed{\boxed{1 \mid 2} \; \boxed{10 \mid 7 \mid 11 \mid 12}} \; 13 \mid 14 \mid 3 \mid \text{-}9 \mid \text{-}8 \mid \text{-}6 \mid \text{-}5 \mid \text{-}4 \mid 15}$$

Then, using this reconstruction to build the ancestor of Asyneuma, we note that intervals [0,1] and [2,13] of Asyneuma are conflicting with respect to $A_{Leg,Tri}$. The resulting set of compatible conserved elements is represented by:

$$A_{Leg,Tri,Asy} = \boxed{0 \; \boxed{\boxed{1 \mid 2} \; \boxed{10 \mid 7 \mid 11 \mid 12} \; 13 \mid 14 \mid 3} \; \text{-}9 \mid \text{-}8 \mid \text{-}6 \mid \text{-}5 \mid \text{-}4 \mid 15}$$

The process continues up the tree until the ancestral gene order at the root of the tree is obtained. Fig. 3 shows the resulting set of conserved intervals. The second phase of the algorithm then starts and the information is propagated down the tree, starting from the root and adding conserved intervals to the children as often as possible. The resulting sets of conserved elements, shown in

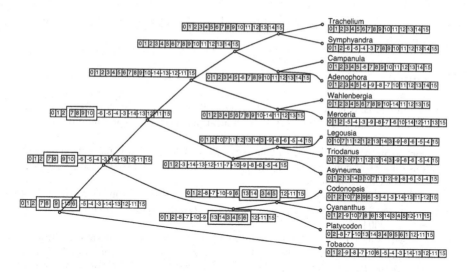

Fig. 4. Reconstructed conserved intervals for internal nodes of the campanulaceae phylogeny, after top-down refinement.

Fig. 4, is often much more refined than those obtained during the first phase of the algorithm. For example, the two ancestors $A_{Leg,Tri}$ and $A_{Leg,Tri,Asy}$ are now pinpointed to single possible permutations, separated by one reversal.

A closer inspection of the ancestral gene orders reconstructed reveals that, although the criterion used for inferring them was not based on a notion of parsimony of genome rearrangements, the distance, in terms of number of rearrangements, between neighboring sets of ancestral conserved intervals is usually very small, and often zero. We observe that most rearrangements that can be deduced from the reconstructed ancestors are reversals, but that a few transpositions and inverted transpositions also occur, for example between $A_{Leg,Tri}$ and Legousia or between Platycodon and its ancestor.

7   Conclusion

This paper presented operations on sets of conserved intervals, as well as associated techniques applied to the reconstruction of ancestral gene orders. The results obtained on a classical data set based on chloroplast genomes are very encouraging.

The next step is to apply our algorithms to the inference of complete ancestral mitochondrial and chloroplast genomes, and eventually to whole nuclear genomes. This would yield a better understanding of the rearrangement processes at work in these genomes. It may also highlight some highly conserved intervals that may correspond to sets of genes with strong positional ties, such as operons in bacteria.

However, the method presented here has still to be validated by using on simulated data. Given phylogenetic trees with known ancestral gene orders, we will test our algorithms on these trees and compare the results to the original ancestral gene orders.

References

1. D.A. Bader, B.M.E. Moret, and M. Yan. A linear-time algorithm for computing inversion distances between signed permutations with an experimental study. *J. Comput. Biol.*, 8(5):483–491, 2001.
2. A. Bergeron, M. Blanchette, A. Chateau, and C. Chauve. Implementation of operations on sets of conserved intervals. Technical report, Computer Science Department, UQAM, To appear.
3. A. Bergeron, J. Mixtaci, and J. Stoye. The reversal distance problem. In O. Gascuel, editor, *Mathematics of phylogeny and evolution*. Oxford University Press, To appear in 2004.
4. A. Bergeron and J. Stoye. On the similarity of sets of permutations and its applications to genome comparison. In *9th Annual International Conference on Computing and Combinatorics (COCOON 2003)*, volume 2697 of *Lecture Notes in Comput. Sci.*, pages 68–79. Springer, 2003.
5. M. Blanchette, T. Kunisawa, and D. Sankoff. Gene order breakpoint evidence in animal mitochondrial phylogeny. *J. Mol. Evol.*, 19(2):193–203, 1999.
6. M. Blanchette, T. Kunisawa, and D. Sankoff. Parametric genome rearrangement. *Gene*, 172(1):GC11–17, 2001.
7. K.S. Booth and G.S. Lueker. Testing for the consecutive ones property, interval graphs, and graph planarity using *PQ*-tree algorithms. *J. Comput. System Sci.*, 13(3):335–379, 1976.
8. G. Bourque and P.A. Pevzner. Genome-scale evolution: Reconstructing gene orders in the ancestral species. *Genome Res.*, 12(1):26–36, 2002.
9. A. Caprara. Formulations and complexity of multiple sorting by reversals. In *3rd Annual International Conference on Research in Computational Molecular Biology (RECOMB 1999)*, pages 84–93. ACM Press, 1999.
10. M.E. Cosner, R.K. Jansen, B.M.E. Moret, L.A. Raubeson, L.S. Wang, T. Warnow, and S.K. Wyman. An empirical comparison of phylogenetic methods on chloroplast gene order data in campanulaceae. In D. Sankoff and J. Nadeau, editors, *Comparative Genomics: Empirical and Analytical Approaches to Gene Order Dynamics, Map Alignment, and the Evolution of Gene Families (DCAF 2000)*, pages 99–212. Kluwer Academic Publishers, 2000.
11. J. Felsenstein. *Inferring phylogenies*. Sinauer Associates, 2003.
12. W. Fitch. Toward defining the course of evolution: Minimum change for a specific tree topology. *Systematic Zoology*, 20:406–416, 1971.
13. S. Hannenhalli and P.A. Pevzner. Transforming men into mice (polynomial algorithm for genomic distance problem). In *36th Annual Symposium on Foundations of Computer Science (FOCS 1995)*, pages 581–592. IEEE Comput. Soc. Press, 1995.
14. S. Hannenhalli and P.A. Pevzner. Transforming cabbage into turnip: polynomial algorithm for sorting signed permutations by reversals. *J. ACM*, 46(1):1–27, 1999.
15. J. A. Hartigan. Minimum mutation fits to a given tree. *Biometrics*, 29:53–65, 1973.
16. B. Larget, D.L. Simon, and J.B. Kadane. Bayesian phylogenetic inference from animal mitochondrial genome arrangements. *J. R. Stat. Soc. Ser. B Stat. Methodol.*, 64(4):681–693, 2002.

17. B.M.E. Moret, A.C. Siepel, J. Tang, and T. Liu. Inversion medians outperform breakpoint medians in phylogeny reconstruction from gene-order data. In *2nd International Workshop on Algorithms in Bioinformatics (WABI 2002)*, volume 2452 of *Lecture Notes in Comput. Sci.*, pages 521–536. Springer, 2001.

18. B.M.E. Moret, J. Tang, and T. Warnow. Reconstructing phylogenies from gene-content and gene-order data. In O. Gascuel, editor, *Mathematics of phylogeny and evolution*. Oxford University Press, To appear in 2004.

19. B.M.E. Moret, S. Wyman, D.A. Bader, T. Warnow, and M. Yan. A new implementation and detailed study of breakpoint analysis. In *6th Pacific Symposium on Biocomputing (PSB 2001)*, pages 583–594, 2001.

20. D. Sankoff. Rearrangements and chromosomal evolution. *Curr. Opin. Genet. Dev.*, 13(6):583–587, 2003.

21. D. Sankoff and M. Blanchette. Multiple genome rearrangement and breakpoint phylogeny. *J. Comput. Biol.*, 5(3):555–570, 1998.

22. D. Sankoff, G. Leduc, N. Antoine, B. Paquin, B.F. Lang, and R. Cedergren. Gene order comparisons for phylogenetic inference: evolution of the mitochondrial genome. *Proc. Natl. Acad. Sci. U.S.A.*, 89(14):6575–6579, 1992.

23. J. Tang and B.M.E. Moret. Scaling up accurate phylogenetic reconstruction from gene-order data. *Bioinformatics*, 19(Suppl. 1):i305–312, 2003.

Sorting by Reversals with Common Intervals

Martin Figeac and Jean-Stéphane Varré

Laboratoire d'Informatique Fondamentale de Lille
Université des sciences et technologies de Lille
59655 Villeneuve d'Ascq Cedex, France
{figeac,varre}@lifl.fr

Abstract. Studying rearrangements from gene order data is a standard approach in evolutionary analysis. Gene order data are usually modeled as signed permutations. The computation of the minimal number of reversals between two signed permutations produced a lot of literature during the last decade. Algorithms designed were first approximative, then polynomial and were further improved to give a linear one. Several extensions were investigated authorizing for example deletion or insertion of genes during the sorting process. We propose to revisit the 'sorting by reversals' problem by adding constraints on allowed reversals. We do not allow to break conserved clusters of genes usually called Common Intervals. We show that this problem is NP-complete. Assuming special conditions, we propose a polynomial algorithm.

Omitted proofs are given as supplementary material at
http://www.lifl.fr/~figeac/supplementary_material/
srac_appendix.pdf

1 Introduction

Prokaryotic and Mitochondrial genomes can be represented by their ordered list of signed genes, called signed permutation. Eukaryotics ones can be transformed into this easy to manipulate representation by using synteny blocks as genes [13]. Evolution modify the arrangement of genes. One of the main rearrangement is the reversal. A reversal reverses a list of genes and changes their sign to the opposite. The sorting by reversals problem (SBR) takes two signed permutations π^2 and π^1 as input and search for a scenario that transforms π^2 into π^1 with a minimum number of reversals. For example, let two genomes represented by their signed permutations: $\pi^1 = +1 +2 +3 +4 +5 +6$ and $\pi^2 = -3 -2 -4 +1 +5 +6$. One minimal scenario with two reversals is:

$$
\begin{aligned}
\pi_2 &= \text{-3 -2 }\underline{\text{-4 +1}}\text{ +5 +6} \\
\pi_2.\rho_1 &= \underline{\text{-3 -2 -1}}\text{ +4 +5 +6} \\
\pi_2.\rho_1.\rho_2 &= \text{+1 +2 +3 +4 +5 +6} = \pi_1
\end{aligned}
$$

We can always rename genes such that one of the two permutations is the identity. Then, the SBR problem becomes searching for a minimal scenario that transforms a signed permutation π into the identity permutation. *i.e.* this is a

I. Jonassen and J. Kim (Eds.): WABI 2004, LNBI 3240, pp. 26–37, 2004.
© Springer-Verlag Berlin Heidelberg 2004

sort. Hannenhalli and Pevzner [10] gave the first polynomial algorithm. Bergeron [4] proposed an $O(n^2)$ algorithm to sort a signed permutation of size n. A linear algorithm was described by Bader and al [2] to compute the minimal reversal distance between two genomes.

A few studies tried to compute more biologically relevant scenario or distance between two genomes. Some of them add evolutionary events as transpositions [16, 8] or deletions and insertions of segments of genes [7]. Others tried to compute more accurate evolutionary distances based on error correction [6, 17]. At last, [14, 1, 3] tried to study the set of allowed reversals. Our focus in this paper is to improve the reversal distance taking natural constraints based on observed data into account.

A second focus on gene order data is to discover shared groups of genes. [15, 11] proposed a stringent motif definition called *common intervals*. We denote by $\pi(i) = \pi_i$ the $i-$th element of a signed permutation π. For $1 \leq l < r \leq n$, $[l, r] = \{l, l+1, \ldots, r\}$ is an *interval* of π. $\pi([l, r]) = \{\pi_i | i \in [l, r]\}$ is the subset of elements of π with respect to the interval $[l, r]$.

Let $\Pi = (\pi^1, \ldots, \pi^k)$ a family of k permutations of n genes (each π^i represents a genome and all the genomes have the same gene content). Without restriction, we assume that $\pi_1 = 1, \ldots, n$, the identity permutation. A subset $c \subseteq \{1, \ldots, n\}$ is called a *common interval* of Π if and only if there exists $1 \leq l_j < r_j \leq n$ for all $1 \leq j \leq k$ such that $c = \pi^1([l_1, r_1]) = \pi^2([l_2, r_2]) = \ldots = \pi^k([l_k, r_k])$. A trivial common interval is the permutation itself. For example, given the above two permutations, common intervals are $\pi^1([1, 6]) = \pi^2([1, 6]) = \{1, 2, 3, 4, 5, 6\}$, $\pi^1([1, 5]) = \pi^2([1, 5]) = \{1, 2, 3, 4, 5\}$, $\pi^1([1, 4]) = \pi^2([1, 4]) = \{1, 2, 3, 4\}$, $\pi^1([2, 4]) = \pi^2([1, 3]) = \{2, 3, 4\}$, $\pi^1([2, 3]) = \pi^2([1, 2]) = \{2, 3\}$ and $\pi^1([5, 6]) = \pi^2([5, 6]) = \{5, 6\}$.

The problem of computing common intervals can be solved in $O(kn+K)$ with K the number of common intervals[1], [11]. In this study, we only deal with the problem of sorting a permutation π into the identity permutation. A common interval c will be always given with respect to π. Then, for sake of clarity, we will denote by $[l, r]$ the common interval $c = \pi([l, r])$.

Comparing gene order of two species leads to a set of Common Intervals. Those intervals are conserved neighboring genes. As they are conserved in the two genomes, they may be conserved during evolution. A minimal rearrangement scenario between two genomes should preserve the set of Common Intervals. We investigate the problem of sorting by reversals taking Common Intervals into account: We set the constraint that common intervals must be conserved during the sorting process.

An intensive work has been done about the study of sorting lists of signed genes [12, 10, 2] and [4, 7, 1]. We don't present a summary on how to compute the minimum number of reversals needed to sort a signed permutation π. We will refer to the HP algorithm [10] to name the sorting algorithm on signed permutations.

---

[1] K, the number of common intervals is at most $\binom{n}{2}$.

We present some definitions and basic facts from the theory of sorting by reversals: Comparing two genomes π and σ, we assume that they share a fixed set of n genes: $\mathcal{G} = \{g_1, g_2, \cdots, g_n\}$. Each genome is then an ordering of this set with sign information. A gene is positively or negatively signed ($\pi_i = +g_i$ or $-g_i$) depending on its strain localization. A reversal $\rho(i, j)$ applied to $\pi = \pi_1 \pi_2 \ldots \pi_n$ produces $\pi.\rho(i, j) = \pi_1 \cdots \pi_{i-1} -\pi_j -\pi_{j-1} \cdots -\pi_i \pi_{j+1} \cdots \pi_n$. A *breakpoint* of π with respect to σ is a pair x, y of elements of $\mathcal{G}$ such that xy appears in π and neither xy nor $-y$-x appears in σ.

Theorem 1. *Given a signed permutation, it always exists a minimal reversal scenario which never creates extra breakpoint [12].*

2   How to Sort Permutations with Common Intervals?

We will assume the following notations. π^u is an unsigned permutation, π^s is a signed permutation and π^p is a partially signed permutation (some genes are not signed and the others are). π is a permutation of one of the previous types. $\mathcal{I}$ is the identity permutation. $\mathcal{I}^s, \mathcal{I}^u$ and $\mathcal{I}^p$ are respectively signed, unsigned and partially signed identity permutations. $d(\pi_1, \pi_2)$ is the minimal number of reversals required to sort π_1 into π_2. $d(\pi)$ is the minimal number of reversals required to sort π into the identity permutation. We introduce some definitions to categorize reversals:

- A reversal $\rho(i, j)$ is *internal* to an interval $[k, l]$ if $k \leq i \leq j \leq l$.
- A reversal $\rho(i, j)$ is *incorporating* an interval $[k, l]$ if $i < k < l \leq j$ or $i \leq k < l < j$.
- A reversal $\rho(i, j)$ is *external* to an interval $[k, l]$ if $j \leq k$ or else $i \geq l$.
- A reversal $\rho(i, j)$ is *interleaving* an interval $[k, l]$ if $\rho(i, j)$ isn't external, nor internal, neither incorporating this interval.

Definition 1. *For a fixed set of intervals, a reversal that doesn't interleave any interval is a* compliant *reversal.*

We can now formally define the problem. Given a signed permutation π^s and C a set of common intervals. We search for $d_C(\pi^s)$, the *minimal number of compliant reversals* (with respect to C) required to transform π^s into the identity permutation. A scenario which transforms π^s into the identity permutation which uses $d_C(\pi^s)$ compliant reversals is called a *minimal compliant sorting scenario* (so-called MCSS).

It is clear that $d_C(\pi^s)$ is at least equal to $d(\pi^s)$. By way of theorem 1, if no common interval has breakpoint, there exists a minimal reversal scenario given by the HP theory that won't disrupt them and thus there exists π^s such that $d_c(\pi^s) = d(\pi^s)$.

Given $\pi^s = $ -2 -3 +1 a signed linear permutation and the associated set of *Common Intervals* $C = \{[1, 2], [1, 3]\}$. Only one minimal sorting scenario transforms π^s into the identity with $d(\pi^s) = 2$: $\pi^s = $ -2 -3 +1 $\rightarrow$ -2 -1 +3 $\rightarrow$ +1 +2 +3.

The first reversal isn't compliant with the common interval $[1,2]$. A MCSS (with five reversals) is: $\pi^s = \underline{-2}\ \underline{-3}\ \underline{+1} \rightarrow \underline{-1}\ \underline{+3}\ \underline{+2} \rightarrow +1\ \underline{+3}\ \underline{+2} \rightarrow +1\ -3\ \underline{+2} \rightarrow +1\ \underline{-3}\ \underline{-2} \rightarrow +1\ +2\ +3$. The theorem follows:

Theorem 2. *For any permutation π and a set of Common Intervals:* $d_C(\pi^s) \geq d(\pi^s)$

Remark 1. We can note that with unsigned permutations, sometimes all minimum scenarios create new breakpoints [9] [2]. As any breakpoint-free sequence of genes is a Common Interval, minimal sorting of unsigned permutations by reversals breaks Common Intervals. As a consequence we have $d_C(\pi^u) \geq d(\pi^u)$.

Definition 2. *Given π a permutation and two Common Intervals $[i,j]$ and $[k,l]$*

- $[k,l]$ *incorporates* $[i,j]$ *and* $[i,j]$ *is included in* $[k,l]$ *if* $k \leq i \leq j \leq l$,
- $[k,l]$ *interlaces* $[i,j]$ *and* $[i,j]$ *interlaces* $[k,l]$ *if* $i \leq k \leq j \leq l$.
- $[k,l]$ *and* $[i,j]$ *overlap each other if one interlaces the other or if one includes the other.*

2.1   Sorting by Reversals With Common Intervals Without Any Interlacing Ones

We start the analysis of the MCSS problem with a sub-problem in which we consider that the given *Common Intervals* do not interlace together. The idea is to process in two steps: First we sort *Common Intervals* such that they will become breakpoint-free, second we apply the HP algorithm to the transformed permutation to sort it. Since there is no breakpoint inside *Common Intervals* after the first step and since the HP algorithm does not create new breakpoints, *Common Intervals* are not broken during the second step.

Lemma 1. *Given π^s a signed permutation and $(\rho_1, ..., \rho_k, \rho_{k+1}, ..., \rho_d)$ a MCSS. If ρ_{k+1} is a reversal internal to a common interval c and ρ_k is a reversal which is not internal to c. Then $(\rho_1, ..., \rho_{k+1}, \rho_k, ..., \rho_d)$ is also a MCSS.*

Proof. Let $\pi^{s'} = \pi^s.\rho_1.\ \dots\ .\rho_{k-1}$. Suppose ρ_{k+1} is internal to c and ρ_k is either incorporating or external to c (it cannot interlace otherwise the scenario is not a MCSS): if ρ_k incorporates c then obviously $\pi^{s'}.\rho_k.\rho_{k+1} = \pi^{s'}.\rho_{k+1}.\rho_k$; if ρ_k is external to c then, as ρ_k and ρ_{k+1} do not overlap together, $\pi^{s'}.\rho_k.\rho_{k+1} = \pi^{s'}.\rho_{k+1}.\rho_k$. ◄

Theorem 3. *Given a set of non interlacing Common Intervals, it always exists a MCSS which sorts each Common Interval before applying incorporating or external reversals.*

---

[2] They don't increase the number of breakpoints, but they create a breakpoint in order to create an adjacency.

Proof. Let $\pi^{s'} = \pi^s.\rho_1. \ \ldots \ .\rho_{k-1}$ and ρ_{k+1} is internal to a Common Interval c. We prove that we have $\pi^{s'}.\rho_k.\rho_{k+1} = \pi^{s'}.\rho_{k+1}.\rho_k$. if ρ_k is internal to c, there is nothing to do. If ρ_k is external to c or incorporating it, as ρ_k and ρ_{k+1} don't overlap we have that $\pi^{s'}.\rho_k.\rho_{k+1} = \pi^{s'}.\rho_{k+1}.\rho_k$. By iteratively applying this process we have a MCSS which sorts each *Common Interval* before applying incorporating or external reversals. Included *Common Intervals* can be sorted first. ◄

This theorem proves that the sketch of algorithm described above may find a Compliant Sorting Scenario. Obviously this algorithm is able to find a compliant sorting scenario if we first sort included Common Intervals. But finding a minimal one (MCSS) requires to carefully sort the *Common Intervals*.

Given $\pi^s = {+}3 \ {+}2 \ {+}5 \ {+}1 \ {+}4$. Common intervals are $[1,2] = \{2,3\}$ and $[1,5] = \{1,2,3,4,5\}$. Assume we first decide to sort the sub permutation ${+}3 \ {+}2$ to ${+}2 \ {+}3$. This is done with three reversals. We obtain ${+}2 \ {+}3 \ {+}5 \ {+}1 \ {+}4$. Five more compliant reversals are needed to obtain the identity permutation. Now if we start to sort ${+}3 \ {+}2$ to ${-}3{-}2$, we obtain ${-}3 \ {-}2 \ {+}5 \ {+}1 \ {+}4$ with two reversals. Four more compliant reversals are needed to obtain the identity permutation. The first scenario requires eight compliant reversals when the second one only requires six compliant reversals. This example shows that we have to carefully choose if the common interval has to be sorted in increasing order and positively signed or else in decreasing order and negatively signed.

We define a *block* of π as an interval $[i,j]$ which does not contain breakpoints and we define a *strip* of π as a maximal block, i.e. a block $[i,j]$ such that there is breakpoints between π_{i-1} and π_i and between π_j and π_{j+1}. A block (strip) is *increasing* if $|\pi_i| < |\pi_j|$, otherwise it is *decreasing*. An increasing (decreasing) block b of a signed permutation π^s is *canonically* signed if all elements of b are positive (negative) [9].

We define the positive identity permutation $\mathcal{I}^s_+ = {+}1 \ \ldots \ {+}n$ and the negative identity permutation $\mathcal{I}^s_- = {-}n \ \ldots \ {-}1$. For a *Common Interval* $c = [i,j]$ relative to π, we define $m = \min(\pi_{i,j})$, $M = \max(\pi_{i,j})$ and ${+}c = \mathcal{I}^s_+([m,M])$ and ${-}c = \mathcal{I}^s_-([m,M])$. ${+}c$ (resp. ${-}c$) is a *canonically signed increasing* (resp. *decreasing*) strip. For example if $\pi^s = {+}5 \ {+}1 \ {-}7 \ {-}3 \ {+}2 \ {-}4 \ {-}6$ and $c = \pi^s([4,6]) = \{2,3,4\}$, then $m = \min(\{2,3,4\}) = 2$, $M = \max(\{2,3,4\}) = 4$, ${+}c = \mathcal{I}^s_+([2,4]) = {+}2 \ {+}3 \ {+}4$ and ${-}c = \mathcal{I}^s_-([2,4]) = {-}4 \ {-}3 \ {-}2$.

We may now propose an exact algorithm to compute $d_C(\pi^s)$ for a signed permutation π^s with respect to a given set of non interlacing *Common Intervals* (algorithm 2.1: SRIC).

Theorem 4. *SRIC computes $d_C(\pi^s)$ reversals for a given set of non interlacing Common Intervals.*

This algorithm runs in $O(2^k n)$ time with n the size of the permutation and k the number of non interlacing *Common Intervals*. In the worst case there is $O(n)$ non interlacing *Common Intervals*.

Algorithm 2.1. Brute Force Algorithm (SRIC)

SRIC: Sorting by Reversals with non Interlacing Common intervals
parameters : π^s, C sorted by increasing size
begin
1.   **If** $C = \emptyset$ **Then**
2.   return $d(\pi^s)$
3.   **Else**
4. $c := C[0]$ /$\star$ $\pi^s := \pi^{s1}.c.\pi^{s2}$ $\star$/
5.   return min $\begin{cases} d(c,\text{+}c) + SRIC(\pi^{s1}.\text{+}c.\pi^{s2}, C \setminus \{c\}) \\ d(c,\text{-}c) + SRIC(\pi^{s1}.\text{-}c.\pi^{s2}, C \setminus \{c\}) \end{cases}$
6.   **End If**
end

2.2 Reducing the Complexity of the Brute Force Algorithm

We will focus on the last step of the SRIC algorithm: $d_C(\pi^1.c.\pi^2) = \min(d_1, d_2)$ with $d_1 = d_C(\pi^1.\text{+}c.\pi^2) + d(c,\text{+}c)$ and $d_2 = d_C(\pi^1.\text{-}c.\pi^2) + d(c,\text{-}c)$

Remark 2. $d(\pi.\rho) - d(\pi) \in \{-1, 0, 1\}$

Remark 3. $d_c(\pi^1.\text{+}c.\pi^2) - d_c(\pi^1.\text{-}c.\pi^2) \in \{-1, 0, 1\}$

Lemma 2. *For c a non incorporating common interval of $\pi = \pi^1.c.\pi^2$, if $d(c,\text{+}c) = d(c,\text{-}c) + 1$ then $d_C(\pi^1.c.\pi^2) = d(c,\text{-}c) + d_C(\pi^1.\text{-}c.\pi^2)$*

Proof. As c is a non including Common Interval, we have $d_C(\pi^1.c.\pi^2) = \min(d_1, d_2)$ with $d_1 = d_C(\pi^1.\text{+}c.\pi^2) + d(c,\text{+}c)$ and $d_2 = d_C(\pi^1.\text{-}c.\pi^2) + d(c,\text{-}c)$

Let $k = d_C(\pi^1.\text{+}c.\pi^2)$ and $l = d_C(\pi^1.\text{-}c.\pi^2)$. Therefore $d_1 = k + d(c,\text{+}c) = k + d(c,\text{-}c) + 1$ and $d_2 = l + d(c,\text{-}c)$. $k = l + \varepsilon$, $\varepsilon \in \{-1, 0, 1\}$ (see remark 3):

- if $k = l - 1$ then $d_1 = d(c,\text{-}c) + 1 + k = d(c,\text{-}c) + l = d_2$.
- if $k = l + 1$ then $d_1 = d(c,\text{-}c) + 1 + k = d(c,\text{-}c) + l + 2 > d_2$.
- if $k = l$ then $d_1 = d(c,\text{-}c) + 1 + k = d(c,\text{-}c) + 1 + l > d_2$.

Thus we have $d_2 \leq d_1$. ◀

Symmetrically we have:

Lemma 3. *For c a non including common interval, if $d(c,\text{+}c) = d(c,\text{-}c) - 1$ then $d_C(\pi^1.c.\pi^2) = d(c,\text{+}c) + d_C(\pi^1.\text{+}c.\pi^2)$*

Definition 3. *For a given Common Interval c,*

- *if $d(c,\text{+}c) = d(c,\text{-}c)$ we will say that c is neutral.*
- *if $d(c,\text{+}c) < d(c,\text{-}c)$ we will say that c is positive.*
- *if $d(c,\text{+}c) > d(c,\text{-}c)$ we will say that c is negative.*

Testing the neutrality of a common interval can be done in linear time. If a common interval isn't neutral, we know how to optimally sort it thanks to lemma 2 and lemma 3.

A signed permutation π^s is a *spin* of a partially signed permutation π^p such that if π_i^p is signed then $\pi_i^s = \pi_i^p$ or if π_i^p is unsigned then either $\pi_i^s = +\pi_i^p$ or $\pi_i^s = -\pi_i^p$ for $1 \leq i \leq n$. For an arbitrary partially signed permutation π^p of order n with k unsigned elements define Π^{ps} as the set of all 2^k spins of π^p.

Lemma 4. *For an arbitrary partially signed permutation π^p, $d(\pi^p) = \min_{\pi^s \in \Pi^{ps}} d(\pi^s)$.*

Proof. we extend proof from [9] for unsigned permutations to partially signed ones: For every spin π^s of π^p, any sorting of π^s mimics a sorting of π^s, hence $d(\pi^s) \geq d(\pi^p)$. Let $\pi^p.\rho_1.\rho_2.\ldots.\rho_d = \mathcal{I}$ be an optimal sorting of π^p with d reversals. Consider the signed permutation $\pi^s = \mathcal{I}_+^s.\rho_d.\ldots.\rho_2.\rho_1$. Since $\pi^s \in \Pi^{ps}$ and $\pi^s.\rho_1.\rho_2.\ldots.\rho_d = \mathcal{I}_+^s$, $d(\pi^s) \leq d(\pi^p)$. Clearly, $d(\pi^p) = \min_{\pi^s \in \Pi^{ps}} d(\pi^s)$. ◄

We define the operation COLLAPSE as follows:

$$\text{COLLAPSE} : \Sigma_n^p \rightarrow \Sigma_{\leq n}^p$$

Σ_n^p is the set of all partially signed permutations of order n and $\Sigma_{\leq n}^p$ is the set of all partially signed permutations of order n or less without any strip (unsigned elements are considered as possibly positive and possibly negative). The function *collapses* the strips of π^s and renames the other elements. For example, COLLAPSE(+3 +1 +2 +4) = +2 +1 +3 (the sub permutation +1 +2 becomes +1, the sub permutation +3 becomes +2 and the sub permutation +4 becomes +3), COLLAPSE(+3 ±1 +2 +4) = +2 +1 +3, COLLAPSE(+3 ±1 -2 +4) = +3 ±1 -2 +4 and COLLAPSE(-3 ±2 +1 +4) = -2 +1 +3. Since [9] has showed that strips of size three or more can be collapsed and since strips of size two are *Common Intervals* which means that their elements cannot be separated, Consequently $d_C(\text{COLLAPSE}(\pi^p)) = d_C(\pi^p)$.

Neutral *Common Intervals* must be sorted to the positive or negative identity: They must be sorted to a canonical increasing or decreasing strip. If we collapse this strip we get one element which sign is either positive or negative. For c a neutral *Common Interval*, we will call $\pm c$ the resulting collapsed unsigned element.

Transforming Common Intervals to unsigned elements results in a partially signed permutation. This partially signed permutation keeps the same set of common intervals and this leads to sort a partially signed permutation without disrupting any *Common Intervals*. Previous results are right with partially signed permutations because they depend on remark 3 which is obviously right for partially signed permutations. If a partially signed *Common Interval* isn't neutral, we know how to optimally sort it. If it is neutral, it depends on the remaining partially signed permutation.

Theorem 5. ASRIC *computes $d_C(\pi^p)$ given a set of non interleaving Common Intervals.*

Algorithm 2.2. ASRIC Algorithm

```
ASRIC
parameters : πᵖ, C sorted by increasing size
begin
  1.    If C = ∅ Then
  2.       return d(πᵖ)
  3.    Else
  4.       c := C[0]   /* πᵖ := πᵖ¹.c.πᵖ² */
  5.       If d(c,+c) < d(c,-c) Then
  6.          return d(c,+c) + ASRIC(COLLAPSE(πᵖ¹.+c.πᵖ²), COLLAPSE(C \ {c}))
  7.       Else If d(c,+c) > d(c,-c) Then
  8.          return d(c,-c) + ASRIC(COLLAPSE(πᵖ¹.-c.πᵖ²), COLLAPSE(C \ {c}))
  9.       Else
 10.          return d(c,+c) + ASRIC(COLLAPSE(πᵖ¹.±c.πᵖ²), COLLAPSE(C \ {c}))
 11.       End If
 12.    End If
end
```

Corollary 1. *ASRIC computes $d_C(\pi^s)$, for π^s a signed permutation, given a set of non interleaving Common Intervals.*

The complexity of this algorithm depends on the complexity of sorting partially signed permutations. We know that sorting by reversals unsigned permutation is NP-complete [5]. As in the worth case a partially signed permutation isn't signed, the algorithm is exponential. But if the total number of unsigned elements is in $O(\log n)$ then sorting a partially signed permutation is a polynomial task [9]. It follows that if the total number of neutral *Common Intervals* is in $O(\log n)$ then the given algorithm is polynomial. Our algorithm runs in $O(n2^{k_0})$ time with k_0 the number of neutral *Common Intervals*. The total number of neutral *Common Intervals* can't be preliminary computed because it depends on the sorting process: At the beginning, Common Intervals are signed and may become partially signed.

For all distances to the positive identity permutation ranging from 25 to 100, we have randomly generated 10 000 signed permutations of size 100. We observed on these randomly generated data that percentage of neutral permutations grows exponentially with their reversal distance. The percentage of neutral permutations never exceeds 50%. We have also computed tests for all permutations of size 1 to 9, and this showed that there is always less than thirty four percent of neutral permutations.

2.3 Sorting by Reversals with All Common Intervals

Lemma 5. *For two interlacing Common Intervals $[i,j]$ and $[k,l]$, there exists four others Common Intervals: $[i, k-1], [k,j], [j+1,l]$ and $[i,l]$.*

Define a *multi-block* a list of consecutive blocks $(b_i, \ldots, b_j)$ such that $|\max(b_k)| + 1 = |min(b_{k+1})|$ or $|\min(b_k)| - 1 = |max(b_{k+1})|$ for $i \leq k < j$.

If $|\max(b_k)| + 1 = |min(b_{k+1})|$ we will call it an increasing multi-block or else a decreasing one. A multi-strip (increasing or decreasing) is a multi-maximal-block. For example, in $\pi = (\text{+1 +2 +3 -6 -5 -4 +9 +8 +7})$ there is a multi-strip composed of three maximal blocks: $\pi([1,3]) = \{1,2,3\}, \pi([4,6]) = \{4,5,6\}$ and $\pi([7,9]) = \{7,8,9\}$.

Lemma 6. *Given m a multi-strip, if m is increasing then $d_C(m, \text{+}m) < d_C(m, \text{-}m)$, or else if m is decreasing then $d_C(m, \text{+}m) > d_C(m, \text{-}m)$.*

Proof. Assume without loss of generality the increasing collapsed multi-strip: $\pm1\pm2\pm3$. $\pm i, i \in \{1,2,3\}$ represents a collapsed Common Interval. The common Interval constraint only allow 1-element-reversal or all-elements-reversal, thus obviously $d_C(\pm1\pm2\pm3, \text{+1+2+3}) < d_C(\pm1\pm2\pm3, \text{-3-2-1})$. And symmetrically for decreasing multi-strip. ◄

Lemma 7. *Given two interlacing Common Intervals $[i,j]$ and $[k,l]$, sorting them without disrupting any Common Interval is optimally done by sorting the three non overlapping Common Intervals: $[i, k-1]$, $[k,j]$ and $[j+1,l]$, and at last by sorting the resulting multi-strip.*

Proof. We first show that this scenario is a compliant one: $[i, k-1]$ is included in $[i,j]$ and thus must be sorted before $[i,j]$. $[k,j]$ is included in $[i,j]$ and in $[k,l]$ and thus must be sorted before $[i,j]$ and before $[k,l]$. $[j+1,l]$ is included in $[k,l]$ and thus must be sorted before $[k,l]$. We have three non overlapping Common Intervals which must be sorted first and this doesn't disrupt any Common Interval. The result of these sorting is an increasing or decreasing multi-strip. Theorem 3 shows that this scenario is optimal because we respect all levels of inclusion. ◄

Lemma 8. *An increasing (respectively decreasing) multi-strip needs the number of negatives blocks (respectively positives) to be sorted without disrupting any Common Interval.*

Proof. A block-strip is constituted of consecutive blocks. The elements of one block are all positives or all negatives. As each block is a Common Interval, none interlacing reversals are allowed. As each group of two consecutive blocks is also a Common Interval, no interleaving reversals are allowed. Thus each block must be sorted without moving it. The only allowed reversal is one which exactly reverses a block. A block-strip is either increasing or decreasing. If it is an increasing strip, by way of lemma 6 all blocks must be positives. It needs the number of negative blocks to make the block-strip a canonical increasing strip. Symmetrical results can be obtained for decreasing strips. ◄

We summarize the section with algorithm 2.3 named SRAC.

Theorem 6. *SRAC computes $d_C(\pi^p)$ for all Common Intervals.*

Corollary 2. *SRAC computes $d_C(\pi^s)$ for all Common Intervals.*

The SRAC algorithm runs in $O(2^{k'}n + kn)$ time for a permutation of n elements with k Common Intervals and k' Common Intervals excluding positives, negatives and overlapping ones. Hence k' is at most equal to n.

Algorithm 2.3. SRAC

```
SRAC: Sorting by Reversals with All Common intervals
parameters : π^p, C sorted by increasing size
begin
  1.    If C = ∅ Then
  2.        return d(π^p)
  3.    Else
  4.        c := C[0]   /* π^p := π^{p1}.c.π^{p2} */
  5.        If c is an increasing multi-strip Then
  6.            d :=number_of_negative_blocks (c)
  7.            return d + SRAC(COLLAPSE(π^{p1}.+c.π^{p2}), C \ {c})
  8.        Else If c is a decreasing multi-strip Then
  9.            d :=number_of_positive_blocks (c)
 10.            return d + SRAC(COLLAPSE(π^{p1}.-c.π^{p2}), C \ {c})
 11.        Else If d(c,+c) < d(c,-c) Then
 12.            return d(c,+c) + SRAC(COLLAPSE(π^{p1}.+c.π^{p2}), COLLAPSE(C \ {c}))
 13.        Else If d(c,+c) > d(c,-c) Then
 14.            return d(c,-c) + SRAC(COLLAPSE(π^{p1}.-c.π^{p2}), COLLAPSE(C \ {c}))
 15.        Else
 16.            return d(c,+c) + SRAC(COLLAPSE(π^{p1}.±c.π^{p2}), COLLAPSE(C \ {c}))
 17.        End If
 18.    End If
end
```

3   Complexity of Sorting by Reversals with Common Intervals

In this paragraph we discuss about a particular case of MCSS: The SROC problem. We consider that the set of given Common Intervals is such that they don't overlap each other. It is a particular case of the general problem: given a set of intervals and a permutation, compute the minimal reversal distance without disrupting any interval. Here we show that this particular case of the main problem is NP-complete. We show it by comparing SROC to USBR, the problem of computing the minimal reversal distance for an unsigned permutation.

We first show that if we are able to solve SROC in exponential time, we are also able to solve USBR in exponential time.

Lemma 9.

$$USBR \leq_P SROC$$

Proof. Let $\pi^u = (\pi^u_1 \ldots \pi^u_n)$, an unsigned permutation. Let $c = (+4\ +1\ +3\ +5\ +2)$ a signed permutation of five elements such that $d\big(c, (+1\ +2\ +3\ +4\ +5)\big) = d\big(c, (-5\ -4\ -3\ -2\ -1)\big) = 5$.

Let $\pi^s = (\pi^s_1 \ldots \pi^s_{5n})$ which is obtained by replacing each π^u_i by the sub permutation $\big(\ +(5(\pi^u_i-1)+4)\ +(5(\pi^u_i-1)+1)\ +(5(\pi^u_i-1)+3)\ +(5(\pi^u_i-1)+5)\ +(5(\pi^u_i-1)+2)\big)$. Given the set of all *Common Intervals*, $C = \{[1,5], [6,10],$

$\ldots, [5n-4, 5n]\}$, we then have $d(\pi^u) = d_C(\pi^s) - 5n$ and therefore USBR can be reduced to SROC in polynomial time. ◄

In a second time we show that if we are able to solve USBR in exponential time, we are also able to solve SROC in exponential time.

Lemma 10.
$$SROC \leq_P USBR$$

Proof. Let $\pi^s = (\pi^s{}_1 \ldots \pi^s{}_n)$, a signed permutation of n elements and k *Common Intervals* from π, called c_i, $1 \leq i \leq k$.

Let π' a copy of π^s. Each non neutral *Common Interval* in π' is replaced by a canonical signed strip. Each neutral *Common Interval* is replaced by an unsigned element representing a canonical increasing or a canonical decreasing strip. We then have π' a partially signed permutation. Each signed element of π' is replaced by a non neutral unsigned *Common Interval*. If π'_j is positively signed then we replace it by the positive unsigned *Common Interval* $(\pi_j \ (\pi_j + 1) \ (\pi_j + 2))$. Or else it is negatively signed and we replace it by the unsigned *Common Interval* $((\pi_j + 2) \ (\pi_j + 1) \ \pi_j)$. The main idea of the algorithm from [9] to sort unsigned permutations is to test all their different spins. Strips of size 3 are a very particular case: $(\pi_j \ (\pi_j + 1) \ (\pi_j + 2))$ is optimally signed positive and $((\pi_j + 2) \ (\pi_j + 1) \ \pi_j)$ is optimally signed negative [9]. At last we have $\pi^u = \pi'$ an unsigned permutation of m elements. The algorithm from [9] allows us to compute $d(\pi^u)$ and to get the optimally signed spin s. Then

$$d_C(\pi) = d(\pi^u) + \sum_{i=1}^{i \leq k} d(c_i, \mathcal{I}_{c_i}),$$

If π'_i is positive, then $\mathcal{I}_{c_i}$ is the positive identity permutation relative to c_i. Or else if π'_i is negative, then $\mathcal{I}_{c_i}$ is the negative identity permutation relative to c_i. Hence, SROC can be reduced to USBR in polynomial time. ◄

Theorem 7. *SROC is NP-complete*

Proof. [5] proves that USBR for unsigned permutations is NP-complete, then we have $SROC \leq_P USBR$ and $USBR \leq_P SROC$. SROC is NP-complete. ◄

Corollary 3. *MCSS is NP-complete*

Proof. As SROC is NP-complete, it gives that the general problem MCSS is also NP-complete (We know that for some types of intervals (without overlapping) it is NP-complete). ◄

4   Conclusion

First we showed a lack of likelihood in minimal reversal scenarios because they break *Common Intervals*. So we presented the problem of finding a minimal reversal scenario that doesn't disrupt *Common Intervals* in order to improve the likelihood of reversal scenarios. Finally we showed that this problem is NP-complete and we proposed an exact algorithm to solve it.

References

1. Y. Ajana, J.F. Lefebvre, E. Tillier, and N. El-Mabrouk. Exploring the set of all minimal sequences of reversals - an application to test the replication-directed reversal hypothesis. *WABI*, 2002.

2. D Bader, B Moret, and M Yan. A linear time algorithm for computing inversion distance between signed permutations with an expermimental study. *Journal of*, 2000.

3. Anne Bergeron, Cédric Chauve, T Hartman, and K St-Onge. On the properties of sequences of reversals that sort a signed permutation. In *Proceedings of JOBIM*, JOBIM, 2002.

4. Anne Bergeron. A very elementary presentation of the hannenhalli-pevzner theory. *CPM*, 2001.

5. Alberto Caprara. Sorting by reversals is difficult. *ACM RECOMB*, 1997.

6. M.E. Cosner, R.K. Jansen B. M. E. Moret, L.A. Raubeson, L. Wang, T. Warnow, and S.K. Wyman. An empirical comparison between bpanalysis and mpbe on the campanulaceae chloroplast genome dataset. *Comparative Genomics*, 2000.

7. N. El-Mabrouk. Sorting signed permutations by reversals and insertions/deletions of contiguous segments. *Journal of Discrete Algorithms*, 2000.

8. N Eriksen. (1+epsilon)-approximation of sorting by reversals and transpositions. *WABI*, 2001.

9. S Hannenhalli and P Pevzner. To cut ... or not to cut (applications of comparative physical maps in molecular evolution. *Seventh Annual ACM-SIAM Symposium on Discrete Algorithms*, pages 304–313, 1996.

10. Sridhar Hannenhalli and Pavel A. Pevzner. Transforming cabbage into turnip: polynomial algorithm for sorting signed permutations by reversals. *STOC*, 1995.

11. Steffen Heber and Jens Stoye. Algorithms for finding gene clusters. *Lecture Notes in Computer Science*, 2149:252, 2001.

12. J. Kececioglu and D. Sankoff. Exact and approximation algorithms for sorting by reversals, with application to genome rearrangement. *Algorithmica*, 13:180–210, 1995.

13. Pavel Pevzner and Glenn Tesler. Transforming men into mice: the nadeau-taylor chromosomal breakage model revisited. *RECOMB*, 2003.

14. David Sankoff. Short inversions and conserved gene clusters. *Bioinformatics*, 18(10):1305–1308, 2002.

15. T. Uno and M. Yagiura. Fast algorithms to enumerate all common intervals of two permutations. *Algorithmica*, 2000.

16. ME Walter, Z Dias, and J Meidanis. Reversal and transposition distance of linear chromosomes. *SPIRE*, 1998.

17. L.S. Wang and T. Warnow. New polynomial time methods for whole-genome phylogeny reconstruction. *to appear in Proc. 33rd Symp. on Theory of Comp.*, 2001.

A Polynomial-Time Algorithm for the Matching of Crossing Contact-Map Patterns

Jens Gramm*

International Computer Science Institute
1947 Center Street, Suite 600, Berkeley, CA 94704
gramm@icsi.berkeley.edu

Abstract. Contact maps are a model to capture the core information in the structure of biological molecules, e.g., proteins. A contact map consists of an ordered set S of elements (representing a protein's sequence of amino acids), and a set A of element pairs of S, called *arcs* (representing amino acids which are closely neighbored in the structure). Given two contact maps (S, A) and (S_p, A_p) with $|A| \geq |A_p|$, the CONTACT MAP PATTERN MATCHING (CMPM) problem asks whether the "pattern" (S_p, A_p) "occurs" in (S, A), i.e., informally stated, whether there is a subset of $|A_p|$ arcs in A whose arc structure coincides with A_p. CMPM captures the biological question of finding structural motifs in protein structures. In general, CMPM is NP-hard. In this paper, we show that CMPM is solvable in $O(|A|^6 |A_p|^2)$ time when the pattern is $\{<, \emptyset\}$-structured, i.e., when each two arcs in the pattern are disjoint or crossing. Our algorithm extends to other closely related models. In particular, it answers an open question raised by Vialette that, rephrased in terms of contact maps, asked whether CMPM for $\{<, \emptyset\}$-structured patterns is NP-hard or solvable in polynomial time. Our result stands in sharp contrast to the NP-hardness of closely related problems. We provide experimental results which show that contact maps derived from real protein structures can be processed efficiently.

1 Introduction

Since the function of biological molecules is highly associated with their three-dimensional structure, structure analysis is an important area of computational biology. Combinatorial models have been developed to capture the structure of molecules, e.g., *contact maps* for protein structure analysis [6] and *arc annotations* for RNA structure analysis [11]. Focusing in the following on proteins, we find that experimentally determined structural information is available for a large number of proteins [2]. One way to organize and analyze these data is to classify proteins according to their structure, a task that in many cases still involves human interaction. Herein, proteins are considered to belong to the same class if they share structural features, even if their primary sequence may not be similar. The resulting question is to determine whether a given protein structure exhibits a given structural feature where such a feature may be formed by amino acids which are not necessarily contiguous in the protein sequence. Here, we address

---

* This research was supported through a postdoc fellowship by the DAAD (German Academic Exchange Association).

I. Jonassen and J. Kim (Eds.): WABI 2004, LNBI 3240, pp. 38–49, 2004.

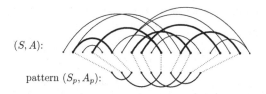

(S, A):

pattern (S_p, A_p):

Fig. 1. Example of a yes-instance of CONTACT MAP PATTERN MATCHING, consisting of two contact maps (S, A) and (S_p, A_p). The occurrence of the pattern in the upper contact map is indicated by bold arcs, the corresponding mapping is indicated by dotted lines.

this kind of pattern matching problem for the model of contact maps and a special class of patterns.

Problem definition. A *contact map* (S, A) consists of an ordered set S and a set $A = \{ (e_l, e_r) \mid e_l, e_r \in S, e_l < e_r \}$. Each pair in A is referred to as an *arc* and, consequently, A is called *arc set*. Given two contact maps (S, A) and (S_p, A_p) with $|S_p| < |S|$, we say that (S_p, A_p) *occurs* in (S_p, A_p) if there is a one-to-one mapping M of S_p to a size-$|S_p|$ subset S' of S such that, for $e, e' \in S_p$, we have both $e < e' \Rightarrow M(e) < M(e')$ and $(e, e') \in A_p \Rightarrow (M(e), M(e')) \in A$. Then, the central problem of this paper is given as follows:

> CONTACT MAP PATTERN MATCHING (CMPM)
> **Given:** Contact maps (S, A) and (S_p, A_p).
> **Question:** Does (S_p, A_p) occur in (S, A)?

An illustration of a yes-instance of CMPM is given in Fig. 1. We refer to (S_p, A_p) as the pattern and we abbreviate $n := |A|$ and $m := |A_p|$. From the order of elements in S, we can in a natural way derive binary relations between arcs. For $a, a' \in A$ with $a = (e_l, e_r)$ and $a' = (e'_l, e'_r)$, we say that $a < a'$ (*a precedes a'*) iff $e_l < e_r < e'_l < e'_r$. We say that $a \sqsubset a'$ (*a is nested* in a') iff $e'_l < e_l < e_r < e'_r$. Finally, we say that $a \lozenge a'$ (*a crosses a'*) iff $e_l < e'_l < e_r < e'_r$. A contact map is $\{<, \lozenge\}$-structured if for each two arcs a and a' either the preceding-relation or the crossing-relation applies, i.e., $a < a'$, $a' < a$, $a \lozenge a'$, or $a' \lozenge a$; other restricted contact map classes are defined analogously.

Previous work. In general, CMPM is NP-complete which can be shown in analogy to [4] where a somewhat different model is used. The model of contact maps received particular attention for computing the similarity of two proteins, formalized as the question for a maximum-size pattern that occurs in both given contact maps [6, 9].

Vialette studied a problem closely related to CMPM in the area of RNA structure comparison [12]. His results imply, among others, that CMPM is NP-hard for $\{\sqsubset, \lozenge\}$-structured patterns, and is solvable in $O(n^2)$ time (in $O(n^2 \log n)$ time) for $\{\sqsubset\}$-structured ($\{\lozenge\}$-structured) patterns. CMPM for $\{<, \sqsubset\}$-structured patterns is solvable in quadratic time [5]. A summary of some results in given in Table 1. The remaining open question in this context, raised by Vialette [12], was – rephrased in terms of contact maps – to determine whether CMPM for $\{\sqsubset, \lozenge\}$-structured patterns is solvable in polynomial time.

Table 1. Overview on results for the complexity of the CONTACT MAP PATTERN MATCHING problem for restricted pattern structures. The result of this paper is marked by an asterisk.

pattern structure	complexity of CMPM
$\{<, \sqsubset, \lozenge\}$	NP-complete [4]
$\{\sqsubset, \lozenge\}$	NP-complete [12]
$\{<, \lozenge\}$	$n^6 m^2$ (*)
$\{<, \sqsubset\}$	$n \log n + nm$ [5]

New Results. In this paper, we answer this open question by giving a polynomial-time algorithm for CMPM with $\{<, \lozenge\}$-structured patterns. The algorithm relies on an involved realization of the dynamic programming principle and has $O(n^6 m^2)$ running time. Due to heuristic improvements, the running time of the algorithm is faster than this worst-case bound suggests. We give experimental results for simulated contact maps and ones derived from real protein structures. Our results show that these datasets can be processed efficiently.

Our result is surprising since it lies at the border of CMPM versions that are still solvable in polynomial time (see Table 1). We show that our algorithm generalizes from contact maps to closely related models, in particular to 2-interval sets as discussed by Vialette [12]. Moreover, we show how our result can be used to give a fixed-parameter algorithm searching for the maximum-size $\{<, \lozenge\}$-structured pattern occurring in two given contact maps or 2-interval patterns, respectively. For 2-interval sets, it was recently established that this question is NP-hard [3]. Notably, our algorithm only poses a constraint on the pattern, no constraint is posed onto the contact map in which we search the pattern. The observation that basic secondary structure elements like alpha helices and anti-parallel beta sheets exhibit $\{<, \lozenge\}$-structured patterns supports the significance of this natural class of pattern structures in proteins. Moreover, in combination with the polynomial-time algorithm for CMPM restricted to $\{<, \sqsubset\}$-structured patterns [5], our algorithm provides an important building stone for further results.

Due to lack of space, most proofs are deferred to the full version of the paper.

2 Additional Notation

As a convention, we use $A_p = \{p_0, \ldots, p_{m-1}\}$ to denote the arc set of the pattern and we use $A = \{a_0, \ldots, a_{n-1}\}$ as the arc set of the contact map in which the pattern is searched. Given an arc $a = (e_l, e_r)$, we use $l(a)$ to refer to the *left endpoint* e_l of a and we use $r(a)$ to refer to the *right endpoint* e_r of a. We assume that both sets are ordered increasingly by their right endpoints and, if some arcs have the same right endpoint, subsequently by their left endpoint; this order, if not existing, can be obtained for A in time $O(|A| \log |A|)$. In the following, we introduce the concepts of *anchors* and *windows*, defined for the arc set of $\{<, \lozenge\}$-structured patterns, which will be essential for stating our algorithm. Given a $\{<, \lozenge\}$-structured contact map (S_p, A_p) and an arc $p \in A_p$, we use anchor(p) to refer to the smallest (w.r.t. the order of A) arc $p' \in A_p$, if it exists, such that $l(p') < l(p) < r(p') < r(p)$. This implies that there is no arc $p'' \in A_p$ with $p'' \lozenge p'$, $p'' \lozenge p$, and $l(p) < r(p'') < r(p')$. If no such arc $p' \in A_p$ exists, then

Fig. 2. Illustration of two concepts for structure patterns: (a) Arc p_j is *anchor* of both arcs p_i and p_{i+1}. (b) The pair of arcs p_{j-1} and p_j comprises the *j-window* with respect to arc p_i.

anchor$(p) := \perp$. The concept of anchors is illustrated in Fig. 2(a). For integer i with $0 \leq i \leq m - 1$, we use anchor(i) to refer to the index of anchor(p_i), i.e., anchor$(i) = j$ iff anchor$(p_i) = p_j$. Given a contact map (S_p, A_p) and two arcs $p_j, p_i \in A_p$ such that $p_{j-1} \between p_i$, and either $p_j \between p_i$ or $p_j = p_i$, we refer to the pair (p_{j-1}, p_j) as the *j-window* of p_i. This concept is illustrated in Fig. 2(b).

As an abbreviation, we use, for an ordered set $A' = \{a_{i_1}, a_{i_2}, \ldots, a_{i_r}\}$ with $A' \subseteq A$, cross$(a_{i_1}, a_{i_2}, \ldots, a_{i_r})$ to denote that all $a_{i_s}, a_{i_t} \in A'$ with $1 \leq s < t \leq r$ satisfy $a_{i_s} \between a_{i_t}$. Given two contact maps (S, A) and (S_p, A_p), let $A' := \{a_{i_1}, a_{i_2}, \ldots, a_{i_r}\}$ and $A'_p := \{p_{j_1}, p_{j_2}, \ldots, p_{j_r}\}$ for some $r < m$, with $a_{i_s} \in A$ and $p_{j_s} \in A_p$ for $1 \leq s \leq r$. We say that A'_p *matches* A' if, for $1 \leq s < t \leq r$, $p_{j_s} \between p_{j_t}$ iff $a_{i_s} \between a_{i_t}$. Intuitively, this means that the arc structures of A' and A'_p coincide. We say that $A_p[j_1, j_2, \ldots, j_r]$ *matches* $A[i_1, i_2, \ldots, i_r]$ if $(S_p, \{p_{j_1}, \ldots, p_{j_r}\})$ occurs in (S, A) while mapping p_{j_1} to a_{i_1}, p_{j_2} to a_{i_2}, $\ldots$, and p_{j_r} to a_{i_r}. Note that the requirement that A'_p matches A' is local, comparing only the arc structure of A'_p and A', whereas the requirement that $A_p[j_1, j_2, \ldots, j_r]$ matches $A[i_1, i_2, \ldots, i_r]$ is global, asking, in addition, for a corresponding occurrence in A of the pattern up to arc p_{j_r}.

We say that a contact map (S, A) is *connected* if there is no bipartition of A into two non-empty subsets A_1 and A_2 such that arcs from A_1 do not cross with arcs from A_2. For an easier exposition, we assume in the following that the arc set of the pattern is *connected*. Further, we assume that in the pattern (S_p, A_p), every element in S_p is endpoint of an arc in A_p.

3 Dynamic Programming Algorithm

In this section we present the algorithm solving CMPM restricted to $\{<, \between\}$-structured patterns. We describe the algorithm in top-down fashion, explaining data structures and an overview in Sect. 3.1 and giving the details of the dynamic programming steps in Sect. 3.2 and 3.3. Sect. 3.4 states the running time and the correctness of the algorithm, Sect. 3.5 points out generalizations of the algorithm to closely related models, and Sect. 3.6 shows an application of this algorithm in a more general problem setting.

3.1 Overview

Central for our algorithm are two sets that we define for every s, $0 \leq s < n$, and for every i, $1 \leq i < m$:

- The set $T_{s,i}$ contains all c, $0 \leq c < n$, such that $A_p[\text{anchor}(i), i]$ matches $A[c, s]$. Thus, assuming that we match p_i with a_s, it stores possible matches for the anchor of pattern arc p_i.

Fig. 3. Given $p_i \in A_p$ with $i > 1$, we distinguish two cases: (Case a) anchor$(p_{i-1}) = $ anchor(p_i) or (Case b) anchor$(p_{i-1}) < $ anchor(p_i).

- The set $S_{s,i}$ contains all 4-tuples $(c, w_1, w_2; j)$ with $0 \le c, w_1, w_2 < n$ and anchor$(i) < j \le i$ such that $A_p[\text{anchor}(i), j-1, j, i]$ matches $A[c, w_1, w_2, s]$. Thus, assuming that we match $p_{\text{anchor}(i)}$ with a_c and p_i with a_s, it stores possible matches for the windows of pattern arc p_i.

In fact, set $S_{s,i}$ is a refinement of set $T_{s,i}$ as the elements of $T_{s,i}$ can be easily obtained from $S_{s,i}$. For the sake of a clearer exposition of our algorithm, we choose to compute these sets separately: First, set $T_{s,i}$ is computed and for every element that is added to $T_{s,i}$, we then compute the corresponding 4-tuples that have to be added to $S_{s,i}$. The algorithm to be presented in the following sections computes the sets, starting with $i = 1$ and computing the sets for a value i with $1 < i < m$ based on the contents of the sets for value $i - 1$. It, thus, follows the classical algorithmic paradigm of dynamic programming.

The algorithm computes sets $T_{s,i}$ and $S_{s,i}$ in the order of ascending values of i. The algorithm has an *initialization phase* that computes the sets $T_{s,1}$ and $S_{s,1}$: Set $T_{s,1}$ contains all $c, 1 \le c < n$, such that $a_c \between a_s$. Since p_0 is necessarily the anchor of p_1 and since (p_0, p_1) constitutes the only window of arc p_1, set $S_{s,1}$ contains all 4-tuples $(c, c, s; 1)$ for which $c \in T_{s,1}$. The *dynamic programming phase* of the algorithm, then, computes $T_{s,i}$ and $S_{s,i}$ for $i > 1$. This computation will be explained in the next two subsections. Having computed sets $T_{s,i}$ and $S_{s,i}$ for all $0 \le s < n$ and all $1 \le i < m$, the algorithm reports that the pattern is found when there exist c^* and s^*, $0 \le c^*, s^* < n$, such that $c^* \in T_{s^*, m-1}$: Then, by definition of $T_{s^*, i}$, $A_p[\text{anchor}(m-1), m-1]$ matches $A[c^*, s^*]$, and thus (S_p, A_p) occurs in (S, A).

3.2 Matching Pattern Anchors

In this subsection, we show how we compute, for given $a_s \in A$ and $p_i \in A_p$, $i > 1$, the set $T_{s,i}$, based on the knowledge of sets $T_{q,i-1}$ and $S_{q,i-1}$ for all $a_q \in A$. The corresponding procedure in pseudocode is given in Fig. 4.

We distinguish two situations concerning p_i which have to be treated differently by the algorithm: Either (Case a) anchor$(p_{i-1}) = $ anchor(p_i), illustrated in Fig. 3(a), or (Case b) anchor$(p_{i-1}) < $ anchor(p_i), illustrated in Fig. 3(b). For each of these cases, we describe in the following how $T_{s,i}$ is computed.

(Case a): For every $a_r \in A$ with $a_r \between a_s$, and for every $c \in T_{r,i-1}$ with cross(a_c, a_r, a_s), we add c to $T_{s,i}$. The resulting set is the final set $T_{s,i}$. Thus, for the computation of $T_{s,i}$ in this case, we iterate over all possible choices of $a_c, a_r \in A$.

---

procedure matchAnchor(a_s; p_i)

Global: Contact maps (S, A) and pattern (S_p, A_p), sets $T_{q,i-1}$ and $S_{q,i-1}$ for all $a_q \in A$.

Input: Arcs $a_s \in A$ and $p_i \in A_p$, $i > 1$.

Output: Sets $T_{s,i}$ and $S_{s,i}$.

Method:

> $T_{s,i} := \emptyset$;
> $S_{s,i} := \emptyset$;
>
> **if** anchor(p_{i-1}) = anchor(p_i) **then**   /*** (Case a) ***/
>> **for all** $a_c, a_r \in A$ with cross(a_c, a_r, a_s) **do**
>>> **if** $\{c\} \in T_{r,i-1}$ **then**
>>>> $T_{s,i} := T_{s,i} + \{c\}$;
>>>> $S_{s,i} := S_{s,i} \cup$ matchWindows(a_c, a_c, a_r, a_s; p_i);
>>>
>>> **end if**
>>
>> **end for**
>
> **else** /* anchor(p_{i-1}) < anchor(p_i) */   /*** (Case b) ***/
>> **for all** $0 \le w_1, w_2, r < n$ **do**
>>> **if** (w_1, w_2, r, s) matches (anchor(p_i) − 1, anchor(p_i), $i - 1, i$) **then**
>>>> $c_r :=$ minAnchor(w_1, w_2, r; anchor(i), $i - 1$);
>>>> **if** (c_r, w_1, w_2; anchor(i)) $\in S_{r,i-1}$ **then**
>>>>> $T_{s,i} := T_{s,i} + \{w_2\}$;
>>>>> $S_{s,i} := S_{s,i} \cup$ matchWindows($a_{c_r}, a_{w_2}, a_r, a_s$; p_i);
>>>>
>>>> **end if**
>>>
>>> **end if**
>>
>> **end for**
>
> **end if**
> **return** ($T_{s,i}, S_{s,i}$);

Fig. 4. Overview in pseudocode on the computation of set $T_{s,i}$ for some $a_s \in A$ and some $p_i \in A_p$ based on the knowledge of sets $T_{q,i-1}$ and $S_{q,i-1}$ for all $a_q \in A$. Procedure matchWindows() that is used to compute $S_{s,i}$ is given in Fig. 5; the call to Function minAnchor() is explained in Sect. 3.2.

(Case b): For every $a_r \in A$ with cross(a_c, a_r, a_s) and every (c, w_1, w_2; anchor(i)) $\in$ $S_{r,i-1}$ with $a_{w_1} < a_s$ but $a_{w_2} \between a_s$, we add w_2 to $T_{s,i}$. The resulting set is the final set $T_{s,i}$. Thus, for the computation of $T_{s,i}$ in this case, we iterate over all possible choices of $a_c, a_{w_1}, a_{w_2}, a_r \in A$.

In fact, this computation can be improved in (Case b), saving the necessity to iterate over possible choices of a_c. Given $a_{w_1}, a_{w_2}, a_r \in A$ and $p_i \in A_p$, we use in Fig. 4 Function minAnchor(w_1, w_2, r; anchor(i), $i - 1$) to return the minimum c – i.e., a_c has minimum right endpoint – such that A_p[anchor($i - 1$), anchor(i) − 1, anchor(i), $i - 1$] matches $A[c, w_1, w_2, r]$. This value of c is then used in the computation of $T_{s,i}$. The call minAnchor(w_1, w_2, r; anchor(i), $i - 1$) needs only constant time since – as will become evident in the following subsection – we can during the computation of set $S_{r,i-1}$ maintain an array that keeps, for every $a_{w_1}, a_{w_2} \in A$, track of the minimum value c having the described property.

In Fig. 4, the calls to Procedure matchWindows() are used to compute set $S_{s,i}$ and will be explained in the following subsection.

3.3   Matching Pattern Windows

In this subsection, we show how we compute, for given $a_s \in A$ and $p_i \in A_p$, $i > 1$, the set $S_{s,i}$, based on the knowledge of sets $S_{q,i-1}$ for all $a_q \in A$. The corresponding pseudocode is given in Figs. 4 and 5.

We compute the set $S_{s,i}$ in "slices", meaning that for given values $a_{c_r}, a_{c_s}, a_r \in A$ for which

$$A_p[\text{anchor}(i-1), \text{anchor}(i), i-1, i] \text{ matches } A[c_r, c_s, r, s], \qquad (1)$$

Procedure matchWindows() computes the set $S'_{s,i} \subseteq S_{s,i}$ of 4-tuples $(c_s, w_1, w_2; j)$ for which $A_p[\text{anchor}(i-1), \text{anchor}(i), j-1, j, i-1, i]$ matches $A[c_r, c_s, w_1, w_2, r, s]$. Iterating over all possible choices of a_{c_r}, a_{c_s}, and a_r, we then compute all "slices" of $S_{s,i}$. In the following, we explain, firstly, how matchWindows() can be called in a more efficient way in the process of computing the sets $T_{s,i}$ (the computation of $T_{s,i}$ was explained in Sect. 3.3). Secondly, we explain how matchWindows() is computed.

Instead of enumerating all possible values of a_{c_r}, a_{c_s}, and a_r in order to find, for given $a_s \in A$ and p_i, all values which satisfy (1), we observe that we already obtain these values in the process of computing $T_{s,i}$. Namely, if (1) is satisfied then Procedure matchAnchor() adds c_s to $T_{s,i}$ while matching $A_p[\text{anchor}(i-1), i-1]$ to $A[c_r, r]$. Therefore, as shown in Fig. 4, we call matchWindows() exactly when c_s is added to $T_{s,i}$, either in (Case a) or in (Case b).

Now we describe how Procedure matchWindows() is computed. With respect to the arc matches determined by (1) – these arc matches are the given arguments of Procedure matchWindows() – we call a pair $(w; j)$ for $a_w \in A$ and $\text{anchor}(i) \leq j < i$ reachable if there exists w_h for every $h = \text{anchor}(i), \text{anchor}(i) + 1, \ldots, j$, such that (1) holds while matching p_h with w_h for all these h. The reachable pairs determine the 4-tuples of the set $S'_{s,i}$ to be constructed, as will be indicated in the following.

We compute the reachable pairs in an inductive way and store them in set $R_{s,i}$. More precisely, the induction starts with pair $(c_s; \text{anchor}(i))$ which belongs to $R_{s,i}$ in any case. Then, we loop through integers j with $\text{anchor}(i) < j < i$ in ascending order. For a given j, we iterate over all pairs $a_{w_1}, a_{w_2} \in A$ with $\text{cross}(a_{c_s}, a_{w_1}, a_{w_2}, a_s)$, and add (a_{w_2}, j) to $R_{s,i}$ if $(a_{w_1}, j-1)$ is already in $R_{s,i}$ and if $(c_r, w_1, w_2; j) \in S_{r,i-1}$. Moreover, we add $(c_s, w_1, w_2; j)$ to $S_{s,i}$. In addition to these 4-tuples added to $S_{r,i-1}$ during the above induction, we add, in any case, $(c_r, r, s; i)$ to $S_{r,i-1}$.

3.4   Running Time and Correctness

The running time and correctness of the algorithm presented in Sect. 3.1 to 3.3 is summarized in the following theorem.

Theorem 3.1. CONTACT MAP PATTERN MATCHING *restricted to* $\{\emptyset, <\}$-*patterns is solvable in* $O(n^6 \cdot m^2)$ *time for a contact map of size n and a pattern of size m.*

For the proof of Theorem 3.1, we employ the following two lemmas (proofs omitted).

Lemma 3.2. *For $i > 1$, if $T_{s,i-1}$ and $S_{s,i-1}$ are computed correctly for $j < i$, then Procedure* matchAnchor() *correctly computes set $T_{s,i}$.* $\qquad\square$

---

procedure matchWindows($a_{c_r}, a_{c_s}, a_r, a_s; p_i$)
Global: Contact maps (S, A) and pattern (S_p, A_p), set $S_{q,i-1}$ for every $a_q \in A$.
Input: Arcs $a_{c_r}, a_{c_s}, a_r, a_s \in A$ and arc $p_i \in A_p$ such that
    $A_p[\text{anchor}(i-1), \text{anchor}(i), i-1, i]$ matches $A[c_r, c_s, r, s]$.
Output: Set $S'_{s,i}$ containing all 4-tuples $(c_s, w_1, w_2; j)$, $j < i$ such that
    $A_p[\text{anchor}(i-1), \text{anchor}(i), j-1, j, i-1, i]$ matches $A[c_r, c_s, w_1, w_2, r, s]$.
Method:

$\quad S'_{s,i} := \emptyset$;
$\quad R_{s,i} := \emptyset$; /* to store reachable pairs */

$\quad$ /* initialize the set $R_{s,i}$ */
$\quad R_{s,i} := R_{s,i} + (c_s; \text{anchor}(i))$;

$\quad$ /* inductive computation of $R_{s,i}$ and $S'_{s,i}$ */
$\quad$ **for all** $j = \text{anchor}(i) + 1$ upto $i - 1$ **do**
$\quad\quad$ **for all** $w_1, w_2 \in A$ **do**
$\quad\quad\quad$ **if** $(\text{anchor}(p_i), p_{j-1}, p_j, p_{i-1}, p_i)$ matches $(a_{c_s}, a_{w_1}, a_{w_2}, a_r, a_s)$ **then**
$\quad\quad\quad\quad$ **if** $(w_1; j-1) \in R_{s,i}$ **and** $(c_r, w_1, w_2; j) \in S_{r,i-1}$ **then**
$\quad\quad\quad\quad\quad R_{s,i} := R_{s,i} + (w_2; j)$;
$\quad\quad\quad\quad\quad S'_{s,i} := S'_{s,i} + (c_s, w_1, w_2; j)$;
$\quad\quad\quad\quad$ **end if**;
$\quad\quad\quad$ **end if**;
$\quad\quad$ **end for**;
$\quad$ **end for**;

$\quad$ /* final step in computation of $S'_{s,i}$ */
$\quad S'_{s,i} := S'_{s,i} \cup \{c_s, r, s; i\}$;

$\quad$ **return** $S'_{s,i}$;

Fig. 5. Overview in pseudocode on the computation of (subsets of) $S_{s,i}$, for $a_s \in A$ and $p_i \in A_p$.

Lemma 3.3. *For $i > 1$, if $T_{s,i-1}$ and $S_{s,i-1}$ are computed correctly and if $T_{s,i}$ is computed correctly, then Procedure* matchWindows() *correctly computes $S_{s,i}$.*   □

Proof (of Theorem 3.1). We prove this theorem by analyzing the algorithm presented in Sect. 3.1 to 3.3. We discuss the algorithm's running time and show its correctness.

Running Time. A set $T_{s,i}$ is computed for n values of s and m values of i. The dynamic programming step to compute a set $T_{s,i}$ for some fixed values of s and i can be done in $O(n^3)$ time as shown in Procedure matchAnchor(), Fig. 4: The worst case is determined by (Case b) in which we iterate over possible choices for $a_{w_1}, a_{w_2}, a_r \in A$. The update of set $S_{s,i}$, for every found match, can be done in $O(n^2 m)$ time as shown in Procedure matchWindows(), Fig. 5: We iterate over possible choices for $a_{w_1}, a_{w_2} \in A$ and $p_j \in A_p$. Altogether, this results in a total running time of $O(n^6 m^2)$.

Correctness. Showing the correctness of the algorithm, basically comes down to showing that sets $T_{s,i}$ and $S_{s,i}$ are computed correctly. We prove this by induction on i, starting with $i = 1$. It is easy to check that sets $T_{s,1}$ and $S_{s,1}$ are computed correctly. For $i > 1$, the correctness of computing $T_{s,i}$ and $S_{s,i}$ follows with Lemmas 3.2 and 3.3.   □

The running time estimate given in Theorem 3.1 is in fact a worst-case estimate. In practice, the six nested loops on A do not all have to consider every element of A as assumed in the worst-case estimation. In a preprocessing step, we can compute, for every $a \in A$, the set of arcs $a' \in A$ for which $a' \between a$. Then, e.g., in (Case a) of Procedure matchAnchor(), we only consider those arcs a_c and a_r for which $a_c \between a$ and $a_r \between a$. In the same way, we can speed-up all loops over A in Procedures matchAnchor() and matchWindows().

3.5 Generalizations

In the following, we discuss the extension of the presented algorithm to two closely related models, namely the one of arc annotations and the one of 2-interval sets. Due to lack of space, we only sketch these extensions, omitting the details.

Arc annotations. Arc annotations are a model mainly used for modelling RNA structures [4,5,8,11]. The main difference between arc annotations and contact maps is that the elements in S carry additional sequence information. Reframing this question in terms of contact maps, for a given alphabet Σ and a given contact map (S, A), we are also given a labeling $\ell : S \longrightarrow \Sigma$, resulting in a *labeled* contact map (S, A, ℓ). Given two labeled contact maps (S, A, ℓ) and (S_p, A_p, ℓ_p), we say that (S_p, A_p, ℓ_p) *occurs* in (S, A, ℓ) if there is a one-to-one mapping M of S_p to a size-$|S_p|$ subset S' of S such that, for $e, e' \in S_p$, we have $e < e' \Rightarrow M(e) < M(e')$, $(e, e') \in A_p \Rightarrow (M(e), M(e')) \in A$, and – in addition to the definition for "usual" contact maps – $\ell_p(e) = \ell(M(e))$. The corresponding CMPM problem is then formulated as in Sect. 1. The labeling can be used, e.g., to capture the primary sequence of the protein or to encode chemical properties of amino acids. In this way, we obtain a problem formulation that incorporates *both* sequence *and* structure information.

We extend our algorithm to labeled contact maps, by requiring, whenever we try to match an element in S with an element in S_p, to check whether their labelings coincide.

2-interval sets. Vialette introduced 2-interval sets in the context of RNA structure analysis. A 2-interval set is given as a pair (I, A) where I, in contrast to S in the definition of contact maps, is given as a set of intervals $[i_l, i_r]$ with positive integers i_l and i_r. Consequently, A is a set of interval pairs, the intervals of one pair being non-overlapping; non-paired intervals can, however, overlap. We specify a partial order on the intervals, saying that two intervals $i = (i_l, i_r)$ and $i' = (i'_l, i'_r)$ satisfy $i \prec i'$ iff $i_r < i'_l$. Apart from this change, we can still employ the definition of CMPM as given in Sect. 1, replacing contact maps by 2-interval sets.

The only change required for the algorithm is to adjust the definition of whether two arcs a and a' are crossing or preceding. For 2-interval sets, two arcs $a = (i, i')$ and $a' = (j, j')$, for intervals i, i', j, and j', are crossing iff $i \prec j \prec i' \prec j'$. Analogously, a and a' are preceding iff $i \prec i' \prec j \prec j'$. Note that, consequently, intervals of a $\{<, \between\}$-structured 2-interval pattern have pairwisely disjoint intervals.

3.6 Applications

In this subsection we discuss how we can use the algorithm presented in Sect. 3.1 to 3.3 in order to solve a related NP-hard problem discussed by Vialette [12] and Blin *et al.* [3]:

the 2-INTERVAL PATTERN problem restricted to $\{<, \between\}$-structured patterns. Given a 2-interval set (S, A) where S is a set of intervals and A is a set of 2-interval pairs, it asks for a maximum size 2-interval set (S_p, A_p) such that A_p is $\{<, \between\}$-structured and occurs in (S, A); as usual, we denote $n = |A|$ and $m = |A_p|$. 2-INTERVAL PATTERN is NP-hard even when restricted to $\{<, \between\}$-structured patterns [3]. With Theorem 3.1 and its generalization outlined in 3.5, we can show that this problem is fixed-parameter tractable with respect to parameter m, i.e., it is solvable in $f(m) \cdot \text{poly}(n, m)$ time:

Theorem 3.4. *The* 2-INTERVAL PATTERN *problem restricted to* $\{<, \between\}$-*structured patterns is solvable in* $O(3^m \cdot n^6 m^2)$ *time.* $\qquad\qquad\square$

The proof of Theorem 3.4 relies on a one-to-one correspondence of $\{<, \between\}$-structures and the Dyck language [7]; similar arguments were already exploited in [1]. Theorem 3.4 directly translates to a special case of the well-known MAXIMUM CONTACT MAP OVERLAP (CMO) problem [6, 9]. Given two contact maps (S_1, A_1) and (S_2, A_2) $(n = \max(|A_1|, |A_2|))$, CMO asks for a maximum-size contact map (S_p, A_p) $(m = |A_p|)$ that occurs both in A_1 and in A_2. Then, CMO restricted to $\{<, \between\}$-structured overlaps requires that (S_p, A_p) is $\{<, \between\}$-structured. Theorem 3.4 implies that CMO restricted to $\{<, \between\}$-structured overlaps is fixed-parameter tractable with respect to $|A_p|$. However, it is open to show that this problem is NP-hard [3].

4 Experimental Results

In this section, we describe experiments that have been conducted with a straightforward C implementation of the algorithm presented in this work. Running times were measured on a Sun SunFire V100 with 1024MB of RAM running Solaris. The contact maps derived from entries of the PDB database are generated based on the distance between the c_α atoms of the protein sequence; the c_α atoms constitute the nodes of the elements of the contact map. Two c_α atoms share an arc if their distance is less than 5.5 Å. A pattern of size m is implanted into a given contact map by randomly choosing the $2m$ endpoints of the pattern among the elements of the contact map.

To give an idea about the practical running times of the algorithm, we took measurements when running our implementation on contact maps into which we implanted a random pattern. We used two kinds of contact maps, in which the pattern was searched, random ones and ones derived from protein domain structures selected from the PDB database such that representatives from different structural classes and architectures [10] were included. Our results are displayed in Fig. 6. We see that the algorithm can process contact maps containing several hundred arcs in acceptable running times (Fig. 6(a)). Considering our results on contact maps which were derived from protein domain structures (Fig. 6(b)), it turns out that they were less difficult to process than random contact maps. In summary, the results show that the algorithm can efficiently process real protein structure data.

To give an idea about a realistic scenario in which the presented algorithm could be applied for structural classification of proteins, we investigated four examples of homologous protein domain superfamilies from the CATH database [10]. We chose, for each of these superfamilies, three protein domain structures from the superfamily and – by

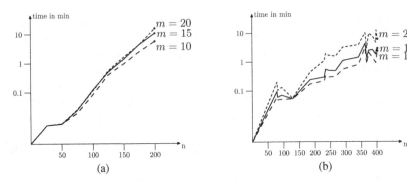

Fig. 6. Performance, displayed on a logarithmic scale, on contact maps of size n into which we implanted a random $\{<, \emptyset\}$-structured contact map pattern of size m. (a) Random contact maps. (b) Contact maps derived from real protein domain structures.

architecture	superfamily	m	#pos	#neg	fn	fp
Trefoil	Lectin	9	27	27	0	2
TIM Barrel	Glycosidases	15	20	20	2	4
3-Layer Sandwich	GroEL	10	17	18	1	1
2-Layer Sandwich	D-Amino Acid Oxidase	17	20	20	0	0

Fig. 7. Using the algorithm for CMPM for classification of protein domain structures, following the classification provided by the CATH database. The length of the chosen pattern is denoted by m, #pos denotes the number of the investigated domains that are in CATH listed as members of the superfamily, #neg denotes the number of investigated non-members. In our classification, we observed fp false positives and fn false negatives.

visual inspection – identified a significant $\{<, \emptyset\}$-structured contact map pattern shared by the three structures. Then, we used this pattern to classify structures, on the one hand of protein domains listed in CATH as members of the superfamily, and of proteins domains listed in CATH as non-members but as 'structural relatives' of members. In difference to CATH, our classification thus *merely* relied on structural features and not on sequence similarity. The results of this experiment are displayed in Fig. 7. The result shows that already using this straightforward approach allows a correct classification in most cases. Making this idea competitive naturally requires an automatic detection of common patterns (e.g., using ideas of Sect. 3.6) and the use of a *set* of characteristic contact map patterns; these extensions remain to be explored.

5   Conclusion

This paper positively answered the previously open question [12] whether matching $\{<, \emptyset\}$-structured patterns in contact maps (or 2-interval sets, respectively) can be done in polynomial time. Constraining ourselves to $\{<, \emptyset\}$-structured patterns may seem a stringent requirement. However, we point out that the algorithm presented in this work gives, together with the quadratic-time algorithm for CMPM restricted to $\{<, \sqsubset\}$-structured patterns [5], the building-stones for a much more general pattern matching framework. It remains an issue of future research to classify those patterns that can now be matched in polynomial time by a combination of these two algorithms. Further, an open question is whether the running time for CMPM restricted to $\{<, \emptyset\}$-structured patterns can be improved. It is to be explored how our results regarding the MAXIMUM CONTACT MAP OVERLAP (CMO) problem (see Sect. 3.6) perform in practice. Finally, it is open to prove or disprove the conjecture that CMO restricted to $\{<, \emptyset\}$-structured overlaps is NP-hard.

Acknowledgements

I thank Nadja Betzler (University of Tübingen, Germany) for providing her implementation which generates contact maps from PDB structure files. Further, I thank Markus Stirnkorb and Jiong Guo (University of Tübingen) and Stéphane Vialette (LRI Orsay, France) for helpful discussions on this topic.

References

1. J. Alber, J. Gramm, J. Guo, and R. Niedermeier. Computing the similarity of two sequences with nested arc annotations. *Theoretical Computer Science* 312:337–358, 2004.
2. H.M. Berman *et al.*. The Protein Data Bank. *Nucleic Acids Research*, 28:235–242, 2000. http://www.rcsb.org/pdb/.
3. G. Blin, G. Fertin, and S. Vialette. New results for the 2-interval pattern problem. In *Proc. of the 15th CPM*, to appear in LNCS, 2004, Springer.
4. P. A. Evans. Finding common subsequences with arcs and pseudoknots. In *Proc. of 10th CPM*, pages 270–280, volume 1645 in LNCS, Springer, 1999.
5. J. Gramm, J. Guo, and R. Niedermeier. Pattern matching for arc-annotated sequences. In *Proc. of the 22nd FSTTCS*, number 2556 in LNCS, pages 182–193, 2002, Springer.
6. D. Goldman, S. Istrail, and C. H. Papadimitriou. Algorithmic aspects of protein structure similarity. In *Proc. of the 40th FOCS*, pages 512–521, 1999, IEEE Computer Society.
7. M. A. Harrison. Introduction to Formal Language Theory. Addison-Wesley, Reading, 1978.
8. T. Jiang, G.-H. Lin, B. Ma, and K. Zhang. The Longest Common Subsequence problem for arc-annotated sequences. In Proc. of the 11th CPM, pages 154–165, volume 1848 in LNCS, Springer, 2000.
9. G. Lancia, R. Carr, B. Walenz, and S. Istrail. 1001 optimal PDB structure alignments: Integer Programming methods for finding the maximum contact map overlap. *Journal of Computational Biology*, 11(1):27–52, 2004.
10. C.A. Orengo, A.D. Michie, S. Jones, D.T. Jones, M.B. Swindells, and J.M. Thornton. CATH – A Hierarchic Classification of Protein Domain Structures. *Structure* 5(8):1093–1108. 1997. http://www.biochem.ucl.ac.uk/bsm/cath/.
11. Z. Wang and K. Zhang. RNA secondary structure prediction. In T. Jiang *et al.* (eds): *Current Topics in Computational Molecular Biology*, pages 345–364, MIT Press, 2002.
12. S. Vialette. On the computational complexity of 2-interval pattern matching problems. *Theoretical Computer Science*, 312(2–3):223–249, 2004.

A 1.5-Approximation Algorithm for Sorting
by Transpositions and Transreversals

Tzvika Hartman[1] and Roded Sharan[2]

[1] Dept. of Computer Science and Applied Mathematics, Weizmann Institute of Science
Rehovot 76100, Israel
`tzvi.hartman@weizmann.ac.il`
[2] International Computer Science Institute, 1947 Center St., Berkeley, CA 94704
`roded@icsi.berkeley.edu`

Abstract. One of the most promising ways to determine evolutionary distance between two organisms is to compare the order of appearance of orthologous genes in their genomes. The resulting genome rearrangement problem calls for finding a shortest sequence of rearrangement operations that sorts one genome into the other. In this paper we provide a 1.5-approximation algorithm for the problem of sorting by transpositions and transreversals, improving on a five years old 1.75 ratio for this problem. Our algorithm is also faster than current approaches and requires $O(n^{3/2}\sqrt{\log n})$ time for n genes.

1 Introduction

When trying to determine evolutionary distance between two organisms using genomic data, one wishes to reconstruct the sequence of evolutionary events that have occurred, transforming one genome into the other. One of the most promising ways to trace the evolutionary events is to compare the order of appearance of orthologous genes in two different genomes [1, 2]. This comparison, which relies on computing global rearrangement events, may provide more accurate and robust clues to the evolutionary process than the analysis of local mutations.

In a genome rearrangement problem, the two compared genomes are represented by permutations, where each element stands for a gene, and the goal is to find a shortest sequence of rearrangement operations that transforms (sorts) one permutation into the other. Previous work focused on the problem of sorting a permutation by reversal operations. This problem was shown to be NP-hard [3]. One of the most celebrated results in this area by Hannenhalli and Pevzner shows that for signed permutations (every element of the permutation has a sign, which represents the direction of the corresponding gene; a reversal reverses the order of the elements it operates on and flips their signs), the problem becomes polynomial [4]. The algorithm is based on representing a permutation using a breakpoint graph (we defer a formal definition to Section 2) which decomposes uniquely into disjoint cycles, and studying the effect of a reversal on its cycle decomposition. There has been less progress on the problem of sorting by transpositions. A transposition is a rearrangement operation in which a segment is cut out of the permutation and pasted in a different location. The complexity of sorting by transpositions is still open, although several 1.5-approximation algorithms are known for it [5–7].

I. Jonassen and J. Kim (Eds.): WABI 2004, LNBI 3240, pp. 50–61, 2004.

A transreversal is a biologically motivated operation that combines a transposition and a reversal: A segment is cut out of the permutation, reversed and pasted in another location. In particular, a reversal is also a transreversal. Transpositions and transreversals capture a large fraction of the genomic rearrangements in evolution. Gu et al. [8] gave a 2-approximation algorithm for sorting signed permutations by transpositions and transreversals. Lin and Xue [9] improved this ratio to 1.75 by considering a third rearrangement operation, called revrev, which reverses two contiguous segments. Both algorithms run in quadratic time.

In this paper we study the problem of sorting permutations by transpositions, transreversals and revrevs. The question of whether the 1.75 known ratio for this problem can be improved, has been open for five years. One of the main difficulties in tackling the complexity of this problem is the vast number of possible configurations that need to be considered when analyzing general linear permutations. We make four contributions toward greatly simplifying the problem. First, we show that the sorting problem is equivalent for linear and circular permutations (Section 2). Second, we reduce the general problem of sorting a circular permutation to that of sorting a permutation with a very simple structure: In its breakpoint graph representation all non-trivial cycles are of length 3 (Section 2). Third, we characterize cycle configurations in the breakpoint graph and show that it suffices to restrict attention to one type of configurations. Fourth, we develop and characterize a novel cycle representation, which allows us to use results on sorting by transpositions only, a well-studied problem, in further eliminating cycle configurations. These characterizations and simplifications are key to our main result: a 1.5-approximation algorithm for sorting by transpositions, transreversals and revrevs (Section 3). Furthermore, our algorithm can be implemented in time $O(n^{3/2}\sqrt{\log n})$, which improves on the quadratic running time of previous algorithms [8, 9]. For lack of space, some proofs are shortened or omitted.

2 Preliminaries

A *signed permutation* $\pi = [\pi_1 \ldots \pi_n]$ on $n(\pi) \equiv n$ elements is a permutation in which each element is labelled by a sign of plus or minus. A *segment* of π is a consecutive sequence of elements $\pi_i, \ldots, \pi_k$ ($k \geq i$). We focus on four rearrangement operations. A *reversal* ρ is an operation that reverses the order of the elements in a segment and flips their signs. If the segment is $\pi_i, \ldots, \pi_{j-1}$ then $\rho \cdot \pi = [\pi_1, \ldots, \pi_{i-1}, -\pi_{j-1}, \ldots, -\pi_i, \pi_j, \ldots, \pi_n]$. Two segments $\pi_i, \ldots, \pi_k$ and $\pi_j, \ldots, \pi_l$ are *contiguous* if $j = k + 1$ or $i = l + 1$. A *transposition* τ exchanges two contiguous (disjoint) segments. If the segments are $A = \pi_i, \ldots, \pi_{j-1}$ and $B = \pi_j, \ldots, \pi_{k-1}$ then $\tau \cdot \pi = [\pi_1, \ldots, \pi_{i-1}, \pi_j, \ldots, \pi_{k-1}, \pi_i, \ldots, \pi_{j-1}, \pi_k, \ldots, \pi_n]$ (note that the end segments can be empty if $i = 1$ or $k = n + 1$). A *transreversal* $\tau\rho_{A,B}$ is a transposition that exchanges segments A and B and also reverses A., i.e., $\tau\rho_{A,B} \cdot \pi = [\pi_1, \ldots, \pi_{i-1}, \pi_j, \ldots, \pi_{k-1}, -\pi_{j-1}, \ldots, -\pi_i, \pi_k, \ldots, \pi_n]$, and $\tau\rho_{B,A} \cdot \pi = [\pi_1, \ldots, \pi_{i-1}, \ldots, -\pi_{k-1}, \ldots, -\pi_j, \pi_i, \ldots, \pi_{j-1}, \pi_k, \ldots, \pi_n]$. A *revrev* operation reverses each of the two segments (without transposing them). Thus, $\rho\rho \cdot \pi = [\pi_1, \ldots, \pi_{i-1}, -\pi_{j-1}, \ldots, -\pi_i, -\pi_{k-1}, \ldots, -\pi_j, \pi_k, \ldots, \pi_n]$.

The problem of finding a shortest sequence of transposition, transreversal and revrev operations that transforms a permutation into the identity permutation is called *sorting* by *transpositions* and *transreversals*[1]. The *distance* of a permutation π, denoted by $d(\pi)$, is the length of the shortest sorting sequence.

Linear vs. Circular Permutations. Key to our approximation algorithm is a reduction from the problem of sorting linear permutations to that of sorting circular permutations (indices are cyclic), on which the analysis is simpler. An operation is said to *operate* on the segments that are affected by it and on the elements in those segments. We say that two operations μ and μ' are *equivalent* if they have the same effect, i.e., $\mu \cdot \pi = \mu' \cdot \pi$ for all π. The following lemma is the basis for the reduction, and is used to prove the subsequent theorem on the equivalence of the sorting problem for linear and circular permutations, similarly to [7].

Lemma 1 *Let x be an element of a circular permutation π, and let μ be an operation that operates on x. Then there exists an equivalent operation μ' that does not operate on x.*

Proof. For reversals, this result was proven by Meidanis et al. [10] and for transpositions by Hartman [7]. For transreversals and revrevs, the claim relies on the observation that a chromosome is equivalent to its reflection, i.e., the reversed sequence of elements with their signs flipped [10] (see the upper part of Figure 1). Consider a permutation with three segments: A, B and C. W.l.o.g. $x \in A$. Then a transreversal that operates on segments A and B and reverses B (resp. A) is equivalent to a revrev that operates on A and C (B and C), since the result is a reflection of the permutation (as illustrated in Figure 1). Similarly, a revrev that operates on A and B (or C) is equivalent to a transreversal that operates on B and C. ∎

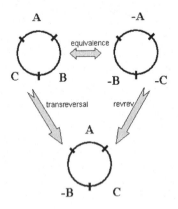

Fig. 1. The equivalence of operations on circular permutations.

---

[1] We do not include revrevs in the problem name, as we provide in the next section a reduction of the problem that allows us to mimic revrevs using transreversals.

Theorem 2 *The problem of sorting linear permutations by transpositions and transreversals is linearly equivalent to the problem of sorting circular permutations by transpositions and transreversals.*

Proof. We show only one direction. Given a linear n-permutation, circularize it by adding an additional element $\pi_{n+1} = x$ and closing the circle. Denote the new circular permutation by π^c. By Lemma 1, any operation on π^c can be mimicked by an operation that does not involve the segment that includes x. Hence, there is an optimal sequence of operations that sorts π^c such that none of them operates on segments that include x. The same sequence can be viewed as a sequence of operations on the linear permutation π, by ignoring x. This implies that $d(\pi) \leq d(\pi^c)$. On the other hand, any sequence of operations on π is also a sequence of operations on π^c, so $d(\pi^c) \leq d(\pi)$. Hence, $d(\pi) = d(\pi^c)$. Moreover, an optimal sequence for π^c implies an optimal sequence for π. ∎

We observe that for circular permutations revrevs and transreversals are equivalent operations. Thus, for circular permutations we can restrict attention to transpositions and transreversals, which are more biologically motivated operations. Moreover, combined with Theorem 2, this observation implies that one can reduce the problem of sorting a linear permutation by transpositions, transreversals and revrevs to that of sorting a circular permutation by transpositions and transreversals only.

The Breakpoint Graph. We follow the construction of Bafna and Pevzner for representing signed permutations [11]. First, a permutation π on n elements is transformed into a permutation $f(\pi) = \pi' = (\pi'_1 \ \ldots \ \pi'_{2n})$ on $2n$ elements, by replacing each positive element i by two elements $2i - 1, 2i$ (in this order), and each negative element by $2i, 2i - 1$. On the extended permutation $f(\pi)$, only operations that cut before odd positions are allowed. This ensures that every operation on $f(\pi)$ can be mimicked by an operation on π. The *breakpoint graph* $G(\pi)$ is an edge-colored graph on $2n$ vertices $\{1, 2, \ldots, 2n\}$. For every $1 \leq i \leq n$, π'_{2i} is joined to π'_{2i+1} by a black edge (denoted by b_i), and $2i$ is joined to $2i + 1$ by a gray edge. Here and in the rest of the paper we identify, in both indices and elements, $2n + 1$ and 1.

It is convenient to draw the breakpoint graph on a circle, such that black edges are on the circumference and gray edges are chords (see Figure 2). Since the degree of each vertex is exactly 2, the graph uniquely decomposes into cycles. A k-cycle is a cycle with k black edges, and it is *odd* if k is odd. k is called the *length* of the cycle. The number of odd cycles in $G(\pi)$ is denoted by $c_{odd}(\pi)$. Gu et al. [8] have shown that for all linear permutations π and operations μ, it holds that $c_{odd}(\mu \cdot \pi) \leq c_{odd}(\pi) + 2$. Their result holds also for circular permutations and can be used to prove the following lower bound on $d(\pi)$:

Theorem 3 ([8]) *For all permutations π, $d(\pi) \geq (n(\pi) - c_{odd}(\pi))/2$.*

Transformation into 3-Permutations. Our goal in this section is to transform the input permutation into a permutation with simple structure, to which we can apply our algorithm and mimic its steps on the original permutation. A permutation is called *simple* if its breakpoint graph contains only k-cycles, where $k \leq 3$. It is called a *3-permutation*

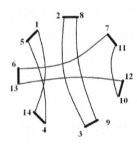

Fig. 2. The circular breakpoint graph of the permutation $\pi = (1 \; -4\;6 \; -5\;2 \; -7 \; -3)$, for which $f(\pi) = (1\;2\;8\;7\;11\;12\;10\;9\;3\;4\;14\;13\;6\;5)$. Black edges are represented as thick lines on the circumference, and gray edges are chords.

if it contains only 1-cycles and 3-cycles. A transformation from π to $\hat{\pi}$ is called *safe* if $n(\pi) - c_{odd}(\pi) = n(\hat{\pi}) - c_{odd}(\hat{\pi})$, i.e., if it maintains the lower bound of Theorem 3. Next, we show how to transform an arbitrary permutation into a 3-permutation using safe transformations (that is, maintaining the lower bound, but not the exact distance). Our starting point is the standard safe transformation into simple permutations (cf. [7]). Hence, it suffices to show how to convert 2-cycles into 3-cycles using safe transformations.

Let C be a 2-cycle and let $b = (\pi'_{2i}, \pi'_{2i+1})$ be one of its black edges. A (C, b)-*padding* extends the original permutation π by adding a new element $\pi_i + 1$, and renaming all elements $j > \pi_i + 1$ by $j + 1$ (the renaming is done on the absolute values of the elements and then their signs are reintroduced, e.g., -3 is renamed to -4). The new element $\pi_i + 1$ has the same sign as π_i, and is located after (resp. before) π_i if it is positive (negative). Finally, the sign of π_i is flipped. The effect on the breakpoint graph is that C is transformed into a 3-cycle (see Figure 3 for an example). Overall, the permutation after the padding has an additional element and one more odd cycle.

Lemma 4 *Every simple permutation π can be transformed into a 3-permutation $\hat{\pi}$ by safe paddings. Moreover, every sorting of $\hat{\pi}$ mimics a sorting of π with the same number of operations.*

Proof. Let π be a simple permutation that contains a 2-cycle C and let $b \in C$. Let $\overline{\pi}$ be the permutation obtained by applying a (C, b)-padding on π. Clearly, $n(\overline{\pi}) = n(\pi) + 1$, and $c_{odd}(\overline{\pi}) = c_{odd}(\pi) + 1$, so the padding is safe. This process can be repeated until a 3-permutation $\hat{\pi}$ is eventually obtained. Since $\hat{\pi}$ is obtained from π by padding new elements, every operation of $\hat{\pi}$ can be mimicked on π by ignoring the padded elements. ∎

In the rest of the paper, we shall restrict attention to circular 3-permutations and often refer to the 3-cycles in our breakpoint graph simply as cycles. In Section 3 we show how to sort a 3-permutation using at most $1.5l$ operations, where l is the lower bound of Theorem 3. By Theorem 2 and Lemma 4 this implies a 1.5-approximation algorithm for sorting arbitrary circular and linear permutations.

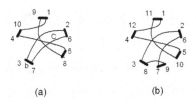

(a)   (b)

Fig. 3. (a) The breakpoint graph of the permutation $\pi = (1, -3, -4, 2, -5)$. (b) The graph of $(1, -3, -5, 4, 2, -6)$, which is obtained by a (C, b)-padding.

Cycle Types. An operation that cuts some black edges is said to *act on* these edges. It is called a *k-operation* if it increases the number of odd cycles by k. An odd cycle is called *oriented* if there is a 2-operation that acts on three of its black edges; otherwise, it is *unoriented*. A *configuration* of cycles is a subgraph of the breakpoint graph that contains one or more cycles. There are four possible configurations of single 3-cycles, which are shown in Figure 4(a-d). It is easy to verify that cycles a and b are unoriented, whereas c and d are oriented. A black edge is called *twisted* if its two adjacent gray edges cross each other as chords in the circular breakpoint graph. A cycle is *k-twisted* if k of its black edges are twisted. For example, in Figure 4 cycle a is 0-twisted and c is 2-twisted.

Observation 5 *A 3-cycle is oriented iff it is 2- or 3-twisted.*

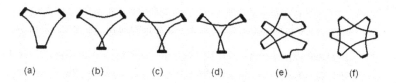

(a)   (b)   (c)   (d)   (e)   (f)

Fig. 4. Configurations of 3-cycles. (a-b) Unoriented 3-cycles. (c-d) Oriented 3-cycles. (e) A pair of intersecting 3-cycles. (f) A pair of interleaving 3-cycles.

Let $b = (i_1, i_2)$ and $b' = (j_1, j_2)$ be two black edges in the breakpoint graph such that i_1, i_2, j_1 and j_2 occur in this order along the circle. Then b_1 and b_2 *induce* two disjoint *arcs* on the circle, one between i_2 and j_1 and the other between j_2 and i_1. Two pairs of black edges are called *intersecting* if they alternate in their order of occurrence along the circle. A pair of black edges intersects with cycle C, if it intersects with a pair of black edges that belong to C. Cycles C and D intersect if there is a pair of black edges in C that intersect with D (see Figure 4e). Two cycles are *interleaving* if their black edges alternate in their order of occurrence along the circle (see Figure 4f). A *1-twisted pair* is a pair of 1-twisted cycles, whose twists are consecutive on the circle in a configuration that consists of these two cycles only. A pair of black edges is said to be *coupled* if they are connected by a gray edge and when reading the edges along the circle they are read in the same direction. (For example, the top edges in Figure 4b are coupled, and so are all pairs of edges in Figure 4a).

The following lemma will be useful in the sequel:

Lemma 6 ([8]) *Let (b_1, b_2) be a pair of coupled black edges. Then there exists a cycle C that intersects with (b_1, b_2).*

3   The Algorithm

A $(0, 2, 2)$-*sequence* is a sequence of three operations, of which the first is a 0-operation and the next two are 2-operations. Since a 2-operation is the best possible in one step, a series of $(0, 2, 2)$-sequences guarantees a 1.5 approximation ratio. A 1-twisted cycle is called *closed* (w.r.t. a configuration) if its two coupled edges intersect with some other cycle in the configuration. A configuration is *closed* if at least one of its 1-twisted cycles is closed; otherwise it is called *open*. In the following we shall consider only closed configurations, since an open configuration implies the existence of a closed one (by Lemma 6). For each possible closed configuration we shall prove the existence of a $(0, 2, 2)$-sequence of operations. First, we deal with interleaving cycle pairs.

Lemma 7 *Let π be a permutation that contains two unoriented, interleaving cycles C and D that do not form a 1-twisted pair. Then π admits a $(0, 2, 2)$-sequence.*

Proof. If both cycles are 0-twisted then a $(0, 2, 2)$-sequence of transpositions is given in [7]. Suppose that C is 0-twisted and D 1-twisted (resp. both are 1-twisted and their twists are not consecutive on the circle). Let a, b and c be the three arcs that are induced by the black edges of C, and let a be the arc that contains the twist of D. First apply a 0-transposition that acts on the black edges of C. This makes D 2-twisted, so it is possible to eliminate it using a 2-transversal. The latter operation makes C 2-twisted (resp. 3-twisted). A 2-transreversal (2-transposition) on C completes the $(0, 2, 2)$-sequence. ∎

In order to deal with intersecting cycles we borrow some of the theory developed in [7] for unsigned permutations. As we show below, some of this theory carries also to our case with some modifications. A useful tool that we will require is the signed canonical labelling[2] of a cycle which we present next.

For a given cycle (of any length), consider the labelling obtained by picking an arbitrary black edge of a cycle, labelling it 1, and labelling the rest of the cycle's edges according to their order of occurrence along the circle. The *signed canonical labelling* of a cycle is the signed permutation obtained by starting with the edge labelled 1 and reading the labels in the order they appear along the cycle, where the signs stand for the direction in which the edge is read: An edge that is visited in the same direction as the edge labelled 1 is positive, and otherwise it is negative (see, e.g., Figure 5). This definition captures the notion of twists in 3-cycles; indeed, a 0-twisted 3-cycle has labelling $(1, 2, 3)$, 3-twisted has labelling $(1, 3, 2)$, etc. Note that a cycle may have more than one possible canonical labelling. A canonical labelling of a 5-cycle is called *oriented* if it starts with $1, b, a$ or $1, -a, -b$ or $1, -b, a$ or $1, b, -a$, where $1 < a < b$. The motivation for this definition comes from the following observation:

Observation 8 *A 5-cycle is oriented iff it has an oriented canonical labelling.*

---

[2] A generalization of the notion of canonical labelling in [6].

Fig. 5. A 5-cycle with signed canonical labelling $(1, -4 - 3 - 2, 5)$.

Lemma 9 *Let C be a 5-cycle that admits a 2-transposition. Then any 5-cycle with the same canonical labelling up to a reversal of one element is also oriented. If in addition the canonical labelling of C is $(1, 4, 2, 3, 5)$ then a 5-cycle with the same canonical labelling up to a reversal of two consecutive elements is also oriented.*

The above lemma is the basis for handling the case of two intersecting 0-twisted cycles, which we present next:

Lemma 10 *Let π be a permutation that contains a closed configuration in which there are two intersecting 0-twisted cycles C and D. Then π admits a $(0, 2, 2)$-sequence.*

Proof. Since C and D are intersecting, C has a pair of coupled edges that do not intersect with D. By Lemma 6 there exists a cycle E that intersects with this pair of edges. The case in which E is 0-twisted was treated in [7]. If E is 1-twisted there are two cases to consider:

1. D and E are non-intersecting. Our starting point is the $(0, 2, 2)$-sequences for configurations of three 0-twisted cycles given in Figure 6, where two of the cycles are non-intersecting, and the third one intersects both. In our case, one of the non-intersecting cycles corresponds to E and is 1-twisted. Depending on the location of the twist in E, it is always possible to apply the first two transpositions shown in Figure 6 to the closed configuration. (The first transposition is applied to the edges shown in the figure, if all are non-twisted, or to a symmetric set of edges). By Lemma 9, the resulting 5-cycle is oriented, which completes the $(0, 2, 2)$-sequence.

2. D and E are intersecting. Consider the $(0, 2, 2)$-sequences for three mutually intersecting 0-twisted cycles given in Figure 7. In our case E is 1-twisted. If all three edges d, e_1 and e_2 that are cut by the first transposition are non-twisted, then we apply the first two transpositions as in Figure 7. By Lemma 9, the resulting 5-cycle, F, is oriented. The same holds for any set of symmetric edges that are non-twisted. The only closed configurations in which no such symmetric set is possible is when some arc induced by a pair of black edges of C contains a single twist. There are three such configurations, for which a $(0, 2, 2)$-sequence is described in Figure 8. ∎

Next, we deal with closed configurations that include two intersecting, 1-twisted cycles. We need the following observation:

Observation 11 *Let π be a permutation that contains a 2-twisted cycle C and a 1-twisted cycle D, such that C and D are intersecting and there is a single non-twist of D in the arc induced by the two twists of C. Then π admits two consecutive 2-operations.*

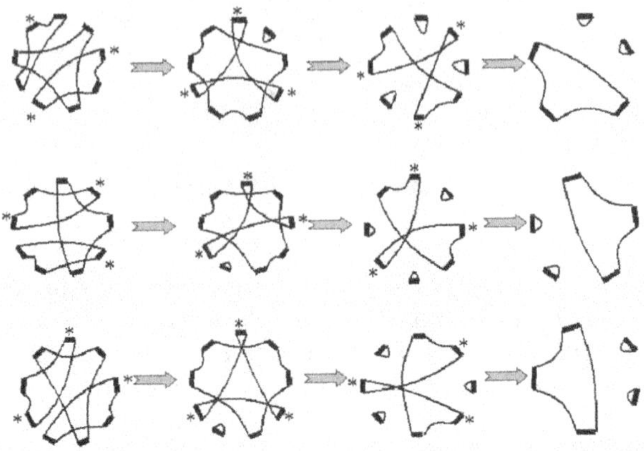

Fig. 6. $(0, 2, 2)$-sequences for three 0-twisted cycles, where two of the cycles are non-intersecting, and a third one intersects both (taken from [7]). At each step the transposition acts on the three black edges marked by a star. For simplicity, every 1-cycle is shown only when it is formed and not in subsequent graphs.

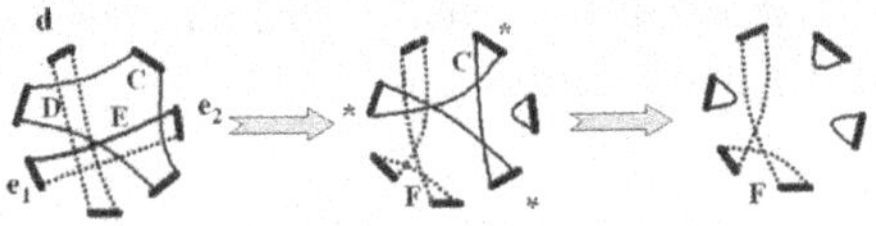

Fig. 7. Three mutually intersecting 0-twisted cycles (taken from [7]). A dashed line represents a path.

Proof. Applying a 2-transreversal on C eliminates it, while making D 2-twisted.■

Lemma 12 *Let π be a permutation that contains a closed configuration with two intersecting, 1-twisted cycles. Then π admits a $(0, 2, 2)$-sequence.*

Proof. There are six possible cases, shown in Figure 9. For cases (a-d), we first apply a 0-reversal that acts on the black edges that are marked by a star. This makes the other cycle 2-twisted, and two additional 2-operations follow from Observation 11. For cases (e-f) we observe that by Lemma 6 there is another cycle that has a black edge in the arc denoted x, and a black edge in one of the other 5 arcs. We apply a 0-reversal that acts on these two edges. If the resulting configuration contains two 2-twisted cycles then the permutation can be shown to admit two 2-operations. Otherwise, two 2-operations follow from Observation 11. ■

The following lemma deals with a closed configuration which involves two intersecting cycles, one of which is 0-twisted and the other 1-twisted. The subsequent lemma deals with 1-twisted pairs of interleaving cycles. The proofs of both lemmas can be found at http://www.icsi.berkeley.edu/~roded/transrev.pdf.

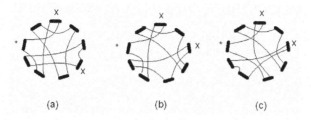

Fig. 8. $(0, 2, 2)$-sequences for some cycle configurations that contain two intersecting 0-twisted cycles. First we apply a 0-transreversal on the three marked edges, such that the segment between the two x's is reversed, resulting in an oriented 3-cycle and a 5-cycle. Next, we eliminate the 3-cycle and are left with an oriented 5-cycle, which allows us to complete the $(0, 2, 2)$-sequence.

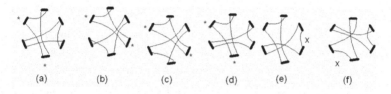

Fig. 9. Closed configurations of two intersecting 1-twisted 3-cycles.

Lemma 13 *Let π be a permutation that contains a 0-twisted cycle, which intersects with the coupled edges of a 1-twisted cycle. Then π admits a $(0, 2, 2)$-sequence.*

Lemma 14 *Let π be a permutation that contains $k \geq 2$ mutually interleaving 1-twisted cycles, such that all their twists are consecutive on the circle and k is maximal with this property. Then π admits a $(0, 2, 2)$-sequence.*

We are now ready to state our main result:

Theorem 15 *There is a 1.5-approximation algorithm for sorting by transpositions and transreversals, which runs in $O(n^{3/2}\sqrt{\log n})$ time.*

Proof. Our algorithm is described in Figure 10. The sequence of operations generated by the algorithm contains only 2-operations and $(0, 2, 2)$-sequences of operations. Therefore, every sequence of three operations increases the number of odd cycles by at least 4 out of 6 possible in 3 steps (as implied from the lower bound of Theorem 3). Hence, the approximation ratio is 1.5.

We sketch the proof of the running time. Steps 1 and 3 can be done in linear time. The number of iterations in Step 2) is linear. Note that identifying pairs of interleaving and intersecting cycles can be done by applying the following query a constant number of times: Given a gray edge find an arbitrary gray edge that intersects it. Thus, the most time-consuming tasks in each iteration are the application of operations to the permutation, and the above query. These two tasks can be performed in $O(\sqrt{n \log n})$ time using the data structure of Kaplan and Verbin [12], as we show next.

Algorithm *Sort* (π)

1. Transform π into a 3-permutation $\hat{\pi}$ (Lemma 4).
2. While $G(\hat{\pi})$ contains a 3-cycle C do:
   (a) If C is oriented, apply a 2-operation.
   (b) Otherwise, find a cycle D that intersects with a coupled pair of C.
   (c) If C and D interleave, apply a $(0, 2, 2)$-sequence (Lemmas 7, 14).
   (d) Else if C or D are 1-twisted, apply a $(0, 2, 2)$-sequence (Lemmas 12, 13).
   (e) Otherwise, apply a $(0, 2, 2)$-sequence (Lemma 10).
3. Mimic the sorting of π using the sorting of $\hat{\pi}$ (Lemma 4).

Fig. 10. Algorithm Sort. After Step 2(a) we assume that all cycles involved in the configuration of C are unoriented. Obviously, if an oriented cycle is involved, then a 2-operation can be applied on it.

For simplicity, we describe the data structure for linear permutations. Consider the breakpoint graph of a linear permutation with n elements. Each gray edge is represented by a pair of vertices, called *mates*. The permutation is partitioned into $\Theta(\sqrt{\frac{n}{\log n}})$ blocks of $\Theta(\sqrt{n \log n})$ vertices each. A splay tree [13] is attached to each block, in which the vertices of the block are maintained according to the order of their mates in the permutation. In [12] it is shown that a reversal can be applied in time $O(\sqrt{n \log n})$. Hence, transpositions (resp. transreversals) can be easily implemented using two (three) reversals in the same time bound. As for queries, we denote the pair of vertices by v_1 and v_2. We may assume that v_1 and v_2 are first elements in their blocks [12], so we need not consider parts of blocks. We scan all blocks that are between v_1 and v_2, and ask if the mate of the leftmost vertex in the block appears in the permutation before v_1, or the mate of the rightmost vertex element appears after v_2. If there is a block that satisfies the condition, then we found an intersecting pair; otherwise, there is no such pair. ∎

Acknowledgments

We would like to thank Ron Shamir for many invaluable discussions, and Elad Verbin for discussions on the data structure. TH was supported in part by an Israel Science foundation grant 309/02 (PI: Ron Shamir) and by ISF grant IS265/02. RS was supported in part by NSF ITR grant CCR-0121555.

References

1. Palmer, J.D., Herbon, L.A.: Tricircular mitochondrial genomes of Brassica and Raphanus: reversal of repeat configurations by inversion. Nucleic Acids Research **14** (1986) 9755–9764
2. Hoot, S.B., Palmer, J.D.: Structural rearrangements, including parallel inversions, within the chloroplast genome of Anemone and related genera. J. Molecular Evooution **38** (1994) 274–281
3. Caprara, A.: Sorting permutations by reversals and Eulerian cycle decompositions. SIAM Journal on Discrete Mathematics **12** (1999) 91–110

4. Hannenhalli, S., Pevzner, P.: Transforming cabbage into turnip: Polynomial algorithm for sorting signed permutations by reversals. Journal of the ACM **46** (1999) 1–27
5. Bafna, V., Pevzner, P.A.: Sorting by transpositions. SIAM Journal on Discrete Mathematics **11** (1998) 224–240
6. Christie, D.A.: Genome Rearrangement Problems. PhD thesis, University of Glasgow (1999)
7. Hartman, T.: A simpler 1.5-approximation algorithm for sorting by transpositions. In: Proc. 14th Annual Symposium on Combinaotrial Pattern Matching (CPM '03), Springer (2003) 156–169
8. Gu, Q.P., Peng, S., Sudborough, H.: A 2-approximation algorithm for genome rearrangements by reversals and transpositions. Theoretical Computer Science **210(2)** (1999) 327–339
9. Lin, G.H., Xue, G.: Signed genome rearrangements by reversals and transpositions: Models and approximations. In: Proc. COCOON '99, Lecture Notes in Computer Science. Volume 1627., Berlin Heidelberg, Springer-Verlag (1999) 71–80
10. Meidanis, J., Walter, M.E., Dias, Z.: Reversal distance of signed circular chromosomes. manuscript (2000)
11. Bafna, V., Pevzner, P.A.: Genome rearragements and sorting by reversals. SIAM Journal on Computing **25** (1996) 272–289
12. Kaplan, H., Verbin, E.: Effficient data structures and a new randomized approach for sorting signed permutations by reversals. In: Proc. 14th Annual Symposium on Combinaotrial Pattern Matching (CPM '03), Springer (2003) 170–185
13. Sleator, D.D., Tarjan, R.E.: Self-adjusting binary search trees. J. Assoc. Comput. Mach. **32** (1985) 652–686

Algorithms for Finding
Maximal-Scoring Segment Sets*
(Extended Abstract)

Miklós Csűrös

Département d'informatique et de recherche opérationnelle
Université de Montréal
C.P. 6128 succ. Centre-Ville, Montréal, Québec, H3C 3J7, Canada
csuros@iro.umontreal.ca

Abstract. We examine the problem of finding maximal-scoring sets of disjoint regions in a sequence of scores. The problem arises in DNA and protein segmentation, and in post-processing of sequence alignments. Our key result states a simple recursive relationship between maximal-scoring segment sets. The statement leads to an algorithm that finds such a k-set of segments in a sequence of length n in $O(nk)$ time. We describe linear-time algorithms for finding optimal segment sets using different criteria for choosing k, as well as an algorithm for finding an optimal set of k segments in $O(n \log n)$ time, independently of k. We apply our methods to the identification of non-coding RNA genes in thermophiles.

1 Introduction

Suppose that $w_1, w_2, \ldots, w_n \in \mathbb{R}$ is an arbitrary sequence of scores with $n > 0$. A *segment* S is a set of consecutive integers: $S = [a, b] = \{a, a + 1, \ldots, b\}$. The *score of a segment* S is the sum of the scores indexed by the segment's elements: $w(S) = \sum_{i \in S} w_i$. A classic example of algorithm design is Jon Bentley's Programming Pearl [1] for finding a segment with maximum score. Such a segment can be found in linear time by scanning the scores once. This paper considers a natural generalization of the maximum-scoring segment problem. Namely, we are interested in finding k disjoint segments with maximum total score. A k-*cover* $\mathcal{C} = \{S_1, \ldots, S_k\}$ is a non-intersecting family of segments. The score of a k-cover $\mathcal{C}$ is the sum of its elements' score: $w(\mathcal{C}) = \sum_{S \in \mathcal{C}} w(S)$. It is useful to define the *indicator vector* $(z_1, \ldots, z_n)$ of a cover $\mathcal{C}$: $z_i = 1$ if $i \in \cup_{S \in \mathcal{C}} S$ and $z_i = 0$ otherwise. Using this notation, $w(\mathcal{C}) = \sum_{i=1}^{n} w_i z_i$. A k-cover is *maximal* if it has maximum score among all k-covers. We define the 0-cover as the empty set with score 0.

A cover may define a segmentation, which alternates high- and low-scoring regions, i.e., segments within and outside the cover. Segmentation methods have been extensively used in the analysis of protein and DNA sequences [2]. Various

---

* Work supported by NSERC grant 250391-02.

I. Jonassen and J. Kim (Eds.): WABI 2004, LNBI 3240, pp. 62–73, 2004.
© Springer-Verlag Berlin Heidelberg 2004

scoring schemes permit the identification of charge clusters and hydrophobic profiles for proteins [3], determination of isochores in DNA sequences [4, 5], discovery of CpG islands [5, 6], and even gene finding [7]. Different methods include maximum likelihood estimation [4], Hidden Markov Models [8, 7], entropy-based [5], and various "moving window" techniques. Segmentation methods are also used to remove low-scoring regions from sequence alignments [9].

Our key result is Theorem 1, which states the incremental nature of maximal covers. This theorem leads to an algorithm that finds a k-cover with maximum score for $k \leq K$ in $O(nK)$ time where K is an upper bound on the cover size. Section 3 describes the algorithms for finding maximal covers using different optimality criteria, as well as an algorithm for finding a maximal k-cover in $O(n \log n)$ time. Section 4 deals with the problem of identifying GC-rich regions in AT-rich genomes, which coincide with non-coding RNA genes in thermophiles. Section 5 discusses related results and concludes the paper.

Theorem 1. *Let $\mathcal{C}_k$ be a maximal k-cover for $k \in [0, n-1]$. There exists a maximal $(k+1)$-cover $\mathcal{C}_{k+1}$ which satisfies one of the following conditions.*

(1) There exists such a segment $[a, b]$ that $\mathcal{C}_{k+1} = \mathcal{C}_k \cup \{[a, b]\}$; or
(2) there exist $a, b, c, d \in [1, n]$ for which $a \leq c < d \leq b$, and $\mathcal{C}_{k+1} = \mathcal{C}_k \cup \{[a, c], [d, b]\} \setminus \{[a, b]\}$.

Theorem 2. *Let $\mathcal{C}_k$ be a maximal k-cover for $k \in [1, n]$. There exists a maximal $(k-1)$-cover $\mathcal{C}_{k-1}$ which satisfies one of the following conditions.*

(1) There exists such a segment $[a, b] \in \mathcal{C}_k$ that $\mathcal{C}_{k-1} = \mathcal{C}_k \setminus \{[a, b]\}$; or
(2) there exist $a, b, c, d \in [1, n]$ for which $a \leq c < d \leq b$, and $\mathcal{C}_{k-1} = \mathcal{C}_k \cup \{[a, b]\} \setminus \{[a, c], [d, b]\}$.

Theorem 1 shows that $\mathcal{C}_{k+1}$ is obtained either (1) by adding a new segment to $\mathcal{C}_k$, or (2) by removing the middle of a segment in $\mathcal{C}_k$. By Theorem 2, the converse is also true: a maximal $(k-1)$-cover can be created from a k-cover by merging two segments, or by removing one. Theorem 2 implies also that *all* maximal covers can be produced by consecutive applications of operations (1) and (2) of Theorem 1. The theorems' proofs are omitted here due to space constraints. The theorems have two immediate consequences. First, Corollary 1 below shows that the score of maximal k-covers is a concave function of k. Secondly, Theorem 1 implies a simple algorithm for calculating successive maximal covers, which we describe in §3.1.

Corollary 1. *Let $1 < k < n$. Let $\mathcal{C}_{k-1}$, $\mathcal{C}_k$, and $\mathcal{C}_{k+1}$ be maximal $(k-1)$-, k-, and $(k+1)$-covers, respectively. Then $w(\mathcal{C}_{k+1}) - w(\mathcal{C}_k) \leq w(\mathcal{C}_k) - w(\mathcal{C}_{k-1})$.*

Proof. Omitted. □

2   Scores Based on Probabilistic Models

2.1   Maximum Likelihood Estimation of Segments

Let $X_1, \ldots, X_n$ be a sequence of independent random letters from an alphabet $\Sigma = \{\sigma_1, \ldots, \sigma_r\}$. The distribution of every X_i is one of two known distributions, specified by the probabilities $p(\sigma_j)$ and $q(\sigma_j)$. A *changed segment* is a segment $[a, b]$ of indices where $\mathbb{P}\{X_i = \sigma_j\} = q(\sigma_j)$ for all $i \in [a, b]$. A segment $[a, b]$ is *unchanged* if $\mathbb{P}\{X_i = \sigma_j\} = p(\sigma_j)$ for all $i \in [a, b]$. Maximum likelihood estimation of changed segments turns into a maximal cover problem. Let $x_i : i \in [1, n]$ be the observed values of X_i. Let $\mathcal{C}$ be a non-intersecting set of hypothetical changed segments. Let $\mathbf{z} = (z_1, \ldots, z_n)$ be the indicator vector for $\mathcal{C}$. The likelihood function is $f(\mathbf{x} \mid \mathbf{z}, \mathbf{p}, \mathbf{q}) = \prod_{i=1}^{n} (p(x_i))^{1-z_i} (q(x_i))^{z_i}$. Define

$$w_i = \log(q(x_i)) - \log(p(x_i)). \tag{1}$$

(Throughout the paper, log denotes natural logarithm.) The log-likelihood can be written as $\log f(\mathbf{x} \mid \mathbf{z}, \mathbf{p}, \mathbf{q}) = \sum_{i=1}^{n} \log p(x_i) + \sum_{i=1}^{n} w_i z_i$. The first term is the log-likelihood of the null hypothesis that there are no changed segments. The second term is the log-likelihood ratio (LLR) of the alternative hypothesis defined by $\mathcal{C}$. Accordingly, a maximal k-cover maximizes the LLR among hypotheses with k changed segments if the scores are set by Eq. (1).

Fig. 1. Maximal k-covers with minimum segment lengths $m_1 = 6$ and $m_0 = 1$. A '+' denotes $w_i = 1$ and '-' denotes $w_i = -1$. An equivalent scoring scheme is realized when $\Sigma = \{0, 1\}$, and $p(0) = 1 - q(0)$ in Eq. (1).

Fu and Curnow [4] examine the problem of finding a k-set of changed segments that maximizes the LLR with restrictions on the minimum lengths of changed and unchanged segments specified by thresholds m_1 and m_0, respectively. Fu and Curnow state a theorem (with an incomplete proof) that is similar to Theorem 1: "Given one set of best k segments [k-cover in our terminology], we can find one set of best $(k + 1)$ segments if it exists, by either adding the best segment which does not overlap with any of the k best segments or by splitting and expanding one of the best k segments." Their claim, however, does not seem to hold in general, as the relationship between maximal covers may be complicated if a minimum length is imposed on the segments. Figure 1 shows an example where more than one segment change between consecutive maximal covers. Alternatively, if segments can be of arbitrary length, then by Theorem 1, there is no need for expansions.

2.2   Selecting the Cover Size: Complexity Penalties

Unless it is warranted by the problem at hand, the reason for restricting segment lengths is to avoid overfitting: the cover $\{[i, i] : w_i > 0\}$ maximizes the likelihood but it hardly captures any meaningful pattern in the data. We suggest that one should instead penalize the cover size. Define the *complexity-penalized score* of a cover $\mathcal{C}$ by $\tilde{w}(\mathcal{C}) = w(\mathcal{C}) - r(|\mathcal{C}|)$ where $r\colon \mathbb{N} \mapsto [0, \infty)$ is a monotone increasing penalty function. The optimal cover has maximum complexity-penalized score. First we describe a penalty based on the minimum description length (MDL). According to the MDL principle [10], one favors the cover $\mathcal{C}$ which minimizes the length of encoding the data and $\mathcal{C}$. Let $\mathbf{z}$ be the indicator vector for $\mathcal{C}$. Given $\mathbf{z}$, every x_i can be encoded in $b(z_i, x_i)$ bits on average, where $b(0, \sigma) = -\log p(\sigma)$ and $b(1, \sigma) = -\log q(\sigma)$. The cover itself can be specified by the endpoints of its segments using $2|\mathcal{C}| \log_2 n$ bits. The total codelength equals $\ell(\mathbf{x}, \mathcal{C}) = \sum_{i=1}^{n} b(z_i, x_i) + 2|\mathcal{C}| \log_2 n = \ell(\mathbf{x}, \emptyset) - \frac{w(\mathcal{C}) - 2|\mathcal{C}| \log n}{\log 2}$. The MDL cover thus maximizes $w(\mathcal{C}) - 2|\mathcal{C}| \log n$, and corresponds to the penalty $r(k) = 2k \log n$. (A more efficient encoding can rely on the fact that there are $\binom{n}{2k}$ possible k-covers. When $k = o(n)$, a k-cover can be encoded in $\log_2 \binom{n}{2k} \approx 2k \log_2 n - 2k \log_2(2k)$ bits. The corresponding penalty equals $r(k) = 2k(\log n - \log(2k))$.)

2.3   A Penalty Based on Statistical Significance

As an alternative to the MDL approach, a penalty can be defined based on statistical significance, measured by the probability that a segment has a large score under the null hypothesis that there are no changed segments. The distribution of the maximum segment score, i.e., the score $w^{(1)}$ of the maximal 1-cover, under the null hypothesis has been extensively studied [11, 12]. Karlin *et al.* [12] prove that $w^{(1)} \to \log n$ almost surely as $n \to \infty$, and that for all x,

$$\lim_{n \to \infty} \mathbb{P}\{w^{(1)} - \log n \leq x\} = \exp(-Ce^{-x}), \qquad (2)$$

where C is independent of n and x, and is defined by a rapidly converging infinite sum. (For the general case of assigning score u_j to every letter σ_j with $\sum_{j=1}^{r} p(\sigma_j) u_j < 0$, $w^{(1)} \to \lambda^{-1} \log n$ where λ is the unique positive solution of $\sum_{j=1}^{r} p(\sigma_j) \exp(\lambda u_j) = 1$. When $u_j = \log \frac{q(\sigma_j)}{p(\sigma_j)}$, $\lambda = 1$ is a solution.) This result provides a means to select α, in order to search for the cover $\mathcal{C}$ that is optimal according to a penalty function $r(k) = \alpha k$. The following theorem characterizes the segment scores in $\mathcal{C}$.

Theorem 3. *Fix $\alpha > 0$, and let $\mathcal{C} = \{[a_1, b_1], \ldots, [a_k, b_k]\}$ be a cover that maximizes $\tilde{w}(\mathcal{C}) = w(\mathcal{C}) - \alpha|\mathcal{C}|$. If $\mathcal{C} \neq \emptyset$, then the following holds. For all $i \in [1, k]$, $w([a_i, b_i]) \geq \alpha$, and there does not exist a, b with $a_i < a \leq b < b_i$ and $w([a, b]) < -\alpha$. For all $i \in [1, k-1]$, if $b_i + 1 < a_{i+1}$, then $w([b_i + 1, a_{i+1} - 1]) \leq -\alpha$. For all $i \in [0, k]$, there does not exist a, b for which $b_i < a \leq b < a_{i+1}$ and $w([a, b]) > \alpha$, where $b_0 = 0$ and $a_{k+1} = n + 1$.*

Proof. Straightforward.   □

By Theorem 3, every changed segment in $\mathcal{C}$ has score at least α and no subsegment of an unchanged segment has a score above α. Consequently, by setting $\alpha = x + \log n$ with an appropriately chosen x we can ensure that every changed segment has significant statistical support, and that a maximal set of such segments is selected. In particular, Eq. (2) implies that for large n, $\mathcal{C}$ is non-empty with probability $1 - \exp(-Cne^{-\alpha})$ under the null hypothesis. Accordingly, for a given [small] $0 < p < 1$, we can use

$$\alpha \geq \log n + \log \frac{C}{-\log(1-p)} \approx \log n + \log \frac{C}{p}, \tag{3}$$

in order to get a non-empty optimal cover with at most p probability. By switching the roles of changed and unchanged segments, a similar argument can be made to measure the statistical support for unchanged segments.

2.4   Two-State Hidden Markov Models

Our last example of penalizing cover size is that of segmentation by a Hidden Markov Model (HMM). Extending the maximum likelihood framework, we impose that the random sequence $X_1, \ldots, X_n$ is generated by a two-state HMM [8, 7]. The two states correspond to changed and unchanged segments. A run of the HMM results in a state sequence $Z_1, \ldots, Z_n$ forming a Markov chain, and the sequence of emitted characters $X_1, \ldots, X_n$. If $Z_i = 0$, then X_i is drawn according to the unchanged segments' distribution $\mathbf{p}$, otherwise it is drawn according to $\mathbf{q}$. The most likely state sequence $\mathbf{z} = (z_1, \ldots, z_n)$ for a given observation sequence $\mathbf{x} = (x_1, \ldots, x_n)$ defines a segmentation of $[1, n]$ into changed and unchanged segments, i.e., segments where $z_i = 1$ vs. segments where $z_i = 0$. Clearly, $\mathbf{z}$ is the indicator vector for a cover. The likelihood function equals $f(\mathbf{x} \mid \mathbf{z}) = \pi(z_1)\left(\prod_{i=1}^{n}(p(x_i))^{1-z_i}(q(x_i))^{z_i}\right)\left(\prod_{i=2}^{n}\tau(z_{i-1} \rightarrow z_i)\right)$, where π are the starting probabilities and τ are the transition probabilities for the states' Markov chain. There exists a well-known method for finding the most likely state sequence, known as the Viterbi algorithm [13], but formulating it as a maximal cover problem enables us to consider further variations with restrictions on the number of state changes (§3.1) or on state durations (§3.2). The LLR of a state sequence $\mathbf{z}$ (viewed as indicator for a cover $\mathcal{C}$) with respect to the null hypothesis that all $z_i = 0$ can be written in the form $\sum_{i=1}^{n} w_i z_i - \alpha|\mathcal{C}|$, where

$$w_i = \log \frac{q(\sigma)}{p(\sigma)} + \log \frac{\tau(1 \rightarrow 1)}{\tau(0 \rightarrow 0)} + \delta_i; \tag{4a}$$

$$\alpha = -\log \frac{\tau(0 \rightarrow 1)}{\tau(0 \rightarrow 0)} - \log \frac{\tau(1 \rightarrow 0)}{\tau(0 \rightarrow 0)} + \log \frac{\tau(1 \rightarrow 1)}{\tau(0 \rightarrow 0)}, \tag{4b}$$

and $\delta_i = 0$ for every $i \in [2, n-1]$, otherwise it hides correction terms: $\delta_1 = -\log \frac{\tau(0 \rightarrow 1)}{\tau(0 \rightarrow 0)} + \log \frac{\pi(1)}{\pi(0)}$ and $\delta_n = -\log \frac{\tau(1 \rightarrow 0)}{\tau(0 \rightarrow 0)}$. Consequently, segmentation by the most likely state sequence in a two-state HMM is an instance of finding an optimal cover using linear complexity penalties.

3 Algorithms

3.1 An Algorithm for Finding a Maximal Cover

By Theorem 1, a maximal $(k + 1)$-cover can be found by updating a maximal k-cover. For each k, one needs to find the segment that can be either added or removed to increase the cover score by the largest amount. The idea is employed by the algorithm MaxCover, which is an adaptation of Bentley's algorithm [1]. (In fact, Bentley credits Joseph Kadane of CMU with the design.)

Algorithm MaxCover
Input: w_i scores for $i \in [1, n]$, K maximum cover size
Output: indicator vector for a k-cover with maximum score for $0 \le k \le K$
C1 initialize $z_i \leftarrow 0$ for $i = 1, \ldots, n$
C2 **for** $k \leftarrow 1, \ldots, K$ **do**
C3   set $i_0 \leftarrow 1$; $w \leftarrow 0$; $S \leftarrow$ null; $w_{\max} \leftarrow 0$
C4   **for** $i \leftarrow 1, \ldots, n$ **do**
C5   **if** $i > 1$ and $z_{i-1} \ne z_i$ **then** $i_0 \leftarrow i$; $w \leftarrow 0$
C6 $w \leftarrow w + w_i$ // *current candidate is $[i_0, i]$ with score w*
C7   **if** $(z_i = 0$ and $w \le 0)$ or $(z_i = 1$ and $w \ge 0)$ **then** $i_0 \leftarrow i + 1$; $w \leftarrow 0$
C8   **else if** $|w| > w_{\max}$ **then** $w_{\max} \leftarrow |w|$; $S \leftarrow [i_0, i]$
C9   **if** $w_{\max} = 0$ **then return** $(z_1, \ldots, z_n)$ **else** set $z_i \leftarrow 1 - z_i$ for all $i \in S$
C10 **return** $(z_1, \ldots, z_n)$

The algorithm scans the scores w_i once for every $k \in [1, K]$ in Lines C4–C8. For every k, the algorithm calculates the maximum increase $w_{\max}$ in cover score that can be achieved by removing a sub-segment or adding a segment (the segment S).

Lemma 1. *The algorithm* MaxCover *finds a cover that has maximum score among covers with at most K segments in $O(nK)$ time.*

Proof. The proof of the running time is straightforward. The proof of correctness is analogous to that of [1]; it is omitted due to space constraints.   □

3.2 Algorithms for Linear Complexity Penalties

Suppose that we want to find the cover $\mathcal{C}$ that maximizes the penalized score $\tilde{w}(\mathcal{C}) = w(\mathcal{C}) - \alpha|\mathcal{C}|$ with some $\alpha \ge 0$. The MDL approach of §2.1 sets $\alpha = 2 \log n$; the statistical significance framework (setting α by Eq. (3)), and the HMM approach of §2.4 also use linear penalty functions.

Let $\mathcal{C}_0 = \emptyset, \mathcal{C}_1, \mathcal{C}_2, \ldots$ be a series of maximal k-covers. By Corollary 1, a cover $\mathcal{C}^*$ maximizing $\tilde{w}$ is the first $\mathcal{C}_k$ for which $w(\mathcal{C}_{k+1}) - w(\mathcal{C}_k) < \alpha$. It is easy to modify MaxCover to find $\mathcal{C}^*$. The only necessary change is in Line C9, where $\mathbf{z}$ needs to be returned if $w_{\max} \le \alpha$. MaxCover then finds $\mathcal{C}^*$ in $O(nK)$ time if it is invoked with $K \ge |\mathcal{C}^*|$. In what follows we develop a faster algorithm.

For all $i \in [1, n]$, define $W^0(i)$ as the maximum of $\tilde{w}$ for covers of $[1, i]$ which do not include i. Define $W^1(i)$ as the maximum of $\tilde{w}$ for covers of $[1, i]$ which do include i.

Lemma 2. *For all $i > 1$, $W^0(i) = \max\{W^0(i - 1), W^1(i - 1)\}$, and $W^1(i) = w_i + \max\{W^0(i - 1) - \alpha, W^1(i - 1)\}$.*

Proof. Straightforward by using the definition.   □

The lemma implies a dynamic programming algorithm. In case of the two-state HMM, the algorithm is equivalent to the Viterbi algorithm [13]. We design a more general method that respects minimum segment length constraints. Specifically, we want to find a cover that maximizes $\tilde{w}$ with the stipulation that changed segments must have lengths at least m_1 and unchanged segments must have lengths at least m_0.

For all $j = 0, 1$, $m \in [1, m_j]$, and $i \in [m, n]$, define $\mathcal{C}_{i,m}^j$ as covers of $[1, i]$ that maximize $\tilde{w}$ while satisfying the requirements for all segment lengths, except for the last one: $\mathcal{C}_{i,m}^0$ is a cover that ends with an unchanged segment of length at least m, and $\mathcal{C}_{i,m}^1$ ends with a changed segment of length at least m.

Lemma 3. *Let* $W_{\text{short}}^0(i) = \tilde{w}(\mathcal{C}_{i,1}^0)$, $W_{\text{long}}^0(i) = \tilde{w}(\mathcal{C}_{i,m_0}^0)$, $W_{\text{short}}^1(i) = \tilde{w}(\mathcal{C}_{i,1}^1)$, *and* $W_{\text{long}}^1(i) = \tilde{w}(\mathcal{C}_{i,m_1}^1)$. *For all* $i > 1$, $W_{\text{short}}^0(i) = \max\{W_{\text{short}}^0(i-1), W_{\text{long}}^0(i-1)\}$ *and* $W_{\text{short}}^1(i) = w_i + \max\{W_{\text{long}}^0(i-1) - \alpha, W_{\text{short}}^1(i-1)\}$. *For all* $i \in [m_0, n]$, $W_{\text{long}}^0(i) = W_{\text{short}}^0(i - m_0 + 1)$, *and for all* $i \in [m_1, n]$, $W_{\text{long}}^1(i) = W_{\text{short}}^1(i - m_1 + 1) + \sum_{j=i-m_1+2}^{i} w_i$.*

Proof. Straightforward by using the definition.   □

Lemma 3 implies a dynamic programming algorithm (referred to as MINLENGTH-COVER), which finds an optimal cover subject to length restrictions. The algorithm runs in $O(n)$ time. The case $\alpha = 0$ is equivalent to the original problem of Fu and Curnow [4], that of finding a segmentation that satisfies the length restrictions.

3.3   A Fast Algorithm for Finding a Maximal Cover

So far we concentrated on computing maximal covers using Theorem 1 or selecting one cover using linear complexity penalties. It is also possible to calculate maximal covers by employing Theorem 2. The main idea is to find the cover that comprises all runs of positive scores and then produce smaller maximal covers consecutively. Below we develop the idea formally. A segment $[i, j]$ is a *positive run* if $w([i, j]) > 0$ and for all $k \in [i, j]$, $w_k \geq 0$. A segment $[i, j]$ is a *negative run* if $w([i, j]) < 0$ and for all $k \in [i, j]$, $w_k \leq 0$. When not all scores are zero, we can decompose $[1, n]$ into an alternating series of maximal negative and positive runs. Let $\mathcal{T} = (T_1, T_2, \ldots, T_m)$ be the resulting series. Let M be the number of positive runs in $\mathcal{T}$. Clearly, the set $\{T \in \mathcal{T} \colon w(T) > 0\}$ is a maximal M-cover. In fact, M is the cover size until which the score of maximal covers increases.

The sequence $\mathcal{T}$ can be calculated in $O(n)$ time. Applying Theorem 2, we produce maximal covers of size less than M one by one. In every step, we need to identify three consecutive segments T_{i-1}, T_i, T_{i+1} that can be merged at the expense of the smallest decrease in the cover score. Such a triple is found by selecting i for which the absolute value $|w(T_i)|$ is minimal. Algorithm MAXCOVER-FAST shown here implements the idea.

Algorithm MAXCOVER-FAST
Input: w_i scores for $i \in [1, n]$, K cover size
F1 Let $\mathcal{T}$ be the sequence of alternating maximal runs
F2 **for** $M = |\{T \in \mathcal{T}: w(T) > 0\}|$ **downto** K **do**
F3   // at this point $\mathcal{T} = ([a_1, b_1], [a_2, b_2], \ldots, [a_m, b_m])$ where $a_{i+1} = b_i + 1$
F4   Choose $[a_i, b_i]$ from $\mathcal{T}$ with minimum $|w([a_i, b_i])|$, $1 < i < m$
F5   Set $\mathcal{T} \leftarrow \mathcal{T} \cup \{[a_{i-1}, b_{i+1}]\} \setminus \{[a_{i-1}, b_{i-1}], [a_i, b_i], [a_{i+1}, b_{i+1}]\}$
F6 **return** the set $\{T \in \mathcal{T}: w(T) > 0\}$

Lemma 4. *Algorithm* MAXCOVER-FAST *finds a maximal K-cover if not all scores are zero, and it is invoked with a K that is not larger than the number M of maximal positive runs. The algorithm can be implemented in such a way that it terminates in $O(n + M \log M)$ time.*

Proof. (Sketch.) An invariant that implies the correctness is that in Line F4, $\mathcal{T}$ alternates segments with positive and negative scores. In order to see that, notice that $\left| w([a_i, b_i]) \right| \leq \min\left\{ \left| w([a_j, b_j]) \right| : j = i \pm 1 \right\}$ in Line F5. Thus, $w([a_{i-1}, b_{i+1}])$, $w([a_{i-1}, b_{i-1}])$, and $w([a_{i+1}, b_{i+1}])$ have the same sign. The algorithm's correctness now follows from Theorem 2. A balanced search tree can be augmented to track the segments in $\mathcal{T}$. Elements of $\mathcal{T}$ are stored at the tree leaves, ordered by the absolute values of the scores. In order to avoid selecting the first or the last segment in Line F4, those two segments are stored with scores $\pm\infty$, preserving only their scores' signs. In addition, leaves are equipped with pointers to preceding and succeeding segments. It is thus possible to perform Line F4 in $O(\log M)$ time, to find neighboring segments in $O(1)$ time, and to update $\mathcal{T}$ in Line F5 in $O(\log M)$ time. Hence the algorithm runs in $O(n + M \log M)$ time.   □

MAXCOVER-FAST can be modified to find an optimal cover $\mathcal{C}^*$ for an arbitrary monotone increasing complexity penalty function. Since maximal covers' scores stop increasing at M, $|\mathcal{C}^*| \leq M$. The algorithm has to track the cover score: at each merging operation in Line F5, the score decreases by $|w([a_i, b_i])|$. All maximal covers of size $\leq M$ are inspected, and the one maximizing $\tilde{w}$ is reported at the end. Consequently, the optimal cover can be found in $O(n \log n)$ time.

4 Non-coding RNA Genes in AT-Rich Thermophiles

A frequently used statistic for DNA sequences is the *GC-content*, which is the relative frequency of G and C in a region. In a recent application, GC-content was used to detect non-coding RNA genes [7,14] in genomes of thermophile Archaebacteria such as *Methanocaldococcus jannaschii*. The optimal growth temperature of thermophile Prokaryotes strongly correlates with the GC-content of transfer and ribosomal RNA genes [15–17]. (For the genome-wide GC-content, however, there does not seem to exist a similar dependence [17].) *M. jannaschii* is a prime candidate for identifying RNA genes on GC-content alone, since while the GC-content of the genome is 31%, known RNA genes have a much higher GC-content of 60–70%. Klein *et al.* [7] trained a two-state HMM in which the states modeled GC-poor and GC-rich regions. They computed the most likely state sequence, in order to select a set of GC-rich segments. After filtering out known genes, they selected the segments with a minimum length of 50, which resulted in nine candidate RNA genes, denoted Mj1–Mj9. They validated four of them by showing that they are transcribed. They identified a fifth gene Mj6a, missed by the HMM, based on sequence similarity. Two candidates (Mj5 and Mj8) are less likely to be RNA genes as they overlap with putative protein coding regions. Schattner [14] also used GC-content and other statistics to identify RNA genes in *M. jannaschii*. He used a moving window, within which the log-likelihood was calculated using essentially the same equations as in §2.1.

We tested our algorithms on *M. jannaschii* (1.66 Mbp, GenBank accession NC_000909.1). Using Eq. (1), we employed the scores $w_i = -0.66$ if the corresponding nucleotide was A or T, and $w_i = 0.72$ for G or C. The scores are based on the genome's overall GC-content, and the 65% GC-content in seven tRNA genes between positions 850000 and 870000. Using MAXCOVER, we computed maximal k-covers: see Fig. 2. The smallest maximal cover that includes all tR-NAs has size $k = 38$. That cover also includes all rRNAs, as well as RNase P RNA and SRP 7S (Signal Recognition Particle) genes. In addition, three novel genes of [7] are also included. The false positive rate can be assessed by the fact that only two intervals overlap with protein-coding genes: Mj5 and Mj8. The maximal 46-cover contains all RNA genes of [7], including Mj6a, not discovered by either the HMM or the sliding windows of [14].

We evaluated different penalty functions r: $r(k) = 2k \log n$ (MDL1) or $r(k) = 2k(\log n - \log(2k))$ (MDL2); Eq. 4b (HMM[1]); and Eq. 3 for significance ($P = 0.1$ and $P = 0.01$). As shown in Fig. 2, the MDL penalties are too severe, and even HMM segmentation stops at $P = 0.01$. There is no need to be very conservative in this case, as the gene candidates identified by the segmentation are further analyzed by different methods. Accordingly, we selected $\alpha = 14$ for the complexity penalty (P-value 0.11 by Eq. (3)), and imposed a minimum segment length of 40. MINLENGTH-COVER finds a 48-cover, which includes all known RNA genes (even Mj6a), five protein-coding genes, and the segment [334439,334485] not identified by either [14] or [7], which is classified as a pseudogene by tRNAscan [18].

We carried out similar experiments with a number of thermophile Prokaryotes. In the maximum likelihood framework of §2.1, one can readily predict the success of gene finding. A linear penalty α set by Eq. (3) can be compared to expected scores of changed segments. A changed segment of length ℓ has expected score $E(\ell) = \ell D$ where D is the relative entropy between the distributions. The threshold $\ell_{\min} = \alpha/D$ thus indicates the minimum detectable gene lengths. By this reasoning, we found that among thermophiles for which whole genome sequences are avaiilable, *N. equitans* [19], *S. tokodaii* [20], *S. solfataricus*, *M. maripaludis* and *P. horikoshii* have low $\ell_{\min}$ values. The analysis of the results is in progress, and is beyond this paper's scope. We summarize here some initial findings. We found that a maximal 58-cover of the *N. equitans* genome (NC_005213.1) includes all tRNAs found by tRNAscan [18] (even pseudogenes),

---

[1] If an HMM is used, the log $\frac{\tau(1 \to 1)}{\tau(0 \to 0)}$ terms are negligible in Eq. (4a).

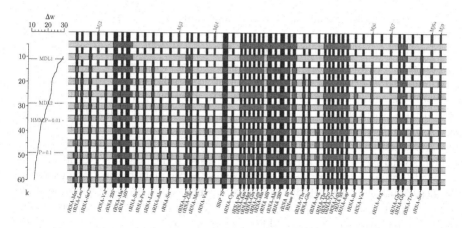

Fig. 2. Segmentation of M. jannaschii. The right-hand side shows maximal k-covers in increments of five. Horizontal grey bands correspond to covers, and darker boxes indicate the segments. Vertical bars show RNA genes with known function (bottom) and those in [7] (top). (Locations are mapped to display coordinates by a piecewise logarithmic function.) The left-hand side plots the score increase $\Delta w(k) = w(\mathcal{C}_k) - w(\mathcal{C}_{k-1})$. The optimal cover for a linear penalty α is the last one with $\Delta w(k) \geq \alpha$.

all rRNAs, four protein-coding genes, and 12 unannotated segments. Seven of the unannotated segments[2] produce no significant BLAST hits in Genbank. The *S. tokodaii* genome (NC_003106.2) contains many repetitive sequences. Its segmentation identifies all annotated RNA genes (including a tRNA that is missed by tRNAscan), and includes 24 transposition-related sequences, and six hypothetical protein-coding genes. There are eight unannotated segments, four of which are repeated more than once in the genome. BLAST finds no similarities between the remaining four[3] segments and known archaeal DNA sequences. We hypothesize that they correspond to non-coding RNA genes. A fifth segment [326016,326321] overlaps in only 11 bp with a hypothetical protein-coding gene. Based on a structural alignment to known RNase P RNA genes in Crenarchaea [21], it is the so far unpublished RNase P RNA gene in *S. tokodaii* (J. W. Brown, personal communication).

5 Discussion

We presented algorithms that calculate optimal covers according to different criteria, in linear or $O(n \log n)$ time for an input of size n. Even a recent review [2] of DNA segmentation methods considered the cover selection problem, based on [4], as one that can only be solved in $O(n^2)$ time. Such a running time may be a serious drawback in the analysis of long DNA sequences.

---

[2] The following segments: [221790,221843], [308960,309006], [339411,339562], [427536,427595], [487434,487490], [487645,487722], [488412,488458].

[3] Segments [1322116,1322329] [2020954,2021005], [2089025,2089145], and [2091849, 2092094].

A related problem, that of finding a maximal *chain* of covers, can also be solved in linear time. A chain of covers is formed by $\mathcal{C}_0, \mathcal{C}_1, \ldots$ where every $\mathcal{C}_k$ is a k-cover, and $\mathcal{C}_k \subset \mathcal{C}_{k+1}$ for all k. A maximal chain of covers $\mathcal{C}_0^*, \mathcal{C}_1^*, \ldots$ is defined recursively: $\mathcal{C}_0^* = \emptyset$, and for every $k > 0$, $\mathcal{C}_k^*$ is a k-cover that has maximum score satisfying $\mathcal{C}_{k-1}^* \subset \mathcal{C}_k^*$. In other words, successive elements are generated using only Case (1) of Theorem 1. Ruzzo and Tompa [6] describe a linear-time algorithm that finds the last $\mathcal{C}_k^*$ in which all segments have positive scores. In the maximal likelihood framework of §2.1, looking for a maximal chain may give unsatisfactory results. Specifically, it may be the case that two changed segments with large scores are separated by a short unchanged segment, and all three get lumped together in one of the covers. Subsequent covers do not change the situation, regardless of the middle segment's score. In Fig. 1, all positive scores are included in $\mathcal{C}_1^*$. Theorem 3 shows that maximal covers may give more sound segmentation results than do maximal chains.

Zhang *et al.* [9] examine the problem of producing pairwise sequence alignments without low-scoring regions. An alignment is a sequence of n columns, each assigned a score. The score of a subalignment, defined by a segment $[a, b] \subseteq [1, n]$ is the sum of its columns scores. Disjoint subalignments thus form a cover. Standard alignment procedures [22] have essentially the same shortcomings as maximal cover chains in that they may include subalignments of arbitrarily low score. In order to avoid such situations, Zhang *et al.* [9] propose that low-scoring regions should be removed from the alignment. In particular, they aim to find a cover $\mathcal{C}$, for which no subsegment of a $S \in \mathcal{C}$ has score less than $-X$ for a threshold $X \geq 0$. They prove that such covers for decreasing values of X form a hierarchy similar to that of maximal covers described by Theorems 1 and 2. They also provide a linear time algorithm implied by the hierarchy that finds such a cover for a given X. In light of Theorem 3, such covers are succinctly characterized by a linear penalty function $r(k) = Xk$. We pointed out the connection between the threshold X and various statistical notions of complexity, as well as the interpretation of the optimal cover as the most likely state sequence in a Markov model. PENALIZED-COVER offers a simple, efficient alternative to the algorithm of [9] for eliminating low-scoring regions from alignments. MINLENGTH-COVER also provides the option of imposing minimum subalignment lengths.

Acknowledgments

This work has benefited from conversations with Balázs Kégl and Stefan Wolf. I am very grateful to James W. Brown for confirming my identification of the *S. tokodaii* RNase P gene, and to Hans-Georg Müller for reading the manuscript. I found a simpler version of Theorem 1 in collaboration with Réka Szabó: it appeared first, along with the MAXCOVER algorithm, in my Masters thesis, written under the direction of Gábor Lugosi at the Technical University of Budapest in 1994.

References

1. Bentley, J.: Programming pearls: algorithm design techniques. Comm. ACM **27** (1984) 865–873
2. Braun, J.V., Müller, H.G.: Statistical methods for DNA sequence segmentation. Statist. Sci. **13** (1998) 142–162
3. Karlin, S., Brendel, V.: Chance and significance in protein and DNA analysis. Science **257** (1992) 39–49
4. Fu, Y.X., Curnow, R.N.: Maximum likelihood estimation of multiple change points. Biometrika **77** (1990) 563–573
5. Li, W., Bernaola-Galván, P., Haghighi, F., Grosse, I.: Applications of recursive segmentation to the analysis of DNA sequences. Comput. Chem. **26** (2002) 491–510
6. Ruzzo, W.L., Tompa, M.: A linear time algorithm for finding all maximal scoring subsequences. In: Proc. 7th Intl. Conf. Intelligent Systems in Molecular Biology, AAAI Press (1999) 234–241
7. Klein, R.J., Misulovin, Z., Eddy, S.R.: Noncoding RNA genes identified in AT-rich hyperthermophiles. Proc. Natl. Acad. Sci. USA **99** (2002) 7542–7547
8. Churchill, G.A.: Stochastic models for heterogeneous DNA sequences. Bull. Math. Biol. **51** (1989) 79–94
9. Zhang, Z., Berman, P., Wiehe, T., Miller, W.: Post-processing long pairwise alignments. Bioinformatics **15** (1999) 1012–1019
10. Barron, A., Rissanen, J., Yu, B.: The Minimum Description Length principle in coding and modeling. IEEE Trans. Inform. Theory **44** (1998) 2743–2760
11. Karlin, S., Altschul, S.F.: Methods for assessing the statistical significance of molecular sequence features by using general scoring schemes. Proc. Natl. Acad. Sci. USA **87** (1990) 2264—2268
12. Karlin, S., Dembo, A., Kawabata, T.: Statistical composition of high-scoring segments from molecular sequences. Ann. Statist. **18** (1990) 571–581
13. Rabiner, L.R.: A tutorial on Hidden Markov Models and selected applications in speech recognition. Proc. IEEE **77** (1989) 257–286
14. Schattner, P.: Searching for RNA genes using base composition statistics. Nucleic Acids Res. **30** (2002) 2076–2082
15. Galtier, N., Lobry, J.: Relationships between genomic G+C content, RNA secondary structures, and optimal growth temperature in Prokaryotes. J. Mol. Evol. **44** (1997) 632–636
16. Wang, H.C., Hickey, D.A.: Evidence for strong selective constraint acting on the nucleotide composition of 16S ribosomal RNA genes. Nucleic Acids Res. **30** (2002) 2501–2507
17. Bao, Q., et al.: A complete sequence of the T. tengcongensis genome. Genome Res. **12** (2002) 689–700
18. Lowe, T.M., Eddy, S.R.: tRNAscan-SE: a program for improved detection of transfer RNA genes in genomic sequence. Nucleic Acids Res. **25** (1997) 955–964
19. Waters, E., et al.: The genome of Nanoarchaeum equitans: insights into early archaeal evolution and derived parasitism. Proc. Natl. Acad. Sci. USA **100** (2003)
20. Kawarabayashi, Y., et al.: Complete genome sequence of an aerobic thermoacidophilic crenarchaeon, Sulfolobus tokodaii strain7. DNA Research **8** (2001) 123–140
21. Brown, J.W.: The ribonuclease P database. Nucleic Acids Res. **27** (1999) 314
22. Smith, T.F., Waterman, M.S.: Identification of common molecular subsequences. J. Mol. Biol. **147** (1981) 195–197

Gapped Local Similarity Search with Provable Guarantees

Manikandan Narayanan[1] and Richard M. Karp[1,2]

[1] Computer Science Division, University of California, Berkeley, CA 94720
{nmani,karp}@cs.berkeley.edu
[2] International Computer Science Institute, Berkeley, CA 94704

Abstract. We present a program *qhash*, based on q-gram filtration and high-dimensional search, to find gapped local similarities between two sequences. Our approach differs from past q-gram-based approaches in two main aspects. Our filtration step uses algorithms for a sparse all-pairs problem, while past studies use suffix-tree-like structures and counters. Our program works in sequence-sequence mode, while most past ones (except QUASAR) work in pattern-database mode.

We leverage existing research in high-dimensional proximity search to discuss sparse all-pairs algorithms, and show them to be subquadratic under certain reasonable input assumptions. Our *qhash* program has provable sensitivity (even on worst-case inputs) and average-case performance guarantees. It is significantly faster than a fully sensitive dynamic-programming-based program for strong similarity search on long sequences.

1 Introduction

Problem and Significance. A gapped local similarity between two sequences is a pair of fixed length substrings, one from each sequence, that align with few mismatches and indels (insertions/deletions). We address the problem of finding all such similarities between two sequences. This is a core problem in bio-sequence similarity search, as it is a variant of the basic local alignment problem [29] with edit distance as the scoring function. Edit distance is simpler than a general scoring function as it treats mismatches and indels via unit costs, nevertheless it is important and very relevant for comparing genomic DNA sequences as discussed next.

Whole genome alignment (wga) is a task necessitated by recent sequencing of several genomes [32]. Current wga programs (see [23] and references therein) rely heavily on exact or near-exact local matches called *seeds* between two genomes to select "anchors" and construct wga. Permitting a few mismatches and indels in seeds make them robust to sequencing errors or short breaks in conserved stretches, and the resulting alignment more sensitive to detecting homologies (see [11,4] and seed-design references therein for related evidence; note that robustness is even more crucial for multi-genome alignment). Current approaches to finding *gapped seeds* first find ungapped seeds and later incorporate gaps via

I. Jonassen and J. Kim (Eds.): WABI 2004, LNBI 3240, pp. 74–86, 2004.

gapped extension [28] or linking of nearby seeds separated by few gaps [7]. One direct approach to finding gapped seeds is enabled by the well-defined problem below that we address. The problem is stated for equal length sequences for convenience in analysis, but can be extended to the case of unequal lengths.

Gapped Local Similarity Problem. Find and output all close substring pairs from two sequences of length n each. A close substring pair consists of two length-m substrings, one from each sequence, with edit distance at most a threshold e^*.

Algorithm Approach. A simple $O(n^2m^2)$ algorithm for the above problem computes edit distances of all substring pairs. A $O(n^2m)$ dynamic-programming [15] and $O(n^2e^*)$ k-differences [25, 22] based algorithms exist for a close problem variant. The QUASAR program [10], which studies this variant, uses q-gram filtration [31] to accelerate strong similarity search on random and practical inputs. We use the same idea, but we do the filtration step very differently using high-dimensional randomized search. Both QUASAR and our filtration are quadratic in worst-case, and subquadratic for strong similarity search on random inputs. But the specific assumptions that make our filtration subquadratic are quite different due to our randomization technique (see next section).

We view q-gram filtration as a high-dimensional search on sparse vectors that represent length-m substrings. A sparse vector is a huge-dimensional binary vector with a small number $m' = (m - q + 1)$ of 1s that indicate all q-grams[1] in a substring (similar to q-gram profiles [31]). The filtration step finds substring pairs that share a high number of common q-grams by solving the next problem.

Sparse All-Pairs Problem. Find and output all close pairs from two groups V and W, each consisting of n sparse high-dimensional vectors. A close pair is a vector each from V and W that share at least a threshold o^* number of common 1s.

A naive algorithm takes $O(n^2m')$ time (assuming quick access to the non-zero columns of each vector). To improve on this without affecting sensitivity, we consider a sparse (o^*, ϵ)-all-pairs problem, where some *approximately* close or *boundary* pairs (that share at least $\frac{o^*}{1+\epsilon}$ common 1s) can also be output. This ϵ-approximate problem enables the use of randomization techniques, which are efficient overall if there aren't too many boundary pairs. We note that Buhler briefly mentions the idea of using the all-pairs problem for filtration [9], but his technique [8] of picking a few random positions wouldn't work for sparse huge-dimensional vectors (as the picked positions would be 0 in most vectors).

Contributions. We make a main contribution to q-gram-based sequence comparison problem and some algorithmic contributions to the sparse (o^*, ϵ)-all-pairs (hereafter called (o^*, ϵ)-SAP) problem, which is the core of our filtration step.

- We present a randomized program *qhash* for gapped local similarity search with provable sensitivity (it finds a user-specified expected fraction, say 90%, of all similarities even on worst-case inputs) and average-case performance guarantees. Our approach differs considerably from past q-gram-based stud-

---

[1] A q-gram of a string is any contiguous length-q substring of the string.

ies as our filtration leverages research in high-dimensional proximity search and works in sequence-sequence mode. Past studies (see survey [26]) use suffix-tree-like structures and counters, and work only in pattern-database mode (QUASAR is sequence-sequence though).

- We carefully analyse the running time of an existing (o^*, ϵ)-SAP algorithm from [16, 13] (neither of which give an *explicit worst-case* time analysis). Our analysis, along the lines of a related near(est)-neighbor algorithm's analysis in [19], yields a time of $\tilde{O}(n^{1+\rho}m' + n^\rho m'(\# \text{ of close pairs}) + m' \ (\# \text{ of boundary pairs}))$, where $\rho < 1$. Note that the $\tilde{O}(.)$ notation hides logarithmic factors.
- We provide a second (o^*, ϵ)-SAP algorithm to improve the number of close pairs term in the above running time. Our second algorithm runs in time $\tilde{O}(n^{2-\delta}m'\delta^{-1} + m'\delta^{-1}(\# \text{ of close pairs}) + m'\delta^{-1}(\# \text{ of boundary pairs}))$, where $0 < \delta < 1$, and is an adaptation of the agglomerative clustering algorithm in [2] and near(est)-neighbor data structures in [21].

The running times above are in expectation over the (o^*, ϵ)-SAP algorithm's internal randomness, and are subquadratic under two main assumptions. First, o^* is at least a constant fraction of m' to make both ρ and δ constants depending only on ϵ. Second, the number of boundary pairs is subquadratic (actually, the first algorithm also requires the number of close pairs to be $o(n^{2-\rho})$). We justify later why these two main assumptions are reasonable for strong similarity search (such as finding gapped seeds for aligning genomes).

Other Related Work. We mention related work not cited so far. Multipattern search algorithms [14] with average-optimal search time for lenient error ranges can solve our problem using $O(n)$ patterns. But such high number of patterns result in quadratic/cubic in n preprocessing time and space, quite infeasible in our setting. High-dimensional sparse nearest-neighbor search was used in the context of a *block* edit distance embedding scheme [24]. We use sparse *all-pairs* in a edit distance filtering scheme to solve a different problem. Sparse all-pairs on certain input distributions is called the bit-vector intersection problem in [20]. Some earlier works (see [27] and references there) use q-gram based measures to study sequence similarity. But we use edit distance, and rely on a q-gram overlap measure for filtration purposes *only*. Our q-gram measure is related to the l_1 distance between q-gram profiles [31]. These earlier measures are related to the (squared) l_2 distance (which is amenable to high-dimensional search too).

2   Algorithms

2.1   Gapped Local Similarity Algorithm

Recall the gapped local similarity problem statement. The algorithm for this problem comprises two steps: find all length-m substring pairs with at least o^* common q-grams in a filtration step, and among them output only those with edit distance at most e^* in a verification step. The o^* is given as follows. A length-m substring has $m' = (m - q + 1)$ q-grams, and two such substrings with an edit distance at most e^* share at least $o^* = (m' - e^* q)$ common q-grams [31] (to account for multiplicity here, count a q-gram occurring x times in one substring and y times in the other as $\min(x, y)$). We do the filtration step using a (o^*, ϵ)-SAP algorithm and the verification step using standard edit distance calculations.

We reduce the filtration step to the (o^*, ϵ)-SAP problem to leverage past research in high-dimensional proximity search [18]. The idea is to first associate each length-m substring with a σ^q-dimensional q-gram profile vector [31] whose ith entry contains the number of copies of q-gram "i" in the substring (σ denotes alphabet size). Then, represent the value in each entry of this profile vector in m'-digit unary to get an expanded profile vector (e.g. in 7-digit, unary$(4) = 1111000$). The expanded profile vectors are of huge dimension $D = \sigma^q m'$, and very sparse with exactly m' 1s. Importantly, the number of common q-grams shared by two substrings is equal to the number of common 1s in their expanded profile vectors. This leads us to the sparse all-pairs problem, in which we further use ϵ-approximation to enable efficient randomization techniques (without affecting sensitivity as noted in Section 1).

The whole algorithm is provably sensitive as the (o^*, ϵ)-SAP algorithm used is provably sensitive. The algorithm's performance depends on the filtration time and filter efficiency. The latter is commonly studied [26] by counting the substring pairs that survive (i.e., share at least $\frac{o^*}{1+\epsilon}$ common q-grams) under a Bernoulli assumption, where each position of one of the sequences is drawn independently and uniformly at random from the σ alphabets. This lemma on the performance for strong similarity search (small e^*) follows from the use of Markov's inequality.

Lemma 1. *The gapped local similarity algorithm runs in time $O(T(n, m', o^*, \epsilon) + m^2(\# \text{ of surviving substring pairs}))$, where $T(n, m', o^*, \epsilon)$ denotes the running time of the (o^*, ϵ)-SAP algorithm. When $q \geq ((c+2) \log_\sigma m + \log_\sigma(1+\epsilon))$ and $e^* = O(m / (\log_\sigma m + \log_\sigma(1+\epsilon)))$, the expected number of surviving substring pairs under a random uniform Bernoulli input assumption is $O(n^2/m^c)$, which is subquadratic when m is an increasing function of n (say $m = \log n$).*

2.2 Approximate Sparse All-Pairs Algorithms

We focus on two (o^*, ϵ)-SAP algorithms in this section. A (o^*, ϵ)-SAP algorithm forms the core of our approach as it is used to implement q-gram filtration.

Preliminaries. We first review basic concepts common to the two algorithms.

Notations and Definitions. Vectors are treated as row vectors, with their positions sometimes referred to as columns. The symbols $\boldsymbol{u}$, $\boldsymbol{v}$, and $\boldsymbol{w}$ denote binary $(0/1)$ vectors of dimension D, unless stated otherwise. The set of indices of all non-zero columns of a binary vector $\boldsymbol{u}$ is denoted by ones$(\boldsymbol{u})$. A sparse vector refers to a binary vector of dimension D that has exactly m' 1s for $m' \ll D$. The algorithms use ones$(\boldsymbol{u})$ as a succinct representation of a sparse vector $\boldsymbol{u}$.

Definition 1. *Consider two binary vectors v, w.*

- *The Hamming distance $\mathrm{h}(v, w)$ is the number of columns in which v and w differ (the distance can be defined over non-binary vectors too).*
- *The overlap or dot-product measure $\mathrm{o}(v, w)$ is the number of columns that have a 1 in both v and w, i.e., $\mathrm{o}(v, w) = |\mathrm{ones}(v) \cap \mathrm{ones}(w)| = v \cdot w$.*
- *The set resemblance $\mathrm{s}(v, w)$ is the ratio of the number of columns that have a 1 in both v and w to the number of columns that have a 1 in either of v or w, i.e., $\mathrm{s}(v, w) = \frac{|\mathrm{ones}(v) \cap \mathrm{ones}(w)|}{|\mathrm{ones}(v) \cup \mathrm{ones}(w)|}$.*

Observation 1. *Let v, w be sparse vectors with $\mathrm{o}(v, w) = o$. Then, $\mathrm{s}(v, w) = \frac{o}{2m' - o}$, and $\mathrm{h}(v, w) = 2(m' - o)$.*

With respect to a $\frac{o^*}{1+\epsilon} : o^*$ overlap measure criterion, vectors (v, w) are called a close pair if $\mathrm{o}(v, w) \geq o^*$, far pair if $\mathrm{o}(v, w) \leq \frac{o^*}{1+\epsilon}$, and boundary pair otherwise. We can now state the problem of this section as below. The equal size assumption in the problem is for convenience in analysis, and can be removed if needed.

Sparse (o^*, ϵ)-All-Pairs (or) (o^*, ϵ)-SAP Problem. Output all close pairs and possibly some boundary pairs, but no far pairs w.r.t the $\frac{o^*}{1+\epsilon} : o^*$ overlap measure criterion from two groups of sparse vectors V and W of size n each.

Source of Subquadraticity. To beat the naive $O(n^2 m')$ algorithm, the two (o^*, ϵ)-SAP algorithms require the two main assumptions in the Contributions section. Both assumptions are reasonable for strong similarity search. Particularly, both hold under the conditions in Lemma 1 (i.e., o^*/m' is at least a constant when q is set at the suggested lower bound, and the *expected* number of boundary pairs is subquadratic for sufficiently large m).

Subquadratic algorithms for an all-pairs problem can be designed from techniques in sublinear algorithms for a related near(est)-neighbor problem [18]. The first algorithm we discuss is based on the nearest-neighbor data structures in [19] and the second on those in [21]. These data structures, which support vectors w.r.t Hamming distance measure, are made to handle sparse vectors of huge dimension w.r.t overlap measure.

From a high-level, both (o^*, ϵ)-SAP algorithms use the idea of random projections to achieve dimensionality reduction, and then do an easier all-pairs computation on the resulting low-dimensional vectors. But the design of each step is quite different between the two.

First Algorithm. The first (o^*, ϵ)-SAP algorithm is an existing algorithm [16, 13], and we provide a careful worst-case analysis of its running time. Explicit worst-case running time is not provided in [16, 13]; [13] discusses running time in practice using algorithm parameters determined from an input-dependent optimization problem. The algorithm that we present and analyse is actually a close variant of the existing algorithm, and follows along the lines of the nearest-neighbor algorithm and analysis in [19].

Random Projection via Minhash. The key ingredient of the algorithm is a random projection achieved via a locality-sensitive hash function [19] which preserves distance information between vectors. The minhash function [6, 12] is used for this purpose.

Definition 2. *Choose a random permutation π of the column indices $1, 2, \ldots,$ D. The minhash of a vector u is defined as the index of the first non-zero column of u in the ordering induced by π, i.e., $\mathrm{minhash}(u) = \arg\min_{i \in \mathrm{ones}(u)}\{\pi(i)\}$.*

The probability over π that two vectors v and w get the same minhash value is exactly their set resemblance $s(v, w)$. We use this fact and Observation 1 for sparse vectors to derive the next lemma.

Lemma 2. *Consider two sparse vectors v, w, and a minhash function $\mathrm{minhash}(.)$ picked at random as in Definition 2. Let $p_1 = s^*$, and $p_2 = \frac{s^*}{1+\epsilon}$, where $s^* = \frac{o^*}{2m'-o^*}$.*

- *If $o(v, w) \geq o^*$, $\mathbf{Pr}_\pi[\mathrm{minhash}(v) = \mathrm{minhash}(w)] \geq p_1$.*
- *If $o(v, w) \leq \frac{o^*}{1+\epsilon}$, $\mathbf{Pr}_\pi[\mathrm{minhash}(v) = \mathrm{minhash}(w)] \leq p_2$.*
- *$p_1 = p_2^\rho$, where the bias $\rho = \left(1 + \frac{\log(1+\epsilon)}{\log(1/s^*)}\right)^{-1}$. If o^*/m' is a constant, $\rho < 1$ is a constant (depending only on ϵ).*

A minhash thus introduces a bias between close and far pairs. We pick k independent random permutations, concatenate the resulting minhash functions, and denote the result by the function $\mathrm{minhash}^k : \{0,1\}^D \to \{1, 2, \ldots, D\}^k$. Note that $\mathrm{minhash}(u)$ for a sparse vector u can be computed in $O(m')$ time as follows. Instead of inefficiently permuting all D column indices, we use a practical approach [6, 16] based on a random linear function $\pi(x) = ax+b \bmod P$, where P is the smallest prime above D, and a, b are random numbers from $\{1, 2, \ldots, D\}$ ([5, 17] analyse the random ordering resulting from this and alternate approaches).

Algorithm Description. The algorithm repeats a single randomized trial many times. To manage the buckets used in the procedure, we resort to either a hash table [19] or sorting [16, 8]. Also, to quickly check if a pair hasn't been output before, the procedure can use a hash table that stores all already output pairs. Note that p_1, p_2 are from Lemma 2.

All-Pairs Procedure I:
input: two groups of sparse vectors, V and W.
output: solves (o^*, ϵ)-SAP on the input.

Set $k = \log_{1/p_2} n$.
repeat $l = 3 \log n / p_1^k$ times **do**
    Pick a random $\mathrm{minhash}^k(.)$ function for this trial as described above.
    for each $v \in V$ **do**   // **Preprocess** V
        Compute $\mathrm{minhash}^k(v)$.
        Store v in the bucket labelled $\mathrm{minhash}^k(v)$.
    for each $w \in W$ **do**   // **Query** W
        Compute $\mathrm{minhash}^k(w)$.
        for each v in the bucket labelled $\mathrm{minhash}^k(w)$ **do**
            Output (v, w) if it's not output before, and $o(v, w) \geq o^*$.

Analysis. The expected running time of the above procedure, as given in the next theorem, is subquadratic under our two main assumptions (about o^*/m' being a constant and the number of boundary pairs being sub-quadratic), and an additional assumption that number of close pairs is $o(n^{2-\rho})$.

Theorem 1. *The All-pairs Procedure I solves the (o^*, ϵ)-SAP problem with high probability in expected running time $O(n^{1+\rho}m'\log^2 n + (m' + n^\rho \log n)$ (# of close pairs) $+ m'\log n$ (# of boundary pairs)), where $\rho < 1$ is a constant depending only on ϵ as o^*/m' is a constant, and space $O(nm' +$ (# of close pairs)).*

Proof. The proof follows along the lines of a proof in [19]. First, computing $l = O(n^\rho \log n)$ minhash$^k(.)$ on all vectors takes $O(lm'kn)$ time. Next, we consider the three types of pairs. The probability that (vectors of) a far pair wrongly map to the same bucket is at most p_2^k, so the expected number of treated far pairs is at most $O(l\, n^2\, p_2^k) = O(ln)$. Similarly, the expected number of treated boundary pairs is at most $O(l\,(\#$ of boundary pairs$)\,p_1^k) = O(\log n\,(\#$ of boundary pairs$))$. Note that a pair is treated by computing its overlap measure in $O(m')$ time.

The probability that a close pair fails to map to the same bucket in all l trials is at most $(1 - p_1^k)^l \simeq \frac{1}{n^3}$, so the error probability over $O(n^2)$ potential close pairs is a low $O(\frac{1}{n})$. Finally, the probability that a close pair maps to the same bucket can't be upper-bounded, since the same close pair could be detected and treated in all l trials in the worst case (e.g. a pair of identical vectors), with the first treatment taking $O(m')$ time and later ones $O(1)$ time each (to find if the pair hasn't been output before). This gives a $O((m' + l)(\#$ of close pairs$))$ term. □

Second Algorithm. The second (o^*, ϵ)-SAP algorithm improves the number of close pairs term in the first algorithm's running time. Our second algorithm is a simple adaptation of the agglomerative clustering algorithm (w.r.t euclidean distance) in [2], which is based on the near(est)-neighbor data structures in [21]. The crux of our adaptation is to rid the running time of its dependence on the huge dimension D, and to transition from euclidean to overlap measure. To do this, we exploit the sparsity of the vectors and simplify the nearest-neighbor data structures in [21] to support only n (and not all 2^D) queries. The rest of the details are the same as the clustering algorithm in [2].

Random Projection via β-Equality Hash. The random projection is now via β-equality hash instead of minhash. The use of β-equality hash to do dimensionality reduction admits a brute-force solution to a later all-pairs step of the algorithm.

Definition 3 ([21]). *Choose a random vector r of dimension D by independently setting each of its D positions to 1 with probability $\frac{\beta}{2}$ and 0 with probability $1 - \frac{\beta}{2}$. The β-equality hash of a vector u is defined as $\text{hash}_\beta(u) = r \cdot u \pmod 2$.*

In the standard "equality test", $\beta = 1$ and the probability over r that two unequal vectors v and w get different hash values is $\frac{1}{2}$. For a general β, this probability is $\frac{1}{2}(1 - (1 - \beta)^{h(v,w)})$, a value dependent on the distance between vectors [21]. We use this probability for an appropriate β and Observation 1 for sparse vectors to derive the next lemma.

Lemma 3. *Consider two sparse vectors v, w, and a β-equality hash function* $\text{hash}_\beta(.)$ *picked at random as in Definition 3 using* $\beta = \frac{1}{2h^*}$, *where* $h^* = 2(m'-o^*)$.
Let $p_1 = \frac{1}{2} - \frac{1}{2}\left(1 - \frac{1}{4(m'-o^*)}\right)^{2(m'-o^*)}$, *and* $p_2 = \frac{1}{2} - \frac{1}{2}\left(1 - \frac{1}{4(m'-o^*)}\right)^{2\left(m' - \frac{o^*}{1+\epsilon}\right)}$.

- *If* $o(v, w) \geq o^*$, $\mathbf{Pr}_r[\text{hash}_\beta(v) \neq \text{hash}_\beta(w)] \leq p_1$.
- *If* $o(v, w) \leq \frac{o^*}{1+\epsilon}$, $\mathbf{Pr}_r[\text{hash}_\beta(v) \neq \text{hash}_\beta(w)] \geq p_2$.
- $p_2 - p_1 > \Delta_p(\epsilon, \frac{o^*}{m'})$, *where the additive bias* $\Delta_p(\epsilon, \frac{o^*}{m'}) = \Theta\left(e^{-\frac{1}{2}} - e^{-\frac{1}{2(1+\epsilon)}}\left(1 + \frac{\epsilon}{1-o^*/m'}\right)\right)$. *If* o^*/m' *is a constant,* $\Delta_p > 0$ *is a constant (depending only on ϵ).*

The next lemma amplifies the constant bias between close and far pairs introduced in Lemma 3 by using many β-equality hash functions. It is a simple consequence of Chernoff bounds.

Lemma 4 ([21]). *Let the function* $\text{hash}_\beta^t : \{0,1\}^D \rightarrow \{0,1\}^t$ *denote the concatenation of t independent β-equality hash functions picked at random as in Lemma 3. Then for sparse vectors v, w,*

- *If* $o(v, w) \geq o^*$, $\mathbf{Pr}_r\left[h\left(\text{hash}_\beta^t(v), \text{hash}_\beta^t(w)\right) > (p_1 + \frac{\Delta_p}{3})t\right] \leq e^{-\frac{2}{9}\Delta_p^2 t}$.
- *If* $o(v, w) \leq \frac{o^*}{1+\epsilon}$, $\mathbf{Pr}_r\left[h\left(\text{hash}_\beta^t(v), \text{hash}_\beta^t(w)\right) < (p_2 - \frac{\Delta_p}{3})t\right] \leq e^{-\frac{2}{9}\Delta_p^2 t}$.

Finally note that a $\text{hash}_\beta(.)$ function on n sparse vectors can be consistently computed in $O(nm')$ time, because we focus only on columns that are non-zero in at least one of the n vectors, and pick and store in a hash table only these (and not all D) columns of the associated random vector r.

Algorithm Description. A single trial of this algorithm uses Lemma 4 to reduce the dimension of vectors from D to t. As any "query" must now map to one of the 2^t possible vectors, we compute the "answers" to all such vectors and store them in a table with 2^t entries. Note that β, Δ_p are from Lemma 3.

All-Pairs Procedure IIa:
input: two sets of sparse vectors, V' and W'.
output: solves (o^*, ϵ)-SAP on the input (first version).
Set $t = \frac{9}{2}\Delta_p^{-2} \log |V'|$.
repeat $g = 3\log_{|V'|} |W'| + 1$ times **do**
    Pick a random $\text{hash}_\beta^t(.)$ function for this trial as in Lemma 4.
    for each binary vector z of dimension t **do**   // **Preprocess** V'
        Create an entry labelled z.
        Store in this entry all $v \in V'$ s.t.
            $h(\text{hash}_\beta^t(v), z) \leq (p_1 + \frac{\Delta_p}{3})t$.
    for each $w \in W'$ **do**   // **Query** W'
        Compute $\text{hash}_\beta^t(w)$.
        for each v in the entry labelled $\text{hash}_\beta^t(w)$ **do**
            Output (v, w) if it's not output before, and $o(v, w) \geq o^*$.

We can't simply pass V, W as inputs V', W' to the above procedure as the running time is dominated by a $2^t|V'| = |V'|^{c_1}$ factor, where $c_1 = (\frac{9}{2}\Delta_p^{-2} + 1) > 2$. We resort to a trick from [2] and pass V after partitioning it into smaller groups.

All-Pairs Procedure IIb:

input: two sets of sparse vectors, V and W.

output: solves (o^*, ϵ)-SAP on the input (improved version).

Set $\delta = \frac{1}{2c_1}$, and partition V into $n^{1-\delta}$ sets, $V_1, V_2, \ldots, V_{n^{1-\delta}}$, of size n^δ each.

for each $i = 1, 2, \ldots, n^{1-\delta}$ **do**

Call All-pairs Procedure IIa with V_i, W as inputs.

Analysis. The expected running time of the above procedure, as given in the corollary below, is subquadratic under our usual two main assumptions (as in the first algorithm). Note that we don't require here the additional assumption (made by the first algorithm) on the number of close pairs. The theorem below analyses Procedure IIa. The proof of this theorem using Lemma 4 is omitted as it follows the same template as the proof of Theorem 1.

Theorem 2. *The All-pairs Procedure IIa solves the (o^*, ϵ)-SAP problem with high probability in expected running time $O(gm'\Delta_p^{-2}\log|V'|(|V'|^{c_1} + |W'|) + gm'(\# \text{ of close pairs}) + gm'(\# \text{ of boundary pairs}))$, and space $O(m'(|V'| + |W'|) + |V'|^{c_1} + (\# \text{ of close pairs}))$.*

We use $g = O(\frac{1}{\delta}) = O(\Delta_p^{-2})$ in the above theorem and union-bound the error probabilities to get the desired running time below.

Corollary 1. *The All-pairs Procedure IIb solves the (o^*, ϵ)-SAP problem with high probability in expected running time $O(n^{2-\delta}m'\delta^{-1}\log n + m'\delta^{-1}(\# \text{ of close pairs}) + m'\delta^{-1}(\# \text{ of boundary pairs}))$, where $0 < \delta < 1$ is a constant depending only on ϵ as o^*/m' is a constant, and space $O(nm' + (\# \text{ of close pairs}))$.*

3   Implementation

The *qhash* program implements the gapped local similarity algorithm (Section 2.1) using the first sparse all-pairs algorithm (Section 2.2). It will be made publicly available at *www.cs.berkeley.edu/~nmani/qhash*. We use the first algorithm, despite the second's improved "close pairs" running time term, because the number of close pairs is likely to be at most linear in practice, and the first is more amenable to input-dependent optimizations described below. A note on All-pairs Procedure I before we proceed. We set $l = \log(1/p_{\mathrm{fn}})/p_1^k$ so that the probability of missing a close pair is at most p_{fn} (i.e., we won't miss more than a p_{fn} fraction of all close pairs in expectation, over the program's internal randomness).

3.1   *qhash* in Practice

We describe optimizations that significantly improve *qhash*'s performance.

Learning k, l from the Input. We empirically estimate the parameters k, l of the first algorithm from the input data. This leads to huge savings in time *without* affecting the sensitivity guarantee. Observe that an increase in k decreases the number of candidate pairs (those with same $\text{minhash}^k(.)$), but could potentially increase l exponentially. The idea then is to pick a k less than the theoretical k so as to minimize an empirical estimate of the total running time. This estimate for a specific k is empirically obtained by quickly observing and counting the candidate pairs in a trial (without actually processing them). Parameters are picked practically in [13] too, but they use a different optimization criteria.

q-Grams with Locations. We can improve q-gram filters by not counting common q-grams that are more than a distance e^* apart in two substrings [30]. This improvement (*not* in the current implementation) could potentially reduce k, l. It can be incorporated in our framework by associating each substring with a set of locational q-grams before calling the sparse all-pairs algorithm. In more detail, we decompose each substring into blocks of size $2e^*$ that overlap by the amount e^*. A q-gram now falls in two such blocks, and we annotate it with these block ids to get two locational q-grams per q-gram.

3.2   Results

We present results on a similarity search for 50-nucleotide long substring pairs that share a strong similarity of 94% ($m = 50$, $e^* = 3$). We set the parameters $p_{fn} = 0.1$ (90% sensitivity), $q = 6$, and $\epsilon = 1$ [2]. Table 1 compares *qhash* against a fully sensitive $O(n^2 m)$ program based on the dynamic-programming (DP) algorithm of [15]. We observe that *qhash* when compared to the other is significantly faster on long sequences and runs in almost the same time on medium-sized sequences. We also observe that the memory requirement of *qhash* is reasonable.

Table 2 shows the effect of empirically estimating k, l by illustrating the trade-off between low and high values of k (see last section). Note that the empirical k, l values in Table 2 are in fact the ones used to obtain the results of Table 1. Additional improvements such as q-grams with location are necessary if *qhash* is to be competitive with the extremely fast BLAST program [1]. We expect such improvements would make *qhash* efficient even for lenient error ranges (i.e., higher e^* where BLAST requires shorter initial seeds to be sensitive).

4   Conclusions

The crux of our work is a new randomized approach to doing q-gram filtration in sequence-sequence mode using high-dimensional proximity search techniques. Our program *qhash* based on this idea has provable guarantees, and is significantly faster than a dynamic-programming-based algorithm for strong gapped local similarity search on long sequences. We suggested future additions that

---

[2] Lemma 1 with $c = 1$ suggests a starting q value, and we reduce it till s^* exceeds 0.4 (to keep l low). The ϵ value doesn't matter much as we learn k, l from the input.

Table 1. Results on a 3 GHz Pentium Linux machine. The test sequences (1) to (4), selected mostly from past search studies, are respectively human–mouse (beta-globin) [8], human–chicken (AC096683) [3], m.pneumoniae–m.genitalium [32], and t.acidophilum–t.volcanium [32]. The overall time of *qhash* includes the time for empirically estimating k, l. The optimistic time estimate of the DP-based program is the product of the sequence lengths (in basepairs) and 1.34×10^{-8} secs, a implementation constant.

Test Sequences No.	Sequence lengths (Kbp–Kbp)	*qhash* program overall time (mins)	DP-based program optimistic time estimate (mins)	Number of close pairs found	*qhash* program memory used (MB)
(1)	20.0–12.6	0.17	0.06	19	6
(2)	161.6–110.4	3.27	3.98	376	32
(3)	816.4–580.1	31.26	105.75	14118	157
(4)	1565.0–1584.8	63.00	553.90	6573	353

Table 2. Learning k, l from the input. The counts refer to candidate pairs observed in *one* randomized trial using the corresponding k. Verifying each such pair would take an estimated 32 μsecs in our machine.

Test Sequences No.	Theoretical Estimation			Empirical Estimation		
	k	l	Cand. Pairs Count	k	l	Cand. Pairs Count
(1)	6	394	98	3	30	2495
(2)	7	930	616	4	71	10413
(3)	8	2192	20244	4	71	322668
(4)	9	5167	7225	4	71	1109426

could make *qhash* competitive in more lenient error ranges. A related open question here is design of random projections w.r.t overlap measure without assuming o^*/m' as a constant.

We can extend any filter technique from pattern-database to sequence-sequence mode using our approach, provided each substring can be associated with a set of the filter's "features" (e.g. it works for gapped q-grams [11]). Our work raises interesting questions about randomized filters, filters with guarantees on the fraction of similarities they miss. Are there other approaches to construct randomized filters (other than taking a traditional filter and giving it a randomized implementation as we did)? Would they have a better filter efficiency over traditional filters due to their flexibility in missing some similarities?

Acknowledgements

We would like to thank Vinod Prabhakaran, Sourav Chatterji and Jayanth Kumar Kannan for many useful discussions, and the latter two for help with much needed computing power. We thank Sridhar Rajagopalan for ideas on how to combine q-gram locations in the scheme. We thank the reviewers for useful suggestions.

References

1. S. Altschul, W. Gish, W. Miller, E. Myers, and D. Lipman. Basic local alignment search tool. *Journal of Molecular Biology*, 215(3):403–410, 1990.
2. A. Borodin, R. Ostrovsky, and Y. Rabani. Subquadratic approximation algorithms for clustering problems in high dimensional spaces. In *Proc. 31st Symp. on Theory of Computing*, pages 435–444, 1999.
3. N. Bray, I. Dubchak, and L. Pachter. Avid: A global alignment program. *Genome Research*, 13(1):97–102, 2003.
4. B. Brejova, D. Brown, and T. Vinar. Vector seeds: An extension to spaced seeds allows substantial improvements in sensitivity and specifity. In *Proc. 3rd Workshop on Algorithms in Bioinformatics*, volume 2812 of *LNCS*, pages 39–54, 2003.
5. A. Broder, M. Charikar, A. Frieze, and M. Mitzenmacher. Min-wise independent permutations. In *Proc. 30th Symp. on Theory of Computing*, pages 327–336, 1998.
6. A. Broder, S. Glassman, M. Manasse, and G. Zweig. Syntactic clustering of the web. In *Proc. 6th Intl. World Wide Web Conf.*, pages 391–404, 1997.
7. M. Brudno and B. Morgenstern. Fast and sensitive alignment of large genomic sequences. In *Proc. IEEE Comp. Soc. Bioinformatics Conf.*, pages 138–147, 2002.
8. J. Buhler. Efficient large-scale sequence comparison by locality-sensitive hashing. *Bioinformatics*, 17(5):419–428, 2001.
9. J. Buhler. *Search Algorithms for Biosequences Using Random Projection*. PhD thesis, University of Washington, 2001.
10. S. Burkhardt, A. Crauser, P. Ferragina, H. Lenhof, E. Rivals, and M. Vingron. q-gram based database searching using a suffix array. In *Proc. 3rd Conf. on Research in Comp. Molecular Biology*, pages 77–83, 1999.
11. S. Burkhardt and J. Karkkainen. Better filtering with gapped q-grams. In *Proc. 12th Symp. on Comb. Pattern Matching*, pages 73–85, 2001.
12. E. Cohen. Size-estimation framework with applications to transitive closure and reachability. *Journal of Computer and System Sciences*, 55(3):441–453, 1997.
13. E. Cohen, M. Datar, S. Fujiwara, A. Gionis, P. Indyk, R. Motwani, J. Ullman, and C. Yang. Finding interesting associations without support pruning. *IEEE Trans. on Knowledge and Data Engineering*, 13(1):64–78, 2001.
14. K. Fredriksson and G. Navarro. Improved single and multiple approximate string matching. In *15th Symp. on Comb. Pattern Matching (to appear)*, 2004.
15. D. Gusfield. *Algorithms on Strings, Trees, and Sequences*, chapter 11.6.5 (Approximate occurrences of P in T). Cambridge Univ. Press, 1997.
16. T. Haveliwala, A. Gionis, and P. Indyk. Scalable techniques for clustering the web. In *Proc. 3rd Intl. Workshop on the Web and Databases*, 2000.
17. P. Indyk. A small approximately min-wise independent family of hash functions. In *Proc. 10th Symp. on Discrete Algorithms*, pages 454–456, 1999.
18. P. Indyk. Nearest neighbors in high-dimensional spaces. In *Handbook of Discrete and Comp. Geometry, 2nd Edition*. CRC Press LLC, Upcoming.
19. P. Indyk and R. Motwani. Approximate nearest neighbors: towards removing the curse of dimensionality. In *Proc. 30th Symp. on Theory of Computing*, pages 604–613, 1998.
20. R. Karp, O. Waarts, and G. Zweig. The bit vector intersection problem. In *Proc. 36th Symp. on Foundations of Computer Science*, pages 621–630, 1995.
21. E. Kushilevitz, R. Ostrovsky, and Y. Rabani. Efficient search for approximate nearest neighbor in high dimensional spaces. In *Proc. 30th Symp. on Theory of Computing*, pages 614–623, 1998.

22. G. Landau and U. Vishkin. Introducing efficient parallelism into approximate string matching and a new serial algorithm. In *Proc. 18th Symp. on Theory of Computing*, pages 220–230, 1986.

23. R. Lippert, X. Zhao, L. Florea, C. Mobarry, and S. Istrail. Finding anchors for genomic sequence comparison. In *Proc. 8th Conf. on Research in Comp. Molecular Biology*, pages 233–241, 2004.

24. S. Muthukrishnan and S. Sahinalp. Simple and practical sequence nearest neighbors with block operations. In *Proc. 13th Symp. on Comb. Pattern Matching*, pages 262–278, 2002.

25. E. Myers. An $O(ND)$ Difference Algorithm and Its Variations. *Algorithmica*, 1(2):251–266, 1986.

26. G. Navarro. A guided tour to approximate string matching. *ACM Computing Surveys*, 33(1):31–88, 2001.

27. P. Pevzner. Statistical distance between texts and filtration methods in sequence comparison. *CABIOS*, 8(2):121–127, 1992.

28. S. Schwartz, W. Kent, A. Smit, Z. Zhang, R. Baertsch R, R. Hardison, D. Haussler, and W. Miller. Human-mouse alignments with blastz. *Genome Research*, 13(1):103–107, 2003.

29. T. Smith and M. Waterman. Identification of common molecular subsequences. *Journal of Molecular Biology*, 147(1):195–197, 1981.

30. E. Sutinen and J. Tarhio. On using q-gram locations in approximate string matching. In *Proc. European Symp. on Algorithms*, pages 327–340, 1995.

31. E. Ukkonen. Approximate string matching with q-grams and maximal matches. *Theoretical Computer Science*, 92(1):191–211, 1992.

32. NCBI Entrez Genomes.
http://www.ncbi.nlm.nih.gov/entrez/query.fcgi?db=Genome.

Monotone Scoring of Patterns with Mismatches
(Extended Abstract)

Alberto Apostolico[1,*] and Cinzia Pizzi[2,**]

[1] University of Padova & Purdue University
axa@dei.unipd.it
[2] University of Padova
cinzia.pizzi@dei.unipd.it

Abstract. We study the problem of extracting, from given source x and error threshold k, substrings of x that occur unusually often in x within k substitutions or mismatches. Specifically, we assume that the input textstring x of n characters is produced by an i.i.d. source, and design efficient methods for computing the probability and expected number of occurrences for substrings of x with (either *exactly* or *up to*) k mismatches. Two related schemes are presented. In the first one, an $O(nk)$ time preprocessing of x is developed that supports the following subsequent queries: for any substring w of x arbitrarily specified as input, the probability of occurrence of w in x within (either exactly or up to) k mismatches is reported in $O(k^2)$ time. In the second scheme, a length or length range is arbitrarily specified, and the above probabilities are computed for all substrings of x having length in that range, in overall $O(nk)$ time. Further, monotonicity conditions are introduced and studied for probabilities and expected occurrences of a substring under unit increases in its length, allowed number of errors, or both. Over intervals of constant frequency count, these monotonicities translate to some of the scores in use, thereby reducing the size of tables at the outset and enhancing the process of discovery. These latter derivations extend to patterns with mismatches an analysis previously devoted to exact patterns.

1 Preliminaries

The problem of extracting unusually frequent or rare patterns from observed sequences arises ubiquitously in applications and has been the subject of much

* Dipartimento di Ingegneria dell' Informazione, Università di Padova, Padova, Italy *and* Department of Computer Sciences, Purdue University, Computer Sciences Building, West Lafayette, IN 47907, USA. Work Supported in part by an IBM Faculty Partnership Award, by the Italian Ministry of University and Research under the National Projects FIRB RBNE01KNFP, and PRIN "Combinatorial and Algorithmic Methods for Pattern Discovery in Biosequences", and by the Research Program of the University of Padova.
** Dipartimento di Ingegneria dell' Informazione, Università di Padova, Via Gradenigo 6/A, 35131 Padova, Italy.

I. Jonassen and J. Kim (Eds.): WABI 2004, LNBI 3240, pp. 87–98, 2004.
© Springer-Verlag Berlin Heidelberg 2004

study in Molecular Biology. This problem may take up different flavors, depending on the assumptions about the source and on the character and structure of the patterns themselves. It is customary to partition the approaches to discovery into two main classes. In the first class, the sample string is tested for occurrences of each and every motif in a family of *a priori* generated, abstract *models* or templates. This is methodologically sound but may pose daunting computational burdens. The second class of approaches assumes that the search may be limited to substrings in the sample or to some more or less controlled neighborhood of those substrings. This may be less firm methodologically but brings about time and space savings. Some hybrid variants consist of postulating or building the models by inference from their incarnations in the sample itself. Except for the case of solid patterns, however, all of the available techniques present some intrinsic exponential buildup that often translates into unbearable computational overhead. We refer to the quoted sample of literature for details. In this paper, we study the approach that consists of extracting from given source x and error threshold k, substrings of x that occur unusually often in x within k mismatches. To quantify "unusually often" for a substring w of x, this is measured by comparing, e.g., the observed frequency and the expected number of occurrences for w with (either exactly, or up to) k mismatches. We study problems related to the efficient computation and representation for these measures.

Throughout the rest of the discussion, a *motif* is a pair (w, k) where w is a string of characters from an alphabet Σ and k is the number of errors or mismatches allowed on w. Thus, the pair (w, k) identifies a family of strings over Σ, whereas the same string belongs to more than one family. We will use $w_{(k)}$ to refer to a string in (w, k). To avoid clutter in notation, we will let the context specify whether k denotes the maximum or exact number of errors. When $k = 0$ we talk of *solid* strings or patterns.

Given a textstring x and a length range $m \pm \delta$ with constant δ, we are interested in particular in the efficient construction of a table $\mathcal{W}(x)$ containing all motifs (w, k) of length between $m - \delta$ and $m + \delta$ and such that w is a substring of x, together with their individual probabilities, and expected number of occurrences. In practical applications such as, e.g., regulatory sequence detection, values of $m \approx 10 - 15$ and $\delta \approx 3 - 6$ are typical. As mentioned, the above expectations combine with frequency counts to yield some of the z-scores in use. Since $\mathcal{W}(x)$ can be quite bulky, we will seek to reduce its size, by limiting it to entries representing local maxima w.r.t. the score.

We use capital letters to denote random strings and variables. In particular, $X = X_1 X_2 X_3 \dots X_n$ denotes a random textstring produced by a source which emits symbols from Σ under i.i.d. assumptions, i.e., the X_i are emitted independently and according to the same distribution. This is denoted by setting $P[X_i = s \in \Sigma] = p_s = p_i \; \forall i$, with obvious meaning.

For an *observed* pattern $y = y_1 y_2 \dots y_m$, the probability of y is $P(y) = p_1 p_2 \dots p_{|y|}$, which is also the expected value of the indicator variable $Z_i | y$, taking value 1 when y occurs beginning at position i in X and 0 otherwise. Thus,

$E[Z_i|y] = E[Z_1|y] = P(y)$. The random variable representing the number of occurrences of y in X is $Z|y = \sum_{i=1}^{n-m+1} Z_i|y$. In the following, and whenever this causes no confusion, we will use $E[y]$ shorthand for $E[Z|y] = \sum_{i=1}^{n-m+1} E[Z_i|y] = (n-m+1)P(y)$. Likewise, we will use $P_k(y)$ for $P(y_{(k)})$, $E_k[y]$ for $E[y_{(k)}]$.

The rest of this paper is organized as follows. In the next section, we give efficient computations for motif probabilities and expected number of occurrence. The monotonicities of these and related parameters are established in Section 3. Section 4 completes our constructions and concludes the paper.

2 String Probabilities and Their Correction Factors

With linear-time preprocessing of a standard prefix computation on the textstring x, it is trivial to compute the probability or expected number of occurrences of any substring of x in constant time. The pre-processing consists of building the array $A[i] = \prod_{h=1}^{i} p_h$, $i = 1, 2, \ldots, n$ with $A[0] = 1$, so that, for instance, for any pair (b, e) of positions, the probability of $\bar{x} = x[b \ldots e]$ is $P(\bar{x}) = \prod_{i=b}^{e} p_i = \frac{A[e]}{A[b-1]}$ and $E[\bar{x}] = A[e]/A[b-1](|x| - e + b)$.

The same computation for patterns having exactly k mismatches with $\bar{x}$ risks to incur exponential cost, since we need to tally all of the $\binom{|\bar{x}|}{k}$ ways to position k mismatches in $\bar{x}$. A less expensive, incremental approach can be built on the notion of *correction factor*. As an example, consider the pattern $y = abaabaa$ on $\Sigma = \{a, b\}$ and let $P(y) = p_a p_b p_a p_a p_b p_a p_a = p_a^5 p_b^2$. A mutation in the first position would change y into $y' = bbaabaa$, with associated probability: $P(y') = p_b p_b p_a p_a p_b p_a p_a = \frac{p_a}{p_a} p_b p_b p_a p_a p_b p_a p_a = \frac{p_b}{p_a} p_a p_b p_a p_a p_b p_a p_a = \frac{p_b}{p_a} P(y)$.

We define $f_a = \frac{p_b}{p_a}$ as the correction factor for the character a. More in general, if a character s is allowed to mutate into any one of the characters in the subset $\Sigma_s \subseteq \Sigma$, we define the Σ_s-*correction factor* for s as $\sum_{s' \in \Sigma_s} p_{s'}/p_s$. Hereafter, we assume for simplicity that $\Sigma_s = \Sigma \setminus \{s\}$ for all characters of Σ and set the correction factor $f_s = \sum_{s' \in \Sigma \setminus \{s\}} p_{s'}/p_s$.

Clearly, if $\hat{p} = P(y)$ is the probability of y, the probability of any string y' differing from y due to the change of a character s is obtained by multiplying $P(y)$ by the correction factor for s. In the example above, the change in probability is the same when the error occurs at any of the positions 1,3,4,6,7. For the remaining positions we would have to multiply $P(y)$ by f_b. In conclusion, if y is a substring of a text x, the probability of occurrence in x of a string y' differing from y in exactly one position is: $P_1(y) = n_a f_a \hat{p} + n_b f_b \hat{p} = \hat{p}(n_a f_a + n_b f_b)$, and $E_1[y] = E[y](n_a f_a + n_b f_b)$.

If we were to compute probabilities and expectations for occurrences of y with exactly k errors, then we would have to compute the above for all possible choices of k positions among m, which entails a complexity of $O(m^k)$. For small Σ, such as in DNA, some improvement is obtainable by distributing errors among the at most $|\Sigma|$ characters rather that the m positions. In our example string, for instance, we have $n_a = 5$ and $n_b = 2$ and the probability of occurrences with 2 mismatches is:

$$\binom{5}{2} f_a^2 + \binom{2}{2} f_b^2 + \sum_{i=1}^{1} \binom{5}{i}\binom{2}{2-i} f_a^i f_b^{2-i} = 10 f_a^2 + f_b^2 + 10 f_a f_b.$$

The dominant term in the calculation for k errors is the one involving all the characters of the alphabet, and it requires $|\Sigma| - 1$ nested cycles totaling $O(k^{|\Sigma|-1})$ time.

Besides being expensive, neither one of these approaches lends itself to iterated computations where, e.g., the probabilities of occurrences with (up to) k mismatches of all m-character substrings of a text are sought. The approach presented next achieves this in $O(k^2)$ per iteration, hence in $O(k^2n)$ time for a text of n characters. It requires an $O(kn)$ time pre-processing of the text, which is described next.

Text Pre-processing. Given a text x of length n, and a fixed number of errors k, we build a $[k \times n]$ matrix A whose generic entry $A[i][j]$ is the correction factor to be applied to the probability of string $x[1 \ldots j]$ with exactly i errors. Array A is readily computed in time $O(kn)$.

Lemma 1. *With $f_{x[j]}$ the correction factor of $x[j]$, the following holds:*

$$A[i][j] = \begin{cases} 1 & \text{if } i = 0 \text{ and } \forall j \\ 0 & \text{if } i \neq 0 \text{ and } j < i \\ f_{x[1]} & \text{if } i = j = 1 \\ A[i][j-1] + A[i-1][j-1]f_{x[j]} & \text{if } i > 0 \text{ and } j > i \end{cases}$$

Proof. We neglect the obvious boundary conditions and concentrate on the last row. The recurrence (see also Figure 1) states that the correction factor to be applied to $x[1 \ldots j]$ when i errors are allowed comes from two tributaries:

1. The symbol at position j is correct and exactly i errors occur in $x[1 \ldots j-1]$
2. There is an error at position j and exactly $i-1$ errors in $x[1 \ldots j-1]$. $\square$

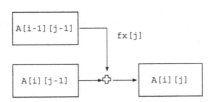

Fig. 1. Computing A[i][j].

Processing. Let $C_k(b, e)$ be the global correction factor to be applied to substring $\bar{x} = x[b \ldots e]$ in order to obtain the probability of $\bar{x}$ when exactly k errors are imposed. Thus, $\hat{p}C_k(b, e) = P(\bar{x}_k)$ and this value depends on string x and on the indices (b, e).

Lemma 2. $C_0(b, e) = 1$. *For $k > 0$,* $C_k(b, e) = A[k][e] - \sum_{i=0}^{k-1} A[k-i][b-1] \cdot C_i(b, e)$.

Table 1. Computing the array A for $x = abccb$ and $k = 4$.

i	a	b	c	c	b
0	1	1	1	1	1
1	f_a	f_a+f_b	$(f_a+f_b)+f_c$	$(f_a+f_b+f_c)+f_c$	$(f_a+f_b+f_c+f_c)+f_b$
2	0	f_af_b	$(f_af_b)+(f_a+f_b)f_c$	$[f_af_b+(f_a+f_b)f_c]+(f_a+f_b+f_c)f_c$	$[f_af_b+(f_a+f_b)f_c+(f_a+f_b+f_c)f_c]f_c]+(f_a+f_b+f_c+f_c)f_b$
3	0	0	$(f_af_b)f_c$	$(f_af_bf_c)+[f_af_b+(f_a+f_b)f_c]f_c$	$f_af_bf_c+[f_af_b+(f_a+f_b)f_c]f_c+(f_a+f_b+f_c)f_c]f_b$
4	0	0	0	$(f_af_bf_c)f_c$	$f_af_bf_cf_c+[f_af_bf_c]+[(f_af_b)+(f_a+f_b)f_c]f_cf_b$

Proof. That $C_0(b, e) = 1$ is obvious. Consider first $C_1(b, e)$. In order to compute it, it suffices to take the correction factor for text $x[1 \ldots e]$ and subtract the errors contributed by positions preceding b. Thus, $C_1(b, e) = A[1][e] - A[1][b - 1]C_0(b, e)$. Consider now the case $k = 2$. By construction, $A[2][e]$ contains the correction factor for $x[1 \ldots e]$ with 2 errors. From this value we need to subtract the following:

- the contribution $A[2][b - 1]$ of 2 errors occurring in positions which precede b;
- the contribution due to 1 error occurring in $x[1 \ldots b-1]$ and 1 error occurring in $x[b \ldots e]$

Thus, the final correction factor is: $C_2(b, e) = A[2][e] - A[2][b - 1] \cdot C_0(b, e) - A[1][b-1] \cdot C_1(b, e)$. Continuing in this fashion we obtain, for general k: $C_k(b, e) = A[k][e] - \sum_{i=0}^{k-1} A[k - i][b - 1] \cdot C_i(b, e)$ which is the formula of the claim. $\square$

Lemma 3. *Once the table A has been built, the computation of the correction factor for any substring $y = x[b \ldots e]$ with $1, 2, ..., k$ errors takes $O(k^2)$ steps.*

Proof. We exhibit an algorithm based on the previous lemma that fulfills the claim.

```
1. array C[0...k]
2. C[0]=1;
3. for i=1 to k
4.     Sum = 0
5.     for j=0 to i-1
6.         Sum = Sum + A[i-j][b-1] C[j]
7.     C[i] = A[i][e] - Sum
```

The values $C[j]$ at row 6 have been already computed in the previous cycles, so that line requires constant time. The total complexity is then charged by the two nested cycles, leading to an $O(k^2)$ algorithm. $\square$

Unlike that of the methods described earlier, this time complexity is independent from both the alphabet size and the pattern length.

Note that in order to compute $C_k(b, e)$ we also compute $C_i(b, e), 0 \le i \le k$, so that in time $O(k^2)$ we actually determine the value for k consecutive correction factors. On average, the time needed to compute a single factor is thus $O(k)$.

In conclusion, after $O(kn)$ time and space pre-processing of the text, it is possible to obtain in $O(k^2)$ time, from input initial and final position of any substring of the text, the correction factor for that substring. Combined with probabilities and length, this also yields the desired probability and expected frequency with mismatches for that substring, at no extra cost. For any fixed length m, the algorithm above supports also the computation of correction factors for all m-character substrings of text x in $O(nk^2)$ time. However, we can achieve a substantial improvement in both time and space for this case. The on-line algorithm presented next proceeds by dynamically adjusting the desired values in a sliding window of any given fixed length m. To avoid clutter in notation we assume $m \geq k$, but this constraint can be removed without penalty.

First, note that a recurrence similar to the one used to compute the correction factor $A[i][j]$ from $A[i-1][j]$ and $A[i-1][j-1]$ can be used to derive the correction factor of a string $w' = x[b \ldots e+1]$ from the correction factor of the string $w = x[b \ldots e]$. Indeed, we have $C_0(b, e+1) = 1$ and, for $k > 0$:
$C_k(b, e+1) = C_k(b, e) + C_{k-1}(b, e)f_{x[e+1]}$.

Lemma 4. $C_0(b+1, e) = 1;$ for $k > 0$, $C_k(b+1, e) = C_k(b, e) - C_{k-1}(b+1, e)f_{x[b]}$.

Proof. Observe that $C_k(b, e) = C_k(e, b)$, i.e., computing the C_k's for a string from left to right or from right to left will not change their values. Therefore, the correction factors of the string w are related to those of the string $w'' = x[b+1 \ldots e]$ by the formula:

$$C_0(b, e) = 1; \quad C_k(b, e) = C_k(b+1, e) + C_{k-1}(b+1, e)f_{x[b]} \quad \text{for } k > 0.$$

which is equivalent to the one in the claim.   □

Triggered by the "universal" initial condition $C_0 = 1$, Lemma 4 enables us to extract in succession the $C_i(b+1, e)$'s values from the $C_i(b, e)$'s, for consecutive values of $i = 1, 2, ..., k$. At that point, we can apply the above Equation, rewritten as:

$$C_k(b+1, e+1) = C_k(b+1, e) + C_{k-1}(b+1, e)f_{x[e+1]}$$

to similarly compute the $C_i(b+1, e+1)$'s.

Clearly, the process takes twice k steps and requires knowledge of as many auxiliary values, hence the computation for all m-character strings completes in $O(nk)$ time and $O(k)$ auxiliary space. By extending this treatment to all substrings of length $m \pm \delta$ would enable us to weigh $\mathcal{W}(x)$ within these bounds. As mentioned, the table at the outset risks to be too bulky, hence we address next monotonicities that shall enable us to neglect part of $\mathcal{W}(x)$ without loss of information.

3   Monotonicities

We begin by making an assumption that limits the skewedness on our probability distributions in exchange for some useful consequences. The assumption is quite reasonable for genomic as well as general applications.

Assumption 1: $p_a \leq \displaystyle\sum_{s \in \Sigma \setminus \{a\}} p_s$

As an immediate consequence of this assumption, we get:

Property 1: $f_a \geq 1 \quad \forall a \in \Sigma;$ **Property 2:** $A[i][j] \geq 0 \quad \forall i, j.$

Property 2 follows from the observation that only positive values are added in our algorithm, once $A[i][j] = 0$ for $i < j$, and $A[1][1]$ is positive.

As seen, a recurrence similar to the one used to compute $A[i][j]$ can be used to compute correction factors of the string $x[b \ldots e]$ from that of $x[b \ldots e - 1]$. Specifically, we have:

$$C_k(b, e) = C_k(b, e - 1) + C_{k-1}(b, e - 1) \cdot f_{x[e]}. \tag{1}$$

This shows that correction factors are always non-negative, and in fact that they are in positive for words longer than the number of errors k.

Property 3: $C_k(w) \geq 0 \quad \forall k, w$ and $C_k(w) > 1 \quad \forall w : |w| \geq k > 0.$

$C_k(b, e)$ is the correction factor to be applied to string $x[b \ldots e]$ in order to allow for *exactly* k errors. It is natural to extend the notion of correction factor to the case where one wants to consider *at most* k errors in a string. The corresponding expression for $x[b \ldots e]$ is obtained by taking the sum of all the correction factors from 0 to k:

$$\bar{C}_k(b, e) = \sum_{i=0}^{k} C_i(b, e) \tag{2}$$

We discuss next some properties of monotonicity of correction factors both for the case of *exactly* k errors (ECF) as well as for the case (UCF) of *up to* k errors. We are specifically interested in the behavior of these factors for a word under each one of the following scenarios:

1. the word length is increased, keeping error number fixed;
2. the number of errors is increased, keeping word size fixed;
3. both word length and number of errors are increased.

We set $w = v \cdot a$ where $w, v \in \Sigma^*$ and $a \in \Sigma$ and study first C_k. We shall make frequent use in our proofs of Equation 1 that is reported below in the crisper version:

$$C_k(w) = C_k(v) + C_{k-1}(v) f_a. \tag{3}$$

Lemma 5. *For $w = va$, $C_k(w) \geq C_k(v)$.*

Proof. From Equation 3, considering Property 1 we have $C_k(w) = C_k(v) + C_{k-1}(v) f_a \geq C_k(v) + C_{k-1}(v)$. By Property 2, $C_{k-1} \geq 0$. We can conclude that $C_k(w) \geq C_k(v) + C_{k-1}(v) \geq C_k(v)$. $\qquad\square$

Lemma 6. *For $w = va$, $C_k(w) \geq C_{k-1}(v)$.*

Proof. By the argument in the previous Lemma, since $C_k(v) \geq 0$ we also have:
$C_k(w) \geq C_k(v) + C_{k-1}(v) \geq C_{k-1}(v)$. □

A counterexample will show that, in general, the correction factor is not monotonically increasing when the number of errors allowed is increased while the length of the string is kept fixed. To see this, assume that the characters of Σ have the same probability. Hence:

$$p_{s_1} = p_{s_2} = \ldots = p_{s_{|\Sigma|}} = p = \frac{1}{|\Sigma|} \text{ and } f_{s_1} = f_{s_2} = \ldots = f_{s_{|\Sigma|}} = f = |\Sigma| - 1$$

In this special case, for a word w we have:

$$C_k(w) = \binom{|w|}{k} f^k.$$

We claim that there is a value $\bar{k}$ for k such that the correction factor is monotonically increasing for $k \leq \bar{k}$ and monotonically decreasing for $k > \bar{k}$. To determine $\bar{k}$, observe that by the definition of $C_k(w)$ we have:

$$C_k(w) < C_{k+1}(w) \implies \binom{|w|}{k} f^k < \binom{|w|}{k+1} f^{k+1} \implies f > \frac{k+1}{|w| - k}$$

Hence $C_k(w) < C_{k+1}(w)$ holds for $k < (|w|f - 1)/(f + 1)$. Combined with its symmetric argument, this leads to conclude that, with $\bar{k} = \lfloor \frac{|w|f-1}{f+1} \rfloor$, we have:

$$\begin{cases} C_k(w) < C_{k+1}(w) & \text{for } k \leq \bar{k} \\ C_k(w) > C_{k+1}(w) & \text{for } k > \bar{k} \end{cases}$$

Thus $C_k(w)$ is bi-tonic in this case. Our counterexample suggests that monotonicity might be preserved within restricted ranges of errors. If we apply the above to almost-evenly distributed DNA strands, for instance, we have $\Sigma = \{a, c, g, t\}$, $f = 3$ and $\bar{k} = \lfloor \frac{3|w|-1}{4} \rfloor$. Hence for words of length 16 we can assume the correction factor to increase monotonically up to 11 errors, which is a very large fraction of the string length for most practical purposes. In addition, we also get that for $12 \leq k \leq 16$ the correction factor is monotonically decreasing.

A more general formula expressing $\bar{k}$ must depend on the distribution and might be hard to come by. However, our next lemma establishes an acceptable lower bound for $\bar{k}$, that corresponds to half the length of the string w.

Lemma 7. $C_k(w) \geq C_{k-1}(w) \qquad \forall k \leq \frac{|w|}{2}$.

Proof. Let $w = w_1 w_2 \ldots w_m$. The inequality holds for $k = 1$, since $C_0(w) = 1$ and:

$$C_1(w) = \sum_{i=1}^{m} f_i \geq \sum_{i=1}^{m} 1 = m \geq 1 = C_0(w).$$

The contribution of each position in this case is the correction factor of the character occupying that position. Hence we obtain a set of $\binom{m}{1} = m$ terms, which may be expressed as $C_1 = (f_1, f_2, \ldots, f_m)$. For $k = 2$, we obtain a set of $\binom{m}{2} = m(m-1)/2$ terms, where each term results from the combination of the characters at two positions of w, say, w_i and w_j, and consists of the product of the corresponding correction factors $f_i f_j$. Specifically, the set of contributions for $k = 2$ is given by $C_2 = (f_1 f_2, f_1 f_3, \ldots, f_1 f_m, f_2 f_3, \ldots, f_2 f_m, \ldots, f_{m-1} f_m)$. Since $\forall i \ f_i \geq 1$, then $f_i f_j = f_j f_i \geq f_i \ \forall i, j$, so that for every term f in C_1 we have at least one element $\bar{f}$ of C_2 such that $\bar{f} \geq f$. This argument propagates from one C to the next for as long as the number of terms increases. But the number of terms is given by the binomial coefficients, hence our condition is preserved only for values of k up to $\frac{k}{2}$. We conclude that $|w|/2$ is always safe as a lower bound for $\bar{k}$.   □

We consider next the modified correction factor $\bar{C}_k(w)$ defined in Equation 2. Interestingly, monotonicity holds here with no restrictions.

Lemma 8. *Let $w = va$. Then, $\bar{C}_k(w) \geq \bar{C}_k(v)$; $\bar{C}_k(w) \geq \bar{C}_{k-1}(w)$; $\bar{C}_k(w) \geq \bar{C}_{k-1}(v)$.*

Proof. From Equation 2 and that of Lemma 5, it immediately follows that:

$$\bar{C}_k(w) = \sum_{i=0}^{k} C_i(w) \geq \sum_{i=0}^{k} C_i(v) = \bar{C}_k(v),$$

leading to the first inequality. We also have from Equation 2:

$$\bar{C}_k(w) = \sum_{i=0}^{k} C_k(w) = C_k(w) + \sum_{i=0}^{k-1} C_i(w) = C_k(w) + \bar{C}_{k-1}(w) \geq \bar{C}_{k-1}(w)$$

which establishes the second inequality. The third inequality follows from the previous two, whence $\bar{C}_k(w) \geq \bar{C}_k(v) \geq \bar{C}_{k-1}(v)$.   □

We now turn to probabilities of strings with errors. By definition, the probability of occurrence for string w when k errors are allowed is given by the product of that string probability and its correction factor for k errors:

$$P_k(w) = P(w) D_k(w)$$

where $D_k(w)$ is either $C_k(w)$ or $\bar{C}_k(w)$ depending on whether we are considering ECFs or UCFs. Our next lemma can be stated in terms of $D_k(w)$ but when $D = C$ the additional assumption that $|w|/2 > k$ is needed.

Lemma 9. *Let $w = va$ ($a \in \Sigma$), then $P_k(w) \leq P_k(v)$, and $P_k(w) \geq P_{k-1}(w)$*

Proof. By definition, $P(w) = P(v) \cdot p_a$ where $w = va$. From Equation 3 and the definition of f_a, it follows:

$$P_k(w) = p_a P(v)(D_k(v) + D_{k-1}(v) f_a)$$
$$= p_a P(v) D_k(v) + D_{k-1}(v) P(v) p_a \times \frac{\sum_{s \neq a} p_s}{p_a}$$

Since $D_k(v) \geq D_{k-1}(v)$, we obtain the inequality:

$$P_k(w) \leq p_a P(v) D_k(v) + P(v) D_k(v) \times \sum_{s \neq a} p_s$$

By the definition of $P_k(v)$ we finally have:

$$P_k(w) \leq p_a P_k(v) + P_k(v) \times \sum_{s \neq a} p_s = P_k(v) \times \sum_{s \in \Sigma} p_s = P_k(v)$$

From Lemma 5 we have: $P_k(w) = P(w)D_k(w) \geq P(w)D_{k-1}(w) = P_{k-1}(w)$. □
Our analysis of monotonicities is summarized in the following

Theorem 1. *Under both ECF and UCF, for any v, z and $w = vz$ and any integer $k < |w|/2$, it is $P_k(w) \leq P_k(v)$ and $P_k(w) \geq P_{k-1}(w)$.*

Proof. By the above discussion and lemmas. □

For words w in a text x such that $|w| = m \pm \delta \ll n = |x|$ the above inequalities translate to expectations, i.e., under the conditions of Theorem 1 we get also that $E_k(w) \leq E_k(v)$ and $E_k(w) \geq E_{k-1}(w)$.

4   Monotone Scores

The degree of surprise associated with the recurrence of a word or motif in a sequence or family of sequences is measured by some *z-score* that takes into account the observed and expected frequencies, perhaps normalized by some parameter such as expectation or higher moments. Basic scores in use are, e.g., $z_1(w) = F(w) - E(w)$, $z_2(w) = F(w)/E(w)$, and

$$z_3(w) = \frac{F(w) - E(w)}{\sqrt{Var(w)}}; \quad z_4(w) = \frac{(F(w) - E(w))^2}{E(w)}$$

where F denotes frequency, E expected frequency and Var variance. The expression and computation of the expected values, moments and related scores of significance depend substantially on the particular notion used. For sequence families, the frequency of a pattern can be defined in at least two ways, depending on whether we count the total number of pattern occurrences or the number of sequences containing each at least one occurrence of that pattern. In this paper, attention is restricted to notions involving the *total number* of occurrences, whether in a single sequence or sequence family (the latter being reduced to a singleton thru concatenation of its members), and using no higher moments. Whereas the first one of our restrictions is not hard to forfeit, the efficient computation of scores involving variance and higher moments have proved to pose serious algorithmic challenges even for solid patterns [3].

In appropriate synergy with frequencies, the monotonicity of expectations extends to the related scores. In particular, such scores are monotone over intervals of constant frequency. Here we limit consideration to this simplest case,

Table 2. Sample table size reductions for exactly (left half) and up-to (right half) k errors.

		$m=12$	$m=13$	$m=14$	$m=15$	$m=16$	$m=17$			$m=12$	$m=13$	$m=14$	$m=15$	$m=16$	$m=17$
k	# entries	34853	34846	34839	34832	34825	34818	k		34853	34846	34839	34832	34825	34818
=	# runs	5015	4994	4987	4982	4977	4974	=		5015	4994	4987	4982	4977	4974
0	avg length	6.93	6.97	6.98	6.99	6.99	6.99	0		6.93	6.97	6.98	6.99	6.99	6.99
	% saving	85.35	85.61	85.67	85.69	85.70	85.71			85.35	85.61	85.67	85.69	85.70	85.71
k	# entries	3326	1108	332	106	42	18	k		34853	34846	34839	34832	34825	34818
=	# runs	543	200	66	24	10	4	=		5538	5192	5051	5006	4989	4980
1	avg length	2.30	2.33	2.36	2.33	2.20	2.00	1		5.92	6.59	6.86	6.95	6.98	6.99
	% saving	21.20	23.92	27.11	30.19	28.57	22.22			78.13	83.29	85.02	85.50	85.63	85.67
k	# entries	13119	8186	3847	1398	474	156	k		34853	34846	34839	34832	34825	34818
=	# runs	1106	6880	605	261	95	32	=		5909	5940	5625	5279	5090	5016
2	avg length	2.27	2.30	2.32	2.30	2.27	2.22	2		4.04	4.84	5.76	6.45	6.80	6.93
	% saving	10.72	15.95	20.72	24.25	25.52	25.00			51.51	65.50	76.81	82.66	84.70	85.38
k	# entries	22949	18075	13156	8230	3960	1522	k		34853	34846	34839	34832	34825	34818
=	# runs	1171	1204	1211	1072	635	263	=		5073	5802	6064	6050	5703	5310
3	avg length	2.20	2.24	2.25	2.26	2.29	2.30	3		2.65	3.26	3.97	4.77	5.67	6.40
	% saving	6.14	8.23	11.55	16.46	20.70	22.54			24.08	37.60	51.69	65.49	76.55	82.33

however, much broader domains of monotonicity can be identified through the interplay of expectation and frequency, and these are under study.

For our application, we add k as a subscript to indicate the number of errors. Thus, e.g., $F_k(w)$ is the number of observed subwords of x at a distance k from w. As there is no substantial difference in our computation whether k is the exact or maximum number of errors we make here no distinction of treatment nor belabor this point further. We only need to discuss the computation of $F_k(w)$ for all words of x of size $|w| = m \pm \delta$. This is done by established techniques in $O(nk)$ or even expected sublinear time (see, e.g., [3]). There are $O(n)$ subwords of length m in x, whence the total computation with δ a constant is $O(n^2 k)$ or expected $O(n^2)$. This information can be organized in the $n \times (2\delta+1)$ table $\mathcal{W}(x)$ at the outset, such that the frequencies of substrings of length $[m - \delta \ldots m + \delta]$ beginning at position i form the i-th column of the table. Looking now at every single column will suffice to firm the intervals of monotonicity for F, whence only one extreme in each class is retained and weighed with expectation and score. The overall cost including expectations is $O(n^2 k)$ or expected $O(n^2)$, depending on the method used.

As an illustration, Table 2 displays the results of computations performed on a sequence of $n = 5,000$ bases randomly generated according to a genomic distribution (specifically: $p_a = p_t = 0.30$; $p_c = p_g = 0.20$, the approximate base composition of yeast nuclear chromosomes), for various values of m and $\delta = 3$. For each triplet (m, k, δ), the substrings of length between $m - \delta$ and $m + \delta$ were considered in succession, and the number of occurrences of each string with exactly (left half of the table) or up-to (right part) k errors were computed. The elimination of entries with zero frequency led to the table sizes reported under "# entries". From these, runs of identical counts corresponding to consecutive extensions of a same substring were identified and compacted each to a single entry. With reference to Table 2, for instance, the leftmost column at $k = 1$ states that of all substrings of length from 9 to 15 only 3,326 had a non-zero F_1 value. Of these, 16% consisted of runs with an average length of 2.3 characters. Using one representative entry per run yields a 21.20% reduction in the size of the table. For comparison, the number of strings having length in this range is 1,073,741,824.

Acknowledgement

We are indebted to one of the Referees for the careful scrutiny of an earlier version of this manuscript.

References

1. APOSTOLICO, A. Pattern discovery and the algorithmics of surprise. In *Artificial Intelligence and Heuristic Methods for Bioinformatics* (2003), P. Frasconi and R. Shamir, Eds., IOS Press, pp. 111–127.
2. APOSTOLICO, A., AND GALIL, Z., Eds. *Pattern matching algorithms*. Oxford University Press, 1997.
3. APOSTOLICO, A., BOCK, M. E., AND LONARDI, S. Monotony of surprise and large-scale quest for unusual words (extended abstract). In *Proc. of Research in Computational Molecular Biology RECOMB* (Washington, DC, April 2002), G. Myers, S. Hannenhalli, S. Istrail, P. Pevzner, and M. Waterman, Eds. Also, *J. Comp. Bio.*, 10:3-4, (July 2003), 283–311.
4. APOSTOLICO, A., AND PARIDA, L. Incremental Paradigms of Motif Discovery. *J. Comput. Bio. 7*, **11**:1, (Jan. 2004), 15–25.
5. BAILEY, T. L., AND ELKAN, C. Unsupervised learning of multiple motifs in biopolymers using expectation maximization. *Machine Learning 21*, 1/2 (1995), 51–80.
6. BRĀZMA, A., JONASSEN, I., UKKONEN, E., AND VILO, J. Predicting gene regulatory elements in silico on a genomic scale. *Genome Research 8*, 11 (1998), 1202–1215.
7. BUHLER, J., AND TOMPA, M. Finding motifs using random projections. *J. Comput. Bio. 9*, 2 (2002), 225–242.
8. HERTZ, G. Z., AND STORMO, G. D. Identifying DNA and protein patterns with statistically sign ificant alignments of multiple sequences. *Bioinformatics 15* (1999), 563–577.
9. JONASSEN, I. Efficient discovery of conserved patterns using a pattern graph. *Comput. Appl. Biosci. 13* (1997), 509–522.
10. KEICH, AND PEVZNER. Finding motifs in the twilight zone. In *Annual International Conference on Computational Molecular Biology* (Washington, DC, Apr. 2002), pp. 195–204.
11. LAWRENCE, C. E., ALTSCHUL, S. F., BOGUSKI, M. S., LIU, J. S., NEUWALD, A. F., AND WOOTTON, J. C. Detecting subtle sequence signals: A Gibbs sampling strategy for multiple alignment. *Science 262* (Oct. 1993), 208–214.

Suboptimal Local Alignments Across Multiple Scoring Schemes

Morris Michael[1,2], Christoph Dieterich[2], and Jens Stoye[1]

[1] Technische Fakultät, Universität Bielefeld, 33594 Bielefeld, Germany
{mmichael,stoye}@techfak.uni-bielefeld.de
[2] Computational Molecular Biology, Max Planck Institute for Molecular Genetics
14195 Berlin, Germany
christoph.dieterich@molgen.mpg.de

Abstract. Sequence alignment algorithms have a long standing tradition in bioinformatics. In this paper, we formulate an extension to existing local alignment algorithms: *local alignments across multiple scoring functions.* For this purpose, we use the Waterman-Eggert algorithm for suboptimal local alignments as template and introduce two new features therein: 1) an alignment of two strings over a set of score functions and 2) a switch cost function δ for penalizing jumps into a different scoring scheme within an alignment.

Phylogenetic footprinting, as one potential application of this algorithm, was studied in greater detail. In this context, the right evolutionary distance and thus the scoring scheme is often not known *a priori*. We measured sensitivity and specificity on a test set of 21 human-rodent promoter pairs. Ultimately, we could attain a 4.5-fold enrichment of verified binding sites in our alignments.

Keywords: Sequence alignment, non-parametric alignment, phylogenetic footprinting, comparative sequence analysis.

1 Introduction

Comparative sequence analysis is a powerful tool in bioinformatics for addressing a variety of issues. Applications range from grouping of sequences (e.g. protein sequences) into families to *de novo* pattern discovery of functional signatures. Thus, sequence comparison aims at detecting "biologically meaningful" similarities between sequences. Considering gene regulation, it has been known for a long time that there is considerable sequence conservation between species in non-protein-coding regions of the genome. Especially, sequence conservation within promoter regions of genes often stems from transcription factor binding sites that are under selective pressure (see [5] for a review). Duret and Bucher [4] give an overview on exploiting sequence conservation across species for the detection of regulatory elements. This concept is commonly referred to as *phylogenetic footprinting.*

Phylogenetic footprinting in a strict sense is carried out on orthologous promoter regions. Local sequence similarities can then be directly interpreted as

I. Jonassen and J. Kim (Eds.): WABI 2004, LNBI 3240, pp. 99–110, 2004.

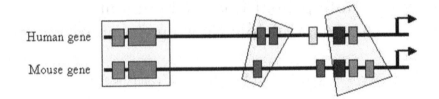

Fig. 1. The concept of phylogenetic footprinting. Local sequence similarities in orthologous promoter regions of genes (light "framed" regions) occur often due to selective pressure on transcription factor binding sites (shaded boxes).

related regions harboring conserved functional binding sites. Selecting suitable sequence pairs and the choice of the right scoring parameters is crucial to the success of the footprinting approach.

Computational Approaches to Phylogenetic Footprinting. If we recall the situation in Figure 1, an appropriate way of detecting local similarities is to retrieve many local alignments from the search space. Waterman and Eggert [14] proposed an extension of the Smith-Waterman [11] local alignment algorithm for finding non-trivial local similarities (non-intersecting suboptimal local alignments). In earlier work, we have employed an implementation of the Waterman-Eggert algorithm successfully in a large-scale study of man-mouse promoter regions [3].

Related Approaches. Heuristic algorithms for large-scale comparison of genomic regions emerged as a new field in computational biology [9]. Recent reviews [12, 2] give a survey of the field and list all available "genome alignment" tools. These software packages are readily applicable to compare whole *syntenic* regions of genomes. BLASTZ [10] is the most similar heuristic solution to the Waterman-Eggert approach since it computes suboptimal local alignments with gaps, which have no constraints on their position. Although, BLASTZ is not guaranteed to find the optimal solution in our setting, it performs well in practice.

Alignment over Several Score Functions. None of the previously mentioned solutions aligns sequence pairs over more than one score function. However, this is desirable in the context of *phylogenetic footprinting*. In maximizing local scores over more than one scoring scheme, sequence properties that can be reflected in the scoring scheme (e.g. local GC-content, bias in substitution patterns) are better captured by the alignment algorithm. As another application, Altschul [1] introduced the idea of multiple score functions for database searches where one does not know *a priori* the right evolutionary distance between two sequences. This is analogous to comparing two promoter regions where we do not know the evolutionary rates of neutrally diverging sequence and sequence elements under selective pressure. In this paper, we propose an extension to the Waterman-Eggert algorithm to meet these demands.

Structure of the Paper. Firstly, we introduce the basic notation and edit operations of local alignment algorithms. Secondly, we formally extend the set of edit operations to alignments over several score functions. Subsequently, we present our implementation of an extended version of the Waterman-Eggert algorithm (as implemented by Huang and Miller [8]), and finally we evaluate the impact of our modifications on the problem of finding regulatory elements by comparative sequence analysis.

2 Definitions and Notation

The empty string is denoted by ε and an alphabet of symbols by $\mathcal{A}$. $\alpha_1\alpha_2\ldots\alpha_n$ is the concatenation of $\alpha_1, \alpha_2, \ldots, \alpha_n$, where α_i can be either a string or a symbol and ε is the neutral element: $\alpha\varepsilon = \alpha = \varepsilon\alpha$. $|w|$ denotes the length and w_i the ith symbol of the string w.

2.1 Basic Definitions

Here, we briefly review the standard terminology of sequence alignment, as we will use it throughout this paper.

Definition 1 (Edit Operation). *An* edit operation *is a pair*

$$(\alpha, \beta) \in (\mathcal{A} \cup \{\varepsilon\}) \times (\mathcal{A} \cup \{\varepsilon\}) \setminus \{(\varepsilon, \varepsilon)\}.$$

It is usually denoted by $\alpha \to \beta$.

   Edit operations describe the step by step transformation of a source string into a target string. Three kinds of edit operations exist:

$\alpha \to \varepsilon$ denotes the *deletion* of the symbol α.
$\varepsilon \to \beta$ denotes the *insertion* of the symbol β.
$\alpha \to \beta$ denotes the *replacement* of the symbol α by β. Here we distinguish between two cases. If $\alpha = \beta$ it is called a *match*, otherwise it is called a *mismatch* or an *exchange*.

A maximal sequence of adjacent insertions and deletions forms a *gap*.

Definition 2 (Alignment). *An* alignment A *of two strings* u *and* v *is a sequence* $(\alpha_1 \to \beta_1, \ldots, \alpha_h \to \beta_h)$ *of edit operations such that* $\alpha_1 \ldots \alpha_h = u$ *and* $\beta_1 \ldots \beta_h = v$.

   An alignment is usually displayed by placing the symbols of the two aligned strings in different lines, where ε is replaced by -.

Example 1. The alignment $(d \to \varepsilon, a \to a, \varepsilon \to i, r \to r, l \to l, i \to i, n \to n, g \to e)$ is displayed as follows:

$$u: \texttt{da-rling}$$
$$v: \texttt{-airline}$$

Definition 3 (Score Function). *A (similarity) score function σ assigns to each edit operation $\alpha \to \beta$ a score $\sigma(\alpha \to \beta)$, where similar pairs of symbols (matches or exchanges) are scored by positive values and dissimilar pairs by negative values. Using affine gap costs, the score of an alignment $A = (\alpha_1 \to \beta_1, \ldots, \alpha_h \to \beta_h)$ is the sum of the scores of all edit operations and an additional cost for each gap: $\sigma(A) = \sum_{k=1}^{h} \sigma(\alpha_k \to \beta_k) + g \cdot \gamma$, where g is the number of gaps in A and $\gamma \leq 0$ denotes the gap open cost.*

Definition 4 (Local Alignment Problem). *Let two strings u and v and a score function σ be given. The* local alignment problem *is to determine an alignment A of u' and v' such that u' and v' are substrings of u and v and the score of A is maximal:*

$$\sigma(u,v) = \max \left\{ \sigma(A) \left| \begin{array}{l} A \text{ is an alignment of } u' \text{ and } v', \\ u' \text{ is a substring of } u, \ v' \text{ is a substring of } v \end{array} \right. \right\}.$$

Each alignment that satisfies this condition is called an optimal local alignment *of u and v, and $\sigma(u,v)$ is called the* optimal local alignment score *of u and v.*

Definition 5 (Suboptimal Alignments). *A local alignment A of u and v is called* suboptimal *if $\sigma(A)$ is smaller than the optimal alignment score of u and v. If we are looking for suboptimal alignments, we want to find them ordered by decreasing score.*

2.2   Several Score Functions

Sequence alignment over several score functions can now be introduced as a direct generalisation of (standard) local alignment. Again, we first define edit operations and score functions before we state the alignment problem.

Definition 6 (Edit Operation over Several Score Functions). *An edit operation over p score functions $\sigma_1, \ldots, \sigma_p$ is a tuple*

$$(\alpha, \beta, i) \in (\mathcal{A} \cup \{\varepsilon\}) \times (\mathcal{A} \cup \{\varepsilon\}) \times \{1..p\} \setminus \{(\varepsilon, \varepsilon, j) \mid j \in \{1..p\}\}.$$

It is denoted by $\alpha \to_i \beta$. Its score is $\sigma_i(\alpha \to \beta)$.

Definition 7 (Alignment over Several Score Functions). *An alignment A of two strings u and v over p score functions $\sigma_1, \ldots, \sigma_p$ is a sequence $(\alpha_1 \to_{i_1} \beta_1, \ldots, \alpha_h \to_{i_h} \beta_h)$ of edit operations such that $\alpha_1 \ldots \alpha_h = u$ and $\beta_1 \ldots \beta_h = v$.*

An alignment over several score functions is displayed by placing the symbols of the two aligned strings in $p+1$ lines. For each edit operation $\alpha \to_i \beta$ the symbol α is placed in the first line and β in the $i+1$st line, where ε is replaced by – and the symbols of each edit operation are in the same column.

Example 2. The alignment $(d \to_2 \varepsilon, a \to_2 a, \varepsilon \to_2 i, r \to_1 r, l \to_1 l, i \to_1 i, n \to_1 n, g \to_2 e)$ is displayed as follows:

```
u:  da-rling
v1:     rlin
v2: -ai     e
```

Definition 8 (Score of an Alignment over Several Score Functions).
*The cost of switching between two score functions σ_1 and σ_2 is determined by a
switch cost function $\delta(\sigma_1 \to \sigma_2)$.*

Using a given switch cost function δ and affine gap costs, the score of an align-
ment $A = (\alpha_1 \to_{i_1} \beta_1, \ldots, \alpha_h \to_{i_h} \beta_h)$ over $\sigma_1, \ldots, \sigma_p$ *is the sum of the scores of
all edit operations, the gap open costs, and all switch costs:*

$$\sigma(A) = \sum_{k=1}^{h} \sigma_{i_k}(\alpha_k \to \beta_k) + \sum_{k=1}^{p} g_k \cdot \gamma_k + \sum_{k=1}^{h-1} \delta(\sigma_{i_{k-1}} \to \sigma_{i_k})$$

*where g_k is the number of gap openings in A scored by σ_k and γ_k denotes the
gap open cost for score function σ_k.*

Note that according to this score function gap open costs are applied where
the gap begins, although the score function may be switched within the gap.

**Definition 9 (Local Alignment over Several Score Functions Prob-
lem).** *Let two strings u and v, p score functions $\sigma_1, \ldots, \sigma_p$ and a switch cost
function δ be given. The* local alignment over several score functions problem *is
to determine an alignment A over $\sigma_1, \ldots, \sigma_p$ of u' and v' such that u' and v' are
substrings of u and v and the score of A is maximal:*

$$\sigma(u, v) = \max \left\{ \sigma(A) \,\middle|\, \begin{array}{l} A \text{ is an alignment over } \sigma_1, \ldots, \sigma_p \text{ of } u' \text{ and } v', \\ u' \text{ is a substring of } u, \ v' \text{ is a substring of } v \end{array} \right\}.$$

Each alignment that satisfies this condition is called an optimal local alignment
of u and v over $\sigma_1, \ldots, \sigma_p$, *and $\sigma(u, v)$ is called the* optimal local alignment score
of u and v over $\sigma_1, \ldots, \sigma_p$.

2.3 Nonintersecting Alignments

In order to avoid redundancies, similar to Waterman and Eggert [14] we consider only nonintersecting alignments. Two alignments are nonintersecting if they
share no replacement $u_i \to v_j$. More formally, we define:

Definition 10 (Projection). *Let two strings u and v, and substrings $u' =
u_{b_u} \ldots u_{e_u}$ and $v' = v_{b_v} \ldots v_{e_v}$ be given. The* projection *of an alignment $A =
(\alpha_1 \to \beta_1, \ldots, \alpha_h \to \beta_h)$ of u' and v' is*

$$\breve{A} = \{(b_u + |\alpha_1 \ldots \alpha_k| - 1, b_v + |\beta_1 \ldots \beta_k| - 1) \mid 1 \le k \le h, \alpha_k \ne \varepsilon \ne \beta_k\}.$$

The projection *of an alignment over several score functions $A = (\alpha_1 \to_{i_1} \beta_1,
\ldots, \alpha_h \to_{i_h} \beta_h)$ of u' and v' is*

$$\breve{A} = \{(b_u + |\alpha_1 \ldots \alpha_k| - 1, b_v + |\beta_1 \ldots \beta_k| - 1, i_k) \mid 1 \le k \le h, \alpha_k \ne \varepsilon \ne \beta_k\}.$$

Definition 11 (Nonintersecting Alignments). *Two local alignments A_1 and
A_2 of u and v over σ or over $\sigma_1, \ldots, \sigma_p$, respectively, are* nonintersecting *if and
only if $\breve{A}_1 \cap \breve{A}_2 = \emptyset$.*

3   Algorithms

Various techniques have been developed to calculate optimal and suboptimal local alignments and have been improved to save resources. Huang and Miller [8] combined some of them to obtain an algorithm that calculates, for a given number K, the K best nonintersecting local alignments. In the following, we first sketch their algorithm, before we extend it to calculate the K best nonintersecting local alignments over p score functions.

3.1   Algorithm of Huang and Miller

Given two sequences u and v of lengths M and N, respectively, in the first phase of Huang and Miller's algorithm, a classical linear-space dynamic programming computation is performed to collect the K highest scores of local alignments of u and v in $O(M \cdot N)$ time and $O(M + N)$ space, together with the start and end position of each such alignment (see Algorithm 1). If some of these scores belong to intersecting alignments, only the highest score of mutually intersecting alignments is stored. Unfortunately, there is no guarantee that the K best scores found this way belong to the overall K best nonintersecting alignments, since some nonintersecting high-scoring alignments may be shaded by even higher scoring intersecting ones. Therefore, additional passes are required after a local alignment has been calculated.

More precisely, right after the first pass, using Hirschberg's [7] technique, the highest-scoring alignment (of length L_1) is computed in $O(L_1)$ space and $O(L_1^2)$ time by calculating a global alignment of the substrings determined by the start and end positions saved with the highest score. Thereby, the used replacements are recorded to be not used again.

Afterwards, the hidden high scoring alignments are discovered by a limited backwards dynamic programming pass determining the region of influence of the calculated alignment, and a forward pass to recompute the score matrix in this region. To determine the region of influence, some additional information is recorded. The possible local alignments are partitioned into equivalence classes. The K classes are stored in a data structure called LIST. An equivalence class S stored in LIST is represented by a tuple (C, F, u, T, B, L, R) where C is the score of the best alignment in S, F is the start position of all alignments in S, u is the end position of an alignment in S that gains score C, and $[T, B] \times [L, R]$ contains the end position of each alignment in S whose score is better than W, the lowest score of the K saved high scoring classes. The region of influence that needs to be recomputed is the part $[T', B] \times [L', R]$ of the score matrix that contains all entire alignments ending in $[T, B] \times [L, R]$ with a score greater than W.

An implementation of LIST must support the following operations:

- `find(f)`: returns the tuple whose $F = f$ or `null` if there is no such one.
- `insert(S)`: adds S to LIST.
- `maxtuple()`: removes a highest scoring tuples in LIST and returns it.
- `minscore()`: returns the lowest score (W) of all tuples in LIST.

Algorithm 1 Calculating alignment start position.

The score for a local alignment ending at (i, j) is usually calculated by

1: $D(i,j) \leftarrow \max\{D(i-1,j), C(i-1,j) + \gamma\} + \sigma(u_i \rightarrow \varepsilon)$
2: $I(i,j) \leftarrow \max\{I(i,j-1), C(i,j-1) + \gamma\} + \sigma(\varepsilon \rightarrow v_j)$
3: $C(i,j) \leftarrow \max\{0, D(i,j), I(i,j), C(i-1,j-1) + \sigma(u_i \rightarrow v_j)\}$

To compute the start position the first line is refined to

1: **if** $D(i-1,j) > C(i-1,j) + \gamma$ **then**
2:    $D(i,j) \leftarrow D(i-1,j) + \sigma(u_i \rightarrow \varepsilon)$
3:    StartD$(i,j) \leftarrow$ StartD$(i-1,j)$
4: **else if** $D(i-1,j) < C(i-1,j) + \gamma$ **then**
5:    $D(i,j) \leftarrow C(i-1,j) + \gamma + \sigma(u_i \rightarrow \varepsilon)$
6:    StartD$(i,j) \leftarrow$ StartC$(i-1,j)$
7: **else** {*tie!*}
8:    $D(i,j) \leftarrow D(i-1,j) + \sigma(u_i \rightarrow \varepsilon)$
9:    StartD$(i,j) \leftarrow \max_{\prec}\{$StartD$(i-1,j),$ StartC$(i-1,j)\}$

StartD(i,j) denotes the start position of a highest scoring local alignment ending at (i,j) with a deletion. Similarly, StartI and StartC denote the start of an alignment ending with an insertion and replacement, respectively.
The other two lines are extended in the same way. If the maximum for $C(i,j)$ is 0, StartC$(i,j) = (i,j)$.
In case of a tie the start position is chosen by an ordering $\prec$ of positions. This way, Huang and Miller showed that two alignments intersect if and only if they have the same start position.

  − `replace(S)`: replaces a lowest scoring tuple in LIST by S.
  − `size()`: returns the number of tuples in LIST.

The LIST is maintained by a function `enter` [8, Figure 3]: `enter(C,F,u,W,l)` first tests if there already is a class S in LIST with the same F. If there is one, its attributes are adjusted. Otherwise, a new class is added. If there are more than 1 classes in LIST, the class with the lowest score is deleted. `enter` returns the new minimum score W.

The steps – calculate alignment, determine its region of influence and search for hidden alignments – are repeated K times. If gaps and mismatches are not penalized too lightly by the used score function, Huang and Miller show that the algorithm takes $O(M \cdot N + \sum_{n=1}^{K} L_n^2)$ time and $O(M + N + \sum_{n=1}^{K} L_n)$ space in the expected case, where L_n is the length of nth reported alignment.

3.2 Extended Algorithm for Alignments over Several Score Functions

An outline of the extended algorithm that calculates the K best local alignments over several score functions $\sigma_1, \ldots, \sigma_p$ is shown in Algorithm 2. It differs from Huang and Miller's algorithm in additional **for** loops (lines 4 and 15) that iterate over the possible score functions, and in the calculations for the scores (lines 5 and 16), the alignment (line 10) and the region of influence (line 12). The function

---

Algorithm 2 Extension of Huang and Miller's alg. for several score functions.

1: $W \leftarrow 0$
2: **for** $i \leftarrow 0$ **to** M **do**
3:      **for** $j \leftarrow 0$ **to** N **do**
4:          **for** $r \leftarrow 1$ **to** p **do**
5:              calculate $C(i, j, r)$ and $\text{StartC}(i, j, r)$.
6:              **if** $C(i, j, r) > W$ **then**
7:                  $W \leftarrow \text{enter}(C(i, j, r), \text{StartC}(i, j, r), (i, j, r), W, K)$
8: **for** $n \leftarrow 1$ **to** K **do**
9:      $S \leftarrow \text{maxtuple}()$
10:     $\text{alignment}(S)$ {*calculates and reports an optimal alignment for the equivalence class S that does not intersect with any already calculated alignment*}
11:     **if** $n \neq K$ **then**
12:         calculate the region of influence $[T', S.B] \times [L', S.R]$.
13:         **for** $i \leftarrow T'$ **to** $S.B$ **do**
14:             **for** $j \leftarrow L'$ **to** $S.R$ **do**
15:                 **for** $r \leftarrow 1$ **to** p **do**
16:                     Calculate $C(i, j, r)$ and $\text{StartC}(i, j, r)$ relating to $[T', S.B] \times [L', S.R]$.
17:                     **if** $C(i, j, r) > W$ and (i, j) in $[S.T, S.B] \times [S.L, S.R]$ **then**
18:                         $W \leftarrow \text{enter}(C(i, j, r), \text{StartC}(i, j, r), (i, j, r), W, K - n)$

---

enter (lines 7 and 18) that maintains the LIST is almost unchanged. It is only adapted to consider the used score function as the third coordinate of start and end positions. Algorithm 3 shows exemplarily how the calculation of $C(i, j, r)$ and $\text{StartC}(i, j, r)$, of the alignment and of the region of interest is extended for several score functions.

The additional **for** loops (Algorithm 2, lines 4 and 15) and the nested loop (Algorithm 3, line 3) needed to compute the score, alignment and region of interest result in an additional factor of p^2 in the time complexity, yielding $O(p^2 \cdot (M \cdot N + \sum_{n=1}^{K} L_n^2))$ in the expected case. Regarding the space complexity, there is an additional factor of p for the intermediate results, yielding $O(p \cdot (M + N) + \sum_{n=1}^{K} L_n)$.

4   Proof of Concept

Now that we have presented our extension of the algorithm, we assess the impact of our modifications. Wasserman *et al.* [13] compiled a small test set of mammalian promoter regions where some binding sites had been verified experimentally. We retrieved 21 well annotated man-rodent sequence pairs from this set and investigated the effect of different parameter settings on the performance of the algorithm. We measured performance based on two factors: *sensitivity* in order to measure the ability of the method to recognize binding sites and *coverage* in order to measure the specificity of detected possible binding sites.

Algorithm 3 Extension of Algorithm 1 for several score functions.

1: $D(i, j, r) \leftarrow \gamma_r$
2: $\text{StartD}(i, j, r) \leftarrow (i, j, r)$
3: **for** $rr \leftarrow 1$ **to** p **do**
4:     **if** $C(i-1, j, rr) + \gamma_{rr} + \sigma(u_i \rightarrow_{rr} \varepsilon) + \delta(\sigma_{rr} \rightarrow \sigma_r) > D(i, j, r)$ **then**
5:         $D(i, j, r) \leftarrow C(i-1, j, rr) + \gamma_{rr} + \sigma(u_i \rightarrow_{rr} \varepsilon) + \delta(\sigma_{rr} \rightarrow \sigma_r)$
6:         $\text{StartD}(i, j, r) \leftarrow \text{StartC}(i-1, j, rr)$
7:     **else if** $C(i-1, j, rr) + \gamma_{rr} + \sigma(u_i \rightarrow_{rr} \varepsilon) + \delta(\sigma_{rr} \rightarrow \sigma_r) = D(i, j, r)$ **then**
8:         $\text{StartD}(i, j, r) \leftarrow \max_{\prec}\{\text{StartD}(i, j, r), \text{StartC}(i-1, j, rr)\}$
9:     **if** $D(i-1, j, rr) + \sigma(u_i \rightarrow_{rr} \varepsilon) + \delta(\sigma_{rr} \rightarrow \sigma_r) > D(i, j, r)$ **then**
10:        $D(i, j, r) \leftarrow D(i-1, j, rr) + \sigma(u_i \rightarrow_{rr} \varepsilon) + \delta(\sigma_{rr} \rightarrow \sigma_r)$
11:        $\text{StartD}(i, j, r) \leftarrow \text{StartD}(i-1, j, rr)$
12:     **else if** $D(i-1, j, rr) + \sigma(u_i \rightarrow_{rr} \varepsilon) + \delta(\sigma_{rr} \rightarrow \sigma_r) = D(i, j, r)$ **then**
13:        $\text{StartD}(i, j, r) \leftarrow \max_{\prec}\{\text{StartD}(i, j, r), \text{StartD}(i-1, j, rr)\}$

Definition 12 (Sensitivity). *The quotient of the number of found binding sites vs. the number of all annotated binding sites. We deem a binding site as found if at least 70% of the site (core region) are covered by an alignment in the lowest employed PAM distance.*

Definition 13 (Coverage). *The length of all alignment parts in the lowest employed PAM distance divided by the arithmetic mean of the two sequence lengths.*

Our test scenario is as follows: The first 10 local alignments are computed for each sequence pair across all combinations of jump costs and scoring functions. Alignment gap open and extension costs are set to 11 and 0.1 times the match score, respectively (see [3]). All scoring matrices are derived from the *HKY model* of sequence evolution [6], which takes single nucleotide frequencies into account, and assume a transition to transversion ratio of 3 : 1. An enumeration of all parameter settings follows below:

1. **Score function sets L (all data in PAM):** $\{1, 5, 15, 20\}, \{1, 5, 10, 15, 20\},$ $\{5, 20, 40, 80\}, \{1, 20\}, \{5, 80\}, \{10, 80\}, \{20, 80\}, \{30, 80\},$ and $\{40, 80\}$.
2. **Switch cost factors F:** 1, 3, 5, 7, 9, 12, 16, 20, 24, and 99999 (no switch in score function within alignment). The switch cost function $\delta(\sigma_1 \rightarrow \sigma_2)$ is then given by $F \cdot |\sigma_1(A \rightarrow \varepsilon) - \sigma_2(A \rightarrow \varepsilon)|$.

Before the systematic evaluation here we present parts of an alignment in order to illustrate how the output of our algorithm may look like. Shown are the 5' untranslated regions of human cardiac actin gene (first row) and mouse alpha-cardiac actin gene (other rows). Switch cost factor is $F = 7$, and the set of score functions is $L = \{1, 5, 15, 20\}$. The annotated binding sites (SRF, SP1, MYF), showing up nicely in the PAM 1 row, are marked by asterisks in the last row:

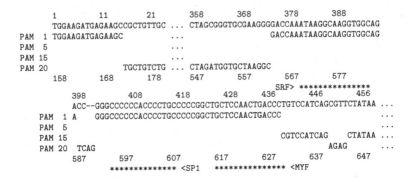

4.1 Influence of Score Function Set

Figure 2 depicts how the performance of the algorithm is affected by the set of
employed score functions. As expected, sets that include small PAM distances
(1 and 5) perform generally better with respect to "specificity", whereas larger
PAM distances are a bit more sensitive. Note that the variance along the x-axis
is substantially larger than for the y-axis. This means that the choice of the
scoring scheme mainly affects "specificity".

4.2 Influence of Switch Costs

Figure 3 demonstrates how switch costs that penalize jumping between different
score functions alter the performance of the algorithm. Evidently, the coverage
drastically increases if switch costs are low ($F \leq 3$) and alignments simply grow
by alternating between score functions. Other than that, we could not observe
any general trend with respect to sensitivity or specificity. The key data on the
test set of 21 promoter pairs is shown in the following table:

Table 1. Table 1 - maximal values

sensitivity	coverage	sensitivity : coverage
87.3 %	**14.3 %**	**4.46**
F=99999; L=30.80	F=20; L=1.5.10.15.20	F=16; L=1.5.10.15.20

5 Discussion

The technique of sequence alignment is vital to bioinformatics. Sequence alignment is used in various fields for tasks as diverse as functional annotation, evolutionary parameter estimation and motif discovery. In this paper, we have presented a versatile algorithm for computing suboptimal local alignments over
multiple score functions. Our implementation does not impose any constraints
or prior assumptions on the position and segmentation of alignments. Consequently, two basic alignment problems are addressed by our solution: 1) Local

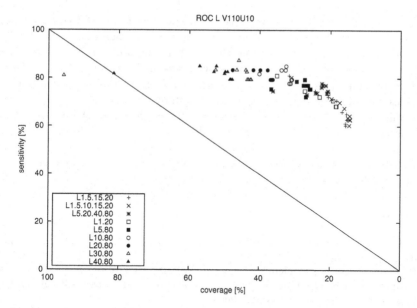

Fig. 2. Performance over all tested score function sets. For each set of PAM distances L with each switch cost factor F a point is plotted. Points with the same value of L are displayed as the same symbol. For a definition of the axis labels see Defs. 12 and 13.

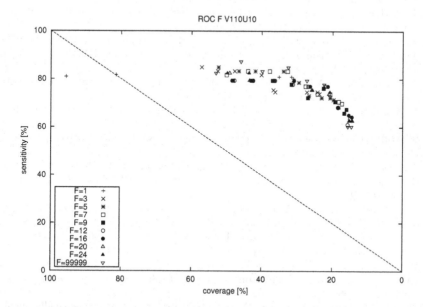

Fig. 3. Performance over all switch cost settings. The same graph as in Fig. 2, but points with the same value of F are displayed as the same symbol.

alignments often show a "mosaic" structure (the alternation of regions of high and low similarity, cf. [15]). We avoid this problem by aligning sequences with more than one scoring functions and thus capture the problem in a better way. 2) Often one does not know the proper scoring scheme in advance. This challenge is solved by employing a set of scoring schemes (e.g. for different evolutionary distances, protein domains, or secondary structures).

In this paper, the algorithm was exemplarily employed for the identification of transcription factor binding sites. For the test data, a 4.5-fold improvement of the sensitivity-to-coverage ratio was attained (see Table 1). This is just an estimation since we do not know how many binding sites escaped experimental validation so far.

References

1. S. F. Altschul. A protein alignment scoring system sensitive at all evolutionary distances. *J. Mol. Evol.*, 36:290–300, 1993.
2. P. Chain, S. Kurtz, E. Ohlebusch, and T. Slezak. An applications-focused review of comparative genomics tools: Capabilities, limitations and future challenges. *Briefings in Bioinformatics*, 4:105–123, 2003.
3. C. Dieterich, B. Cusack, H. Wang, K. Rateitschak, A. Krause, and M. Vingron. Annotating regulatory DNA based on man-mouse genomic comparison. *Bioinformatics*, 18(Suppl 2):S84–S90, 2002. (Proceedings of ECCB 2002).
4. L. Duret and P. Bucher. Searching for regulatory elements in human noncoding sequences. *Curr. Opin. Struct. Biol.*, 7:399–406, 1997.
5. R. C. Hardison. Conserved noncoding sequences are reliable guides to regulatory elements. *Trends Genet.*, 16:369–372, 2000.
6. M. Hasegawa, Y. Iida, T. Yano, F. Takaiwa, and M. Iwabuchi. Phylogenetic relationships among eukaryotic kingdoms inferred from ribosomal RNA sequences. *J. Mol. Evol.*, 22:32–38, 1985.
7. D. S. Hirschberg. A linear space algorithm for computing maximal common subsequences. *Commun. ACM*, 18:341–343, 1975.
8. X. Huang and W. Miller. A time-efficient, linear-space local similarity algorithm. *Adv. Appl. Math.*, 12:337–357, 1991.
9. W. Miller. Comparison of genomic DNA sequences: solved and unsolved problems. *Bioinformatics*, 17:391–397, 2001.
10. S. Schwartz, W. J. Kent, A. Smit, Z. Zhang, R. Baertsch, R. C. Hardison, D. Haussler, and W. Miller. Human-mouse alignments with BLASTZ. *Genome Res.*, 13:103–107, 2003.
11. T. F. Smith and M. S. Waterman. Identification of common molecular subsequences. *J. Mol. Biol.*, 147:195–197, 1981.
12. A. Ureta-Vidal, L. Ettwiller, and E. Birney. Comparative genomics: genome-wide analysis in metazoan eukaryotes. *Nat. Rev. Genet.*, 4:251–262, 2003.
13. W. W. Wasserman, M. Palumbo, W. Thompson, J. W. Fickett, and C. E. Lawrence. Human-mouse genome comparisons to locate regulatory sites. *Nature Genetics*, 26:225–228, 2000.
14. M. S. Waterman and M. Eggert. A new algorithm for best subsequence alignments with application to tRNA-rRNA comparisons. *J. Mol. Biol.*, 197:723–728, 1987.
15. Z. Zhang, P. Berman, T. Wiehe, and W. Miller. Post-processing long pairwise alignments. *Bioinformatics*, 15:1012–1019, 1999.

A Faster Reliable Algorithm to Estimate the p-Value of the Multinomial llr Statistic

Uri Keich and Niranjan Nagarajan

Department of Computer Science, Cornell University, Ithaca, NY-14850, USA
{keich,niranjan}@cs.cornell.edu

Abstract. The subject of estimating the p-value of the log-likelihood ratio statistic for multinomial distribution has been studied extensively in the statistical literature. Nevertheless, bioinformatics laid new challenges before that research by often concentrating its interest on the "thin tail" of the distribution where classical statistical approximation typically fails. Hence, some of the more recent development in this area have come from the bioinformatics community ([5], [3]).
Since algorithms for computing the exact p-value have an exponential complexity, the only generally applicable algorithms for reliably estimating the p-value are lattice based. In particular, Hertz and Stormo have a dynamic programming algorithm whose complexity is $O(QKN^2)$, where Q is the size of the lattice, K is the size of the alphabet and N is the size of the sample. We present a new algorithm that is practically as reliable as Hertz and Stormo's and has a complexity of $O(QKN \log N)$. An interesting feature of our algorithm is that it can guarantee the quality of its estimated p-value.

1 Introduction

The subject of goodness-of-fit tests in general and of using the (generalized) log-likelihood ratio (*llr*) statistic, in particular, is of great importance in applications of statistics. In many applications, an important question to answer is how unlikely is it that an observed sample came from a particular multinomial distribution (H_0)? But in order to answer this question, we first need to quantify the similarity level between the observed sample distribution and the null distribution. The llr statistic, G^2 (defined below) is a popular measure as it is provably optimal under some conditions. Indeed, it is so popular that it has several other names which are more commonly used in the information theory and bioinformatics community: entropy distance, relative entropy, information content, Kullbak-Leibler divergence etc., all of which (upto a factor of N) stand for $I = G^2/2$, where

$$I = \sum_k X_k \log\left(X_k/(N\pi_k)\right) \ ^1,$$

for a null multinomial distribution $\pi = (\pi_1, \ldots, \pi_K)$ and a random sample $X = (X_1, \ldots, X_K)$ of size $N = \sum_k X_k$. Note that $I = 0$ if and only if the empirical

---

[1] One can readily show that $G^2 = 2I$ is a generalized llr (e.g. [12]).

I. Jonassen and J. Kim (Eds.): WABI 2004, LNBI 3240, pp. 111–122, 2004.
© Springer-Verlag Berlin Heidelberg 2004

distribution is identical to π which is to be expected from something that is supposed to measure the distance (not in a metric sense) between these distributions.

The question of how unlikely is it that the particular sample $n = (n_1, \ldots, n_K)$ of size $N = \sum n_k$ came from π can then be translated to the following p-value that we need to compute:

$$P_{H_0}\left(I \geq \sum_k n_k \log \frac{n_k}{N \pi_k}\right),$$

or more generally, given an observed score s, what is $P_{H_0}(I \geq s)$? The latter question has been studied extensively in the statistical literature and has several types of estimates and means of computation which we survey below before describing our own novel technique.

The first type of estimates are in the form of universal upper and lower bounds such as the following one from Hoeffding [7]:

$$c_0 N^{-(K-1)/2} \exp(-s) \leq P(I \geq s) \leq \binom{N+K-1}{K-1} \exp(-s), \tag{1}$$

where c_0 is a positive absolute constant which can be taken to be $1/2$. Kallenberg has provided sharper (and more complicated) bounds [8] but he added that "...the bounds are not intended as direct numerical approximations of the involved probabilities ...they are useful because they have the right order of magnitude" and this will be relevant to us later on.

We can obtain asymptotically correct estimates based on the result that keeping π fixed and letting the sample size $N \to \infty$,

$$P_{H_0}(G^2 \geq s) \longrightarrow \chi^2_{K-1}(s)$$

(e.g. [12]). The rate of convergence and various corrections have been studied and are discussed in [4]. While the χ^2 approximation is a valid asymptotic result, in a typical application N is fixed and as s approaches the tail of the distribution the approximation can be quite poor. For example, for a null distribution of $\pi_i = i/10$ with $i = 1, \ldots, 4$ and $N = 40$, the p-value of $s = 120$ is roughly 7.8e-27 while the χ^2 approximation yields 7.7e-26, a factor of 10 off (and all else being equal, as s grows this will become worse).

Algorithmically, the simplest approach to computing the p-value is by naively enumerating all possible empirical distributions. However, the number of possible distributions grows like $\binom{N+K-1}{N}$, thus giving us a $O(N^{K-1})$ algorithm. Aware of these problems, Baglivo et al. [1] designed a polynomial time algorithm to approximate the p-value using a lattice. In principle the lattice can also be used to guarantee the quality of the approximation. However Baglivo et al.'s Algorithm employs the DFT (discrete Fourier transform) [10] and is therefore prone to exceedingly large numerical errors which originate from the inherent roundoff errors that would accompany any implementation of the DFT[6].

In bioinformatics I is heavily used in the context of evaluating the quality of an ungapped multiple sequence alignment [13] as in the popular motif-finder programs Meme [2] and Consensus [5]. Other usages were recently surveyed in [3]. Given the typical size of bioinformatics data, we are often forced to deal with exceedingly small p-values for which the χ^2 approximation breaks down. Thus, it should be of no surprise that some of the advancements in this area came from this community. Hertz and Stormo [5] provide a dynamic programming algorithm which, similar to Baglivo et al., uses a lattice approximation of the p-value. Both algorithms have a complexity of $O(QKN^2)$, where Q is the size of the lattice. However, Hertz and Stormo's algorithm has a much better handle of the numerical errors and by and large their algorithm is accurate to the mesh of the lattice. A slight modification of this algorithm is implemented as part of Meme's statistical evaluation of its results (version 3.0.3).

More recently Bejerano [3] introduced a new branch and bound algorithm to find the *exact* p-value. Since this approach does not use a lattice and is also a numerically stable algorithm, in general, it yields the most accurate result. However, it is only suitable for small Ks as it exhibits an exponential behavior in K. For $K = 4$ it has a runtime of the order of N^2 and in general it seems to have a runtime function that has the order of N^{K-2}. In another recent work Rahmann [11] apparently re-discovered Hertz and Stromo's dynamic programming method but in addition the paper also includes a clearer exposition of the problem and the algorithm. The paper also makes the important observation that in order to preserve accuracy, Q has to be increased linearly with N (assuming fixed π).

In this paper, we present a new algorithm that yields a lattice approximation of the p-value, $P_{H_0}(I \geq s)$ in $O(QKN \log N)$ time. We start with Baglivo et al.'s Algorithm and modify it using a technique we recently developed in [9] to control the numerical errors in FFT (fast Fourier transform) based convolutions. An interesting feature of our algorithm is that it provides a fairly reliable (and useful) upper bound on the numerical error in our estimate for the p-value.

2   Baglivo et al.'s Algorithm

Instead of computing the pmf (probability mass function) of I, Baglivo et al. suggest that we compute the pmf of the lattice valued random variable I_Q which approximates I, where

$$I_Q = \sum_k \text{round} \left[\delta^{-1} X_k \log(X_k/(N\pi_k)) \right]^2.$$

Here $\delta = \delta(Q) = I_{\max}/(Q-1)$ is the mesh size, $I_{\max} = N \log \pi_{\min}^{-1}$ is the maximal entropy and $\pi_{\min} = \min\{\pi_k\}$. By estimating p_Q, the pmf of I_Q, we can use

$$\sum_{\lceil s/\delta + K/2 \rceil} p_Q(j) \leq P(I \geq s) \leq \sum_{\lfloor s/\delta - K/2 \rfloor} p_Q(j).$$

to get a good estimate of $P(I \geq s)$ (assuming that the lattice is fine enough.)

---

[2] Note that due to rounding effects I_Q might be negative but we shall ignore this as the arithmetic we perform is modulo Q. The concerned reader can redefine $\delta = I_{\max}/(Q - 1 - \lceil K/2 \rceil)$.

In order to compute p_Q Baglivo et al.'s Algorithm starts by computing the DFT of p_Q, $\Phi = Dp_Q$:

$$\Phi(l) = \sum_{j=0}^{Q-1} p_Q(j) e^{i\omega_0 jl} \qquad \text{for } l = 0, 1, \ldots, Q-1,$$

where $\omega_0 = 2\pi/Q$. Once we have Φ, we can recover p_Q by applying D^{-1}, the inverse-DFT:

$$p_Q(j) = (D^{-1}\Phi)(j) = \frac{1}{Q} \sum_{l=0}^{Q-1} \Phi(l) e^{-i\omega_0 lj}.$$

At first glance this seems like a page out of the adventures of Baron Munchausen since after all we need p_Q in order to compute Φ to begin with. However, there is an alternative way to compute Φ as we outline next. As is explained in [1], we know that

$$\Phi(l) = \frac{1}{P(X_+ = N)} \sum_{x \in \mathbb{Z}^{+K} : \sum x_j = N} \prod_{j=1}^{K} p_j(x_j) e^{i\omega_0 l s_j(x_j)} = \frac{\psi_K(N, l)}{P(X_+ = N)},$$

where X_+ is a Poisson $\lambda = N$ random variable, p_k is the Poisson $\lambda = N\pi_k$ pmf and $s_k(y) = \text{round}[\delta^{-1} y \log(y/N\pi_k)]$ (the contribution to I_Q from the k-th letter appearing y times). It is not difficult to check that ψ_k actually satisfies the following recursive formula [1]:

$$\psi_k(n, l) = \sum_{x=0}^{n} p_k(x) e^{il\omega_0 s_k(x)} \psi_{k-1}(n - x, l). \qquad (2)$$

Thus using (2) $\Phi(l)$ can be recovered in $O(KN^2)$ steps for each l separately and hence $O(QKN^2)$ steps overall. Finally, using an FFT implementation of DFT [10] they get an estimate of p_Q in an additional $O(Q \log Q)$ steps (which should typically be absorbed in the first term). However, as we mentioned earlier, the algorithm as it is has a serious limitation in that numerical errors introduced by the use of the FFT can quickly become dominant in the calculations. An example of this phenomena can be observed with the parameter values, $Q = 8192$, $N = 100$, $K = 20$ and $\pi_i = 1/20$, where Baglivo et al.'s Algorithm yields a *negative* p-value for $P(I \geq 40)$.

3   Our Algorithm 1.0

To reduce the often unacceptable level of numerical errors in Baglivo et al.'s Algorithm we follow [9] and apply an exponential shift to p_Q. A simple example can help explain the idea. Let $p(x) \propto e^{-x}$ for $x \in \{0, 1, \ldots, 255\}$. In Figure 1 we compare p with $q = \widetilde{D^{-1}}(\widetilde{D}p)$, where $\widetilde{D}$ and $\widetilde{D^{-1}}$ are the machine implemented FFT and inverse FFT operators. As can be seen, while theoretically equal, in practice the two differ significantly. Now, if we apply an exponential shift to p prior to invoking the FFT operators then we get $\max_x |\log q_\theta(x)/p(x)| < 1.78 \cdot 10^{-15}$, where $\theta = 1$, $p_\theta(x) = p(x)e^{\theta x}$ and $q_\theta(x) = \left(\widetilde{D^{-1}}(\widetilde{D}p_\theta)\right)(x) \cdot e^{-\theta x}$. So p is recovered almost up to machine precision

($\varepsilon_0 \approx 2.2 \cdot 10^{-16}$). The reason this works is that by applying the correct exponential shift we "flatten" p so that the smaller values are not overwhelmed by the largest ones during the computation of the fourier transforms.

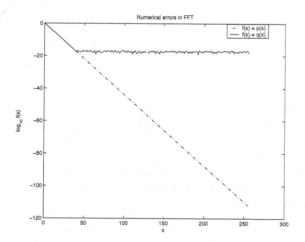

Fig. 1. The destructive effects of numerical roundoff errors in FFT.

This figure illustrates the potentially overwhelming effects of numerical errors in applications of FFT. $p(x) \propto e^{-x}$ for $x \in \{0, 1, \ldots, 255\}$ is compared with what should (in the absence of numerical errors) be the same quantity: $q = \widetilde{D^{-1}}(\widetilde{D}p)$, where $\widetilde{D}$ and $\widetilde{D^{-1}}$ are the machine implemented FFT and inverse FFT operators, respectively. This dramatic difference all but vanishes when we apply the correct exponential shift prior to applying D.

Needless to say this exponential shift will not always work. However, we do know that "to first order" our p-value behaves like e^{-s} (with fixed N and K) as is evident from (1). This suggests that we would benefit from applying an exponential shift to p_Q. Let

$$p_\theta(j) = \frac{p_Q(j)e^{\theta\delta j}}{M(\theta)},$$

where $M(\theta) = Ee^{\theta I_Q}$ is the MGF (moment generating function) of I_Q. Figure 2 shows an example of the flattening effect such a shift has on p_Q. Note that for a given θ, $M(\theta)$ can be reliably estimated in $O(KN^2)$ steps by replacing $e^{il\omega_0 s_k(x)}$ with $e^{\theta s_k(x)}$ in (2) and essentially repeating Baglivo's et al. procedure (but for a single value of θ).

The discussion so far implicitly assumed that we know p_Q which of course we do not. Nevertheless, we can compute $\Phi_\theta = Dp_\theta$ by slightly modifying the algorithm of Baglivo et al. All we need to do is replace the Poisson pmfs p_k with a shifted version $p_{k,\theta}(x) = p_k(x)e^{\theta s_k(x)}/M(\theta)^{\pi_k}$ and also replace ψ_k with the obvious $\psi_{k,\theta}$.

$$\psi_{k,\theta}(n, l) = \sum_{x=0}^{n} p_{k,\theta}(x)e^{il\omega_0 s_k(x)}\psi_{k-1,\theta}(n - x, l). \tag{3}$$

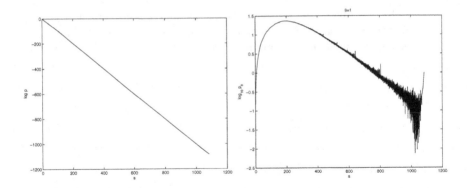

Fig. 2. How can an exponential shift help?

The graph on the left is that of $\log p_Q(s/\delta)$ where $N = 400$, $Q = 8192$ and $\pi_k = k/15$ for $k = 1, \ldots, 5$. The graph on the right is of the log of the shifted pmf, $\log p_\theta(s/\delta)$ where $\theta = 1$. Note the dramatic flattening effect of the exponential shift (keeping in mind the fact that the scales of the y-axes are different).

This allows us to compute $\widetilde{\Phi}_\theta(l)$, an estimate of[3] $\Phi_\theta(l) = \psi_{K,\theta}(N, l)/P(X_+ = N)$ in the same $O(KN^2)$ steps for each fixed l. We then compute an estimate $\widetilde{p_Q}$ of p_Q based on $p_Q(j) = (D^{-1}\Phi_\theta)(j)e^{-\theta\delta j}M(\theta)$. It is important to note that should the need arise (for example, for estimating very small p-values in the tail of the distribution) we can just as well estimate $\log p_Q(j) = \log p_\theta(j) - \theta\delta j + \log M(\theta)$ which significantly extends the range of p-values that can be computed by our algorithm. Finally, the p-value is estimated by $\sum_{j \geq s/\delta} \widetilde{p_Q}(j)$ (or the logarithmic version of that summation).

We are still left with the question of which θ to use. Equation (1) suggests $\theta = 1$ and indeed it typically yields the widest range of js for which $\widetilde{p_Q}(j)$ provides a "decent" approximation of $p_Q(j)$. However, for a given $s_0 = j_0\delta$ there would typically be a better choice of θ. Intuitively, we want to center p_θ about s_0 and that is the case with $\theta_{s_0} = \mathrm{argmin}_\theta [-\theta s_0 + \log M(\theta)]$. The minimization procedure can be carried out numerically[4] by using, for example, Brent's method [10] which would cost us another $O(KN^2)$ (but is outside the main loop on l). Finally, the next claim whose technical proof is outlined in the appendix gives an upper bound on our error.

Claim 1
$$|\widetilde{p_Q}(j) - p_Q(j)| \leq C(NK + \log Q)\varepsilon_0 e^{-\theta\delta j + \log M(\theta)}, \qquad (4)$$

where C is some small universal constant and ε_0 is the machine precision.

Note that for $s_0 = \delta j_0$ this upper bound is exactly minimized for θ_{s_0}, thus giving us another justification for our choice of θ.

Although (4) can be used to measure the quality of our approximation, we would like to emphasize a different "quality-control" method that works well in practice.

---

[3] Due to unavoidable numerical errors we cannot expect to recover $\Phi_\theta(l)$ precisely.

[4] A crude approximation of θ_{s_0} would typically suffice for our purposes.

As observed in [9] Im $\widetilde{p}_\theta$ is not 0 only because of numerical errors. Thus, $\varepsilon_{Im} = \max_j \left| \text{Im } \widetilde{p}_\theta(j) \right|$ is typically an indicator of the level of noise in $\widetilde{p}_\theta$. For example, with $N = 400$, $Q = 8192$, and $\pi_k = k/15$ for $k = 1,\ldots,5$ we applied our algorithm 1.0 with $\theta = 1$ to find that setting a noise threshold of Re $\widetilde{p}_\theta(j) > 10^3 \varepsilon_{Im}$ correctly recovers all of the non vanishing entries of p_θ at 9-digit accuracy[5].

4   Our Algorithm 2.0

Algorithm 1.0 fixed the problem of numerical errors that plagued Baglivo et al.'s Algorithm but its runtime complexity of $O(QKN^2 + Q \log Q)$ is essentially the same as that of Hertz and Stormo. An advantage of Baglivo et al.'s Algorithm, however, is that it has a stingier memory requirement that scales as $O(Q + N)$ as opposed to $O(QN)$ for Hertz and Stormo. An important observation that helps us to improve on the runtime of our algorithm is the fact that (3) can be expressed as a convolution between the vectors $p_{k\theta l}(x) = p_{k,\theta}(x)e^{il\omega_0 s_k(x)}$ and $\psi_{k-1,\theta}(\cdot, l)$. A naively implemented convolution requires $O(N^2)$ steps and hence that factor in the overall complexity. Alternatively, an FFT-based convolution, justified by the equation $(D(u * v))(j) = (Du)(j)(Dv)(j)$ [10], would only require $O(N \log N)$ steps cutting down the overall complexity to $O(QKN \log N + Q \log Q)$ [6].

Simply implementing (3) using FFT, however, reintroduces the severe numerical errors we worked hard to get rid of. The following example illustrates what is happening: for $\theta = 1$ one can easily verify that $p_{k,\theta}(x) \approx e^{-\lambda_k + x}/\sqrt{2\pi x}$. Computing $Dp_{k,\theta}(x)$ therefore faces essentially the same problem (only mirrored) as the one demonstrated in our example of FFT applied to e^{-x}. The solution is therefore to apply a negative exponential shift to $p_{k\theta l}$ and $\psi_{k-1,\theta}(\cdot, l)$ (i.e. multiply by $e^{-\theta_2(k)x}$).

The problem of choosing $\theta_2(k)$ is more involved than that of choosing θ. To begin with we have to choose a shift for each $k = 1,\ldots,K$. In addition, in each case we have to worry about simultaneously shifting three vectors: $p_{k\theta l}$, $\psi_{k-1,\theta}(\cdot, l)$ and their convolution $\psi_{k,\theta}(\cdot, l)$. We propose the following solution:

$$\theta_2(k) = \text{argmin}_{\theta'} \left[\theta' \sum_{i=1}^k \lambda_i + \log M_k(-\theta') \right], \tag{5}$$

where M_k is the MGF of $q_{k,\theta}(x) = (\psi_{k-1,\theta}(\cdot, 0) * p_{k,\theta})(x)$ for $x = 0,\ldots,2N$ [7]. The intuition behind this choice of $\theta_2(k)$ is that it guarantees that the mean of $q_{k,\theta}$ is at $\sum_{i=1}^k \lambda_i$ which, loosely speaking, says that the distribution of $\sum_{i=1}^k X_i$ has maximal resolving power (relative to numerical noise) about its mean $\sum_{i=1}^k \lambda_i$.

We currently do not have theoretical bounds on the error that arises due to the use of $\theta_2(k)$. We instead tested our algorithm on a wide range of parameters and compared

---

[5] Except for $p_\theta(Q-1)$ which has only 4-digits accuracy.

[6] While Hertz and Stormo make a passing remark that they can also use FFT-based convolution to cut the complexity to $O(QKN \log N)$ it seems unsubstantiated to us given that (17) in [5] is not strictly a convolution.

[7] Note that $q_{k,\theta}(x) = \psi_{k,\theta}(x, 0)$ for $x = 0,\ldots,N$.

Table 1. Range of test parameters.

Parameter	Values
K	4, 10, 20
N	50, 100, 200, 400
π	$Uniform, Sloped, Blocked$
s	$\frac{i}{21} * I_{max}$ $i \in [1..20]$

Uniform refers to the distribution where $\pi_j = 1/K$, Sloped refers to the case where $\pi_j = j/(K * (K+1)/2)$, and Blocked refers to the case where

$$\pi_j = \begin{cases} 3/(4 * \lfloor K/4 \rfloor) & j \le \lfloor K/4 \rfloor \\ 1/(4 * (K - \lfloor K/4 \rfloor)) & otherwise \end{cases}$$

our results with those obtained using Hertz and Stormo's algorithm (which is provably accurate) to get an estimate of the errors that arise in our algorithm. The range of parameters is given in Table 1. With Q set to 16384 and the other parameters exhaustively varying over the sets specified we found that our algorithm agreed with Hertz and Stormo's algorithm to at least 9 decimal places in all cases. This was found to be true even when we ran an experiment where we choose values of s much closer to I_{max} (using an interval halfing process on the range $[(\frac{20}{21} * I_{max})..I_{max}]$ to get 8 values of s) and let the other parameters vary as before. This gives us reasonable confidence in the belief that our methodology for choosing $\theta_2(k)$ works. We are currently working on a formal justification for this observation. In terms of complexity, the main loop now takes $O(QKN \log N)$. The other terms add $O(KN^2 + Q \log Q)$ to the runtime but this should be small compared to the runtime cost of the main loop[8] thus giving us a $O(QKN \log N)$ algorithm.

5   Comparison to Other Algorithms

For estimating a single p-value the complexity of version 2.0 of our algorithm is an improvement over Baglivo et al.'s Algorithm which has a time complexity of $O(QKN^2)$. More importantly, our algorithm offers much better control over the accumulation of numerical errors. For example, when $N = 50$, $K = 10$ and $\pi_i = 1/10$, Baglivo et al.'s Algorithm is able to recover the p-value for only 8 out of the 20 s values that we test on (where we only require correctness to 1 decimal place.) In contrast, our algorithm recovers the p-value accurately to at least 10 decimal places in all cases. In addition, our algorithm has a built-in quality control mechanism and should the p-value be too small for machine representation (not uncommon in bioinformatics applications) we can give the result in terms of log(p-value).

Bejerano's algorithm is more accurate than ours and is faster for $K \le 4$. However even for mildly large Ks it becomes impractical (in particular this is the case with $K = 20$) as it grows exponentially with K, presumably like $O(N^{K-2})$. Hertz and

---

[8] As observed in [11], in order to preserve the bound on the distance between p_Q and our real subject of interest, p_I, (the pmf of I), Q has to grow linearly with N.

Stormo are overall our closest competitors but their complexity is $O(QKN^2)$. In addition, if their algorithm is implemented explicitly as written [5], then it tends to suffer from intermediate underflows. For example, when applied to $N = 200$, $Q = 16384$, and $\pi \equiv 1/20$, all entries of p_Q less than 10^{-135} are estimated as 0. These intermediate errors can be eliminated if we switch to performing arithmetic on $\log p_Q$ instead of on p_Q, but that results in a non-trivial constant sitting in front of the $O(QKN^2)$. Alternatively, one can speed up the log arithmetic by using tables, as in Meme's (v3.0.3) implementation of Hertz and Stormo's algorithm, although that has the potential of introducing uncomfortably large numerical errors. For example, for $N = 50$, $\pi_i = i/10$ $i = 1, \ldots, 4$ and $Q = 10^4$, Meme's implementation seems to estimate the p-value of $s = 6$ as 0.0027 whereas the correct answer is 0.0095. An additional advantage of our method over Hertz and Stormo's algorithm (that it shares with Baglivo et al.'s Algorithm) is that the computation for each value of l can be carried out separately, incurring a space requirement that is $O(Q + N)$ whereas for Hertz and Stormo it is $O(QN)$.

We implemented our algorithm and Hertz and Stormo's algorithm in Matlab to compare the accuracy of the two algorithms. As a by-product, we also measured the running time of the two algorithms in Matlab and found that ours was on average between 10 and 100 times faster than Hertz and Stormo's algorithm on the range of parameters that we tested. In the case where $N = 400$ and $K = 20$, while Hertz and Stormo's algorithm took nearly a day and a half to compute the p-value, our algorithm took less than 7 minutes to do so. For more accurate runtime comparisons we have also written C programs that implement the two alogrithms. We have also worked on optimizing the C code for producing a fair comparison of the two algorithms (and there is still some scope for improvement, especially in the FFT implementations.) This is important because we found that while the code for Hertz and Stormo's algorithm shows little speedup when we turn on C compiler optimizations, the code for our algorithm runs twice as fast with even minor improvements to the code. The asymptotic behavior of the two algorithms is however clear even for small values of N. Figure 3 shows the behavior with increasing N for a fixed choice of the other parameter values (the graph looks the same for other choices of the parameter values too.)

Finally, while our algorithm was designed for finding a single p-value it turns out that in practice it can be easily adapted to reliably estimate p_Q in its entirety. The latter task is also performed by Hertz and Stormo's algorithm. In some cases our algorithm already does that. For example, setting $s = 500$ for $N = 400$, $Q = 8192$ and $\pi = i/15$ $i = 1, \ldots, 5$ we get a reliable estimate for the entire p_Q (relative error $< 10^{-6}$). More generally, in the cases that we have tried we can reliably recover the entire range of values of p_Q using as little as 2-3 different ss, or equivalently, θs (recall that each estimate has a quality control factor which allows us to choose the estimate which has better error gaurantees). Using version 2.0 of our algorithm this approach can still be significantly cheaper than running Hertz and Stormo's algorithm. We should also point out that Baglivo et al.'s comment regarding their algorithm being easily parallelizable is still valid for our algorithm since (3) can be computed separately for each $l = 0, 1, \ldots, Q - 1$.

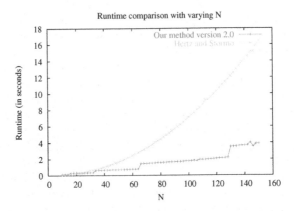

Fig. 3. Runtime comparison of Hertz and Stormo's algorithm and version 2.0 of our method. The parameter values used in this comparison are $K = 20$, $\pi_j = j/(K * (K + 1)/2)$, $s = 20$ and $Q = 1024$. Note that the discontinuities in the curve for our method are due to the fact that our implementation of FFT works with arrays whose sizes are powers of 2.

6   Future Work

There are many questions related to our work that deserve further study. For example, the theoretical and experimental question of how many different θs are needed to recover p_Q in its entirety is wide open. Another interesting area of study involves finding a theoretical grounding for our methodology for computing $\theta_2(k)$. On the practical side, the code for our algorithm is still being optimized, but we hope to make it available soon for the general public.

Finally, the most exciting prospect resulting from this work is that of combining it with our previous work [9] to obtain a fast and accurate system to compute the p-value of the entropy score of a multiple sequence alignment of biosequences. Note that the current work addresses the issue of computing the p-value of the entropy score of one column whereas our previous work focuses on computing the p-value of the sum of the entropy scores from all columns assuming that we have p_Q.

Acknowledgements

The first author would like to thank Gill Bejerano for the numerous discussions that we had on this subject and for generously sharing his code with us.

References

1. J. Baglivo, D. Olivier, and M. Pagano. Methods for exact goodness-of-fit tests. *Journal of the American Statistical Association*, 87(418):464–469, 1992.
2. T.L. Bailey and C. Elkan. Fitting a mixture model by expectation maximization to discover motifs in biopolymers. In *Proceedings of the Second International Conference on Intelligent Systems for Molecular Biology*, pages 28–36, Menlo Park, California, 1994.

3. G. Bejerano. Efficient exact value computation and applications to biosequence analysis. In M. Vingron, S. Istrail, P.A. Pevzner, and M.S. Waterman, editors, *Proceedings of the Seventh Annual International Conference on Computational Molecular Biology (RECOMB-03)*, pages 38–47, Berlin, Germany, 2003. ACM Press.
4. N. Cressie and T.R.C. Read. Person's χ^2 and the loglikelihood ratio statistic g^2: A comparative review. *International Statistical Review*, 57(1):19–43, 1989.
5. G.Z. Hertz and G.D. Stormo. Identifying DNA and protein patterns with statistically significant alignments of multiple sequences. *Bioinformatics*, 15:563–577, 1999.
6. K.A. Hirji. A comparison of algorithms for exact goodness-of-fit tests for multinomial data. *Communications in Statistics-Simulation and Computations*, 26(3):1197–1227, 1997.
7. W. Hoeffding. Asymptotically optimal tests for multinomial distributions. *Annals of Mathematical Statistics*, 36:369–408, 1965.
8. W.C.M. Kallenberg. On moderate and large deviations in multinomial distributions. *Annals of Statistics*, 13(4):1554–1580, 1985.
9. U. Keich. Efficiently computing the p-value of the entropy score. *Journal of Computational Biology*, in press.
10. W.H. Press, S.A. Teukolsky, W.T. Vetterling, and B.P. Flannery. *Numerical recipes in C. The art of scientific computing*. Cambridge University Press, second edition, 1992.
11. S. Rahmann. Dynamic programming algorithms for two statistical problems in computational biology. In Gary Benson and Roderic D. M. Page, editors, *Proceedings of the Third International Workshop on Algorithms in Bioinformatics (WABI-03)*, volume 2812 of *Lecture Notes in Computer Science*, pages 151–164, Budapest, Hungary, 2003. Springer.
12. J.A. Rice. *Mathematical Statistics and Data Analysis*. Duxbury Press, second edition, 1995.
13. G.D. Stormo. DNA binding sites: representation and discovery. *Bioinformatics*, 16(1):16–23, 2000.

Appendix: Outline of the Proof of Claim 1

Proof. The following lemma can be readily derived from the results in [9] (see Lemmas 1-3, (20) & (21)). For $\alpha \in \mathbb{C}$ we denote by $\tilde{\alpha}$ its machine estimator and define $e_\alpha = \tilde{\alpha} - \alpha$. For $\alpha, \beta \in \mathbb{C}$, we define

$$e_{\alpha+\beta} = \widetilde{\tilde{\alpha} + \tilde{\beta}} - (\alpha + \beta),$$

and similarly for $e_{\alpha\beta}$.

Lemma 1. *If* $|e_\alpha| < c_\alpha \varepsilon_*$ *and* $|e_\beta| < c_\beta \varepsilon_*$, *then*

$$|e_{\alpha+\beta}| \le (\max\{c_\alpha, c_\beta\} + 1)\varepsilon_*(|\alpha| + |\beta|)$$
$$|e_{\alpha\beta}| \le (c_\alpha + c_\beta + 5)\varepsilon_*(|\alpha\beta|).$$

Using this lemma one can use (2) to prove by induction on k that

$$|\psi_k(n, l) - \widetilde{\psi}_k(n, l)| \le cNk\varepsilon_*\psi_k(n, 0).$$

Note that when computing $\psi_k(n, 0)$ we only deal with positive numbers. It follows that

$$|\widetilde{\Phi}_\theta(l) - \Phi_\theta(l)| \le CNK\varepsilon_*|\Phi_\theta(0)| = CNK\varepsilon_*,$$

since $\Phi_\theta(0) = 1$. Using this result, Lemma 4 from [9], the fact that for $x \in \mathbb{C}^Q$, $\|D^{-1}x\|_\infty \le \frac{1}{Q}\|x\|_1$ and the triangle inequality we get

$$
\begin{aligned}
\|D^{-1}\Phi - \widetilde{D^{-1}\widetilde{\Phi}}\|_\infty &\le \|D^{-1}(\Phi - \widetilde{\Phi})\|_\infty + \|(D^{-1} - \widetilde{D^{-1}})\widetilde{\Phi}\|_\infty \\
&\le \frac{1}{Q}\|\Phi - \widetilde{\Phi}\|_1 + \frac{(C\log Q)\varepsilon_*}{Q}\|\widetilde{\Phi}\|_1 \\
&\le CNK\varepsilon_* + (C\log Q)\varepsilon_*.
\end{aligned}
$$

Claim 1 now follows by applying the inverse exponential shift (i.e. by multiplying $e^{-\theta\delta j + \log M(\theta)}$) to this error bound.

Adding Hidden Nodes to Gene Networks
(Extended Abstract)

Benny Chor and Tamir Tuller[*]

School of Computer Science, Tel-Aviv University, Tel-Aviv 69978, Israel
{bchor,tamirtul}@tau.ac.il

Abstract. Bayesian networks are widely used for modelling gene networks. We investigate the problem of *expanding* a given Bayesian network by adding a hidden node – a node on which no experimental data are given. Finding a good expansion (a new hidden node and its neighborhood) can point to regions where the model is not rich enough, and help locate new, unknown variables that are important for understanding the network. We study the computational complexity of this expansion, show it is hard, and describe an EM based heuristic algorithm for solving it. The algorithm was applied to synthetic datasets and to yeast gene expression datasets, and produces good, encouraging results.

Keywords: Bayesian networks, minimum description length, network expansion, gene network, EM, maximum likelihood, compression.

1   Introduction

The problem of inferring relationships among components of complex systems is a central problem in computational learning theory, with applications to systems biology, medical diagnosis, econometrics, and other disciplines. Stated in mathematical terms (with appropriate objective function) the problem is almost always computationally hard. In addition to the computational hardness, a large number of samples may be required in order to generate a good description of the system. In many cases, experts in the field have a very good idea on the structure and on many relevant parameters of the correct solution. This suggests an incremental approach: Start with the expert's solution, and when the system's size exceeds some threshold, use computational methods to expand it further. Such incremental grow may alleviate both problems mentioned above in many practical scenarios. In this work we explore this approach in the context of *Bayesian networks* (BNs).

Our input is a dataset and a network purported to explain it. We look for a way to expand the given network by a *hidden node* - a node representing a random variable, for which we have no data. Expanding the network means adding the new node, connecting it to the existing network, possibly modifying nodes and connections within a small bidirectional radius of the new node, and updating the probability tables. To tell a good extension from a bad one, we use the

---

[*] Corresponding author.

I. Jonassen and J. Kim (Eds.): WABI 2004, LNBI 3240, pp. 123–134, 2004.

natural and well known minimum description length (MDL) measure [9]. MDL "encourages" models for which the data have high likelihood, but penalized models with long descriptions. This way, overfitting the data is discouraged. We show that even this simpler problem of extending a given network is computationally hard, and design a heuristic based on the EM paradigm to perform this task. We tested our approach on a number of small synthetic networks, as well as on genetic networks with 100 nodes (genes), based on DNA microarrays gene expression data. For each network, we found a number of hidden nodes whose addition to the network significantly reduces the MDL score. This suggests small perturbations of the original models that better explain the data. We emphasize that our problem is different from the construction of a BN with some hidden variable "from scratch", since in our problem we start from a given network, and want to alter (edit) just a small portion of it. Network extension by hidden node was previously considered, for example in [7]. However, the essence of our work is not in finding "good" hidden node *per se*, but in analyzing a given model by identifying areas where the network need to be edited to better explain the data.

The use of BNs to analyze gene expression data was initiated by Friedman *et. al.*, and become a useful tool for inferring cellular networks [2, 3]. Segal *et. al.* introduced a related model, "module networks" [10], which is similar to a BN, except that genes are grouped into modules. Both models were used successfully to extract meaningful biological phenomena. The computational complexity of learning a BNs was studies by Hôffgen and by Chickering *et. al* . [5, 1], who proved it is NP-hard. In many practical cases the *sample complexity* is also a bottleneck: The amount of available data is insufficient to determine a correct model, this is markedly true for DNA microarray data, where the number of genes is typically two orders of magnitude greater than the number of samples. For example, Friedman and Yakhini [4] show that in order to find a BN with confidence δ and with entropy distance ε from the "true" BN, $O\left((1/\varepsilon)^{4/3}\log(1/\varepsilon)\log(1/\delta)\log\log(1/\delta)\right)$ samples are requires. (This expression hides constants which may be a factor larger than the number of nodes in the BN.) Tanay and Shamir [11] have considered computational expansion of genetic networks. There are several differences between their approach and ours. They consider combinatorial networks while ours are Bayesian. But most importantly, in our case the goal is to add a variable that is unknown (hidden), on which we have no data, whereas in [11] the variable is known, in the sense that there are samples of it.

2   Model and Definitions

In this section we first define the MDL score of a model (BN) and data. Our goal is to locate a hidden node that, when added to the network, will improve the MDL score. We provide bounds on the possible improvement.

2.1   The MDL Score for a BN

The minimum description length (MDL) criteria was used in 1978 by Rissanen [9] as a fundamental principle to model data. MDL scores a BN by two components:

The total number of bits needed to describe the model, $DL(BN)$, and the log likelihood of the data given that model, $-\log(Pr(D|BN))$.

$$MDL\_SCORE(D, BN) = DL(BN) - \log(Pr(D|BN)) \qquad (1)$$

An overly complicated model will be penalized by the $DL(BN)$ component, so by minimizing MDL we avoid overfitting of the model to the data.

BNs models dependencies among random variables and are especially useful when the "direct dependencies" are *sparse*. This model can be described by a directed acyclic graph $G = (V, E)$, which contains a node $x_i \in V$ for each random variable, and a directed edge $(x, y) \in E$ if y depends on x. In our case, the random variables are discrete. Every random variable is associated with a table, which describes its probability distribution, given the values of its parents.

The description of a BN for discrete random variables includes the description of a directed graph, of length $DL_{graph}(BN)$, and a probability table for every node, x_i, of length in the graph, $DL_{table}(x_i, pa_i, BN)$ (see figure 2). Let n denote the number of random variables (nodes in the graph). Let pa_i denote the set of variables that are parents of x_i in the BN. Then $DL_{graph}(BN)$ equals

$$DL_{graph}(BN) + \sum_{i=1}^{n} DL_{table}(x_i, pa_i, BN) \ .$$

Let $||x||$ be the number of possible assignments, or the range, that a variable, x, can assume. The graph G can be encoded by using $\log(n)$ bits to denote the number of parents of each node, and $\log \binom{n}{|pa_i|}$ bits to name x_i's parents. So $DL_{graph}(BN)$ equals

$$\sum_{i=1}^{n} \left(\log(n) + \log \binom{n}{|pa_i|} \right) \ .$$

We now consider the number of bits needed to encoded the probability table of node x_i. The number of entries is $(||x_i|| - 1) \cdot \prod_{j \in pa_i} ||x_j||$. Let N denote the number of samples. The measure of likelihood will not benefit from being more refined than the samples, so it would make no sense to represent probabilities with accuracy greater than $1/N$. Therefore, $\log(N)$ bits suffice to represent each entry in the probability table. By the law of large numbers, $\log(N)/2$ per entry suffices. Therefore

$$\sum_i DL_{table}(x_i, pa_i, BN) = \sum_i \frac{1}{2}(||x_i|| - 1) \cdot \prod_{j \in pa_i} ||x_j|| \log(N) \qquad (2)$$

When dealing with gene expression datasets, the number of measurements is usually low. Fore many genes (random variables x_i), even under severe discretization (say three or four values per variable), not all the combinations of pa_i appear in the dataset. Again, by the definition of likelihood, missing combinations give no contribution, so in such cases we can represent the probability table of a node x_i more succinctly by not counting missing combinations.

2.2 Bounds on the Change in Likelihood
When Adding a Hidden Node

When a hidden node is added to a Bayesian network, the "model description" component, $DL(BN)$, almost always increases. Our goal is to *decrease* the MDL score, $MDL\_SCORE(D, BN) = DL(BN) - \log(Pr(D|BN))$. The only hope to do it, then, is by sufficiently *increasing* the likelihood component $\log(Pr(D|BN))$. In this subsection we give upper and lower bounds on the amount of possible decrease in $-\log(Pr(D|BN))$. These bounds depend on the initial BN and the data.

Lemma 1 deals with upper bounds on the possible increase in log likelihood by adding a hidden node and connecting it to a subset of nodes, S, in a general BN. These bounds are important for evaluating the improvement in the likelihood by adding a hidden node.

The entropy of a discrete random variable, x, is $H(x) = -E(\log(p_x)) = -\sum_j p(x = j) \cdot \log(p(x = j))$. The log likelihood of N independent samples of this random variable is $\log L(x_i) = \sum_j \log(p(x = j))$. For large enough N we get, by the law of large numbers, that $-\log L(x)$ converges to $N \cdot H(x)$. So bounds on the entropy of the data, given the model, correspond to bounds on the log likelihood. We can get a similar result when x is a vector and the mutual distribution of its components are described by some model. If the entropy of the data, given the model, is low, then we can describe the data (given the model) with fewer bits (compression), and this implies that the likelihood of the data (given the model) is high.

The compressibility, or the improvement in the likelihood, is always non-negative, since we can choose the parameters of the probability tables of hidden node's children such that they are independent of the hidden node. However when the initial model describes the data perfectly, the compressibility will be zero. Thus, a general lower bound on possible improvement in likelihood depends on how well the initial model described the data.

When we consider an upper bound on the compressibility, if the initial BN has optimal parameters for its structure, the maximal increase in the log likelihood when adding a hidden node will be attained when the initial BN has no edges. This network does not capture any of the dependencies between the nodes, while the network with the hidden node may capture some of them.

Lemma 1. *Consider a network of n disconnected nodes, $x_1, .., x_n$, and an acyclic network created by adding a single hidden node as a neighbor of these n nodes in this disconnected network. For any data D composed of N samples, the increase in the log likelihood component in the DL score for the second network, compared to the original one, is at most*

$$N \cdot \min\left(\frac{1}{2}(\sum_{i,k:i\neq k} I(x_i, x_k)), max_{x_k}(\sum_i H(x_i) - H(x_k))\right)$$

The next Lemma suggests that a good compressibility of a set of n random variables in a Bayesian network implies a good compressibility of some subset of these n variables. Our heuristic is based on this lemma.

Lemma 2. *Let S be a set of n random variables, $x_1, .., x_n$ in a BN. Suppose each of these variables can attain k different values. If adding a hidden node h as a neighbor of all these variable can improve the log likelihood of the N data samples by $N \cdot C_n$ bits, than there are $n - \ell$ variable out of those n variables such that connecting only them to a hidden node yields compression by at least $N \cdot C_{n-\ell} \geq N \cdot (C_n - \ell \cdot \log(k))$ bits.*

For lack of space, the proofs of the last two lemmata are deferred to the full version of this paper.

3   Computational Hardness

As we mentioned earlier, the problem of constructing a optimal BN for a given dataset is known to be NP-hard. Furthermore it is easy to show that most of edge editing problems in a BN (changing edges and correspondingly tables) are NP-hard. By a reduction from clique, one can show that finding a subset of variables that is "heavy" in terms of pairwise mutual information, $\sum_{i,j \in S} I(x_i, x_j) \geq B$, is NP hard in general. The previous lemmata suggest that such "heavy" subsets are good candidates to connect to a hidden node in order to decrease the MDL score. In this section we explore some BN modification problems related to adding a hidden node, and show NP hardness.

Our problem has some variants, differing according to the types of editing operations that are allowed on neighbors of the hidden node. They are editing the probability table, and adding/removing/reversing edges. Another variant depends on how far from the hidden node these operations can take place (starting from distance 1 and up to the network diameter). Every such variant, if it has a reasonable degree of freedom, can be proved to be NP-hard. Here, we give the proof's sketch for the NP-hardness of one specific variant, we deal with another variant in the full version of this paper.

Theorem 1. *The following problem is NP-hard:*
Input: A Bayesian network, BN, with n nodes (variables). Data with N conditions. A real number, m.
Question: Is there a way to decrease the MDL score of the Bayesian Network parameters and the conditions by m bits by the following local modification: Adding one hidden node, connecting it to some neighborhood by outgoing edges, and then possibly modifying tables of nodes with bidirectional distance at most 2 from the hidden node. Edge editing of the existing nodes is not allowed.

Proof (sketch): We start with an undirected 3-regular graph $G = (V, E)$. We first transform it to a DAG, G', with all old nodes and also new ones. Each new node v_e corresponds to an edge e in the original graph. If $e = (u, v)$ then there are two directed edges to v_e - one from v, one from u (see figure 1).

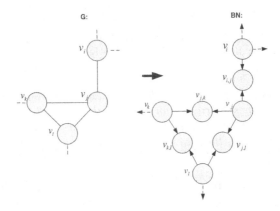

Fig. 1. A reduction from Vertex Cover on three regular graphs to our problem.

The new graph G' has $|V| + |E|$ nodes, and this will be the BN we'd like to modify. First we describe the data: There are N samples, where N is an even number to be determined later. Exactly $N/2$ of the samples have the form $(1, 1, ..., 1)$, and $N/2$ of the form $(0, 0, ..., 0)$. The probability tables of all "edge nodes" v_e are the same. If its parents have both 0, v_e emits a 0 with probability p_0, and 1 with probability $1 - p_0$. Likewise for $(1, 1)$ which leads to 1 with probability p_1, and 1 with probability $1 - p_1$. The tables of the original nodes are all identical (recall they have no incoming edges), they emit 0 with probability p_0, and 1 with probability p_1. For this data, the optimal network has a central "hidden" node that has outgoing edges to all other nodes. The hidden node is 0 with probability $\frac{1}{2}$ and 1 with probability $\frac{1}{2}$, and it totally determines all other nodes. However, the local modification we consider cannot change the topology and connectivity that much. What we allowed is adding a hidden node, and modifying the probability table of its children and grandchildren. The choice of N and of p_0, p_1 in the probability table is such that there is a modification with MDL score $\leq m$ if and only if G has vertex cover of size $\leq k$.

Note that the combination that the value of the parents are different (*i.e.* $(0, 0, 1)$ or $(1, 1, 0)$) are not supported by the data. One of the goals of the modifications is to remove these "anomalies". The best set of children of the hidden node will be the nodes of the vertex cover, the hidden node determine the value of these nodes and they, in turn, determine the value of all edge nodes. The original nodes outside the vertex cover are left fatherless. Therefore, their value will be independent of the other nodes in the BN, thus reducing the overall likelihood. However, it is not worthwhile to connect the hidden node to these orphans, since the increase in the description length (table size) will be larger than the log likelihood gain. Therefore, the overall MDL score is directly related to the VC size of the original graph.

For lack of space, the detailed proofs of the last and the following theorems are deferred to the full version of this paper.

Theorem 2. *The following problem is NP-hard:*
Input: A Bayesian network, BN, with n nodes (variables). Data with N condi-tions. A real number, m.
Question: Is there a way to decrease the MDL score of the Bayesian Network pa-rameters and the conditions by m bits by the following local modification: Adding one hidden node, connecting it to some neighborhood by outgoing or incoming edges, and then possibly modifying tables and the edges between nodes with bidi-rectional distance at most 1 from the hidden node. (Edge editing is allowed here.)

4   Algorithm

We first develop an EM heuristic algorithm that, given all the directed connec-tions of the hidden node to/from the network, determines the probability tables of the hidden node and of the other nodes the BN. This reduces our problem to finding a good way of connecting the hidden node to the Bayesian network. For a small k, it is feasible to exhaustively search all size k neighborhoods a hidden node. For larger k, we use a heuristic for choosing good candidate k-sets. By Lemma 2, good but large candidate neighbor sets for the hidden node have a smaller subset that yields, by itself, a large decrease in the MDL score. Our algorithm keeps a pool of tens to a few hundreds best smaller neighborhoods. To go to larger neighborhoods, we examine unions of sets from this pool. Suppose we decided upon a subset of nodes that we want to be connected to the hidden node. We then greedily examine the orientation of the edges connecting to the hidden node. In addition, we allow modifications (insert/delete/reverse) to the edges connecting neighbors of the hidden node. Here, we also employ a greedy (hill climbing) search. The final output is an ordered list of a small number of the best sets generated this way.

4.1   Updating Parameters, Given the Hidden Node Edges

Assume that the connectivity of the hidden node to the Bayesian network is given (all edges to/from the hidden node, and their directionality). We want to optimize the probability tables parameters of the hidden node and its neighbors. We used *EM* algorithm for solving this optimization problem. For lack of space, the details of this algorithm are deferred to the full version of this paper.

4.2   Local Graph Modification

Suppose we have a candidate for the hidden node and its neighbors. We now consider two type of local edge modification (between the hidden node neighbor's edges, and the edges going to/from the hidden node). Each step have to retains the acyclicity of the graph.

1. Deleting an edge between two neighbors of the hidden node.
2. Reversing the direction of an edge between two neighbors of the hidden node, or between the hidden node and one of its neighbors.

Deleting edges enables a decrease in the description length of the model, by removing variable's dependencies which are induced by the hidden node. The second operation may improve the likelihood (after an appropriate update to the table) by optimizing the structure of the hidden node's neighborhood. We used hill climbing method in order to find the best structure of the connectivity in the hidden node's neighborhood, we continue performing edge modification operation as long as an operation which improve the DL is found.

4.3 Finding Hidden Nodes with Large Neighborhoods

Exhaustive search over all possible neighbors sets, and the directed connectivity of the set members of size k (checking all $\binom{n}{k}$ hidden node neighbors' candidates) is feasible only if k is small. For large k we used an heuristic method for choosing good candidate set for the neighborhood of the hidden node. By Theorem 2, a good candidate set for the hidden node's neighbors set often has a subset that, by itself, gives a large decrease of the MDL score.

We use this idea in the following heuristic algorithm for finding hidden nodes of large in/out degree by checking good candidate sets. The algorithm use dynamic programming, and doesn't necessarily guarantee the optimal solution. Let $MDLD(S)$ be the decrease in the MDL due to adding a hidden node as a neighbor to the set S, and let t_k be some threshold. The algorithm seeks a set of size k which is a union of some subsets $S_1, .., S_j$ of size smaller than k, (which were found in previous iterations of the algorithm), such that $\sum_{i=1}^{j} MDLD(S_i) \geq t_k$. Formally the algorithm has the following steps.

1. Let r_k denote some natural number. For every size, k, we save the r_k sets with the largest decrease in the MDL due to adding a hidden node as a neighbor to this set.
2. A set of size k is a candidate to be the neighbors set of the hidden node only if it satisfies the following condition: There are sets $S_1, .., S_j$ which satisfy:
   (a) $\forall_i |S_i| < k$.
   (b) $\forall_i S_i$ is among the $r_{|S_i|}$ sets of size $|S_i|$ chosen previously.
   (c) $|\bigcup_{i=1}^{j} S_i| = k$.
   (d) $\sum_{i=1}^{j} MDLD(S_i) \geq t_k$
3. Calculate explicitly the decrease in the DL due to adding a hidden node as a neighbor to the sets which was found in the second step, and keep the r_k best sets.

5 Empirical Results

In this section we report on the performance of our method on synthetic and real inputs. We start with a description of the test bench for the algorithm. We continue with synthetic toy problems that help gain intuition on the problem. Then, we check the ability of our method to recognize hidden variables when the input is yeast gene expression data and a BN that was generated from it. We remark that the variables in our networks are all ternary (have 3 values).

5.1   Synthetic Networks

We started with a number of six node BNs (graphs and tables). We sampled the network, chose one node to hide, and eliminated this node's entries from the data. We then rebuilt a five node BN, based on the "truncated" data. The input to our algorithm was this five node net, and the truncated data. The algorithm expanded the network - adding a sixth, hidden, node and updating the probability tables. When there are enough samples, adding a hidden node with connectivity close to the original usually got the highest improvement in the MDL score. This verifies that our method is indeed capable of discovering regions that are influent by hidden variables.

We now describe two specific DAGs on which the experiment of the previous paragraph was carried out, both depicted graphically in figure 3: The DAGs are the outgoing star, and the path. In different runs we chose different parameters for the probability tables of the nodes. We sampled the network, added a Gaussian noise to the samples, and hid node 0, namely eliminated this node's entries from the data. We then built the best of all five node Bayesian networks, based on the "truncated" data, using exhaustive search over the topologies. The input to our algorithm was this five node net and the truncated data. The algorithm expanded the network by adding a sixth node, connecting it, and completing the probability tables. Table 1 summarize our result for the case of the outgoing star: For every parameters set of the original network probability tables, the average values of mutual information over pairs of leaves, $\overline{I(x,y)}$, and over triplets of leaves, $\overline{I(x,y,z)}$, before adding noise, were calculated. Note that for ternary variables, the maximum value the mutual information ($I(x,y)$ or $I(x,y,z)$) can attain is $\log_2 3$. Likewise, we computed the average decrease in the DL score, obtained when the hidden node is added with either four neighbors, $\overline{MDL(x,y,z,w)}$, or five neighbors, $\overline{MDL(x,y,z,w,t)}$. Before adding the hidden node, the MDL score of the BNs and data was computed. The noise was Guassian noise with mean 0 and variance either 2 or 4, while the "distance" between the signals was 10 (*i.e.* variables attain values $0, 10, 20$). The signal plus noise was then: discretecized, bringing it back to the original range $(0, 10, 20)$. It can be seen that if the mutual information of pairs and triplets in the original model is large enough, then adding a hidden node in the right place yield a decrease in the MDL score.

For comparison, we applied our method on a "true" a number of five node Bayesian network with similar statistical parameters (ternary variables and specification of noise), and data which were sampled from it. The average decrease in the MDL, which was obtained by our method when the hidden node have four neighbors was -1383, and -1482 for five neighbors. So in this case the algorithm indeed detects that a hidden node is *not* called for.

In the case of the path we got no improvement in the MDL score when we tried to add a hidden node, this is due to the facts that an edge between two nodes in a Bayesian network gives the same likelihood (with the right parameters) as in the case that this two nodes are connected via a hidden node (with shorter parameter's description length). This is also the case when we try to add a

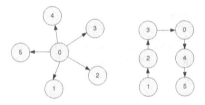

Fig. 2. The original Bayesian networks of the toy problem.

Table 1. The results of applying our algorithm on a toy problem, the outgoing star (figure 2).

Noise variance	Original MDL	$\overline{I(x;y)}$	$\overline{I(x;y;z)}$	$\overline{MDL(x,y,z,w)}$	$MDL(x,y,z,w,t)$
2	1400	0.62	0.59	−121	332
4	1450	0.62	0.59	−125	330
2	1470	0.5959	0.5677	−141	325
4	1480	0.5959	0.5677	−158	291
2	2500	0.11	0.08	−234	−98
4	2550	0.11	0.08	−250	−120

hidden node which is connected only to one node, the parameters of that node can be optimize to give the same likelihood (with shorter parameter's description length). when the hidden node have larger set of neighbor. A proof is omitted because of a lack in space.

5.2   Genomic Network

We tested our method on a 99 nodes Bayesian network of Pe'er *et. al* . [8], which was generated from the yeast gene expression data of Rosseta's compendium (Hughes *et. al* . [6]). The MDL score of this BN and data is 45,000 bits. We applied the algorithm to the full BN and data, and looked for different ways to add a hidden node. We found a vast number of possible ways for improving the MDL score of the net. About 3%, or 4,700 out of the $\binom{99}{3}$ possible neighborhoods of size 3 of the hidden node, and about 0.5%, or 1,800, out of the $\binom{99}{4}$ possible neighborhoods of size 4 of the hidden node gave a decrease in the MDL score. The distribution of the improvement in MDL scores for neighborhoods of size 4 that improved the MDL score is given on figure 3. It can be seen that approximately 1.5% of those neighborhoods gave improvements of 500 bits or more.

In addition to exhaustively checking all the neighborhood of size 3 and 4, we also applied the algorithm from section 3.3 to get good candidates for larger neighbor sets. One such outcome is shown in figure 4: This is a 6 node neighborhoods that improves the MDL score by 1000 bits (starting from the same 99 nodes BN with MDL score 45,000).

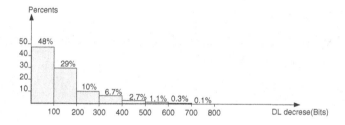

Fig. 3. The distribution of the improvement in the MDL score for size 4 neighborhoods, when adding a hidden node to the 99 nodes network.

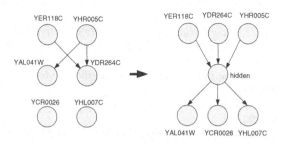

Fig. 4. An example of decreasing the MDL score by adding a hidden node. Node are labels by genes' names.

6 Conclusions

Bayesian and related networks seem like a successful tool for studying gene networks [3, 8, 10, 2]. Studying ways of expanding such networks is therefore an important research direction. In this work, we investigated the problem of extending a given network by adding *hidden node* to it. Our approach can help an expert to focus attention on region or regions in his or her own model, that may require additional parameters,additional experimental data or, more generally, further investigation. We showed that the problem is computationally hard, and developed a heuristic, based on EM, for solving it. We applied this to a number of synthetic and real networks, with encouraging results. In our biological networks, the improvements in the MDL score are not likely to be an artifact, and probably represent a real biological phenomena.

It is easy to see that there are cases where adding more than one hidden node at a time is superior to adding the hidden nodes one by one. In this work we chose to deal with just one hidden node at a time for simplicity and complexity reasons. Our method can be generalized, in principle, to this more general case as well. Furthermore it will be interesting to generalized this method for other probabilistic models, not just Bayesian networks.

Acknowledgements

Thanks to Nir Friedman for helpful discussions, and to Dana Pe'er for sending us her BNs and the Rosseta dataset.

References

1. D. M. Chickering. Learning bayesian network is NP-complete. *In D. Fisher and H.J. Le, editors, Learning from data: AI and statistic V, New York NY. Springer-Verlag*, pages 121 – 130, 1996.
2. N. Friedman. Inferring cellular networks using probabilistic graphical models. *Science*, pages 799 – 805, 2004.
3. N. Friedman, M. Linial, I. Nachman, and D. Pe'er. Using bayesian network to analyze expression data. *Journal of Computational Biology*, (7):601–620, 2000.
4. N. Friedman and Z. Yakhini. On the sample complexity of learning bayesian networks. *In Proc. 12th Conf. on Uncertainty in Artificial Intelligence, Portland, OR*, pages 274 – 282, 1996.
5. K. L. Höffgen. Learning and robust learning of product distributions. *In COLT*, pages 77 – 83, 1993.
6. T. R. Hugehes and M. J. Marton. Functional discovery via a compendium of expression profiles. *Cell*, 102(1):109–26, 2000.
7. C. K. Kwoh and D. F. Gillies. Using hidden nodes in byesian networks. *Artificial Intelligence*, 88:1–38, 1996.
8. D. pe'er, A. Regev, G. Elidan, and N. Friedman. Inferring subnetworks from perturbed expression profiles. *Bioinformatics*, 1(1):1–9, 2001.
9. J. Rissanen. Modeling by shortest data description. *Automatica*, 14:465 – 471, 1978.
10. E. Segal, M. Shapira, D. Pe'er, D. Botstein, D. Koller, and N. Friedman. Module networks: identifying regulatory modules and their condition-specific regulators from gene expression data. *Nature Genetics*, 34:166–176, 2003.
11. A. Tanay and R.Shamir. Computational expansion of genetic networks. *ISMB*, pages 1–9, 2001.

Joint Analysis of DNA Copy Numbers and Gene Expression Levels

Doron Lipson[1,*], Amir Ben-Dor[2], Elinor Dehan[3], and Zohar Yakhini[2]

[1] Computer Science Dept., Technion, Israel
dlipson@cs.technion.ac.il
[2] Agilent Laboratories, Palo Alto, CA
{amir_ben-dor,zohar_yakhini}@agilent.com
[3] Functional Genomics Unit, Sheba Medical Center, Israel
ed486@endeavor.med.nyu.edu

Abstract. Genomic instabilities, amplifications, deletions and translocations are often observed in tumor cells. In the process of cancer pathogenesis cells acquire multiple genomic alterations, some of which drive the process by triggering overexpression of oncogenes and by silencing tumor suppressors and DNA repair genes. We present data analysis methods designed to study the overall transcriptional effects of DNA copy number alterations. Alterations can be measured using several techniques including microarray based hybridization assays. The data have unique properties due to the strong dependence between measurement values in close genomic loci. To account for this dependence in studying the correlation of DNA copy number to expression levels we develop versions of standard correlation methods that apply to genomic regions and methods for assessing the statistical significance of the observed results. In joint DNA copy number and expression data we define significantly altered submatrices as submatrices where a statistically significant correlation of DNA copy number to expression is observed. We develop heuristic approaches to identify these structures in data matrices. We apply all methods to several datasets, highlighting results that can not be obtained by direct approaches or without using the regional view.

1 Introduction

Alterations in DNA copy number (DCN) are characteristic of many cancer types and are thought to drive some cancer pathogenesis processes. Alterations include large chromosomal gains and losses as well as smaller scale amplifications and deletions. Genomic instability triggers the overexpression or activation of oncogenes and the silencing of tumor suppressors and DNA repair genes. Alterations in DCN have been initially measured using local fluorescence in situ hybridization-based techniques. These evolved to a genome wide technique called Comparative Genomic Hybridization (CGH, see [8]), now commonly used for the identification of chromosomal alterations in cancer [10,2]). In this genome-wide cytogenetic method differentially labelled tumor and normal DNA are co-hybridized to normal metaphases. Fluorescence level ratios between the two

---

* Corresponding author

I. Jonassen and J. Kim (Eds.): WABI 2004, LNBI 3240, pp. 135–146, 2004.
© Springer-Verlag Berlin Heidelberg 2004

labels allow the detection of chromosomal amplifications and deletions. This method has, however, a limited resolution (10-20 Mbp) which maks it impossible to predict the borders of chromosomal changes or to identify changes in copy numbers of single genes and small genomic regions. In a more advanced method termed *array CGH* (aCGH) the tumor and normal DNA are co-hybridized to a microarray of thousands of genomic clones of BAC, cDNA or oligonucleotide probes ([13, 11, 6]). The use of aCGH allows the determination of changes in DCN of relatively small chromosomal regions. When using oligonucleotide arrays the resolution can, in theory, be finer than single genes.

The development of high resolution mapping of DCN alterations and the progress of expression profiling technologies enable the study of the effects of chromosomal alterations on cellular processes and how these are mediated through altered expression of genes residing in altered regions. By measuring DNA copy numbers and mRNA expression levels on the same set of samples we gain access to the relationship of copy number alterations to how they are manifested in altering expression profiles. In [12] the authors used (metaphase slides) CGH to identify large scale amplifications in 23 metastatic colon cancer samples and performed expression profiling on the same samples. They observed some correlation between DCN and expression levels but on overall effect of alterations on transciction. In [14] an opposite observation is reported, for breast cancer samples. That is: a strong global correlation between copy number changes and expression level variation is observed. Similarly, Hyman et al [7] studied copy number alterations in 14 breast cancer cell lines and identified 270 genes with expression levels that are systematically attributable to gene amplification. The statistics used by both latter studies is based on simulations and takes into account single gene correlations but not local regional effects. Recently, Linn et al studied expression patterns and genome alterations in DFSP and discovered common 17q and 22q amplifications that are associated with elevated expression of resident genes [9].

Our purpose in this paper is to provide algorithmic and statistical methods to rigorously support data analysis designed to improve our understanding of copy number to transcription relationships (specifically in aCGH data). Regions of high correlation are potentially related to the tumor pathogenesis. More specifically, genes affected by changes in DCN potentially play a role in driving tumor differentiation. Note that the correlation between expression data vectors and their corresponding (same gene) DCN data vectors should behave completely random if all the variation in the DCN vector arises due to experimental errors. We are therefore mostly interested in detection of statistically significant correlations. These might not show up when low resolution and global data analysis approaches are employed. For example, low penetrance (not all cells in the sample) and low prevalence (not all samples in the study) alterations might effect expression below the 2-fold mark and only in some of the samples, but in a significant manner when a genomic region is considered. In addition – the detection of regions that manifest a significant correlation can aid in detecting actual low penetrance alterations in high resolution even if the DCN data, alone, do not support such discoveries.

Throughout the paper we use C and E to denote the DNA copy number and gene expression data matrices, where the (i, j)-th entry of each matrix represent the data for the ith gene in the jth sample. We abbreviate "DNA copy number" and "gene expression" as DCN and GE respectively.

In Section 2 we describe methods to quantify the correlation observed for any pair of rows $C(i, \cdot)$ and $E(i, \cdot)$. In Section 3 we extend these to account for the regional character of DCN alterations. In Section 4 we discuss submatrices of affected samples and genes. Conclusions are discussed in Section 5.

Datasets. We demonstrate our methods on two joint DCN-GE breast cancer datasets. The first, described in [14] is of 6,095 genes across 41 samples (4 cell lines, 37 primary breast tumors). The second, from [7] is of 13,824 genes across 14 cell-line samples. All datasets are measured on cDNA microarrays.

2   Correlation Scoring Methods

Consider a single gene g and let $u = u_g$ and $v = v_g$ denote the corresponding DCN and GE data vectors of g. Let n denote the number of samples (length of u and v). In this section we present several approaches for scoring g by looking for dependencies between u and v.

2.1   Pearson Product Moment Correlation

The most common measure of the dependence between two vectors u, v is the Pearson correlation coefficient:

$$r(u, v) = \frac{\sum (u - \bar{u})(v - \bar{v})}{\sqrt{\sum (u - \bar{u})^2}\sqrt{\sum (v - \bar{v})^2}}. \tag{1}$$

r measures the degree to which u and v maintain a *linear* relationship. It may therefore be less suitable when the DCN and GE values follow some non-linear relationship. Nonetheless, previous large-scale DCN-GE comparative studies [14] used Pearson correlation as a sole scoring method to evaluate dependence.

2.2   Separating-Cross Correlation

A different methodology for comparing gene copy measurements with gene expression levels, such as the one described in [7] utilizes user chosen thresholds for classifying DCN measurements as deleted or amplified and for classifying GE measurements as under-expressed or over-expressed. Such methods do not rely on any assumption of linearity or on the value of the mean; however, they are somewhat dependent on the specific choice of thresholds. The separating-crosses scores we now introduce are a generalized approach to threshold-based analysis of the dependence between two vectors.

We can view the two vectors u and v as n points (u_i, v_i) in the plain. An axis parallel cross $t = t_{x,y}$, centered at (x, y), partitions the plain into four quadrants, denoted by A_t, B_t, C_t, and D_t (See figure 2). We denote by a_t the number of points (u_i, v_i) that belong to the quadrant A_t. The other quadrants counts b_t, c_t and d_t are defined similarly. Clearly, $a_t + b_t + c_t + d_t = n$.

Roughly speaking, u and v are correlated if there exist a cross t such that both a and d are large (compared with b and c). More generally, assume we are given a function of the quadrants counts (such a function is called a *cross-function*), $f(a, b, c, d)$. We are interested in the maximal obtainable value of f:

$$F(u, v) = \max_t \{ f(a_t, b_t, c_t, d_t) \}. \tag{2}$$

The function F is called a *separating cross score function*.

Let π denote the ranks of the samples with respect to the vector u. That is, $u(\pi^{-1}(1)) < \cdots < u(\pi^{-1}(n))$. For example, for $u = (2, 1.5, 9, 0.4)$, $\pi = (3, 2, 4, 1)$. Similarly, we denote by τ the samples permutation induced by v. Since cross-functions (and thus score functions) depend only on quadrants counts and not on the actual locations of the points, we have $F(u, v) = F(\pi, \tau)$. Thus, for every function $f(\pi, \tau, t)$, we can compute $F(\pi, \tau)$ by examining $(n - 1)^2$ possible crosses. We describe one cross score function, the *Maximal Diagonal Product* (MDP). Consider the separating-cross-function $DP(\pi, \tau, t) = a_t \cdot d_t$, which we call *Diagonal Product* (DP), and the corresponding score function:

$$MDP(\pi, \tau) = \max_t \{ DP(\pi, \tau, t) \}. \tag{3}$$

A useful attribute of the MDP score is that it provides a distinction between samples that contribute to the maximum score (points within A_t and D_t) and those that do not (points within B_t and C_t). We make use of this attribute in identifying affected samples in Section 4. The combinatorial nature of this score allows rigorous calculation of its statistical properties, the discussion of which is beyond the scope of this paper.

3   Regional Analysis

Some cancer related alterations in genomic DNA have direct effect on mRNA levels, possibly leading to downstream functional deficiencies. These alterations are most likely *localized* in one or more of the following aspects:

1. The alteration in genomic DNA is limited to certain chromosomal segments.
2. The expression of all genes within a specific genomic segment may not be effected to the same extent.
3. Not all samples contain identical or similar genomic alterations.
4. Within specific samples, alterations occur with varying levels of penetrance.

Previous work on DCN-GE expression relationship consider only correlation between the gene-expression levels of single genes and their respective DNA copy number measurements. Pollack et al [14] study the global behavior of DCN-GE correlation and show that the distribution of Pearson correlation values between DCN and GE differs from the expected distribution. They report 54 amplified genes with moderately or highly elevated expression levels. Hyman et al [7] also demonstrate global single gene correlations and identify 270 genes with expression levels significantly influenced by changes in DCN .

CGH based studies show that chromosomal alterations frequently apply to long stretches of the genome that may span a large number of genes. The expression pattern of a gene that is affected by such an aberration is expected to correlate not only with the copy number of its own coding DNA but also with the DCN measurements of neighboring genes. We therefore expect that analysis that takes regional effects into account to yield better results that might offset the negative effects of noise in the data or low penetrance. Both [14] and [7] did not account for such regional considerations. In this section we suggest a framework for considering local correlation between genomic alteration and variance in gene expression levels, that accounts for regional effects.

Given a gene g_i we define its k-neighborhood as the continuous sequence of genes indexed by $\Gamma_k(i) = (i - k, ..., i + k)$. A straightforward approach to quantifying the correlation of the gene's GE vector $E(i, \cdot)$ with the DCN vectors in its neighborhood $\Gamma_k(i)$ is by calculating the average correlation of $E(i, \cdot)$ to each of the respective DCN vectors:

$$r(i, \Gamma_k(i)) = \frac{1}{2k + 1} \sum_{j=i-k}^{i+k} r(i, j), \tag{4}$$

where $r(i, j)$ is any correlation measure between the vectors $E(i, \cdot)$ and $C(j, \cdot)$.

Alternative approaches to regional correlation include the correlation of $E(i, \cdot)$ to the vector of (weighted or uniform) average DCN in $\Gamma_k(i)$, or the product of the p-values of the respective correlations.

Permuted Data. When performing analyses that take gene order into account we compare results to a null model that assumes neighboring genes are independent of each other. To this end, we also perform our analysis on gene-permuted matrices E' and C' where the same permutation was applied to the rows of both matrices. We expect regional effect results to be dependent on the original chromosomal order of the genes.

Computing p-Values. To identify regions where DCN and GE correlate beyond the extent expected for the consistent DCN values we perform the simulation analysis outlined below. The general idea is to evaluate a locus dependent p-value for chromosomal regions. Correlations in regions where a very consistent DCN measurements are observed need to cross much higher thresholds to be significant since distributions expected at random in such regions have larger variation (weaker smoothing effect of averaging because of the consistent DCN values). Consider a neighborhood $\Gamma_k(i)$ and fix L, the size of simulation applied. We randomly draw $L-1$ expression vectors (rows of E) indexed by $i_1, i_2, ..., i_{L-1}$, and for each compute $r_l = r(i_l, \Gamma_k(i))$, the correlation of the random expression vector to the neighborhood $\Gamma_k(i)$. To the correlation $r^* = r(i, \Gamma_k(i))$, actually observed at i, we assign its rank ρ amongst $r_1, r_2, ..., r_{L-1}, r^*$, a number between 1 and L. The p-value for the region correlation observed at i is $pV(i) = \rho/L$.

3.1   Results

We applied the above locus dependent p-value calculations to investigate copy number to expression correlations in [14]. Figure 1 depicts the cumulative distribution of $pV(i)$, where i ranges over all genes in the dataset. As expected, randomly permuting the dataset yields a straight line that can be used as reference (curve E), while significant single gene correlations (i.e. $r(i, i)$, curve C) are overabundant at all p-values. Significant correlations are even more abundant when computed for neighborhoods of size $k = 2$ and $k = 10$ (curves B and A, respectively). Note that these results depend on both the chromosomal order (as the gene-permuted data yields a lower abundance of significant correlation scores than single gene correlations, curve D) and on direct DCN to GE correlations (due to the method of calculating $pV(i)$). The region-dependent $pV(i)$ scores enable the identification of loci where the gene expression levels significantly correlate with the DCN measurements with greater statistical confidence. For illustration, consider a threshold of $pV(i) \leq 0.001$. A random dataset of 6000 genes is expected to contain 6 genes with this score whereas single gene correlations yield 164 such genes (FDR = 3.7%). Averaged correlation against $\Gamma_2(i)$ yields 214 significant loci (FDR = 2.8%) and working with $\Gamma_{10}(i)$ yields 289 significant loci (FDR = 2.1%). Thus, using region-based analysis delivers almost 80% more loci where DCN-GE correlation may be identified with high confidence. Furthermore, additional regions of correlation are thus detected (details in the supplement [1]).

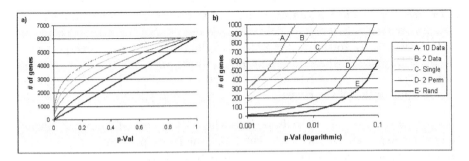

Fig. 1. a) Cumulative distribution of values of $pV(i)$ (p-value of the average correlation $r(i, \Gamma(i))$), for neighborhoods of size $k = 10$ (A), $k = 2$ (B), single genes (C), and for gene-permuted data with a neighborhood of size $k = 2$ (D). Randomly permuted data yields a straight line, as expected (E). b) Zoom of lower values of the same curves at logarithmic scale. Note the different values attained at $pV(i) = 10^{-3}$.

4   Genomic-Continuous Submatrices

In the previous section we mentioned that genomic alterations are often localized to a subset of the samples as well as to a specific chromosomal segment. In this section we discuss the task of detecting both the genomic segment in which an aberration has occurred, the affected samples and the transcriptional effect of the aberration. To this end we define a model of significantly altered genomic-continuous submatrices and present two algorithmic approaches for detecting them. We then apply the suggested methods to the two breast cancer datasets.

4.1   Definition

For a given pair of DCN and GE matrices C, E over an ordered set of genes G and a set of samples S we define a *genomic-continuous submatrix* (GCSM) as $M = G' \times S'$ where $G' \subset G$ is a continuous segment of genes and $S' \subseteq S$ is a subset of samples, and its *complement submatrix* $\overline{M} = G' \times \{S - S'\}$. Let $C(M)$ and $E(M)$ denote the projections of the matrices C and E on the subsets G' and S' (the DCN and GE *submatrices* corresponding to M).

Under our biological model, a genomic alteration in a given chromosomal segment and a given sample should affect most of the DCN measurements in this segment but only some of the respective GE measurements (since changes in expression depend on other factors that determine regulation). Informally, we say that a given GCSM M is *significantly amplified* if:

- *Most* DNA copy values in the set $C(M)$ are positive.
- *Some* genes $g_i \in G'$ have higher expression values $\{E(i,j) : s_j \in S'\}$ compared to $\{E(i,j) : s_j \notin S'\}$.

More formally, let us define a score that reflects the degree to which M is significantly amplified. First, we define a score $F(M; C)$ that reflects overabundance of positive values in $C(M)$ in comparison to $C(\overline{M})$ using the hypergeometric distribution. Let $N = |C(M \cup \overline{M})|$ and $n = |C(M)|$. Let K and k be the number of positive values in $C(M \cup \overline{M})$ and $C(M)$, respectively. Given N, n, K the hypergeometric probability of finding k or more positive values in $C(M)$ is:

$$F(M; C) = HG(N, K, n, k) = \sum_{i=k}^{N} \frac{\binom{n}{i}\binom{N-n}{K-i}}{\binom{N}{K}}. \tag{5}$$

Similarly, we define a score $F(M; E)$ that reflects the overabundance of genes in G' that are significantly differentially expressed, comparing S' and $S - S'$, in the correct direction (higher in S' than in $S - S'$). A TNoM (Threshold Number of Misclassifications) score may be assigned to each gene according to its performance as a S' versus $S - S'$ classifier [5,3]. Rigorous p-values can be calculated for TNoM. If the probability, for a single gene, of obtaining a score of s or better under the null model is $p(s)$ then the number of genes with scores s or better, amongst the $|G'|$ genes examined, is Binomial$(n, p(s))$ distributed.

Let $n(s)$ denote the number of genes with such scores actually observed in the data. Let $\sigma(s)$ be the tail probability of the Binomial$(n, p(s))$ distribution, at $n(s)$. $F(M; E)$ is then defined to be $\max_{0 \leq s \leq |S'|} - \log(\sigma(s))$. For a more detailed description of differential expression overabundance please see [4].

Under the null model, DCN and GE vectors are completely uncorrelated. A total score for an amplification in M is:

$$F(M; C, E) = - [\log_{10} F(M; C) + \log_{10} F(M; E)] \qquad (6)$$

The above discussion addresses amplifications only. However, any deletion in a subset S' is equivalent, under F, to an amplification in $S - S'$.

4.2 Algorithmic Approach

Locating a partition of samples that maximizes TNoM overabundance for a given set of genes is by itself a difficult task that has been approached by heuristic methods [4]. The problem of locating a partition that maximizes a combined hypergeometric and TNoM overabundance score is clearly at least as hard, and consequently we resort to heuristic approaches for locating significantly altered GCSMs. Note that due to the fact that we are looking for continuous segments only, all possible segments may be enumerated in $O(n^2)$, where n is the number of genes. The difficulty remains in determining which subset S' maximizes $F((G' \times S'); C, E)$ for a given segment G'. We suggest two algorithmic approaches:

Max-hypergeometric Algorithm. As the definition of the score of a GCSM M is composed of two parts, a reasonable heuristic approach to locating high-scoring GCSMs is to select the sample partitions that maximize one part of the score – the hypergeometric score – for each possible segment, and for these partitions to calculate the combined score. For a given segment G' the calculation of $\max_{S' \subseteq S}[-\log(F((G' \times S'); C))]$ may be performed in $(O(|S|))$ time by ordering the samples according to decreasing number of positive entries, as described in the following pseudo code:

Algorithm 1 MHA - Max-Hypergeometric Algorithm.

Input: C, E, t - a significance threshold, l - maximum segment length.
Output: A list of high scoring GCSMs, L.
for all segments $G' \subset G$ of length $\leq l$ **do**
    For each sample $s_i \in S$ let $p_i = \#$ of positive entries in $C(G', s_i)$.
    Order the samples s.t. $p_{\pi(1)} \geq ... \geq p_{\pi(|S|)}$.
    $maxScore = \max_{1 \leq i < |S|} F((G', \{S_{\pi(1)}, ..., S_{\pi(i)}\}); C, E)$.
    if $maxScore > t$ **then**
        Add $M = (G', S')$ to L.

Consistent Correlation Algorithm. One shortcoming of MHA is that it depends on a sufficiently strong pattern in the DCN measurements alone in order to detect high-scoring GCSMs. However, in Section 3 we argued that in some cases significant correlation between DCN and GE patterns is indicative of a chromosomal aberration even when the DCN signal *per se* is weak. The second algorithmic approach relies on DCN -GE correlations for locating candidate partitions S'. To this end, we make use of a helpful attribute of the MDP correlation score. Recall from Section 2.2 that for a given gene g_i the score $\text{MDP}(i)$ defines a cross-threshold t for which the product $A_t \cdot D_t$ is maximized (see Figure 2). It is therefore straightforward to separate the samples that contribute to the score $\text{MDP}(i)$ – those within A_t or D_t – from those that do not (within B_t or C_t). Taking into consideration the chromosomal neighborhood of g_i we can increase our confidence that g_i's expression level in a specific sample is affected by the aberration. Consider, for example, a sample s that falls in D_t when computing $\text{MDP}(i,j)$ for $E(i)$ and all relevant $C(j)$. The probability of such an event occurring at random decreases exponentially with k.

For a gene g_i and a sample $s \in S$ we define the *sample MDP score* of s as:

$$\text{SMDP}(s,i) = \frac{1}{2k+1} \sum_{j=i-k}^{i+k} \left\{ \left[1_{s \in A_t(i,j)} \text{MDP}(i,j) \right] - \left[1_{s \in D_t(i,j)} \text{MDP}(i,j) \right] \right\},$$

where $A_t(i,j)$ and $D_t(i,j)$ are the sets of samples that fall into quadrants A_t and D_t for the threshold t that attains the MDP for $E(i)$ and $C(j)$. Note that $-\text{MDP}(i, \Gamma_k(i)) \leq \text{SMDP}(s,i) \leq \text{MDP}(i, \Gamma_k(i))$ and extrema are attained if s falls in either A_t or D_t in all of the crosses.

The above method allows us to rank the samples $s_i \in S$ according to increasing odds that they have been affected by an amplification. As before, this ranking suggests $O(|S|)$ partitions to be evaluated. In practice, we may opt to run the algorithm on a filtered set of genes $\tilde{G} \subset G$ that pass some minimal regional correlation threshold, in accordance with the statistics mentioned in Section 3.

Algorithm 2 CCA - Consistent Correlation Algorithm.

Input: C, E, $\tilde{G}$ - a subset of genes, k - the neighborhood size, t - a significance threshold, l - maximum segment length.
Output: A list of high scoring GCSMs, L.
for all genes $g_i \in \tilde{G}$ **do**
    For each sample $s_j \in S$ calculate $p_i = \text{SMDP}(s_j, i)$.
    Order the samples s.t. $p_{\pi(1)} \geq ... \geq p_{\pi(|S|)}$.
    for all segments $G' \subset G$ of length $\leq l$ s.t. $g_i \in G'$ **do**
        $maxScore = \max_{1 \leq i < |S|} F((G', \{S_{\pi(1)}, ..., S_{\pi(i)}\}); C, E)$.
        if $maxScore > t$ **then**
            Add $M = (G', S')$ to L.

Analysis. Note that the two algorithms are appropriate for two different types of high-scoring GCSMs. MHA is optimal when $F(M; C)$ is a dominant factor of the total score, i.e. when the DCN measurements alone point to a chromosomal aberration. CCA is appropriate when there is a strong correlation between $E(M)$ and $C(M)$ suggesting that both $F(M;C)$ and $F(M;E)$ have a significant part in the total score. In the latter case we expect that a chromosomal alteration has a significant effect on transcriptional activity. A third extremal case, for which neither algorithm is appropriate, is the case in which $F(M;E)$ alone is a dominant factor of the total score. A biological interpretation of this event is co-regulation of neighboring genes, unlinked to chromosomal aberration (as in an operon). This type of effect is not in the scope of this study and may be overlooked by both algorithms.

4.3 Results

We applied both algorithms to detect high-scoring GCSMs in the joint DCN-GE breast tumor data of Pollack et al [14], using a threshold of $F(M;C,E) > 25$. A considerable number of significant GCSMs was detected both by MHA and CCA spanning genomic segments of 5-52 loci, or 0.25-4Mbp. The two algorithms locate similar but not identical high-scoring GCSMs. MHA produced no GCSMs with scores above the given threshold for either randomly permuted data or the same dataset where the genomic order was randomly permuted, verifying that the high scores attained were due to regional phenomena (see Figure 3). A distribution with somewhat lower values is obtained when MHA is run on a the dataset where only the genomic order of the expression vectors was randomly permuted, suggesting that the DCN matrix on its own is accountable for a large fraction of the high scores obtained for the complete data. CCA does not produce any results on random data since, by definition, it initiates a search only when the regional correlation score is statistically significant. Genomic aberrations were located in various chromosomal segments, including 1p, 1q, 3p, 8p, 8q, 13q, 17q and 20q reported in [14] and in several additional locations. These include a GCSM with $F(M) = 60.3$ in 17q (28-32Mbp), and an a GCSM with $F(M) = 50.2$ in 11q (69-73Mbp). Genes affected by altered GCSMs include TP53, FGFR1 and ERBB2, known to be involved in breast cancer (see supplement [1] for a complete list of high-scoring GCSMs, and some affected genes). Similar results were obtained for the joint DCN-GE breast cancer cell-line dataset of Hyman et al [7]. Our method validates the novel alterations in 9p13 (GCSM score 27.1) and in 17q21.3 (GCSM score 37). The same aberrations were located in a subset of the samples of [14], with GCSM scores of 32.5 and 53 respectively. Figure 4 depicts a significant alteration in 17q11 that is significantly associated with altered expression levels of 5 resident genes (data from [14]).

5 Conclusion

The advanced stage of expression profiling technologies, coupled with continued development of aCGH and other technologies for measuring DCN alterations enable a more comprehensive study of the molecular profile of cancer. Specifically, these technology advances enable the generation of joint data sets where

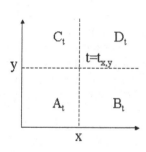

Fig. 2. Notation for separating crosses.

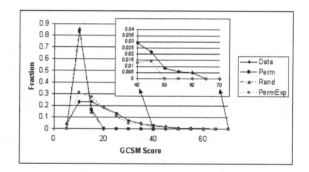

Fig. 3. Distribution of GCSM scores obtained by MHA on data of [14] (Data), on randomly permuted data (Rand), and on randomly permuted gene order (both DCN and GE - Perm; only GE - PermExp).

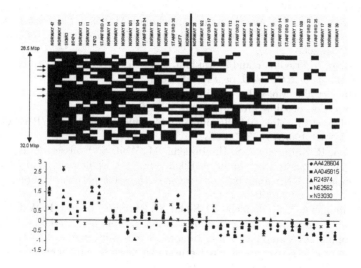

Fig. 4. A significantly amplified GCSM identified in 17q11 in breast tumor data [14] (score of 60.3). Each column represents a sample. Samples are partitioned into two subsets (S' and $S - S'$) by a vertical line. Top panel depicts the DCN submatrix ($C(M)$) for the respective genomic segment (28.5–32.0 Mb), where dark entries represent positive values. Lower panel depicts GE profiles of 5 resident genes (respective rows in $E(M)$) that are significantly differentially expressed between S' and $S - S'$. The positions of these genes in the genomic segment are indicated by arrows in the top panel.

DCN and GE values are measured on the same set of samples and genes. Using these data to understand cancer pathogenesis mechanisms related to alterations and their transcriptional effects will require the development of adequate data analysis methodology and tools. In this paper we provide a first statistical and algorithmic treatment of this methodology.

We describe methods for assessing the correlation of DCN to GE for single genes, taking into account regional effects. We also develop a statisitical and algorithmic framework for identifying altered regions with a correlated transcriptional effect. Applying our methods to published breast cancer data we provide stronge evidence of the transcriptional effects of altered DCN and identify chromosomal segments and sample subsets where these effects are more pronounced. Specifically, we identify regions where a statistically significant alteration affects more than one resident gene.

In future work we intend to improve the algorithmics and to further develop rigorous statistics for biologically meaningful structures in DCN-GE data.

References

1. Supplement data available at http://bioinfo.cs.technion.ac.il/cghexp/.
2. B.R. Balsara and J.R. Testa. Chromosomal imbalances in human lung cancer. *Oncogene*, 21(45):6877–83, 2002.
3. A. Ben-Dor, L. Bruhn, N. Friedman, I. Nachman, M. Schummer, and Z. Yakhini. Tissue classification with gene expression profiles. In *Proceedings of RECOMB*, pages 54–64, 2000.
4. A. Ben-Dor, N. Friedman, and Z. Yakhini. Class discovery in gene expression data. In *Proceedings of RECOMB*, pages 31–8, 2001.
5. M. Bittner et al.. Molecular classification of cutaneous malignant melanoma by gene expression profiling. *Nature*, 406(6795):536–40, 2000.
6. I. Hedenfalk et al.. Molecular classification of familial non-BRCA1/BRCA2 breast cancer. *PNAS*, 100(5):2532–7, 2003.
7. E. Hyman, P. Kauraniemi, S. Hautaniemi, M. Wolf, S. Mousses, E. Rozenblum, M. Ringner, G. Sauter, O. Monni, A. Elkahloun, O.P. Kallioniemi, and A. Kallioniemi. Impact of DNA amplification on gene expression patterns in breast cancer. *Cancer Research*, 62:6240–5, 2002.
8. O.P. Kallioniemi, A. Kallioniemi, D. Sudar, D. Rutovitz, J. Gray, F. Waldman, and D. Pinkel. Comparative genomic hybridization: a rapid new method for detecting and mapping DNA amplification in tumors. *Semin Cancer Biol*, 4(1):41–46, 1993.
9. S.C. Linn et al.. Gene expression patterns and gene copy number changes in DFSP. *Amer J of Pathology*, 163(6):2383–2395, 2003.
10. F. Mertens, B. Johansson, M. Hoglund, and F. Mitelman. Chromosomal imbalance maps of malignant solid tumors: a cytogenetic survey of 3185 neoplasms. *Cancer Research*, 57(13):2765–80, 1997.
11. D. Pinkel et al.. High resolution analysis of DNA copy number variation using comparative genomic hybridization to microarrays. *Nat Gen*, 20(2):207–211, 1998.
12. P. Platzer et al.. Silence of chromosomal amplifications in colon cancer. *Cancer Research*, 62(4):1134–8, 2002.
13. J.R. Pollack, C.M. Perou, A.A. Alizadeh, M.B. Eisen, A. Pergamenschikov, C.F. Williams, S.S. Jeffrey, D. Botstein, and P.O. Brown. Genome-wide analysis of DNA copy-number changes using cDNA microarrays. *Nature Genetics*, 23(1):41–6, 1999.
14. J.R. Pollack et al.. Microarray analysis reveals a major direct role of DNA copy number alteration in the transcriptional program of human breast tumors. *PNAS*, 99(20):12963–8, 2002.

Searching for Regulatory Elements of Alternative Splicing Events Using Phylogenetic Footprinting

Daichi Shigemizu[1] and Osamu Maruyama[2]

[1] Graduate School of Systems Life Sciences, Kyushu University
Hakozaki 6-10-1, Higashi-ku, Fukuoka 812-8581, Japan
3s103005y@sls.kyushu-u.ac.jp
[2] Faculty of Mathematics, Kyushu University
Hakozaki 6-10-1, Higashi-ku, Fukuoka 812-8581, Japan
om@math.kyushu-u.ac.jp

Abstract. We consider the problem of finding candidates for regulatory elements of alternative splicing events from orthologous genes, using phylogenetic footprinting. The problem is formulated as follows: We are given orthologous sequences $P_1, \ldots, P_a$ and $N_1, \ldots, N_b$ from $a + b$ different species, and a phylogenetic tree relating these species. Assume that for $i = 1, \ldots, a$, P_i is known to have some alternative splicing events, although N_j does not have any alternative splicing events. Our objective is to find all sets of substrings $s_1, \ldots, s_a$ of $P_1, \ldots, P_a$ such that $s_1, \ldots, s_a$ are similar to each other to some extent, and such that any substrings of $N_1, \ldots, N_b$ are not similar to $s_1, \ldots, s_a$. To this aim, we have modified the phylogenetic footprinting algorithm given by Blanchette et al. to solve our problem. We report the results of our preliminary computational experiments on several sets of orthologous genes of the five species, *H.sapiens*, *M.musculus*, *D.melanogaster*, *C.elegans*, and *A.thaliana*. It is interesting that many of the substrings selected by our algorithm from the coding sequences of *H.sapiens* are substrings in the intronic sequences flanking the alternatively spliced exons of the coding sequence. This result implies that regulatory elements of alternative splicing events would be located in intronic sequences flanking the alternatively spliced exons.

1 Introduction

One of the most important problems in computational biology is finding regulatory elements of *alternative splicing events* [1, 2]. Alternative splicing is a mechanism of producing different proteins from the same ORF by splicing alternative exons on the ORF. It is estimated that up to 60% of all human genes are alternatively spliced [3], and alternative splicing is recognized to be possibly one of the major mechanisms causing the variety of proteome. Great efforts have been made to compile databases, including information on identification, annotation and prediction of alternatively spliced genes and their alternatively spliced isoforms. For example, the following databases are available: Alternative Splicing

I. Jonassen and J. Kim (Eds.): WABI 2004, LNBI 3240, pp. 147–158, 2004.

Database (ASDB) [4], Alternative Splicing Database of Mammals (AsMamDB) [5], SpliceDB [6], Putative Alternative Splicing Database (PALS db) [7], Intron Information System (SIS) [8], Alternative Splicing Annotation Project (ASAP) [9], Manually Annotated Alternatively Spliced Events (MAASE) [3]. Thus, it is a quite interesting issue to develop a computational method for finding convincing candidates for regulatory elements of alternative splicing events from information on alternative splicing in available databases.

Blanchette *et al.* [10] has formulated a motif-finding problem, called the *substring parsimony problem*, whose instance is orthologous sequences from n different species and a phylogenetic tree relating the species. They have developed a dynamic programming algorithm to solve the problem. The software of the algorithm is called *FootPrinter* [11] and available at http://bio.cs.washington. edu /software.html. We refer the algorithm in [10] as the footprinter algorithm. The algorithm is designed based on the idea of *phylogenetic footprinting* [12], a technique that identifies regulatory elements by finding unusually well conserved regions in a set of orthologous noncoding DNA sequences from multiple species. This is based on the observation that functional sequences tend to evolve much more slowly than nonfunctional sequences as they are subject to selective pressure. Phylogenetic footprinting exploits this difference in mutation rates.

Assume that we are given orthologous sequences $P_1, \ldots, P_a$ and $N_1, \ldots, N_b$ from $a + b$ different species, and a phylogenetic tree relating these species, and that for $i = 1, \ldots, a$, P_i is known to have some alternative splicing events, although N_j does not have any alternative splicing events. P_i and N_j are said to be *positive* and *negative*, respectively. It would be an interesting approach to find sets of substrings $s_1, \ldots, s_a$ of $P_1, \ldots, P_a$, using the phylogenetic footprinting technique, such that $s_1, \ldots, s_a$ are similar to each other to some extent, and such that each of the substrings of $N_1, \ldots, N_b$ are not similar to any of the substrings $s_1, \ldots, s_a$. If exist, $s_1, \ldots, s_a$ are good candidates for regulatory elements of alternative splicing events.

However, the footprinter algorithm is not designed to cope with the case where orthologous sequences are categorized into two groups. In this paper, we have formulated a two-categorical version of the substring parsimony problem and modified the footprinter algorithm for solving the new problem. We have also carried out preliminary computational experiments on orthologous genes of *H.sapiens*, *M.musculus*, *D.melanogaster*, *C.elegans*, and *A.thaliana*. It is interesting that many of the substrings selected from the coding sequences of *H.sapiens* are substrings in the intronic sequences flanking the alternatively spliced exons of the coding sequence, which implies that regulatory elements of alternative splicing would be located in the intronic sequences flanking alternatively spliced exons. This result agrees with the observation in [13] that intronic sequences flanking alternatively spliced exons are conserved between human and mouse.

This paper is organized as follows: In Section 2, we describe the formulation of our problem. In the next section, we explain how to modify the footprinter algorithm for solving our problem. In Section 4, we give the results of our computational experiments on the five species.

2   Problem Formulation

Before describing our problem, the substring parsimony problem [10] is explained, since our problem is derived from the problem. Let $\Sigma = \{A, C, G, T\}$, and let k be a positive integer. For strings s and t of the same length, the Hamming distance between s and t is denoted by $d(s, t)$. Let $S_1, \ldots, S_n$ be a set of orthologous sequences from n different species. We define a phylogenetic tree T relating these n species as follows: A phylogenetic tree $T = (V, E, \phi)$ be a rooted tree where V is the set of nodes, E is the set of edges, and ϕ is a function from the leaves of T to the power set of Σ^k. The n leaves of T are denoted by $l_1, \ldots, l_n$. For the leaf l_i of the i-th species, $\phi(l_i)$ is set to the set of all substring of length k in S_i. For a tree $T' = (V', E', \phi')$, V' and E' are also denoted by $V(T')$ and $E(T')$, respectively. For a subset W of V, a *node-labeling function of W over Σ^k* is a function $f : W \to \Sigma^k$ such that if $v \in W$ is a leaf of T then $f(v)$ should be an element of $\phi(v)$. We denote the set of all node-labeling functions of W over Σ^k by F_W. Blanchette *et al.* [10] have formulated the *substring parsimony problem*, which is the problem of finding all the sets of substrings $s_1, \ldots, s_n$ of $S_1, \ldots, S_n$, respectively, such that the sum, over all edges e, of the Hamming distance between the labels of the nodes connected by e is at most a threshold α. Formally, this problem can be stated as follows:

Substring Parsimony Problem
Instance: a phylogenetic tree $T = (V, E, \phi)$ relating n different species with orthologous sequences $S_1, \ldots, S_n$, and integers k and α.
Problem: find the following set:

$$\left\{ (f(l_1), f(l_2), \ldots, f(l_n)) \;\middle|\; f \in F_{V(T)} \text{ and } \sum_{(u,v) \in E(T)} d(f(u), f(v)) \leq \alpha \right\},$$

where $l_1, l_2, \ldots, l_n$ are the leaves of T.

In [10], the *parsimony score* of a set of substrings $s_1, \ldots, s_n$ of $S_1, \ldots, S_n$, respectively, in T is defined as the minimum of $\sum_{(u,v) \in E(T)} d(f(u), f(v))$ over all the node-labeling functions f in $F_{V(T)}$ such that $f(l_i) = s_i$ for $i = 1, \ldots, n$. Note that this problem does not require that T should be rooted. However, this requirement should be satisfied in our new problem, which we will describe soon, because the solutions depend on what node is the root.

Here, we consider the case where given orthologous sequences are categorized into two classes: *positive* sequences $P_1, \ldots, P_a$ and *negative* sequences $N_1, \ldots, N_b$, where $a + b = n$. If a sequence is positive then it has some alternative splicing events. If a sequence is negative then it does not have any alternative splicing events. To embed this information in a phylogenetic tree, we modify it as follows: A phylogenetic tree $T = (V, E, \phi)$ is extended into $T = (V, E, \phi, \lambda)$ where λ is a function from the leaves in V to $\{+1, -1\}$. A leaf v is said to be *positive* if $\lambda(v) = +1$, and *negative* if $\lambda(v) = -1$. The subtree rooted at a node v of T is the subtree consisting of v and its descendants, denoted by T_v. A subtree T_v is called *negative* if all the leaves of T_v are negative. For a non-root node v, the parent of v is denoted by $p(v)$. A negative subtree T_v is called *maximal* if $T_{p(v)}$, the subtree rooted at $p(v)$, is not negative. The set of the roots of the maximal negative subtrees in T is denoted by $\Omega = \{\omega_1, \omega_2, \ldots, \omega_m\}$. Fig. 1 shows an example of Ω.

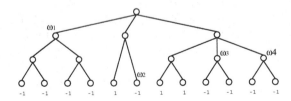

Fig. 1. An example of Ω. The integers attached to the leaves v are the values of $\lambda(v)$. We can see that this tree has four maximal negative subtrees, whose root nodes are denoted by $\omega_1, \omega_2, \omega_3$, and ω_4.

Let α' be a nonnegative integer. For each maximal negative subtree with the root ω_i, we construct a set of root labels $f(\omega_i)$ such that the sum, over all edges e of the subtree, of the Hamming distance between the labels of the nodes connected by e is at most α'. We denote this set by L_i. For each $\omega_i \in \Omega$, L_i is defined as follows:

$$L_i = \left\{ f(\omega_i) \,\middle|\, f \in F_{V(T_{\omega_i})} \text{ and } \sum_{(u,v) \in E(T_{\omega_i})} d(f(u), f(v)) \leq \alpha' \right\}.$$

We can consider the strings in L_i to be *representative* strings of the orthologous sequences assigned to the leaves of the maximal negative subtree.

Next, we modify T into a new tree $T_{\alpha'}$, using $L_1, \ldots, L_m$ and subtrees $T_{\omega_1}, \ldots, T_{\omega_m}$, which is defined as follows: For each $i = 1, \ldots, m$, T_{ω_i} is removed from $T = (V, E, \phi, \lambda)$ except ω_i, and ϕ and λ are modified such that $\phi(\omega_i) = L_i$ and $\lambda(\omega_i) = -1$. Notice that ω_i is changed from the root of a maximal negative subtree of T to a leaf of $T_{\alpha'}$. These leaves $\omega_1, \ldots, \omega_m$ are set to be negative, and the other leaves of $T_{\alpha'}$ remain positive.

Now we can formulate our problem. Informally speaking, the main difference between the substring parsimony problem and our problem, called the *binary substring parsimony problem*, is that in our problem, L_i, a set of representative length k strings of the orthologous sequences assigned to the leaves in a maximal negative subtree of the whole tree T, is assigned to the root node ω_i of the subtree, and then the Hamming distance between a label of the parent node of ω_i and an arbitrary string in L_i should be greater than or equal to a threshold β. We adopted this rigid constraint on negative orthologous sequences, which corresponds to the third condition of the following formal statement of our problem, because this work is the first trial of searching for candidates for regulatory elements of alternatively splicing events using the footprinting technique. Thus the found substrings would be more convincing if exist.

Binary Substring Parsimony Problem
Instance: a tree $T = (V, E, \phi, \lambda)$, and non-negative integers α, α', β.
Problem: Find all $(f(l_1), f(l_2), \ldots, f(l_a)) \in (\Sigma^k)^a$ such that

1. $f \in F_{V(T_{\alpha'}) - \{\omega_1, \ldots, \omega_m\}}$,

2. $$\sum_{\substack{(u,v) \in E(T_{\alpha'}) \text{ s.t. } v \notin \Omega}} d(f(p(v)), f(v)) \leq \alpha,$$

3. $\min_{t \in L_i} d(f(p(\omega_i)), t) \geq \beta$, for each $\omega_i \in \Omega$,

where m is the number of the maximal negative subtrees of T, and $l_1, l_2, \ldots,$ l_a are all the leaves in T with $\lambda(l_i) = +1$.

3 Algorithms

In this section, we describe an algorithm to solve the binary substring parsimony problem. The algorithm uses the footprinter algorithm as a subroutine to find the set of feasible labels at the root node of maximal negative subtrees in the given phylogenetic tree. In addition, we use another dynamic programming algorithm, which is obtained by slightly modifying the footprinter algorithm.

1. Find all the maximal negative subtrees of T, denoted by $T_{\omega_1}, \ldots, T_{\omega_m}$.
2. For $i = 1, \ldots, m$, solve the substring parsimony problem whose instance consists of T_{ω_i}, the length k of labels to be searched, and the upper bound of parsimony scores, α', and determine L_i, the set of feasible labels at the root of T_{ω_i}, for each $i = 1, \ldots, m$.
3. Generate $T_{\alpha'}$ from T, $T_{\omega_1}, \ldots, T_{\omega_m}$ and $L_1, \ldots, L_m$.
4. Find all sets of substrings $s_1, \ldots, s_a$ of positive sequences $P_1, \ldots, P_a$ from $T_{\alpha'}$, satisfying the conditions 1, 2, and 3 of the binary substring parsimony problem.

Fig. 2. An overview of our procedure to solve the binary substring parsimony problem.

3.1 Overview

An overview of our procedure to solve the binary substring parsimony problem is given in Fig. 2. Notice that the task of Step 1 in Fig. 2 can be computed in $O(n)$ time. The task of Step 2 can be solved by applying the footprinter algorithm to each maximal negative subtree of T. The computation time for one maximal negative subtree is $O(\tilde{n} \cdot \min(lk(3k)^{\alpha'/2}, k(4^k + l))))$, where $\tilde{n}$ is the number of leaves in the maximal negative subtree, and l is the average length of input sequences. (It is assumed in [10] that a string of length k fits in a single computer word.) Thus, the total computation time of Step 2 is $O(b \cdot \min(lk(3k)^{\alpha'/2}, k(4^k + l))))$. Recall that b is the number of negative leaves of T. The task of Step 3 can be done in $O(m)$ time, and the task of Step 4 can be solved by a dynamic programming obtained by slightly modifying the footprinter algorithm, which is describe in the next subsection.

3.2 Dynamic Programming

It is not hard to see Step 4 in Fig. 2 can be solved by the following dynamic programming algorithm: Let u be a node in $V(T_{\alpha'})$.

$$
W_u[s] = \begin{cases}
0 & \text{if } u \text{ is a positive leaf and } s \in \phi(u), \\
+\infty & \text{if } u \text{ is a positive leaf and } s \notin \phi(u), \\
0 & \text{if } u \text{ is a negative leaf } \omega_i \text{ and } \min_{t \in L_i} d(s,t) \geq \beta, \\
+\infty & \text{if } u \text{ is a negative leaf } \omega_i \text{ and } \min_{t \in L_i} d(s,t) < \beta, \\
\displaystyle\sum_{v \in C(u)} X_{(u,v)}[s] & \text{otherwise,}
\end{cases}
$$

where $C(u)$ is the set of children of u, and

$$
X_{(u,v)}[s] = \begin{cases}
W_v[s] & \text{if } v \in \{\omega_1, \dots, \omega_m\}, \\
\min_{t \in \Sigma^k} (W_v[t] + d(s,t)) & \text{otherwise.}
\end{cases}
$$

The original of this dynamic programming algorithm is the footprinter algorithm. The difference between them is that in our algorithm, if the node u is a leaf, the returned value depends on whether u is positive or negative. In the same way as the footprinter algorithm, the node-labeling functions of the solutions can be recovered by tracing back the recurrence, from the root, denoted by r, down to the leaves, for each entry $s \in \Sigma^k$ of W_r such that $W_r[s] \leq \alpha$.

3.3 Bounding

A few techniques to improve the time complexity of the substring parsimony problem, which are called *d-bounding*, *sibling bounding*, and *parent bounding*, are shown in [10]. The technique of d-bounding is a divide-and-conquer method. Note that d is the upper bound of a parsimony score in [10]. This bounding technique is based on the following fact. Let u be an internal node of T. Consider the subtree rooted at u, denoted by T', a string $s \in \Sigma^k$, and an arbitrary node-labeling function f such that $f(u) = s$. We can see that if $\sum_{(u,v) \in E(T')} d(f(u), f(v)) > d$ then $\sum_{(u,v) \in E(T)} d(f(u), f(v)) > d$. Thus, f can be pruned whenever the condition holds, since this case never provide feasible solutions to the problem on T. Sibling bounding is also a divide-and-conquer method applied to the scores of siblings. These bounding techniques are applicable to our naive dynamic programming algorithm solving the binary substring parsimony problem.

However, parent bounding, which is also a divide-and-conquer method using the total score over all edges in the whole tree but the edges used in sibling bounding, can not be available to the binary substring parsimony problem. The reason is as follows: A requirement for using this bounding method is that the solutions should be independent of the node at which the tree is rooted. But, the solutions to the binary substring parsimony problem are not independent of the node at which the tree is rooted.

We then describe how to modify the naive dynamic programming algorithm using d-bounding and sibling bounding. For a non-negative integer p and an edge (u, v), a table $B^p_{(u,v)}$ is defined as the set of strings s of length k such that $X_{(u,v)}[s] = p$. We also define R^p_v as the set of strings of length k such that $W_v[s] = p$. If v is a leaf in $T_{\alpha'}$ with $\lambda(v) = -1$, we have

$$
B^p_{(u,v)} = \begin{cases} \left\{ s \in \Sigma^k \, \middle| \, \min_{t \in L_i} d(s, t) \geq \beta \right\} & \text{if } p = 0, \\ \emptyset & \text{otherwise.} \end{cases}
$$

Note that for each $s \in \Sigma^k$, it can be computable in $O(k \cdot l \cdot (3k)^{\alpha'/2})$ time whether s is in $B^0_{(u,v)}$ or not, because L_i can be shown to contain $O(l \cdot (3k)^{\alpha'/2})$ strings, which is derived from the discussion in Section 2.3 of [10]. For the other edges (u, v), we have $B^0_{(u,v)} = R^0_v$ and for $p \geq 0$,

$$
B^{p+1}_{(u,v)} = R^{p+1}_v \cup \bigcup_{t \in B^p_{(u,v)}} N(t) - \bigcup_{p-1 \leq j \leq p} B^j_{(u,v)},
$$

where $N(t) = \{s \in \Sigma^k \mid d(s, t) = 1\}$. This is d-bounding for the binary substring parsimony problem.

Sibling bounding is applicable to even a node having a sibling which is a negative leaf. We are also going to compute the entries of the $B_{(u,v^*)}$ tables for all $v^* \in C(u)$ in parallel, using the following bound: an entry s of $X_{(u,v)}[s] = p$ can be used as a seed of expansion if

$$
\alpha \geq p + \max_{\substack{w \in C(u) \\ w \neq v}} \begin{cases} X_{(u,v)}[s] & \text{if } X_{(u,v)}[s] \text{ has been computed,} \\ p + 1 & \text{otherwise.} \end{cases}
$$

Notice that this is formally the same as the rule for sibling bounding in [10].

The total running time is $O(b \cdot \min(lk(3k)^{\alpha'/2}, k(4^k + l))) + O(m 4^k lk(3k)^{\alpha'/2}) + O((a+m) \cdot \min(lk(3k)^{\alpha/2}, k(4^k + l)))$. If $\alpha = \alpha'$, it turns to be $O(m 4^k lk(3k)^{\alpha/2} + n \cdot \min(lk(3k)^{\alpha/2}, k(4^k + l)))$, which is just slightly larger than the running time of the footprinter algorithm.

4   Experiments

In this section, we report the results of our preliminary computational experiments using the algorithm we describe in the previous section.

Recall that an instance of the algorithm consists of the followings: (i) a set of species, (ii) a phylogenetic tree relating the species, (iii) a set of orthologous genes of the species, (iv) for each of the orthologous genes, information on whether alternative splicing events of the gene exist or not. In this work, we use the five species, $H.sapiens$, $M.musculus$, $D.melanogaster$, $C.elegans$, and $A.thaliana$. In SpliceDB[4] it can be recognized that, each of the five species has at least one gene that has alternative splicing events. A phylogenetic tree relating these five species is generated by ClustalW[14] with 18SrRNAs of the species. The result is given in Fig. 3. To find sets of orthologous genes among these species, we have used the InParanoid database [15] (http://inparanoid.cgb.ki.se/index.html), which is a database of pairwise orthologous sequences among 7 species, including the five species we use here, and *E.coli* and *S.cerevisiae*. The coding sequences of the species are obtained through the web site of Ensembl [16] (http://www.ensembl.org/) except *A.thaliana*. The coding sequences of *A.thaliana* can be obtained through Entrez Gene at http://www.ncbi.nlm.nih.gov/. Whether a gene has some alternative splicing events or not has been determined using the UniProt database [17] (http://www.expasy.uniprot.org/). If there is the key word "alternative splicing" in the section of Alternative Products of the gene, the gene is determined to have some alternative splicing events. Otherwise, the gene is temporarily determined not to have any alternative splicing events.

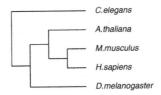

Fig. 3. The phylogenetic tree relating *H.sapiens*, *M.musculus*, *D.melanogaster*, *C.elegans*, and *A.thaliana*.

Table 1. A set of orthologous sequences related to Arginyltransferase 1. The column labeled with λ shows the existence of alternative splicing events in each genes. If the value is +1 then the gene has alternative splicing, and otherwise it does not.

Species	Swiss-Prot ID	λ
H.sapiens	ATE1_HUMAN	+1
M.musculus	ATE1_MOUSE	+1
D.melanogaster	ATE1_DROME	+1
C.elegans	P90914	-1
A.thaliana	Q9C776, ATE1_ARATH	-1

4.1   Arginyltransferase 1

The first set of orthologous sequences of the five species is related to arginyltransferase 1. The Swiss-Prot IDs of the sequences and the information on the existence of alternative splicing events for each of them are given in Tab. 1. Notice that *A.thaliana* has two orthologous sequences, Q9C776 and ATE1_ARATH, to the others in InParanoid [15]. Thus, we have executed our program on two different sets of orthologous sequences, where one includes only Q9C776 and another includes only ATE1_ARATH. We have searched for good node-labeling functions with several different parameter sets. A summary of the result in the case where Q9C776 is included in the orthologous sequence set, is given in Tab. 2,

Table 2. The column labeled with # shows the number of feasible labels at the root node for each parameter set.

$(k, \alpha, \alpha', \beta)$	#	$(k, \alpha, \alpha', \beta)$	#	$(k, \alpha, \alpha', \beta)$	#	$(k, \alpha, \alpha', \beta)$	#
(5, 0, 0, 1)	0	(6, 0, 0, 1)	7	(7, 0, 0, 1)	157	(7, 1, 0, 2)	0
(5, 0, 0, 2)	0	(6, 0, 0, 2)	0	(7, 0, 0, 2)	0		
(5, 0, 0, 3)	0	(6, 0, 0, 3)	0				

which shows the parameter sets with which we have executed our program and the number of feasible labels at the root node.

The 7 feasible labels at the root node in the result on the parameter set $(k, \alpha, \alpha', \beta) = (6, 0, 0, 1)$ is as follows: CCCCAA, CCGTGT, GACGGT, GAGGCC, GCGTGC, GCTAGG, and GGTCCG. Note that GCTAGG is overlapped with CUGCUA and UGCUA, which are highly over-represented hexamer and pentamer sequences in the downstream intron regions adjacent to the brain-specific alternative exons [1]. In addition, CCCCAA and GAGGCC are substrings of 3' splice sites of 25 brain specific alternative exons used in [1]. It is quite interesting that, although our approach and data are different from those in [1], similar sequences are found.

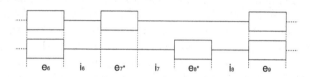

Fig. 4. Known isoforms of ATE1_HUMAN. The type of alternative splicing is mutually exclusive. The j-th region occupied by at least one exon in the coding sequence is denoted by e_j. If e_j is included in an alternative exon, an asterisk is attached to e_j. The exonic regions $e_1, \ldots, e_5, e_{10}, \ldots, e_{13}$ are omitted from this figure, which are constitutive exons. The remaining regions are introns, which are denoted by i_j.

We have done a further analysis to count the occurrences of each of the 7 hexamer sequences in the exonic and intronic regions of the coding sequence of *H.sapiens*. The information of the known isoforms of the sequence, given in Fig. 4, is available in the database Ensemble [16]. Notice that these hexamer sequences are also substrings of the coding sequences of the five genes because α is set to zero. Tab. 3 shows the number of occurrences of each of the 7 hexamer sequences in the exons e_j and introns i_j. Interestingly, most of the occurrences are located in the introns. Furthermore, i_8, which flanks e_8^*, has many occurrences. Recall that e_8^* is an alternatively splice exon. Thus, regulatory elements of alternative splicing would be located in the intronic sequences flanking alternatively spliced exons. In addition, this result agrees with the observation shown in [13] that intronic sequences flanking alternatively spliced exons are conserved between human and mouse. Therefore, some of the 7 hexamer sequences would be regulatory elements of alternatively splicing of the gene ATE1. Furthermore, e_9, e_{11}, and e_{12} are currently constitutive, but they might be actually alterna-

Table 3. The column labeled e_j (i_j) shows the number of the occurrences of each of the 7 pentamer sequences in the region e_j (i_j). The column labeled U (D) is the last (first) 200 bases of the upstream (downstream) intronic sequence flanking this coding regions. If the value of a cell is zero, it is omitted.

	U	e_1	i_1	e_2	i_2	e_3	i_3	e_4	i_4	e_5	i_5	e_6	i_6	e_7^*	i_7	e_8^*	i_8	e_9	i_9	e_{10}	i_{10}	e_{11}	i_{11}	e_{12}	i_{12}	e_{13}	D
GCTAGG					2		1		1								9		7				9		6	1	
CCGTGT		1					3				4						3		4				2		2		
GCGTGC							1		1		2						2								2		
CCCCAA					1		3		1		2						2		9				8		8		
GGTCCG									1								2										
GAGGCC	1						1		1		4		3				8		5				16		11		
GACGGT																	2						2		1		
total	1	1			3		9		5		12		3				28		25				37		30	1	

tive exons, since the introns i_9, i_{11}, and i_{12}, flanking e_9, e_{11}, and e_{12}, respectively, have relatively many occurrences. If ATE1_ARATH is used instead of Q9C776, the results are quite similar to the previous case.

4.2   Protein Arginine N-Methyltransferase 1

The second set of orthologous sequences of the five species is related to protein arginine N-methyltransferase 1. The Swiss-Prot IDs of *H.sapiens*, *M.musculus*, *D.melanogaster*, and *C.elegans* are ANM1_HUMAN, ANM1_MOUSE, Q9VGW7, and Q9U2X0. *A.thaliana* has three entries in Swiss-Prot which are orthologous to the others. They are O82210, Q9SU94, and O81813. We have obtained results on three different sets of orthologous sequences with various parameter sets. For example, the feasible labels at the root node for the orthologous sequence set including O82210 with the parameter set $(k, \alpha, \alpha', \beta) = (5, 0, 0, 1)$ is as follows: AACGC, ACACC, ACGCC, ACGCG, AGACC, AGGAC, CCAGA*, CCCCC, CCGAC, CGCAG, CTGAT*, GACAG, GACCC*, GAGGC*, GAGGG, GGACC*, GGATG, GGGCC, GGGGT, GGGTA*, GGGTG, TAAGT*, TACGC, TATTC*, TCCCC*, and TGAGG*. An asterisk indicates that the sequence with it is a substring of 5' or 3' splice sites of the 25 brain specific alternative exons used in [1]. Thus, possibly, alternative exons would have specific patterns of 5' and 3' splice sites. We need further analysis on this issue. We have also counted the occurrences of each of the pentamer sequences in each of the exons and introns of the orthologous sequence of *H.sapiens*. Three isoforms of the gene of ANM1_HUMAN are known, given in Fig. 5. Most of the occurrences are also located in the introns, as shown in the previous result. Furthermore, it is interesting that the intron i_1, which flanks the two alternative exons e_2^* and e_3^*, has the most occurrences (21%) of the pentamer sequences (data not shown). Thus, this result also supports the hypothesis that the intronic sequences flanking an alternatively spliced exon have regulatory elements of alternative splicing.

4.3   Trifunctional Purine Biosynthetic Protein Adenosine-3

The third set of orthologous sequences of the five species is related to trifunctional purine biosynthetic protein adenosine-3. The Swiss-Prot IDs of *H.sapiens*,

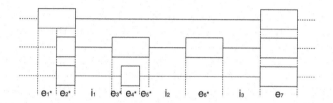

Fig. 5. Known isoforms of ANM1_HUMAN. The exonic regions $e_8, \ldots, e_{15}$ are omitted from this figure, which are constitutive exons.

M.musculus, *D.melanogaster*, *C.elegans*, and *A.thaliana* are PUR2_HUMAN, PUR2_MOUSE, Q9VM53, Q20143, and PUR2_ARATH, respectively. PUR2_HUMAN and PUR2_MOUSE are known to have alternative splicing events, and the others are not. The feasible label at the root node for the orthologous sequence set with the parameter $(k, \alpha, \alpha', \beta) = (5, 0, 0, 1)$ is only CCCCT. As in the previous results, the upstream intron flanking an alternative exon having alternative 3' splice sites in the coding sequence of PUR2_HUMAN is one of the regions having the most occurrences of the pentamer sequence (data not shown).

5 Concluding Remarks

Through our preliminary computational experiments, we have attained the hypothesis that regulatory elements of alternative splicing are located in the intronic sequences flanking alternatively spliced exons. Currently, we are planning to execute computational experiments on all the sets of orthologous genes of several species. Through such an exhaustive search, we can expect to detect convincing regulatory elements of alternative splicing events.

Acknowledgments

We thank Hideki Hirakawa for his valuable advice in arranging biological data. We also thank Akiko Matsuda for her help in generating phylogenetic trees. This work was supported in part by Research for the Future Program of JSPS, and Grant-in-Aid for Scientific Research on Priority Areas (C) and Young Scientists (B) of MEXT.

References

1. Brudno, M., Gelfand, M., Spengler, S., Zorn, M., Dubchak, I., Conboy, J.G.: Computational analysis of candidate intron regulatory elements for tissue-specific alternative pre-mRNA splicing. Nucl. Acids Res. **29** (2001) 2338–2348
2. Sakai, H., Maruyama, O.: Extensive search for discriminative features of alternative splicing. In: Pacific Symposium on Biocomputing 9. (2004) 54–65

3. Zheng, C., Nair, T., Gribskov, M., Kwon, Y., Li, H., , Fu, X.D.: A database designed to computationally aid an experimental approach to alternative splicing. In: Pacific Symposium on Biocomputing 9. (2004) 78–88
4. Dralyuk, I., Brudno, M., Gelfand, M., Zorn, M., Dubchak, I.: ASDB: database of alternatively spliced genes. Nucl. Acids Res. **28** (2000) 296–297
5. Ji, H., Zhou, Q., Wen, F., Xia, H., Lu, X., Li, Y.: AsMamDB: an alternative splice database of mammals. Nucl. Acids Res. **29** (2001) 260–263
6. Burset, M., Solovyev, I.S.V.: SpliceDB: database of canonical and non-canonical mammalian splice sites. Nucl. Acids Res. **29** (2001) 255–259
7. Brett, D., Hanke, J., Lehmann, G., Haase, S., Delbruck, S., Krueger, S., Bork, J.R.P.: EST comparison indicates 38% of human mRNAs contain possible alternative splice forms. FEBS Lett. **474** (2000) 83–86
8. Croft, L., Schandorff, S., Clark, F., Burrage, K., Arctander, P., Mattick, J.: ISIS, the intron information system, reveals the high frequency of alternative splicing in the human genome. Nat. Genet. **24** (2000) 340–341
9. Lee, C., Atanelov, L., Modrek, B., Xing, Y.: ASAP: the alternative splicing annotation project. Nucl. Acids Res. **31** (2003) 101–105
10. Blanchette, M., Schwikowski, B., Tompa, M.: Algorithms for phylogenetic footprinting. J. Comput. Biol. **9** (2002) 211–223
11. Blanchette, M., Tompa, M.: FootPrinter: a program designed for phylogenetic footprinting. Nucl. Acids Res. **31** (2003) 3840–3842
12. Tagle, D., Koop, B., Slightom, M.G.J., Hess, D., Jones, R.: Embryonic epsilon and gamma globin genes of a prosimian primate (Galago crassicaudatus) nucleotide and amino acid sequences, developmental regulation and phylogenetic footprints. J. Mol. Biol. **203** (1988) 439–455
13. Sorek, R., Ast, G.: Intronic sequences flanking alternatively spliced exons are conserved between human and mouse. Genome Research **13** (2003) 1631–1637
14. Thompson, J., Higgins, D., Gibson, T.: CLUSTAL W: improving the sensitivity of progressive multiple sequence alignment through sequence weighting, positions-specific gap penalties and weight matrix choice. Nucl. Acids Res. **22** (1994) 4673–4680
15. Remm, M., Storm, C., Sonnhammer, E.: Automatic clustering of orthologs and in-paralogs from pairwise species comparisons. J. Mol. Biol. **314** (2001) 1041–1052
16. Birney, E., Andrews, D., Bevan, P., Caccamo, M., Cameron, G., Chen, Y., Clarke, L., Coates, G., Cox, T., Cuff, J., Curwen, V., Cutts, T., Down, T., Durbin, R., Eyras, E., Fernandez-Suarez, X.M., Gane, P., Gibbins, B., Gilbert, J., Hammond, M., Hotz, H., Iyer, V., Kahari, A., Jekosch, K., Kasprzyk, A., Keefe, D., Keenan, S., Lehvaslaiho, H., McVicker, G., Melsopp, C., Meidl, P., Mongin, E., Pettett, R., Potter, S., Proctor, G., Rae, M., Searle, S., Slater, G., Smedley, D., Smith, J., Spooner, W., Stabenau, A., Stalker, J., Storey, R., Ureta-Vidal, A., Woodwark, C., Clamp, M., Hubbard, T.: Ensembl 2004. Nucl. Acids Res. (2004) D468–D470
17. Apweiler, R., Bairoch, A., Wu, C., Barker, W., Boeckmann, B., Ferro, S., Gasteiger, E., Huang, H., Lopez, R., Magrane, M., Martin, M., Natale, D., Yeh, C.O.N.R.L.: UniProt: the universal protein knowledgebase. Nucl. Acids Res. **32** (2004) D115–D119

Supervised Learning-Aided Optimization of Expert-Driven Functional Protein Sequence Annotation

Lev Soinov[1], Alexander Kanapin[2], and Misha Kapushesky[3]

[1] Algorithms and methods
EBI, Wellcome Trust Genome Campus, Hinxton, Cambridge, CB10 1SD, UK
[2] InterPro and SwissProt data retrieval and encoding
EBI, Wellcome Trust Genome Campus, Hinxton, Cambridge, CB10 1SD, UK
[3] Calculations and programming
EBI, Wellcome Trust Genome Campus, Hinxton, Cambridge, CB10 1SD, UK

Abstract. The aim of this work is to use a supervised learning approach to identify sets of motif-based sequence characteristics, combinations of which can give the most accurate annotation of new proteins. We assess several of InterPro Consortium member databases for their informativeness for the annotation of full-length protein sequences. Thus, our study addresses the problem of integrating biological information from various resources. Decision-rule algorithms are used to cross-map different biological classification systems in order to optimise the process of functional annotation of protein sequences. Various features (e.g., keywords, GO terms, structural complex names) may be assigned to a sequence via its characteristics (e.g., motifs built by various protein sequence analysis methods) with the developed approach. We chose SwissProt keywords as the set of features on which to perform our analysis. From the presented results one can quickly obtain the best combinations of methods appropriate for the description of a given class of proteins.

Introduction

Availability of a wide variety of effective protein sequence analysis methods calls for an evaluation of their comparative performance and for development of approaches to integrated cross-method consistent annotation. The InterPro Consortium was formed in order to address this problem [1]. The InterPro database is a single resource collecting sequence pattern data from PROSITE (regular expressions and profiles) [2], Pfam (hidden Markov models) [3], PRINTS (fingerprints) [4], and from several other databases-participants. InterPro is a manually curated database, in which the curation process is supported by various automated procedures.

One of the most straightforward approaches to characterizing a novel sequence is to compare it to the already annotated in InterPro proteins. While this potentially can produce high-quality functional predictions, the motif-focused nature of InterPro complicates the interpretation of such analyses, because in most cases it is impossible to find a single InterPro entry corresponding to the combination of motifs found in a given sequence. On the other hand, there are various systems for direct functional annotation of full-length protein sequences, such as GeneOntology (GO) [5] or SwissProt [6]. The quality of annotation that one can get based on similarity to InterPro entries/motifs can thus be improved by combining these two annotation paradigms. A

I. Jonassen and J. Kim (Eds.): WABI 2004, LNBI 3240, pp. 159–169, 2004.

correspondence between the annotation specific for a group of previously characterized full-length protein sequences and the domain/repeat architecture of a given sequence could help to achieve a more complete functional description of the protein
encoded by the sequence.

The analytical system presented in this work demonstrates an approach to linking
keyword-based functional annotation in SwissProt to motif-architecture data from a
subset of the oldest InterPro member databases using Machine Learning. In particular,
the constructed classification system uses motif information to assign SwissProt keywords to proteins. We chose to use SwissProt keywords as the testing ground for our
method. We could have chosen another annotation source, for example, GeneOntology, however, SwissProt keywords present one of the few independent from InterPro
manually curated annotation resources. While GO IDs are assigned to only 67% of
SwissProt proteins (to 3% of these they are assigned manually and to 97% via an
automated system through InterPro), 83% of SwissProt proteins are covered by SwissProt keywords, which are therefore a better candidate (data from SwissProt release
43.1, UniProt 1.7). The approach taken here can, however, be easily extended to GO
in the future.

In this study we use supervised algorithms that are well suited to the situation
where source data of diverse types need to be analysed together. A number of studies
related to automated annotation of novel proteins have been previously reported
[7,8,9,10], however, while our system can predict SwissProt keyword assignments to
proteins, our main purpose is to identify those motif-based characteristics whose
combinations can be used to achieve the most accurate annotation of new sequences.
The problem we aim to address is the inference of such combinations – something
that is nearly impossible to do manually. The idea of using features selected by an
induction algorithm as inputs to the subsequent classification procedure has proven to
be effective in various classification schemes. We show in this work that selected
feature combinations, together with a formal index of their informativeness (information content of these combinations useful for the purposes of protein functional annotation), may serve as an aid in expert-driven curation of protein data.

Data

When using supervised classification algorithms it is necessary to present the source
data (known as *the training set* in machine learning terms) in the form of examples of
correct classifications. These examples consist of sets of characteristics (*features*) that
are used to form predictor functions (*classifiers*) that assign a certain property (class
label) to a previously unseen example correctly.

All the motif-characteristics (called "signatures" in InterPro terminology) coming
from the InterPro member databases form the set of features, while the corresponding
SwissProt keywords make up the labels for the protein sequences. In order for our
algorithms to make use of the diverse data described, the data need to be represented
in a suitable, consistent format – we chose to use the bit vector representation, mapping the collection of InterPro signatures to a vector of 0's and 1's: each feature is set
to 1 if the corresponding InterPro signature (i.e., motif) is found in the protein sequence and to 0 otherwise.

Two different protein classifications, InterPro and SwissProt keywords, were
mapped. Only proteins annotated simultaneously via InterPro signatures and Swiss-

Prot keywords were used as examples of correct classification. Judging from the distribution of the number of keywords annotating the proteins, those keywords that are very general (the distribution tails off around keywords matching ≥ 200 proteins) were not used, as they are irrelevant for specific functional annotation. This threshold is based on the characteristic distribution of protein numbers in InterPro entries and, although its choice is somewhat arbitrary, it takes into account the majority of SwissProt keywords, corresponding to reasonably small sets of proteins (it would not make sense to take into account a keyword that is assigned to nearly all proteins). Please, note that the decision on assigning a keyword to a protein is taken individually for any given keyword. This implies a separate classification problem for every keyword. As a result, 592 SwissProt keywords were selected and, respectively, 592 separate training sets were created (the chosen keywords are available as supplemental data).

Methods

To form the training sets, we used filter methods for feature subset selection and ROC curves analysis for finding optimal misclassification costs in our cost-sensitive classification scheme [11]. Classifiers were created in the form of decision rules and were used further to identify InterPro signatures most useful for the keyword-based functional annotation of protein sequences. Finally, the relative informativeness of InterPro methods to the purposes of the annotation process was estimated.

Training Set Construction

The classification problem here is the two-class problem: given a SwissProt keyword we must decide whether or not to assign it to a given protein. Therefore, the training set should consist of proteins to which this keyword is assigned (*positive examples*) and those to which it isn't (*negative examples*). Such a training set would, however, be highly imbalanced: the number of positive examples would often be much smaller than the number of negative ones because the number of proteins that are not labelled by any one keyword is much greater than of those that are. Previous works on this subject have largely ignored the imbalance problem, allowing the majority class (negative examples) to dominate over the information present in the minority class, leading to the creation of trivial classifiers (e.g., assign/not assign a keyword regardless of sequence characteristics of a given protein) [8]. The issue of the training set imbalance is particularly important in our scenario, since it is the positive examples that are in the minority and they are the ones of primary interest. Therefore, to account equally for properties of both positive and negative classes, the disproportion of the example types comprising the training set should be reduced.

Feature Subset Selection

It would be natural to form the training set by selecting proteins matched by the given keyword as positive examples and all others as negative. However, such a selection would saturate the training set with a lot of negative examples that are irrelevant for classification purposes. The learning procedure should take into account the cases possessing similar sets of features, where the non-trivial problem of discrimination between the two classes exists. Indeed, cases where protein sequences contain motifs that occur in positive examples, but in combinations that result in negative decisions, are the ones that are hard to identify. On the other hand, proteins associated with signatures coming only from negative examples are classified as negatives automatically. Selecting as negative examples those proteins that match any of the same motifs, as do the positive example proteins reduced the imbalance of the training set significantly, discarding from 70% to 99% of negative instances per set as irrelevant. We also ranked all features according to the amount of mutual information shared with the label within each training set, keeping only 100 top-ranked features with mutual information more than zero as the most informative [12]. These data preparation steps are described schematically in Diagram 1.

ROC Curves for Cost-Sensitive Classification

Despite the preliminary filtering and feature subset selection steps, the positive/negative imbalance turns out to still be significant for most of the training sets. Therefore, we used a cost-sensitive learning scheme in order to prevent the occurrence of classifiers that are either trivial or highly biased towards the majority class. Positive and negative examples were given different weights, and ROC curves were constructed for each of the training sets by varying the weight ratio. While ROC curves are usually used to compare the performance of two classification methods against each other [11], we use this method here for the selection of optimal parameters for our classification scheme. In the absence of any *a priori* assumptions about misclassification costs we used the most north-western point of each ROC curve to choose the weights' ratio, thus maximizing the sum of TP (true positive) and TN (true negative) rates.

Classification Procedure

The mapping between InterPro motif architectures and SwissProt keywords was defined as a set of *decision rules* that specify which combinations of motifs describe the same properties of protein sequences as a given keyword. See Table 1 for an example of one such rule-set, consisting of four rules. To construct a list of accurate decision rules, decision tree building algorithms may be used as the first step, and the trees would then be transformed into lists of rules. This approach is not always the best one to follow, since different branches of a decision tree have different accuracy rates and cover different subsets of the training set, thus requiring additional analysis for each derived rule. Alternatively, the decision tree learning procedure can be repeated iteratively, selecting the best branch (i.e., the best decision rule) at each step and discarding those examples that are covered by this branch. Doing so, a hierarchical system of decision rules can be constructed, forming the desired classifier. The advantage of this method is that, although each rule is a part of a whole classifier, it can be considered and evaluated separately just by presenting those rules that are higher in the hierarchy as additional clauses to the one considered. The method used here was the one implemented in the open source machine-learning package WEKA. We used WEKA's J48.PART as the core algorithm within our cost-sensitive learning scheme [13].

Table 1. Decision rules concerning keyword with ID 1031305. For instance, the first rule here says that if the PRINTS motif with ID PR00499 is not found in the sequence of a given protein and if the Pfam motif with ID PF04382 is not found either, then this keyword should **not** be assigned to that protein. Further, if this rule does not cover the given sequence, following rules can be tried sequentially. The last rule covers all the remaining cases.

Rule	If		Then
1	PR00499 is not found PF04382 is not found	**and**	Keyword with ID 1031305 is **not assigned**
2	Not covered by Rule 1 PS50001 is not found PF04382 is found SM00150 is not found	**and** **and** **and**	Keyword with ID 1031305 is **assigned**
3	Not covered by Rules 1-2 SM00150 is found	**and**	Keyword with ID 1031305 is **assigned**
4	Not covered by Rules 1-3		Keyword with ID 1031305 is **not assigned**

Informativeness Index

Clearly, features that repeatedly appear in classification rules of high accuracy should be more informative and useful for annotation than others. Therefore, these are the features we seek to identify as optimal for accurate full-sequence annotation and for which we assessed their informativeness, as described below. The decision rules' accuracy was measured by applying them directly to the training set.

To measure the informativeness of the found feature combinations we identified for each SwissProt keyword the InterPro signatures that were selected by the classification algorithm for constructing the corresponding decision rules. An index chosen as the measure of relative informativeness was calculated as follows: since the generated decision rules are organised into a hierarchical structure (the more examples a rule covers and the more accurate the rule is, the higher it is in the hierarchy), we assign higher values to signatures of higher rules. Each signature's index is defined as the number of times it participates in all positive (ones saying that the keyword should be assigned) rules for a given keyword. Obviously, when annotating a new protein, only positive rules play a useful role. Thus, in the example in Table 1, all signatures have index 2, because they all participate in all positive rules (note that Rule 3 includes Rules 1-2, and, hence, all signatures composing those rules). These index counts were then converted into relative percentage contributions. These percentage contributions should not be confused with the actual information content of the features. Informativeness indices provide a good approximation of information load per feature and, more importantly, the means for ranking features according to the information carried by them. However, these indices have a different meaning than information bits per feature and can be used only as relative estimates.

As sets of signatures most relevant for keyword-based functional annotation were identified, we proceeded to estimate relative contributions of InterPro methods to the annotation of protein sequences in InterPro. We calculated the average informativeness indices (per method per entry) associated with a given InterPro method by, firstly, considering those SwissProt keywords that were assigned to proteins covered by the given entry and, secondly, calculating the proportion of signatures provided by each InterPro member database among the signatures within the rules generated for these keywords (Diagram 2 presents a graphical depiction of these procedures). These indices are a relative measure of how much useful information in the annotation context is delivered by the different methods. We performed this analysis for entries that contain signatures of more than one InterPro method and whose proteins are characterised by the considered set of 592 keywords: 1035 entries. The results of our analysis are presented in Table 3.

Results

Decision Rules

All the decision rules obtained for the considered set of 592 keywords and their accuracy rates on the training sets are available as supplemental data. 3065 rules were generated with average accuracy of 71% on the training data; for 1782 rules it was ≥90% and 1497 had 100% accuracy. While the primary aim of this work has been to discover optimal (for keyword-based functional annotation of proteins) combinations of features comprising the constructed rules, it is worthwhile to remark here that these rules can themselves be used for annotation with fairly high accuracy. For 24 keywords no positive rules were obtained (all rules were negative or trivial, *i.e.,* "never assign the keyword") – rules generated for the 568 remaining keywords were used in the subsequent analysis.

Table 2. Excerpt from supplemental data, *Method informativeness for keyword-based annotation.*

Keywords		Relative Informativeness of InterPro Member Databases						
ID	Keyword	PROSITE Patterns	PROSITE Profiles	Pfam	PRINTS	ProDom	SMART	TIGRFAMs
100050	Amino-acid biosynthesis	8.93%	7.14%	60.71%	0.00%	0.00%	14.29%	8.93%
1441138	Pentaxin	100.00%	0.00%	0.00%	0.00%	0.00%	0.00%	0.00%
189923	Tumor antigen	33.33%	0.00%	16.67%	16.67%	0.00%	16.67%	16.67%
1140070	Acetylcholine receptor inhibitor	16.67%	0.00%	33.33%	33.33%	16.67%	0.00%	0.00%
17790019	Potassium channel	6.79%	15.43%	19.75%	58.02%	0.00%	0.00%	0.00%

Table 3. Relative informativeness per method per InterPro entry for different InterPro member databases, average and maximum among the 1035 selected InterPro entries.

Member DB	Average informativeness index	Maximum informativeness index
PROSITE patterns	2.52	100.00
PROSITE profile	0.70	50.00
Pfam	4.78	100.00
PRINTS	2.01	100.00
ProDom	1.13	33.33
SMART	1.03	50.00
TIGRFAMs	0.83	100.00

Method Informativeness for Keyword-Based Annotation

For each of the 568 keywords, the relative informativeness index of each InterPro method was calculated as an intermediate step of the calculations described in the "Informativeness index" subsection of Methods. The informativeness indices of InterPro methods for the annotation of proteins characterised by a given SwissProt keyword are given in the supplemental data. For example, as Table 2 shows (an excerpt from the full supplemental data table), proteins characterised by the keyword "Potassium channel" are best annotated by motifs coming from PRINTS, while for those to which "Pentaxin" is assigned, "PROSITE Patterns" turns out to be the most effective source of information.

General Method Informativeness

Table 3 contains the calculated informativeness indices of InterPro methods, per InterPro method per entry. The informativeness index is a relative measure of useful information contribution towards keyword-based functional annotation and may be used to assess only the comparative performance of different InterPro methods. The average and maximum informativeness indices are given for each InterPro member database. The meaning of the values in this table is as follows, taking PRINTS as an example: its average informativeness index of 2.01 means that every PRINTS-signature, contributing to an InterPro entry, on average carries 2.01% of information (the meaning of "information" here is explained in the "Informativeness index" subsection of Methods) towards annotation of proteins of this entry. At the same time, for some entries, a PRINTS signature can carry up to 100.00% of information useful for annotation of the respective proteins.

Figure 1 contains plots of the cumulative distributions of numbers of keywords across the entire range of method informativeness for all the InterPro methods surveyed. In general, the lower the graph for a given method, the higher the number of keywords about which the signatures derived by that method carry a significant share of information. This plot complements the data in Table 3 and clearly shows that Pfam dominates over the other member databases in terms of average informativeness towards keyword-based functional annotation of proteins.

Discussion

In the majority of cases accurate functional annotation of uncharacterised protein sequences is possible only by integrating different methods and types of information (cf. Figure 1). Often a human expert may find it difficult or impossible to identify the optimal combinations of data sources for the most efficient annotation. Supervised techniques are frequently considered to be the most effective approach when diverse types of data are to be analysed together [12]. In this study we considered the methods used by the member databases involved in the InterPro Consortium [1] and employed supervised learning algorithms as a means of assessing the informativeness of different sequence analysis methods that are used for protein classification. We also identified those combinations of methods that are optimal for keyword-based functional annotation of proteins.

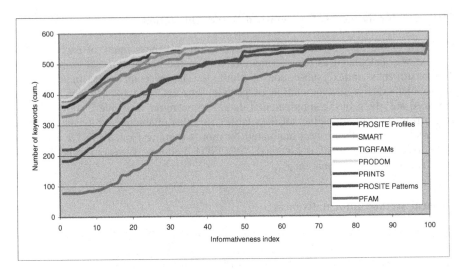

Fig. 1. Plots of the cumulative distributions of numbers of keywords across the range of informativeness indices (from 0 to 100%). Pfam is clearly the most informative method on average.

The features (InterPro signatures) selected by the used classification algorithm and the combinations in which they appear in the constructed classifiers represent subsets of InterPro methods that are most efficient in protein annotation. InterPro signatures that are not included in the rules could require additional consideration or could be seen as carriers of information about proteins' functional properties that are not described by the considered keywords.

Our analysis underscores the significance of all the methods involved in InterPro. However, it shows that they should be treated differently when classifying different groups of proteins. The results indicate that several subsets of proteins annotated with SwissProt keywords can be characterised using combinations of only a few of the available methods. These combinations are not trivial and can be easily extracted manually by searching through the lists of signatures. In fact, our approach complements expert annotation and does not contradict it. The process of integration of different sequence analysis methods would benefit from the fusion of expert opinion and automated assessment of the results. As for the properties of the considered methods, some of them are important for the isolation of vast groups of functionally related proteins, while others are critical for providing highly specific classifications. It is also interesting that the best coverage and specificity of classification can be achieved for different groups of proteins by using different methods (see Table 3 and classification results for the keywords, excerpted in Table 2, full data in the supplement). For instance, informativeness indices of signatures comprising the decision rules for the keyword "Pentaxin" indicate the effectiveness of PROSITE patterns [14] in identifying the group of proteins characterised by this keyword. As a family with a distinct function (lipoprotein ligand-binding), according to InterPro it can be described by a number of other methods, such as PRINTS, Pfam, ProDom and SMART. Therefore, our study confirms that PROSITE patterns are particularly strong in this case [14], because the functionality of these proteins is defined by the presence of a binding site.

Our results show that, on average, Pfam, PROSITE patterns and PRINTS are the most useful sources of information towards keyword-based functional annotation. Pfam has a broad area of application, while at the same time SMART, TIGRFAMs and PROSITE profiles are highly informative only within certain groups. Indeed, (cf. Figure 1) for the majority of keywords the informativeness of the latter methods is comparatively low, while they, in combinations with other methods, can characterise some other keywords fully. It is interesting to note that we did not find, among the selected 1035 entries, any, whose keywords can be fully characterised by SMART, ProDom or PROSITE profile signatures (Table 3). Also we did not find any SwissProt keywords assigned to proteins that were fully described (via generated decision rules) solely by these methods. We would suggest, then, that methods with small average informativeness indices should not be used for protein functional annotation on their own, however, in combinations with the other methods could augment the annotation with specific details.

Conclusions

Supervised classification methods have proven to be effective for the prediction of properties of uncharacterized objects on the whole and for feature subset selection problems specifically. Classification results may be approached as providers of optimal feature sets for subsequent analysis. Our study demonstrates the effectiveness of this approach and provides computationally supported guidance in selecting best methods for expert driven annotation of proteins and protein families.

Acknowledgements

We are thankful to Dr. Maria Krestyaninova of InterPro and Dr. Alvis Brazma for useful discussions and comments. Lev Soinov and Misha Kapushesky would like to acknowledge that they are supported by a grant from the Wellcome Trust (the Bio-Map project).

Supplemental Data

Please see http://www.ebi.ac.uk/~ostolop/skk04 for additional data for this work.

References

1. Mulder NJ, Apweiler R, Attwood TK, Bairoch A, Barrell D, Bateman A, Binns D, Biswas M, Bradley P, Bork P, Bucher P, Copley RR, Courcelle E, Das U, Durbin R, Falquet L, Fleischmann W, Griffiths-Jones S, Haft D, Harte N, Hulo N, Kahn D, Kanapin A, Krestyaninova M, Lopez R, Letunic I, Lonsdale D, Silventoinen V, Orchard SE, Pagni M, Peyruc D, Ponting CP, Selengut JD, Servant F, Sigrist CJ, Vaughan R, Zdobnov EM. The InterPro Database, 2003 brings increased coverage and new features. *Nucl. Acids. Res.* (2003). Jan 1;31(1):315-8.

2. Falquet L, Pagni M, Bucher P, Hulo N, Sigrist CJ, Hofmann K, Bairoch A. The PROSITE database, its status in 2002. *Nucleic Acids Res.* (2002) Jan 1;30(1):235-8.
3. Bateman A, Birney E, Durbin R, Eddy SR, Howe KL, Sonnhammer EL. The Pfam protein families database. *Nucleic Acids Res.* (2000) Jan 1;28(1):263-6.
4. Attwood TK, Croning MD, Flower DR, Lewis AP, Mabey JE, Scordis P, Selley JN, Wright W. PRINTS-S: the database formerly known as PRINTS. *Nucleic Acids Res.* (2000) Jan 1;28(1):225-7.
5. Harris MA, Clark J, Ireland A, Lomax J, Ashburner M, Foulger R, Eilbeck K, Lewis S, Marshall B, Mungall C, Richter J, Rubin GM, Blake JA, Bult C, Dolan M, Drabkin H, Eppig JT, Hill DP, Ni L, Ringwald M, Balakrishnan R, Cherry JM, Christie KR, Costanzo MC, Dwight SS, Engel S, Fisk DG, Hirschman JE, Hong EL, Nash RS, Sethuraman A, Theesfeld CL, Botstein D, Dolinski K, Feierbach B, Berardini T, Mundodi S, Rhee SY, Apweiler R, Barrell D, Camon E, Dimmer E, Lee V, Chisholm R, Gaudet P, Kibbe W, Kishore R, Schwarz EM, Sternberg P, Gwinn M, Hannick L, Wortman J, Berriman M, Wood V, de la Cruz N, Tonellato P, Jaiswal P, Seigfried T, White R; Gene Ontology Consortium. The Gene Ontology (GO) database and informatics resource. *Nucleic Acids Res.* (2004) Jan 1;32, Database issue:D258-61.
6. Boeckmann B, Bairoch A, Apweiler R, Blatter MC, Estreicher A, Gasteiger E, Martin MJ, Michoud K, O'Donovan C, Phan I, Pilbout S, Schneider M. The Swiss-Prot protein knowledgebase and its supplement TrEMBL in 2003.*Nucleic Acids Research* (2003) Jan 1;31(1):365-70.
7. Jensen LJ, Gupta R, Staerfeldt HH, Brunak S. Prediction of human protein function according to Gene Ontology categories. *Bioinformatics* (2003) Mar 22;19(5):635-42.
8. Bazzan AL, Engel PM, Schroeder LF, Da Silva SC. Automated annotation of keywords for proteins related to mycoplasmataceae using machine learning techniques. *Bioinformatics* (2002) Oct;18 Suppl 2:S35-43.
9. Kretschmann E, Fleischmann W, Apweiler R. Automatic rule generation for protein annotation with the C4.5 data mining algorithm applied on SWISS-PROT. *Bioinformatics* (2001) Oct;17(10):920-6.
10. Pavlidis P, Weston J, Cai J, Noble WS. Learning gene functional classifications from multiple data types. *J Comput Biol.* (2002) 9(2):401-11.
11. Provost F, Fawcett T, Kohavi R. Building the Case Against Accuracy Estimation for Comparing Induction Algorithms. ICML-98.
12. Witten I, Frank E. Data Mining-Practical Machine Learning Tools and Techniques with JAVA Implementations, Morgan Kaufmann, 1999.
13. WEKA. (http://www.cs.waikato.ac.nz/~ml/weka).
14. Hulo N, Sigrist CJ, Le Saux V, Langendijk-Genevaux PS, Bordoli L, Gattiker A, De Castro E, Bucher P, Bairoch A. Recent improvements to the PROSITE database. *Nucleic Acids Res.* (2004) 32, 134-7.
15. Mulder NJ, Apweiler R, Attwood TK, Bairoch A, Bateman A, Binns D, Biswas M, Bradley P, Bork P, Bucher P, Copley R, Courcelle E, Durbin R, Falquet L, Fleischmann W, Gouzy J, Griffith-Jones S, Haft D, Hermjakob H, Hulo N, Kahn D, Kanapin A, Krestyaninova M, Lopez R, Letunic I, Orchard S, Pagni M, Peyruc D, Ponting CP, Servant F, Sigrist CJ; InterPro Consortium. InterPro: An integrated documentation resource for protein families, domains and functional sites. *Brief Bioinform.* (2002) Sep;3(3):225-35.

Diagrams

Diagram 1

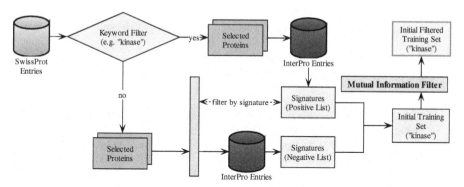

Caption: Data preparation steps for an example keyword (kinase), demonstrating various filters and selection procedures aimed at creating the initial training set to be then subjected to ROC curves analysis for obtaining optimal weight ratios of positive and negative example classes.

Diagram 2

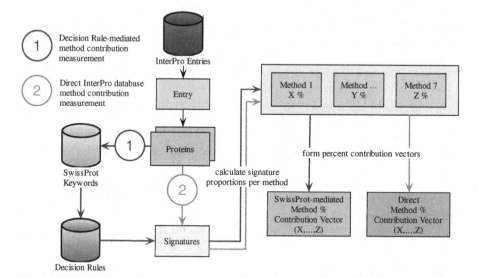

Caption: Calculation of the informativeness index vectors of the InterPro methods, (1) via the supervised learning classification scheme and (2) via the direct count of method contribution per InterPro entry.

Multiple Vector Seeds for Protein Alignment

Daniel G. Brown

School of Computer Science, University of Waterloo
Waterloo ON N2L 3G1 Canada
browndg@math.uwaterloo.ca

Abstract. We present a framework for improving local protein alignment algorithms. Specifically, we discuss how to extend local protein aligners to use a collection of *vector seeds* [3] to reduce noise hits. We model picking a set of vector seeds as an integer programming problem, and give algorithms to choose such a set of seeds. A good set of vector seeds we have chosen allows four times fewer false positive hits, while preserving essentially identical sensitivity as BLASTP.

1 Introduction

Pairwise alignment is one of the most important problems in bioinformatics. Here, we continue an exploration into the seeding and structure of local pairwise alignments, and show that a recent strategy for seeding nucleotide alignments can be expanded to protein alignment. Heuristic protein sequence aligners, exponplified by BLASTP [1], find almost all high-scoring alignments. However, the sensitivity of heuristic aligners to moderate-scoring alignment can still be poor. In particular, alignments with BLASTP score between 40 and 60 are commonly missed by BLASTP, even though many are of truly homologous sequences. We focus on these alignments, and show that a change to the seeding strategy gives success rates comparable to BLASTP with far fewer false positive hits.

Specifically, multiple spaced seeds [10], and their relatives, vector seeds [3], can be used in local protein alignment to reduce the false positive rate in the seeding step of alignment by a factor of four. We present a protocol for choosing multiple vector seeds that allows us to find good seeds that work well together. Our approach is based on solving a set-cover integer program whose solution gives optimal thresholds for a collection of seeds. Our IP is prone to overtraining, so we discuss how to reduce the dependency of the solution on the set of training alignments. The problem we are trying to solve is NP-hard to approximate to any constant factor, so we present heuristics to solve it, though most instances are of moderate enough size to use integer programming solvers.

Our successful result here contrasts with our previous work [3], in which we introduced vector seeds. There, we found that using only one vector seed would not substantially improve BLASTP's sensitivity or selectivity. The use of multiple seeds is the important change in the present work. This successful use of multiple seeds is similar to what has been reported recently for pairwise

I. Jonassen and J. Kim (Eds.): WABI 2004, LNBI 3240, pp. 170–181, 2004.

nucleotide alignment [9, 13, 12], but the approach we use is different since protein aligners require extremely high sensitivity.

Our results confirm the themes developed by us and others since the initial development of spaced seeds. The first theme is that spaced seeds help in heuristic alignment because the very surprisingly conserved regions that one uses as a basis for building an alignment happen more independently in true alignments than for unspaced seeds. In protein alignments, there are often many small regions of high conservation, each of which has a chance to have a hit to a seed in it. With unspaced seeds, the probability that any one of these regions is hit is low, but when a region *is* hit, there may be several more hits, which is unhelpful. By contrast, a spaced seed is likely to hit a given region fewer times, wasting less runtime, and will also hit at least one region in more alignments, increasing sensitivity.

The second theme is that the more one understands how local and global alignments look, the more possible it is to tailor alignment seeding strategies to a particular application, reducing false positives and improving true positives. Here, by basing our set of seeds on sensitivity to true alignments, we choose a set of seed models that hit diverse types of short conserved alignment subregions. Consequently, the probability that one of them hits a given alignment is high, since they complement each other well.

2   Background: Heuristic Alignment and Spaced Seeds

Since the development of heuristic sequence aligners [1], the same approach has been commonly used: identify short, highly conserved regions, and build local alignments around these "hits." This avoids the use of the Smith-Waterman algorithm [11] for pairwise local alignment, which has $\Theta(nm)$ runtimes on input sequences of length n and m. Instead, assuming random sequences, the expected runtime is $h(n, m) + a(n, m)$, where $h(n, m)$ is the amount of time needed to find hits in the two sequences, and $a(n, m)$ is the expected time needed to compute the alignments from the hits. Most heuristic aligners have $h(n, m) = \Theta(n + m)$, while $a(n, m) = \Theta(nm/k)$ for some large constant k. (Even when we align sequences with true homologies, most hits are between unrelated positions, so the estimation of the runtime need not consider whether the sequences are related.) It is the speedup factor of k that is important here.

Most heuristic aligners look at the scores of matching characters in short regions, and use high-scoring short regions as hits. For example, BLASTP [1] hits are three consecutive positions in the two sequences where the total score, according to a BLOSUM or PAM scoring matrix, of aligning the three letters in one sequence to the three letters of the other sequence is at least +13. Finding such hits can be done easily, for example by making a hash table of one sequence and searching positions of the hash table for the other sequence, in time proportional to the length of the sequences and the number of hits found.

To generalize BLASTP's hits, we defined *vector seeds* [3]. A vector seed is a pair (v, T), where $v = (v_1, \dots, v_k)$ is a vector of position multipliers and

T is a threshold. Given two sequences A and B, let $s_{i,j}$ be the score in our scoring matrix of aligning the $A[i]$ to $B[j]$. If we consider position i in A and j in B, we then get an hit to the vector seed at those positions when $v \cdot (s_{i,j}, s_{i+1,j+1}, \ldots, s_{i+k-1,j+k-1}) \geq T$. In this framework, BLASTP's seed is $((1, 1, 1), 13)$.

Vector seeds generalize the earlier idea of spaced seeds [10] for nucleotide alignments, where both scores and the vector are 0/1 vectors and where T, the threshold, equals the number of 1s in v. A spaced seed requires an exact match in the positions where the vector is 1, and the places where the vector is 0 are "don't care" positions.

Spaced seeds have the same expected number of junk hits as unspaced seeds. For unrelated random noise DNA sequences, this is $nm4^{-w}$, where w is the number of ones in the seed (its *support*). Their advantage comes because more distinct internal subregions of a given alignment will match a spaced seed than the unspaced seed, because the hits are more independent. The probability that an alignment of length 64 with 70% conservation matches a good spaced seed of support 11 can be greater than 45%, because there are likely to be more subregions that match the spaced seed than the unspaced seed; by contrast, the default BLASTN seed, which is eleven consecutive required matches, hits only 30% of alignments.

Spaced seeds have three advantages over unspaced seeds. First, their hits are more independent, which means that it is more likely that a given alignment has at least one hit to a seed; fewer alignments have many. Second, the seed model can be tailored to a particular application: if there is structure or periodicity to alignments, this can be reflected in the design of the seeds chosen. For example, in searching for homologous codons, they can be tailored to the three-periodic structure of such alignments [5, 4]. Finally, the use of multiple seeds allows us to boost sensitivity well above what is achievable with a single seed, which for nucleotide alignment can give near-100% sensitivity in reasonable runtime [9].

Keich et al. [8] have given an algorithm for a simple model of alignments to compute the probability that an alignment hits a seed; this has been extended by both Buhler et al. [5] and Brejova et al. [4] to more complex sequence models. Several authors [10, 9, 3, 5, 6, 12] have also proposed using multiple seeds, and given heuristics to choose them. This problem was recently given a theoretical framework by Xu et al. [13].

3 Choosing a Good Set of Seeds

Spaced seeds have made substantial impact in nucleotide alignments, but less in protein alignment. Here, we show that they have use in this domain as well. Specifically, multiple vector seeds, with high thresholds, give essentially the sensitivity of BLASTP with four times fewer noise hits. Slightly fewer alignments are hit, but the regions of alignment hit by the vector seeds are all of the same good ones as hit by the BLASTP seed and a few more. In other words, BLASTP hits more alignments, but the hits found by BLASTP and not the vector seeds are mostly in areas unlikely to be expanded to full alignments.

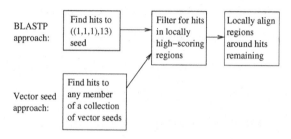

Fig. 1. Flowchart contrasting BLASTP's approach to heuristic sequence alignment to the one proposed here. The only difference is in the initial collection of hits. The smaller collection of hits found with vector seeds gives as many hits to true alignments that survive to the third stage as does BLASTP, yet far fewer noise hits must be filtered out.

We adapt a framework for identifying sets of seeds introduced by Xu *et al.* [13]. We model multiple seed selection as a set cover problem, and give heuristics for the problem. For our purposes, one advantage of the formulation is that it works with explicit alignments: since real alignments may not look like a probabilistic model, we can pick a set of seeds for sensitivity to a collection of true alignments. Unfortunately, this also gives rise to problems, as the thresholds may be set high due to overtraining for a given set of alignments.

3.1   Background Rates

One important detail that we need before we begin is to the background hit rate for a given vector seed. We noted previously [3] that this can be computed, given a scoring matrix. Namely, from the scoring matrix, we can compute the distribution of letters in random sequences implied by the matrix; this can then be used to compute the distribution of scores found in unrelated sequences. Using this, we can compute the probability that unrelated sequences give a hit to a given seed at a random position, or indeed the probabiltiy distribution on the score for a given seed vector at a random position.

For the default BLASTP seed, the probability that two random unrelated positions have a hit is quite high, 1/1600. Because of this high level of false positives, BLASTP must filter hits further, in hopes of throwing out hits in unrelated sequences. Specifically, BLASTP rapidly examines the local area around a hit, and if this region is not also well conserved, the hit is thrown out. Sometimes, this filtering throws out all of the hits found in some true alignments, and thus BLASTP misses them, even though they hit the seed.

Our goal in using vector seeds here is to reduce the false positive rate, while still hitting the overwhelming majority of alignments, and hitting them in places that are highly enough conserved as to make a full alignment likely. A flowchart of our proposal, and the approach of BLASTP, is in Figure 1.

For a set Q of vector seeds, we say that its false positive rate is the probability that *any* seed in Q has a hit to two random positions in unrelated sequences.

This is not equal to the sum of the false positive rates for all seeds in Q, since hits to one seed may overlap hits to another. However, we will use this approximation in our optimization.

3.2   An Integer Program to Choose Many Seeds

Here, we give an integer program to find the set of seeds that hits all alignments in a given training set with overall lowest possible false positive rate. Since our IP trivially encodes the Set-Cover problem, it is NP-hard to solve, or even to approximate to a sublogarithmic factor. However, for moderate-sized training sets, we can solve it in practice, or use simple heuristics to get good solutions.

Given a set of vector seeds $Q = \{(v_1, T_1), \ldots, (v_n, T_n)\}$, we say that they hit a given alignment a if any member of Q hits the alignment. Our goal in picking such a set will be to minimize the false positive rate of the set Q, given that we hit all alignments in a collection A. We note that this optimization goal is the alternative to the goal of Xu *et al.* [13], which maximized seed sensitivity when a maximum number of spaced seeds is allowed. This alternative goal is appropriate for protein alignment, however: we want to achieve extremely high sensitivity, as close to 100% as possible.

The Integer Program. Suppose we are given a collection of alignments $A = \{a_1, \ldots, a_m\}$ and a set of seed vectors $V = \{v_1, \ldots, v_n\}$. We will choose thresholds $(T_1^*, \ldots, T_n^*)$ for the vectors of V such that the vector seed set $Q^* = \{(v_1, T_1^*), \ldots, (v_n, T_n^*)\}$ hits all alignments in A and the false positive rate of Q^* is as low as possible. The T_i^* may be ∞, which corresponds to not choosing the vector v_i at all.

Each alignment a must be hit, so one of the thresholds must be low enough to hit a. To verify this, we compute the best-scoring hit for each vector v_i in each alignment a_j; let the score of this hit be $T_{i,j}$. If we choose T_i^* so that it is at most $T_{i,j}$, then the vector seed (v_i, T_i^*) will hit alignment a.

To model this as an integer program, we have a collection of integer variables $x_{i,T}$ for each possible threshold value for seed v_i. The scores at a position come in a small range of integers, the possible set of reasonable thresholds form a small range. Variable $x_{i,T}$ is 1 when the threshold for seed vector x_i is at most T; for each vector v_i, its threshold chosen is the smallest T where $x_{i,T} = 1$.

To compute the false positive rate, we let $r_{i,T}$ be the probability that a random place in the background model has score exactly T according to seed x_i. We add these up for all of the false hits above the chosen thresholds. Our integer program is as follows:

$$\min \quad \sum_{i,T} x_{i,T} r_{i,T}, \text{ such that} \tag{1}$$

$$\sum_i x_{i,T_{i,j}} \geq 1 \text{ for all alignments } a_j \tag{2}$$

$$x_{i,T} \geq x_{i,T-1} \text{ for all thresholds } T \text{ above the minimum} \tag{3}$$

$$x_{i,T} \in \{0,1\} \text{ for all } i \text{ and } T. \tag{4}$$

The second constraint requires that when we choose a given threshold, we take all hits scoring at least that threshold. Our framework is quite general: given *any* collection of alignments and the sensitivity of a collection of seeds to the alignments, one can use this IP formulation to choose thresholds to hit all alignments while minimizing false positives.

NP-Hardness. Our problem of computing the best set of seeds is NP-hard, and NP-hard to even approximate to within a logarithmic factor. We can straightforwardly represent an intance of Set-Cover as an instance of this problem.

An instance of Set-Cover is a ground set S and a collection T of subsets of S; the goal is the smallest cardinality subset of T whose union is S. An instance of Set-Cover is transformed into an instance of our problem by making an alignment for each element of S whose positional scores are 0/1 and a different seed vector for each member of T, each of which has the same number of 1s, and each of which has only one allowed threshold, equal to its number of ones. For a given element s of S, we add a region to its representing alignment for each of the members of T that contains s, which only matches the seed for that member of T. The construction is straightforward.

As a result, minimizing the total false positive rate while still matching all alignments will correspond to picking the smallest cardinality set of seeds, or correspondingly, the smallest Set-Cover. Also, the non-approximability result for Set-Cover [7] translates to this domain: the problem is not approximable to within a logarithmic factor of optimality unless $P = NP$.

Expansions of the Framework. In our experiments, we use the vector seed requirement as a threshold; one could use a more complicated threshold scheme to focus on hits that would be expanded to full alignments. That is, our minimum threshold for $T_{i,j}$ could be the highest-scoring hit *that is expanded to a full alignment* of seed vector v_j in alignment a_i. We could also have a more complicated way of seeding alignments, and still, as long as we could compute false positive rates, we could require that all alignments are hit and minimize false positive rates.

Also, we can limit the total number of vector seeds used in the true solution (in other words, limit the number of vectors with finite threshold). We do this by putting an upper bound on $\sum_i x_{i,T}$ for the maximum threshold T. In practice, one might want an upper bound of 4 or 8 seeds, as each chosen seed requires a method to identify hits, and one might not want to have to use too many such methods in the goal of keeping fewer indexes of a protein sequence database, for example.

Finally, we might want to not allow seeds to be chosen with very high threshold. The optimal solution to the problem will have the thresholds as on the seeds as high as possible while still hitting each alignment. This allows overtraining: since even a tiny increase in the thresholds would have caused a missed alignment, we may easily expect that in another set of alignments, there may be alignments just barely missed by the chosen thresholds. This is particularly possible if thresholds are allowed to get extremely high and only useful for a single alignment. This overtraining happened in some of our experiments, so we lowered the maximum so that they were either found in a fairly narrow range (+13 to +25), or set to ∞ when a seed was not used. As one way of also addressing overtraining, we considered lowering the thresholds obtained from the IP uniformly, or just lowering the thresholds that have been set to high values.

We note one simplification of our formulation: false hit rates are not additive. Given two spaced seeds, a hit to one may coincide with a hit to the other, so the background rate of false positives is lower than estimated by the program. When we give such background rates later, we will distinguish those found by the IP from the true values.

Solving the IP and Heuristics. To solve this integer program is not necessarily straightforward, since the problem is NP-hard. In our experiments, we used sets of approximately 400 alignments, and the IP has been able to solve directly quickly, using standard algorithms.

Straightforward heuristics also work well for the problem, such as solving the LP relaxation and rounding to 1 all variables with values close to 1, until all alignments are hit, or setting all variables with fractional LP solutions to 1 and then raising thresholds on seeds until we start to miss alignments.

We finally note that a simple greedy heuristic works well for the problem, as well: start with low thresholds for all seed vectors, and repeatedly increase the threshold whose increase most reduces the false positive rate, until no such increase can be made without missing an alignment. This simple heuristic performed essentially comparably to the integer program in our experiments, but since the IP solved quickly, we used its results.

One other advantage to the IP formulation is that the false-positive rate from the LP relaxation is a lower bound on what can possibly be achieved; the simple greedy heuristic offers no such lower bound.

4   Experimental Results

Here, we present experiments with multiple seed selection in the context of protein alignments. One feature of our analysis is that we study not merely which alignments are hit by a seed, but *where* they are hit. Most false hits occur in regions where, surrounding the false hit, there is no particularly high score. This is because they occur by chance, and the expected score of unrelated sequence positions is negative. By contrast, many true hits happen in high-scoring local sub-regions of an alignment. To reduce runtime, heuristic aligners only extend hits found in these good local regions. This improves speed, but means that a true alignment, even if it is hit by a seed, may still be missed if the hit was in a bad part of the alignment.

Thus, we examine the scores in all ten amino acid-long window around each hit, and characterize each alignment by both whether it is hit at all, *and* by how good the best region around a hit to a seed, or a collection of seeds, is. For moderate-scoring alignments, an ensemble of eight vector seeds gives comparable performance to BLASTP, finding hits in the same good local regions, while giving five times fewer false positives. Since the sensitivity of BLASTP in these moderate-scoring alignments is poor, this offers either a much faster way of equalling BLASTP's performance, or the possibility of allowing higher false positive rate and achieving far greater sensitivity.

4.1 BLAST Alignments

We begin by exploring several sets of alignments generated using BLASTP. Our target score range for our alignments is BLASTP score between +40 and +60 (BLOSUM score +112 to +168). These moderate-scoring alignments can happen by chance, but also are often true. Alignments below this threshold are much more likely to be errors, while in a database of proteins we used, such alignments are likely to happen to a random sequence by chance only one time in 10,000, according to BLASTP's statistics.

We begin by identifying a set of BLASTP alignments in this score range. To avoid over-representing certain families of alignments in our test set, we did an all-versus-all comparison of 8654 human proteins from the SWISS-PROT database[2]. (We note that this is the same set of proteins and alignments we used in our previous vector seed work [3].) We then divided the proteins into families so that all alignments with score greater than 100 are between two sequences in the same family, and there are as many families as possible. We then chose ten sets of alignments in our target score range such that in each set, we chose at most 8 from proteins in each family; we divided these ten sets into five training sets and five testing sets.

We then considered the set of 35 vector patterns of length at most 7 that include three or four 1s (the *support* of the seed). We used this collection of vector patterns as we have seen no evidence that non-binary seed vectors are preferable to binary ones for proteins, and because it is more difficult to find hits to seeds with higher support than four, due to the high number of needed hash table keys.

We computed the optimal set of thresholds for these vector seeds such that every alignment in a training set has a hit to at least one of the seeds, while minimizing the background rate of hits to the seeds and only using at most ten vector patterns. Then, we examined the sensitivity of the chosen seeds for a training set to its corresponding test set. The results are found in Table 1. Some seed sets chosen showed signs of overtraining, but others were quite successful, where the chosen seeds work well for their training set as well, and have low false positive rate.

We took the best seed set with near-100% sensitivity for both its training and testing data and used it in further experiments. This seed set is shown in Figure 2. We note that this seed set has five times lower false positive rate than does BLASTP, while still hitting all of its testing alignments but four (which is not statistically significant from zero). We also considered a set of thresholds where we lowered the higher thresholds slightly to allow more hits, and possibly avoid overtraining on the initial set of alignment. These altered thresholds are shown as well in Figure 2, and give a total false positive rate of 1/6900. (This set of thresholds also all 402 test alignments for that instance.)

Table 1. Hit rates for optimal seed sets for various sets of training alignments, when applied to an unrelated test set. With a training set of approximately 400 alignments, we find a set of seeds that typically hits approximately 99% of alignments. We used the seed set from the third of these experiments for further experimentation.

Training alignments	False positive rate	Test alignment fraction hit
404	1/12,600	407/423 (96%)
403	1/9200	401/407 (98.5%)
409	1/8000	398/402 (99.0%)
394	1/10,700	395/400 (98.75%)
415	1/9500	410/416 (98.5%)

Table 2. Seeds and thresholds chosen by integer programming for 409 test alignments. The false positive rate of these 5 seeds is 1/8000, or five times lower than for BLASTP's seed. We also consider the set of thresholds where we lower the values discovered by the IP by 1, except for the already low thresholds below 20; this gives false positive rate 1/6900, or more than four times lower than BLASTP.

Seed vector	IP Threshold	Lowered threshold
$(1,0,0,0,1,1,1)$	21	20
$(1,0,0,1,0,1,1)$	21	20
$(1,0,1,0,0,1,1)$	20	20
$(1,0,1,1,0,0,1)$	19	19
$(1,0,1,1,0,1)$	23	22
$(1,0,0,1,1,1)$	18	18
$(1,1,1,0,0,1)$	20	20
$(1,1,0,1,1)$	21	20

4.2   All Alignments

We then considered a larger set of alignments in our target range of good, but not great scores, to verify if the advantage of multiple seeds still holds. We used the Smith-Waterman algorithm to compute all alignments between pairs of a 1000-sequence subset of our protein data set, and computed how many of them were not found by BLASTP. Only 970 out of 2950 alignments with BLOSUM62 score between +112 and +168 had been identified by BLASTP, even though alignments in this score range would have happened by chance only 1 time in 10000 according to BLASTP's statistics.

Almost all of these 2950 alignments, 2942, had a hit to the BLASTP default seed. Despite this, however, only 970 actually built a successful BLASTP alignment. Our set of eight seeds had hits to 1939 of the 1980 that did not build a BLASTP alignment and to 955 of the 970 that did build a BLASTP alignment, so at a first glance, the situation does not look good. However, the difference between having a hit and having a hit in a good region of the alignment is where we are able to show substantial improvement.

The discrepancy between hits and alignments comes because the BLASTP seed can have a hit in a bad part of the alignment, which is filtered out. Typically,

Table 3. Hits in locally good regions of alignments. Shown are the fraction of 2950 alignments found with a hit to a seed that is found in a region of length 10 whose total subalignment score is above a given threshold. The default BLASTP seed hits more alignments than a collection of seeds whose overall false positive rate is much lower. However, when we filter for hits found in a locally good region, as does BLASTP, the two strategies are comparable: approximately 48% of alignments have a hit in a moderately high-scoring region for both approaches. The set of seeds, whose total false positive rate is 40% lower than the $((1, 1, 1), 15)$ vector seed vastly outperforms it. If we desire still higher sensitivity, the set of seeds with slightly higher thresholds gives performance essentially identical to BLASTP once we start looking for even modestly good regions of alignments, while still having four times fewer noise hits.

Subregion score threshold	Seed $((1,1,1),13)$	Multiple seeds	Multiple seeds Thresholds raised	Seed $((1,1,1),15)$
Any region	99.3%	96.5%	97.0%	91.8%
+25	77.8%	73.2%	75.2%	63.9%
+30	47.7%	46.7%	47.6%	41.1%
+35	24.3%	24.3%	24.4%	23.0%
+40	13.0%	13.0%	13.0%	12.8%

such hits occur in a region where the source of positive score is quite short, which is much more likely with an unspaced seed than with a spaced seed. We looked at all of the regions of length 10 amino acids of alignments that included a hit to a seed (either the BLASTP seed or one of the multiple seeds), and assigned the best score of such a region to that alignment; if no ungapped region of length 10 surrounded a hit, we assumed it would certainly be filtered out. The data are shown in Table 3, and show that of the alignments hit by the spaced seeds, they are hit in regions that are essentially identical in conservation to where the BLASTP seed hits them. For example, 47.7% of the alignments, contain a 10-amino acid region around a hit to the $((1, 1, 1), 13)$ seed with BLOSUM score at least +30, while 46.7% contain such a region surrounding a hit to one of the multiple seeds with higher threshold. If we use the lower thresholds that allow slightly more false positives, their performance is actually slightly better than BLASTP's.

Table 3 also shows that the higher-threshold seed $((1, 1, 1), 15)$, which has worse false positive rate $(1/5700)$ than our ensembles of seeds, performs substantially worse: namely, only 64% of the alignments have a hit to the single seed found in a region with local score above +25, while 73% of the alignments have a hit to one of the multiple seeds with this property. This single seed strategy is clearly worse than the multiple seed strategy of comparable false positive rate, and the optimized seeds perform comparably to BLASTP in identifying the alignments that actually have a core conserved region.

Our experiments show that multiple spaced seeds can have an impact on local alignment of protein sequences. Using many spaced seeds, which we picked by optimizing an integer program, we find seed models with a comparable chance of finding a good hit in a moderate-scoring alignment than does the BLASTP seed, with four to five times fewer noise hits. The difficulty with the BLASTP seed is that it not only has more junk hits, and more hits in overlapping places, it also has more hits in short regions of true alignments, which likely to be filtered and thrown out.

5   Conclusions and Future Work

We have shown the first true success in using spaced seeds for protein homology search detection. Our result shows that using multiple vector seeds can give sensitivity to good parts of local protein alignments essentially comparable to BLASTP, while reducing the false positive rate of the search algorithm by a factor of four to five.

Our set of vector seeds is chosen by optimizing an integer programming framework for choosing multiple spaced seeds when we want 100% sensitivity to a collection of training alignments. Using this framework, we identified a set of seeds for moderate-scoring protein alignments whose total false positive rate in random sequence is four times lower than the default BLASTP seed. This set of seeds had hits to slightly fewer alignments in a test set of moderate-scoring alignments found by the Smith-Waterman algorithm than found by BLASTP; however, the BLASTP seeds hit subregions of these alignments that were actually slightly worse than hit by the spaced seeds. Hence, given the filtering used by BLASTP, we expect that the two alignment strategies would give comparable sensitivity, while the spaced seeds give four times fewer false hits.

Future Work. The most important future work for protein alignments with multiple seeds is to try them out in a full local aligner; to that end, we are currently studying the expansion of existing heuristic protein aligners to use spaced and multiple seeds. Recently, it has been shown that the use of a surprisingly small number of spaced seeds can give near complete sensitivity to high-scoring alignments in the case of nucleotide alignments [9], but we do not expect that this will extend to the case of protein alignments. The difficulty is that protein alignments include more and shorter highly conserved regions, which is why it is hard to do better than the default unspaced seed.

Finally, we are interested in other appplications of alignment, both inside bioinformatics and outside, where our two goals of hit independence and designing hit structures are compatible with the domain. For example, we are interested in how to improve RNA alignment, to improve protein alignments in programs like PSI-BLAST, and whether spaced seeds have application in local search in music databases.

Acknowledgments

The author would like to thank Ming Li for introducing him to the idea of spaced seeds, and the bioinformatics group at the University of Waterloo for being a good sounding board for ideas. This work is supported by the Natural Science and Engineering Research Council of Canada, and by the Human Frontier Science Program.

References

1. S. F. Altschul, W. Gish, W. Miller, E. W. Myers, and D. J. Lipman. Basic local alignment search tool. *Journal of Molecular Biology*, 215(3):403–410, 1990.
2. A. Bairoch and R. Apweiler. The SWISS-PROT protein sequence database and its supplement TrEMBL in 2000. *Nucleic Acids Research*, 28(1):45–48, 2000.
3. B. Brejova, D. Brown, and T. Vinar. Vector seeds: an extension to spaced seeds allows substantial improvements in sensitivity and specificity. In *Proceedings of the 3rd Annual Workshop on Algorithms in Bioinformatics (WABI)*, pages 39–54, 2003.
4. B. Brejova, D. Brown, and T. Vinar. Optimal spaced seeds for homologous coding regions. *J. Bioinf. and Comp. Biol.*, 1:595–610, January 2004.
5. J. Buhler, U. Keich, and Y. Sun. Designing seeds for similarity search in genomic DNA. In *Proceedings of the 7th Annual International Conference on Computational Biology (RECOMB)*, 2003. 67-75.
6. K.P. Choi and L. Zhang. Sensitive analysis and efficient method for identifying optimal spaced seeds. *J. Comp and Sys. Sci.*, 68:22–40, 2004.
7. D. Hochbaum. Approximating covering and packing problems. In D. Hochbaum, editor, *Approximation algorithms for NP-hard problems*, pages 94–143. PWS, 1997.
8. U. Keich, M. Li, B. Ma, and J. Tromp. On spaced seeds for similarity search. *Discrete Appl. Math.*, 138:253–263, 2004.
9. Ming Li, Bin Ma, Derek Kisman, and John Tromp. Patternhunter II: Highly sensitive and fast homology search. *Journal of Bioinformatics and Computational Biology*, 2004. To appear.
10. B. Ma, J. Tromp, and M. Li. PatternHunter: faster and more sensitive homology search. *Bioinformatics*, 18(3):440–445, March 2002.
11. T.F. Smith and M.S. Waterman. Identification of common molecular subsequences. *J. Mol. Biol.*, 147:195–197, 1981.
12. Y. Sun and J. Buhler. Designing multiple simultaneous seeds for DNA similarity search. In *Proceedings of the 8th Annual International Conference on Computational Biology (RECOMB)*, 2004. 76–84.
13. J. Xu, D. Brown, M. Li, and B. Ma. Optimizing multiple spaced seeds for homology search. In *Combinatorial Pattern Matching, 15th Annual Symposium*, 2004. To appear.

Solving the Protein Threading Problem
by Lagrangian Relaxation

Stefan Balev[*]

Laboratoire d'Informatique du Havre, Université du Havre
25 rue Philippe Lebon BP 540, 76058 Le Havre cedex, France
Stefan.Balev@univ-lehavre.fr

Abstract. This paper presents an efficient algorithm for aligning a query amino-acid sequence to a protein 3D structure template. Solving this problem is one of the main steps of the methods of protein structure prediction by threading. We propose an integer programming model and solve it by branch-and-bound algorithm. The bounds are computed using a Lagrangian dual of the model which turns out to be much easier to solve than its linear programming relaxation. The Lagrangian relaxations are computed using a dynamic programming algorithm. The experimental results show that our algorithm outperforms the commonly used methods. The proposed algorithm is general enough and can be easily plugged in most of the threading tools in order to increase their performance.

Keywords: protein threading, protein structure prediction, sequence-structure alignment, integer programming, dynamic programming, Lagrangian relaxation and duality, subgradient optimization.

1 Introduction

The problem of determining the three-dimensional structure of a protein given its one-dimensional sequence is one of the grand challenges confronting the computational biology today [1, 2]. The knowledge of the 3D structure of the proteins is essential for understanding their biological functions. The experimental methods of structure determining are still expensive and slow and cannot cope with the explosion of sequences becoming available. That is why the progress of the molecular biology depends on the availability of reliable and fast computational structure prediction methods.

One of the most promising computational approaches to the problem is protein threading [3]. It is based on the assumption that the number of the possible protein structures is limited. Each structure defines an equivalence class and the problem reduces to classification of the query sequence into one of these classes. Threading is a complex and time consuming computational technique consisting of the following main steps [3]: (i) constructing a database of structure templates; (ii) choosing a score function evaluating each possible alignment of a

[*] This work was partially supported by the GénoGRID project (ACI GRID 2002-2004) and the French-Bulgarian Project RILA'2003 (programme d'actions integrées).

query sequence and a structure template; (iii) finding the best (with respect to the score function) alignment of the query sequence and each of the structure templates in the database; (iv) choosing the most appropriate template based on the (normalized) scores of the optimal alignments found on the previous step.

This paper deals with the third step of the above procedure, which is the most time consuming one, because of the huge number of the possible query-to-template alignments. Below we present a general mathematical model of protein threading problem, compatible with most of the existing approaches [3–10].

A *structure template* is an ordered set of m blocks. Block i has length l_i, $i = 1, \ldots, m$ and represents a sequence of contiguous amino-acid positions. Blocks usually correspond to conserved elements of the 2D structure (α-helices and β-sheets). The template also contains information about the pairwise interactions between residues belonging to the blocks. For our purposes it is convenient to generalize this information to a set E of interactions (links) between blocks. In this way we obtain the so-called *generalized contact map graph*, whose vertices are the blocks and whose edges are the interactions between them.

An *alignment* or *threading* of a query sequence of length N amino-acids and a template is covering of segments of the query sequence by the template blocks. The blocks must preserve their order and are not allowed to overlap. A threading is completely determined by the starting positions of all blocks. In order to simplify our notations, we will use relative positions [4]. If the absolute position of block i is j, then its relative position is $j - \sum_{k=1}^{i-1} l_k$. In this way the possible (relative) positions of each block are between 1 and $n = N+1-\sum_{i=1}^{m} l_i$. Formally, the set of possible threadings can be represented as $T = \{\pi = (\pi_1, \ldots, \pi_m) : 1 \leq \pi_1 \leq \ldots \pi_m \leq n\}$. It is easy to see that the number of possible threadings (the search space size of the threading problem) is $|T| = \binom{m+n-1}{m}$, which is a huge number even for small threading pairs.

The *scoring function* incorporates all biological and physical knowledge on the problem. It describes the degree of compatibility between sequence residues and their corresponding positions in the structure template. The choice of an adequate scoring function is essential for the quality of the threading method. It is complex matter which is beyond the scope of this paper. For our purposes we only assume that the scoring function is additive and can be computed by considering no more than two blocks at a time. These assumptions allow to represent the score function in the following way. Let

$$c_{ijl}, \ i = 1, \ldots, m-1, \ 1 \leq j \leq l \leq n \tag{1}$$

be the score of the segment (called loop) between blocks i and $i+1$ when block i is on position j and block $i+1$ is on position l. To simplify our notations, we suppose that the scores of the loops before the first block and after the last block are incorporated in c_{1jl} and $c_{m-1,j,l}$. The coefficients c_{ijl} can also incorporate the scores of putting block i on the jth position and the scores generated by the interaction between blocks i and $i+1$. Let $R \subseteq E$ be the set of remote links (interactions between non-adjacent blocks) and let

$$c_{ijkl}, \ (i, k) \in R, \ 1 \leq j \leq l \leq n \tag{2}$$

be the score generated by the interaction between blocks i and k when block i is on position j and block k is on position l. Using these notations, the score of threading π is

$$\varphi(\pi) = \sum_{i=1}^{m-1} c_{i\pi_i \pi_{i+1}} + \sum_{(i,k)\in R} c_{i\pi_i k\pi_k} \tag{3}$$

and the threading problem can simply be stated as

$$\min\{\varphi(\pi) \; : \; \pi \in T\} \; . \tag{4}$$

It has been proved that the above problem is NP-hard [11]. Moreover, it is MAX-SNP-hard [12], which means that there is no arbitrarily close polynomial approximation algorithm, unless P = NP. In spite of these discouraging theoretical results, several threading algorithms have been developed, including a divide-and-conquer approach by Xu et al. [5] and a dedicated branch-and-bound algorithm by Lathrop and Smith [4]. Recently two algorithms based on Mixed Integer Programming (MIP) models have been proposed by Andonov et al. [7, 8] and Xu et al. [6]. It is interesting to note that all these algorithms perform surprisingly well on real life instances, and especially when a query is threaded to itself. Probably it is due to the nature of the scoring functions used, which strongly attract the blocks to their optimal positions and guide the search in the huge space of possible threadings. Nevertheless, the above algorithms can solve only moderate size instances in reasonable time and are still very computationally expensive.

The problem (4) is solved many times in the threading process. The query is aligned to each template from the database. In order to normalize the scores, some methods need to thread a large set of queries to each template from the database [10]. The designers of score functions need millions of alignments in order to tune their parameters. That is why a really efficient threading algorithm is needed. In the following sections we present an algorithm which to the best of our knowledge outperforms the existing methods and hopefully meets the efficiency requirements. Our algorithm is based on a MIP model. The advantage of MIP models proposed in [8, 6] is that their linear programming (LP) relaxations give the optimal solution for most of real life instances. Their drawback is their huge size (both number of variables and number of constraints) which makes even solving the LP relaxation slow. Our model suffers from the same drawback but instead of solving its LP relaxation, we consider a Lagrangian relaxation which turns out to be much easier to solve.

2   Optimization Algorithm

A classical approach to attack a hard optimization problem like (4) is to consider a relaxation which is easier to solve than the original problem and provides tight lower bound on the optimal objective value. A *relaxation* of (4) is an optimization problem of the form $\min\{\varphi'(\pi) \; : \; \pi \in T'\}$, where $T \subseteq T'$ and $\varphi(\pi) \geq \varphi'(\pi)$ for $\pi \in T$.

2.1   Relaxation of the Feasible Set

Let us suppose for a moment that there are no remote interactions between the blocks ($R = \emptyset$). In this case the problem (4) may be easily solved in $O(mn^2)$ time using the dynamic programming recurrence

$$F(i + 1, l) = \min_{1 \leq j \leq l} \{F(i, j) + c_{ijl}\}, \; i = 1, \ldots, m - 1, \; l = 1, \ldots, n \quad (5)$$

with $F(1, j) = 0$, $j = 1, \ldots, n$. The remote links are what makes the problem difficult. In order to take them into account, let α_{ik} and β_{ik}, $(i, k) \in R$ be the positions of blocks i and k. Using these notations, we can reformulate the problem (4) in the following way

$$\text{Minimize} \quad \sum_{i=1}^{m-1} c_{i\pi_i\pi_{i+1}} + \sum_{(i,k)\in R} c_{i\alpha_{ik}k\beta_{ik}} \quad (6)$$

$$\text{subject to} \quad 1 \leq \pi_1 \leq \cdots \leq \pi_m \leq n \quad (7)$$

$$\alpha_{ik} = \pi_i, \; \beta_{ik} = \pi_k, \quad (i, k) \in R . \quad (8)$$

The constraints (8) make the problem difficult and we will relax a part of them. Let C be a set of blocks covering all links in R, that is, if $(i, k) \in R$ then $i \in C$ or $k \in C$. We partition the set of links into two subsets R^α and R^β, such that if $(i, k) \in R^\alpha$ then $i \in C$ and if $(i, k) \in R^\beta$ then $k \in C$ (if the both ends of some link are in C, then we break the tie arbitrarily). We relax the constraints $\beta_{ik} = \pi_k$ for the links in R^α and $\alpha_{ik} = \pi_i$ for the links in R^β. In this way each link has a fixed end and a free end. In order to make the relaxation tighter, we impose some extra constraints on the order of the free ends of the links and replace (8) by the following set of constraints

$$\alpha_{ik} = \pi_i, \quad (i, k) \in R^\alpha \quad (9)$$

$$\beta_{ik} = \pi_k, \quad (i, k) \in R^\beta \quad (10)$$

$$1 \leq \alpha_{ik} \leq \beta_{ik} \leq n, \quad (i, k) \in R \quad (11)$$

$$\beta_{ik_1} \leq \beta_{ik_2}, \quad (i, k_1) \in R^\alpha, \; (i, k_2) \in R^\alpha, \; k_1 < k_2 \quad (12)$$

$$\alpha_{i_1k} \leq \alpha_{i_2k}, \quad (i_1, k) \in R^\beta, \; (i_2, k) \in R^\beta, \; i_1 < i_2 . \quad (13)$$

Figure 1 shows an example of the original and the relaxed constraints. It is easy to show that the problem of minimizing (6) subject to constraints (7), (9)-(13) is a relaxation of the original threading problem (4). In the rest of this section we give an algorithm to solve it.

Consider a block $i \in C$ and let $(i, k_1), \ldots, (i, k_p)$, $i < k_1 < \ldots k_p$ be all the links in R^α with beginning i. Let $G^\alpha(i, j, s, l)$ be the minimal score contribution of the links $(i, k_1), \ldots, (i, k_s)$ when block i is on position j and the position of block k_s is *at most* l. Then the minimal contribution of the links $(i, k_1), \ldots, (i, k_p)$ when block i is on position j is given by $G^\alpha(i, j, p, n)$. These contributions can be computed using the recurrences

$$G^\alpha(i, j, s, j) = G^\alpha(i, j, s - 1, j) + c_{ijk_sj} \quad (14)$$

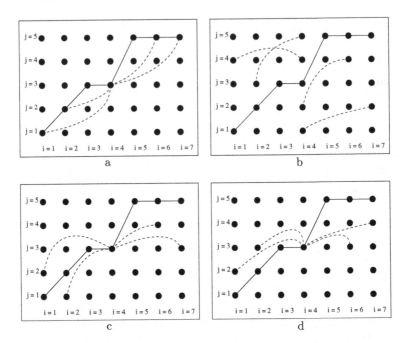

Fig. 1. Example of threading instance with $m = 7$ blocks and $n = 5$ free positions. Columns correspond to blocks and rows to positions. The set of remote links is $R = \{(1,4),(2,4),(4,6),(4,7)\}$. It is covered by the block set $C = \{4\}$. The solid lines connect the points representing the block positions as defined by the π-variables. The dashed lines connect the points representing the block positions as defined by α- and β-variables. (a) solution satisfying constraints (7) and (8); (b) solution satisfying constraints (7) only; (c) solution satisfying constraints (7) and (9)-(11); (d) solution satisfying constraints (7) and (9)-(13).

for $s = 1,\ldots,p$, $j = 1,\ldots,n$, and

$$G^\alpha(i,j,s,l) = \min\{G^\alpha(i,j,s-1,l) + c_{ijk_sl}, G^\alpha(i,j,s,l-1)\} \qquad (15)$$

for $s = 1,\ldots,p$, $1 \leq j < l \leq n$, with $G^\alpha(i,j,0,l) = 0$, $1 \leq j \leq l \leq n$.

Symmetrically, consider a block $k \in C$ and let $(i_1,k),\ldots,(i_q,k)$, $i_1 < \ldots i_q < k$ be all the links in R^β with end k. Let $G^\beta(s,j,k,l)$ be the minimal score contributions of the links $(i_1,k),\ldots,(i_s,k)$ when block k is on position l and the position of block i_s is at most j. Then the minimal contribution of the links $(i_1,k),\ldots,(i_q,k)$ when block k is on position l, $G^\beta(q,l,k,l)$, can be computed using the recurrences

$$G^\beta(s,1,k,l) = G^\beta(s-1,1,k,l) + c_{i_s1kl} \qquad (16)$$

for $s = 1,\ldots,q$, $l = 1,\ldots,n$, and

$$G^\beta(s,j,k,l) = \min\{G^\beta(s-1,j,k,l) + c_{i_sjkl}, G^\beta(s,j-1,k,l)\} \qquad (17)$$

for $s = 1,\ldots,q$, $1 < j \leq l \leq n$, with $G^\beta(0,j,k,l) = 0$, $1 \leq j \leq l \leq n$.

Let $H(i,j)$, $i \in C$, $j = 1, \ldots, n$ be the contribution of all remote links with fixed end i when block i is on position j ($H(i,j)$ is the sum of the corresponding G^α and G^β) and let $H(i,j) = 0$ for $i \notin C$, $j = 1, \ldots, n$. To solve the relaxation it is sufficient to apply the recurrences (5) modified in the following way

$$F(i+1,l) = \min_{1 \le j \le l} \{F(i,j) + c_{ijl} + H(i+1,l)\}, \; i = 1, \ldots, m-1, \; l = 1, \ldots, n \quad (18)$$

with $F(1,j) = H(1,j)$, $j = 1, \ldots, n$. The time complexity of computing G^α and G^β is $O(rn^2)$, where $r = |R|$, which gives an overall complexity $O((m+r)n^2)$ of the described algorithm. In the worst case r is of order m^2, hence the complexity is $O(m^2 n^2)$. However, for real-life instances in most of the cases r is of order m, hence the complexity is $O(mn^2)$.

Lathrop and Smith [4] use similar algorithm to compute the lower bounds in their branch-and-bound algorithm. The difference is that in their case C is the set of all blocks participating in remote links and each link (i,k) is considered twice – as belonging to R^α and to R^β. The coefficients c_{ijkl} of each of the two copies of the link are divided by two.

The quality of the bounds produced by the relaxation described above depends on the choice of the set C. There are many ways to choose a set of blocks covering all the remote links. Intuitively, a smaller set C involves more restrictions on the positions of the free ends of the links and hence a tighter relaxation. On the other hand, the problem of finding a minimal vertex subset covering all the edges of a graph is NP-hard (see [13], p. 144). That is why we use a simple greedy algorithm to determine the set C: find the block with maximal degree, put it in C and delete all links in which this block participates; repeat this procedure until there are no more links.

2.2   Lagrangian Relaxation and Dual

To make the relaxation presented in the previous section stronger, we use a standard optimization technique called Lagrangian relaxation. For a general presentation of this method and proofs of the facts used in this section, the reader is referred to any good integer optimization textbook, for example [13]. The main idea of Lagrangian relaxation is to take into account the relaxed constraints by adding in the objective function terms which penalize the violation of these constraints. In order to use Lagrangian relaxation we need to present our problem as MIP. We introduce binary variables x_{ijl}, $i = 1, \ldots, m-1$, $1 \le j \le l \le n$ and y_{ijkl}, $(i,k) \in R$, $1 \le j \le l \le n$. The variable x_{ijl} is one iff block i is on position j and block $i+1$ is on position l and $y_{ijkl} = 1$ iff block i is on position j and block k is on position l. The connection between these variables and π, α, and β notations from the previous section is the following:

$$\pi_i = j \text{ and } \pi_{i+1} = l \text{ iff } x_{ijl} = 1 \quad (19)$$

$$\alpha_{ik} = j \text{ and } \beta_{ik} = l \text{ iff } y_{ijkl} = 1 \; . \quad (20)$$

Using these variables the protein threading problem can be rewritten as

$$\text{Minimize} \quad \sum_{i=1}^{m-1} \sum_{1 \le j \le l \le n} c_{ijl} x_{ijl} + \sum_{(i,k) \in R} \sum_{1 \le j \le l \le n} c_{ijkl} y_{ijkl} \tag{21}$$

$$\text{subject to} \quad \sum_{1 \le j \le l \le n} x_{1jl} = 1 \tag{22}$$

$$\sum_{l=1}^{j} x_{i-1,l,j} - \sum_{l=j}^{n} x_{ijl} = 0, \ i = 2, \ldots, m-1, \ j = 1, \ldots, n \tag{23}$$

$$\sum_{l=j}^{n} x_{ijl} - \sum_{l=j}^{n} y_{ijkl} = 0, \ (i,k) \in R, \ j = 1, \ldots, n \tag{24}$$

$$\sum_{j=1}^{l} x_{k-1,j,l} - \sum_{j=1}^{l} y_{ijkl} = 0, \ (i,k) \in R, \ l = 1, \ldots, n \tag{25}$$

$$x_{ijl} \in \{0,1\}, \ i = 1, \ldots, m-1, \ 1 \le j \le l \le n \tag{26}$$

$$y_{ijkl} \in \{0,1\}, \ (i,k) \in R, \ 1 \le j \le l \le n \ . \tag{27}$$

The constraints (22) and (23) define a feasible threading in the terms of x-variables. They correspond to (7). The constraints (24) and (25) connect the x-variables and the y-variables. They correspond to (8).

Andonov et al. [7,8] and Xu et al. [6] propose similar MIP models for the protein threading problem. They use the standard branch-and-bound algorithm with LP bounds (constraints (26) and (27) are replaced by $0 \le x_{ijl} \le 1$ and $0 \le y_{ijkl} \le 1$) to solve them. In this paper we explore the specific structure of the above model and define a Lagrangian relaxation which can be easily solved by the algorithm from the previous section. The advantage of this approach is that we obtain tighter bounds with less computational efforts.

As in the previous section, we relax the constraints

$$\sum_{j=1}^{l} x_{k-1,j,l} - \sum_{j=1}^{l} y_{ijkl} = 0, \ (i,k) \in R^{\alpha}, \ l = 1, \ldots, n \tag{28}$$

$$\sum_{l=j}^{n} x_{ijl} - \sum_{l=j}^{n} y_{ijkl} = 0, \ (i,k) \in R^{\beta}, \ j = 1, \ldots, n \ . \tag{29}$$

We associate Lagrangian multipliers λ_{ik}^{l} to the constraints (28) and λ_{ik}^{j} to (29). The constraints imposing an order on the free ends of the links (12) and (13) can also be expressed as linear constraints on the y-variables and added to the model. In this way we obtain a *Lagrangian relaxation* which can be shortly written as

$$z_{\text{LR}}(\lambda) = \min\{cx + dy + \lambda(Cx + Dy) \ : \ (x,y) \in X\} \ , \tag{30}$$

where: x is the vector of the variables x_{ijl}; y is the vector of the variables y_{ijkl}; c is the vector of the coefficients c_{ijl}; d is the vector of the coefficients c_{ijkl};

(C, D) is the matrix of the relaxed constraints (28) and (29); λ is the vector of the Lagrangian multipliers λ_{ik}^l and λ_{ik}^j; and X is the relaxed set of feasible threadings.

It is well known that $z_{\text{LR}}(\lambda)$ is a lower bound on the optimal objective value of the original problem *for any* $\lambda \in \mathbb{R}^{rn}$. In the previous section we have shown how to solve the Lagrangian relaxation for $\lambda = 0$. The same algorithm can be used to solve the relaxation for any λ if the score coefficients are modified appropriately (c is replaced by $c + \lambda C$ and d is replaced by $d + \lambda D$).

The problem of finding the tightest among the Lagrangian relaxations,

$$z_{\text{LD}} = \max_{\lambda \in \mathbb{R}^{rn}} z_{\text{LR}}(\lambda) \ , \tag{31}$$

is called *Lagrangian dual*. There are two commonly used methods to solve the Lagrangian dual – column generation and subgradient optimization. In our algorithm we use the second one since the experiments have shown that it performs better and diverges faster for our problem. Below we give a short description of the subgradient algorithm used.

Initialization: Let $\lambda^0 = 0$, $\theta_0 = 1$, $t = 0$.

Iteration t: Solve the Lagrangian relaxation for $\lambda = \lambda^t$. Let (x^t, y^t) be the solution found. Let $s^t = -Cx^t - Dy^t$ (it is known that s^t is a subgradient of $z_{\text{LR}}(\lambda)$ for $\lambda = \lambda^t$). If $s^t = 0$ (that is, if (x^t, y^t) is feasible for the original problem) then stop ((x^t, y^t) is optimal solution of the original problem). Otherwise let $\theta_{t+1} = \theta_t \rho$. If $\theta_{t+1} < \varepsilon$ then stop. Otherwise let $\lambda^{t+1} = \lambda^t + \theta_{t+1} s^t$, $t = t + 1$.

The parameters $0 < \rho < 1$ and ε determine the decrease of the subgradient step and when to stop. In our implementation $\varepsilon = 0.01$ and ρ is chosen so that the algorithm stops after 500 iterations. This procedure guarantees that λ^t converges to the optimal value of λ when $t \to \infty$ (provided that θ_0 and ρ are sufficiently large). In practice, we have no guarantee to find the real optimum. Moreover, the sequence $z_{\text{LR}}(\lambda^0), z_{\text{LR}}(\lambda^1), \ldots$ is not monotonically increasing and that is why the lower bound provided by the subgradient optimization is $\max_t \{z_{\text{LR}}(\lambda^t)\}$ which is not always z_{LD}.

2.3   Branch-and-Bound Algorithm

We use a best-first branch-and-bound algorithm to solve the protein threading problem. The lower bounds at each node are computed using Lagrangian dual as described in the previous section. Since $z_{\text{LR}}(\lambda^t)$ is a lower bound on the optimal objective value for each t, the subgradient optimization can be stopped prematurely if at some iteration t, $z_{\text{LR}}(\lambda^t)$ is greater or equal to the current record. In addition, each subgradient iteration provides a feasible solution of the original problem (the one determined by the values of the x^t-variables). The objective value of this solution is compared to the current record and the record is updated if necessary.

The splitting of a problem in a given node of the branch-and-bound tree into subproblems is done by restricting the possible positions of the blocks. Let in a given node the possible positions of block i be between l_i and u_i,

$i = 1, \ldots, m$. The dynamic programming recurrences can be easily modified to take into account these restrictions, or alternatively, the score coefficients corresponding to impossible positions can be set to $+\infty$. Consider the solution (π, α, β) of the relaxation of the subproblem of this node. Let (i, k) be the link for which the difference $|c_{i\pi_i k\pi_k} - c_{i\alpha_{ik} k\beta_{ik}}|$ is maximal. If this maximum is zero then the solution is feasible for the original problem and we are done. Otherwise, suppose that $(i, k) \in R^\beta$, then $\pi_i \neq \alpha_{ik}$. Let $\gamma = \lfloor \frac{\pi_i + \alpha_{ik}}{2} \rfloor$. Then we split into two subproblems by setting $l_i' = l_i$, $u_i' = \gamma$ for the first of them and $l_i'' = \gamma + 1$, $u_i'' = u_i$ for the second one. In the case $(i, k) \in R^\alpha$ we split on block k in similar way.

3   Experimental Results

In order to evaluate the performance of our algorithm and to test it on real problems, we integrated it in the structure prediction tool FROST [9, 10]. In our experiments we used the structure database of FROST, containing about 1200 structure templates, as well as its score function. FROST uses a specific procedure to normalize the alignment scores. For each template in the database this procedure selects 5 groups of about 200 sequences each one. The lengths of the sequences in each group are equal. Each of the about 1000 sequences is aligned to the template. The values of the score distribution function F in the points 0.25 and 0.75 are approximated by this empirical data. When a "real" query is threaded to this template, the raw alignment score S is replaced by the *normalized distance* $\frac{F(.75) - S}{F(.75) - F(.25)}$. This procedure involves about 1,200,000 alignments and is extremely computationally expensive. On the other hand, the procedure needs to be repeated after each change of the score function parameters. The implementation of Lathrop and Smith's algorithm [4] used by FROST is not able to produce the required alignments in reasonable time. That is why FROST uses a heuristic steepest-descend-like alignment algorithm to compute the distributions. Even using this approximate algorithm, the computing of score distributions of all templates takes a couple of months on a 16 PC cluster.

To test the efficiency of our algorithm we used the data from 9,136 threadings made in order to compute the distributions of 10 templates. Figure 2 presents the running times for these alignments. The optimal threading was found in less than one minute for all but 34 instances. For 32 of them the optimum was found in less than 4 minutes and only for two instances the optimum was not found in one hour. However, for these two instances the algorithm produced in one minute a suboptimal solution with a proved objective gap less than 0.1%.

It is interesting to note that for 79% of the instances the optimal solution was found in the root of the branch-and-bound tree. This means that the Lagrangian relaxation produces a solution which is feasible for the original problem. The same phenomenon was observed in [6–8] where integer programming models are solved by linear relaxation. The advantage of our method is that the Lagrangian relaxation is solved much faster by our dedicated algorithm than the linear relaxation by general purpose simplex method. For comparison, the times

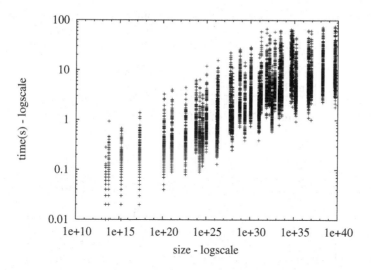

Fig. 2. Running times of 9,136 threading instances as a function of the search space size. The experiment is made on 1.8GHz Pentium PC with 512MB RAM.

to solve instances of size of order 10^{38} reported in [8] are more than one hour, while instances of that size are solved in about 15 seconds by our algorithm.

We compared our algorithm (L) to two other algorithms used by FROST – a steepest-descent heuristic (H) and an implementation of the branch-and-bound algorithm from [4] (B). The comparison was made over 952 instances (the sequences threaded to the template 1ASYA when computing its score distribution). Each of the three algorithms was executed with a timeout of 1 minute per instance. We compare the best solutions produced during this period. The results of this comparison are summarized in Table 1. For the smallest instances (the first line of the table) the performance of the three algorithms is similar, but for instances of greater size our algorithm clearly outperforms the other two. It was timed out only for two instances, while B was timed out for all instances. L finds the optimal solution for all but 2 instances, while B finds it for no instance. The algorithm B cannot find the optimal solution for any instance from the fourth and fifth lines of the table even when the timeout is set to 2 hours. The percentage of the optima found by H degenerates when the size of the problem increases. Note however that H is a heuristic algorithm which produces solutions without proof of optimality. Table 2 shows the distributions computed by the three algorithms. The distributions produced by H and especially by B are shifted to the right with respect to the real distribution computed by L. This means that for example a query of length 638AA and score 110 will be considered as significantly similar to the template according to the results provided by B, while in fact this score is in the middle of the score distribution.

The use of our algorithm made possible to compute the exact score distributions of all templates from the FROST database for the first time. An experiment

Table 1. Comparison between three algorithms: branch-and-bound using Lagrangian relaxation (L), heuristic steepest-descent algorithm (H), and branch-and-bound of Lathrop and Smith (B). The results in each row are average of about 200 instances.

query length	m	n	$\|T\|$	average time(s) L H B	opt(%) L H B
342	26	4	3.65e03	0.0 0.1 0.0	100 99 100
416	26	78	1.69e24	0.6 43.6 60.0	100 63 0
490	26	152	1.01e31	2.6 53.8 60.0	100 45 0
564	26	226	1.60e35	6.4 56.6 60.0	100 40 0
638	26	300	1.81e38	12.7 59.0 60.0	99 31 0

Table 2. Distributions produced by the three algorithms.

query length	distribution (L) $F(.25)$ $F(.50)$ $F(.75)$	distribution (H) $F(.25)$ $F(.50)$ $F(.75)$	distribution (B) $F(.25)$ $F(.50)$ $F(.75)$
342	790.5 832.5 877.6	790.5 832.6 877.6	790.5 832.5 877.6
416	296.4 343.3 389.5	299.2 345.4 391.7	355.2 405.5 457.7
490	180.6 215.2 260.4	184.5 219.7 263.4	237.5 290.4 333.0
564	122.6 150.5 181.5	126.3 157.5 187.9	183.3 239.3 283.4
638	77.1 109.1 142.7	87.6 118.5 150.0	154.5 197.0 244.6

on about 200 query proteins of known structure shows that using the new algorithm improves not only the running time of the method, but also its quality. When using the exact distributions, the sensitivity of FROST (measured as the percentage of correctly classified queries) is increased by 7%. Moreover, the quality of the alignments produced by our algorithm (measured as the difference with the VAST alignments) is also about 5% better compared to the quality of the alignments produced by the heuristic algorithm.

4   Conclusion

In this paper we presented an efficient algorithm for threading a query sequence to a structure template. The algorithm is based on standard combinatorial optimization techniques adapted to the problem. The integration of this algorithm in the structure prediction tool FROST allows to reduce the time of computing the score distributions of the templates, a procedure which must be repeated after each modification of the score function, from several months to a couple of days. The experiments show that the algorithm helps not only to improve the running time, but also the prediction accuracy. The model used is based on several assumptions widely adopted by the protein threading community. It can be easily plugged in most of the existing threading methods. The performance of the algorithm can be further improved by tunning some of its parameters, such as the choice of the link covering set C and the subgradient optimization parameters. Another open question is the adaptation of the algorithm to local query-to-structure alignment needed to predict the structure of multi-domain proteins.

References

1. Head-Gordon, T., Wooley, J.C.: Computational challenges in structural and functional genomics. IBM Systems Journal **40** (2001) 265–296
2. Lengauer, T.: Computational biology at the beginning of the post-genomic era. In Wilhelm, R., ed.: Informatics: 10 Years Back - 10 Years Ahead. Volume 2000 of Lecture Notes for Computer Science. Springer (2000) 341–355
3. Lathrop, R., Rogers Jr., R., Bienkowska, J., Bryant, B., Buturovic, L., Gaitatzes, C., Nambudripad, R., White, J., Smith, T.: Analysis and algorithms for protein sequence-structure alignment. In Salzberg, S., Searls, D., Kasif, S., eds.: Computational Methods in Molecular Biology. Elsevier Science (1998) 227–283
4. Lathrop, R., Smith, T.: Global optimum protein threading with gapped alignment and empirical pair potentials. J. Mol. Biol. **255** (1996) 641–665
5. Xu, Y., Xu, D., Uberbacher, E.C.: An efficient computational method for globally optimal threading. Journal of Computational Biology **5** (1998) 597–614
6. Xu, J., Li, M., Lin, G., Kim, D., Xu, Y.: RAPTOR: optimal protein threading by linear programming. Journal of Bioinformatics and Computational Biology **1** (2003) 95–118
7. Andonov, R., Yanev, N.: Solving the protein threading problem in parallel. In: HiCOMB 2003 – Second IEEE International Workshop on High Performance Computational Biology. (2003)
8. Andonov, R., Balev, S., Yanev, N.: Protein threading: From mathematical models to parallel implementations. INFORMS Journal on Computing (2004) To appear.
9. Marin, A., Pothier, J., Zimmermann, K., Gibrat, J.F.: FROST: a filter-based fold recognition method. Proteins **49** (2002) 493–509
10. Marin, A., Pothier, J., Zimmermann, K., Gibrat, J.F.: Protein threading statistics: an attempt to assess the significance of a fold assignment to a sequence. In Tsigelny, I., ed.: Protein structure prediction: bioinformatic approach. International University Line (2002)
11. Lathrop, R.: The protein threading problem with sequence amino acid interaction preferences is NP-complete. Protein Engineering **7** (1994) 1059–1068
12. Akutsu, T., Miyano, S.: On the approximation of protein threading. Theoretical Computer Science **210** (1999) 261–275
13. Nemhauser, G.L., Wolsey, L.A.: Integer and Combinatorial Optimization. Wiley (1988)

Protein-Protein Interfaces: Recognition of Similar Spatial and Chemical Organizations

Alexandra Shulman-Peleg[1],[*], Shira Mintz[2],
Ruth Nussinov[2],[3],[**], and Haim J. Wolfson[1]

[1] School of Computer Science, Raymond and Beverly Sackler
Faculty of Exact Sciences
Tel Aviv University, Tel Aviv 69978, Israel
shulmana@post.tau.ac.il
[2] Sackler Inst. of Molecular Medicine, Sackler Faculty of Medicine
Tel Aviv University, Tel Aviv 69978, Israel
[3] Basic Research Program, SAIC-Frederick, Inc
Lab. of Experimental and Computational Biology
Bldg. 469, Rm. 151, Frederick, MD 21702, USA

Abstract. Protein-protein interfaces, which are regions of interaction between two protein molecules, contain information about patterns of interacting functional groups. Recognition of such patterns is useful both for prediction of binding partners and for the development of drugs that can interfere with the formation of the protein-protein complex. We present a novel method, Interface-to-Interface (I2I)-SiteEngine, for structural alignment between two protein-protein interfaces. The method simultaneously aligns two pairs of binding sites that constitute an interface. The method is based on recognition of similarity of physico-chemical properties and shapes. It assumes no similarity of sequences or folds of the proteins that comprise the interfaces. Similarities between interfaces recognized by I2I-SiteEngine provide an insight into the interactions that are essential for the formation of the complex and can be related to its function. Its high efficiency makes it suitable for large scale database searches and classifications.

Web server: http://bioinfo3d.cs.tau.ac.il/I2I-SiteEngine

1 Introduction

Most of the cellular processes are governed by association and dissociation of protein molecules. The understanding of such processes can shed light on the mechanism of molecular recognition. A *protein-protein interface* is defined by a pair of regions of two interacting protein molecules that are linked by non-covalent bonds. Analysis and classification of protein-protein interfaces [1, 2] is

---

[*] To whom correspondence should be addressed.

[**] The publisher or recipient acknowledges right of the U.S. Government to retain a nonexclusive, royalty-free license in and to any copyright covering the article. Funded in part by the NCI under contract NO1-CO-12400.

I. Jonassen and J. Kim (Eds.): WABI 2004, LNBI 3240, pp. 194–205, 2004.
© Springer-Verlag Berlin Heidelberg 2004

the first step in deciphering the driving forces stabilizing molecular interactions. Recognition of certain interface binding organizations shared by different protein families suggests their important contribution to the formation and stability of protein-protein complexes. This may constitute targets for drug discovery and assist in predicting side effects. Furthermore, similar interfaces suggest not only similar binding organizations, but also similarity in binding partners and function and may provide hints for potential drug leads that will mimic them.

While sequence patterns have been widely used for functional annotations [3], many functionally similar binding sites are sequence order independent [4]. In such cases methods that assume sequence order may be inapplicable. A sequence order independent method was used by Keskin et al. [5] for clustering of all the known structures of protein complexes. The geometric hashing procedure used in their method considered only the geometric constraints between the interface C_α atoms. However, side chains play an important role in the interaction between two molecules. Current methods that do consider side chain atoms align single binding sites and do not consider their interacting partner [4, 6–8].

This paper presents a novel method, Interface-to-Interface (I2I)-SiteEngine, to recognize similarities between protein-protein interfaces independent of the sequence or the fold of the proteins that comprise them. In addition to geometric considerations used in previous alignment methods, this method takes into account biological considerations in the form of physico-chemical properties of the interacting atoms (both backbone and side-chain). The novelty of I2I-SiteEngine is in recognition of patterns of interacting functional groups shared by a pair of interfaces. Extending the algorithmic approach of our previous method for comparison of small molecule binding sites, SiteEngine [8], the current method performs a simultaneous alignment of *two* binding sites that constitute an interface. Such a simultaneous alignment not only improves the performance of the algorithm, but also may provide more significant biological results. The method introduces a hierarchical scoring scheme which is similar to SiteEngine, but is applied simultaneously to both sides of the interface. First, using a low-resolution representation by chemically important surface points, it performs efficient scoring and filtering of all possible solutions, while retaining the correct ones. Then, as the number of potential solutions is reduced to a smaller subset, the resolution of the molecular representation is increased, leading to more precise calculations. These compare the similarity of the surfaces as well as of local shapes of the chemically similar regions. In this paper we focus on the algorithmic improvements that are introduced by I2I-SiteEngine to specifically treat pairs of binding sites that interact with each other.

We apply the method on a *pilot* dataset to define clusters that contain similar interfaces. Some of the clusters include similar interfaces comprised by proteins with different structural folds. We analyze these clusters and show biological applications which emphasize the importance of such classifications and of procedures to search them.

2   Method

Problem Definition. We define an *interface* as an unordered pair of interacting binding sites (A and B), that belong to two non-covalently linked protein molecules. Two interfaces are considered to be similar, if the binding sites that comprise them share similar physico-chemical properties and shapes. Given two interfaces $I=(A, B)$ and $I'=(A', B')$ the goal is to find the best alignment between them. Specifically, let S denote a set of scoring functions that are used to measure the similarity of the aligned properties. The problem which is heuristically solved by I2I-SiteEngine can be formalized as follows: find a rigid transformation T (rotation and translation) that maximizes the value of $S(I, T(I'))$.

In addition, we assume that the correspondence between the binding sites of the two complexes is unknown, meaning that the binding site A, can be aligned either to A' or to B'. Since the algorithmic procedures that are applied in both cases are the same, the description below refers only to the first option. However, both alignments are considered by the method and the solution that provides the highest score is selected.

Below we present the main stages of the algorithm that include: (1) representation; (2) calculation of the candidate transformations, T, by a matching algorithm (3) scoring the solutions by the set of functions S.

Representation. Efficient, biologically significant, representation of each binding site is crucial for the recognition of functional similarities between unrelated proteins. As depicted in Figure 1, each binding site is represented by the surface of its binding region and by the set of its important functional groups.

The interacting surface is defined by a set of its solvent accessible surface points [9] that are located less than 4Å from the surface of the other protein. Following the definition of Schmitt et al [6], each amino acid of a protein is represented as a set of its important functional groups, localized by pseudocenters, according to the interactions in which it may participate. Each surface point is assigned a

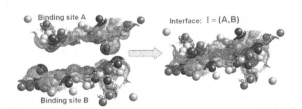

Fig. 1. An interface as a pair of interacting binding sites. Pseudocenters are represented as balls and the surface patches as dots. Hydrogen bond donors are blue, acceptors - red, donors/acceptors - green, hydrophobic aliphatic - orange and aromatic - white.

physico-chemical property according to the functional group to which it belongs and only surface exposed pseudocenters are retained. Surface points that are represented by the same pseudocenter constitute a physico-chemical surface patch. A *patch center* is the surface point nearest to the center of gravity of the patch. The average curvature of the surface patch is estimated by the solid angle shape function [10,11] computed at the *patch center*. This parameter of shape is assigned to the corresponding original pseudocenter and is used throughout the algorithm for fast shape comparisons. In addition, a set of all the *patch centers* of the interface is used in a subsequent stage as a low resolution representation of the surfaces.

The Matching Algorithm. In this stage we compute all of the candidate transformations, T, that superimpose one interface onto the other. When considering protein-protein interfaces, we are supplied with valuable information regarding the functional groups of two binding sites that interact with each other. Utilizing this information increases the speed and the quality of the alignment. Therefore, in the matching stage, for each binding site we consider only *interacting* pseudocenters which have a *complementary* physico-chemical property (with which it can interact) at the other binding site. Specifically, hydrogen bond donors are complementary to acceptors, while hydrophobic aliphatic and aromatic pseudocenters can interact only with similar features. Assuming that at least three such pseudocenters must be present in each interface, we define each triplet of interacting pseudocenters as an *I-triangle*.

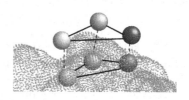

Fig. 2. Interacting triangle (I-triangle). Pseudocenters are represented as balls (colored as in Figure 1) and the surfaces of the two molecules as green dots. Dotted lines represent the interactions.

The flow of the matching algorithm is presented in Figure 3. Given two interfaces $I=(A, B)$ and $I'=(A', B')$ the first stage is to recognize the interacting triangles of each interface. This is achieved by a supplementary hashing procedure that stores all triplets of pseudocenters from the binding sites B and B'. These hash tables are used to check each triplet of pseudocenters from the binding sites A and A' whether it can form three interaction thus creating an I-triangle. Specifically, each triplet of pseudocenters of A (A') is used to access the hash

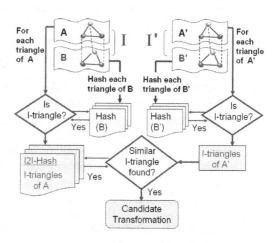

Fig. 3. Overview of the matching algorithm.

table of B (B') to check whether there are three centers in B (B') that have complementary properties at suitable spatial locations.

Specifically, each triplet of pseudocenters of A (A') is used to access the hash table of B (B') to check whether there are three centers in B (B') that have complementary properties at suitable spatial locations. If a triangle of A is recognized as an I-triangle, it is stored in the main matching hash table, which we denote I2I-Hash. If a triangle of A' is recognized as an I-triangle it is used to access the I2I-Hash of A to look for a similar I-triangle (similar physico-chemical properties, curvatures and side lengths[1]). If found, the two I-triangles are used to define a candidate transformation (rotation and translation) that superimposes one interface onto the other.

The keys to all the hash tables are the three parameters of the side lengths of a triangle and an additional index, that encodes the properties of the nodes. Nodes that can function both as hydrogen donors and acceptors are encoded twice, once as a donor and once as an acceptor. The hash tables of B and B' are accessed with the index of complementary properties and additional distance constraints are imposed to ensure that the matched nodes are located close enough to interact with each other.

The Scoring Functions. In this stage we evaluate the candidate transformations that superimpose one interface onto another and select the *one* that provides the best alignment of physico-chemical properties and shapes. Here we extend the scoring procedures presented in the SiteEngine method [8] to a simultaneous comparison between two pairs of binding sites.

Surface Scoring. We use two scoring functions to compare shapes and physico-chemical properties of the surfaces superimposed by each candidate transformation. Both scoring schemes apply each candidate transformation to a certain set of surface points and compare the physico-chemical properties as well as the shapes of the corresponding regions of the two interfaces. The first score, *Fast Low-Resolution Score*, applies this concept to a set of *patch centers*, which provide a low-resolution representation of the surfaces of the interface. The computation of this score is extremely fast and it efficiently estimates the potential similarity of the regions superimposed by a candidate transformation. The high ranking solutions are further clustered according to the RMSD between the pseudocenters under the candidate transformation. After the number of candidate solutions is reduced, the second score, the *Overall Surface Score*, performs a thorough comparison of the overall surfaces aligned by each transformation using a higher level of resolution of molecular representation.

1:1 Correspondence Score. For each retained candidate transformation, we determine a one-to-one correspondence between the two sets of pseudocenters of the two interfaces. A set of pseudocenters of an interface $I=(A, B)$ is the union of the sets of pseudocenters of binding sites A and B that constitute it. The correspondence is obtained by calculating the maximum weight match in a weighted

---

[1] The side lengths of the triangles are similar up to a user defined threshold (3.0Å in this paper).

bipartite graph [12,13], which represents the largest set of pairs of similar pseudo-centers. The construction of a weighted *bipartite graph* for comparison between two protein-protein interfaces $I=(A,B)$ and $I'=(A',B')$ is performed in the following way: (1) Each pseudocenter from binding sites A, B, A' and B' defines a node. (2) Assuming that a candidate transformation aligns a binding site A to A' and a binding site B to B', edges of a *bipartite graph* can only connect nodes of A to A' and nodes of B to B'. Following these restrictions, an edge is added between each pair of pseudocenters that have similar spatial locations and physico-chemical properties. (3) Each edge is assigned a weight that reflects the differences in distance and shape between the nodes. The maximum weight match [13] in this graph, provides a 1:1 correspondence between subsets of pseudocenters of the two interfaces. Due to restriction on the creation of the edges we obtain two separate 1:1 correspondences: one between subsets of pseudocenters of A and A', and another between subsets of B and B'. The obtained 1:1 correspondence is used for two purposes: One is to improve each candidate transformation by the Least-Squares Fitting method [14]. The other is to score this transformation by summing the similarity scores between the matched pseudocenters and their corresponding surface regions.

Final Ranking. Each candidate transformation is assigned a score, $S(I,T(I'))$, which is the weighted sum of all the score functions described above. A candidate transformation with the highest score is selected. When searching a dataset of interfaces for those that are similar to a specific interface of interest, the score of each comparison is normalized by the score of the query compared to itself. We denote this score as *Match Score* and it represents how much of the binding pattern of interest was found to match during the search.

Complexity and Running Times. The complexity of the algorithm is $O(N \cdot K \cdot m)$, where N is the maximal number of I-triangles of an interface, K is the maximal number of I-triangles retrieved in a hash table query, and m is the number of pseudocenters in the largest binding site (selected within A, B, A', B'). Theoretically, when matching triplets of arbitrary nodes, $N \cdot K$ is $O(m^6)$. However, in practice, since we consider only I-triangles and limit their side lengths to be within a predefined range, $N \cdot K$ is proportional to m^2 and the overall running time of the algorithm is $O(m^3)$. For a typical interface in our *pilot* dataset $m \sim 70$. The mean overall CPU time, measured for the comparisons of this study is 28 seconds (3.0 GHz Xeon processors, 4GB memory). For each comparison the running time includes the application of the method twice for the selection of the highest ranking correspondence between the binding sites. It does not include the preprocessing of the surfaces and grids, which is done offline in a matter of seconds. Sample running times of specific algorithm executions are provided in Table 1. The I2I-SiteEngine method, as well as additional details regarding its implementation and default parameters are available on line at our web server http://bioinfo3d.cs.tau.ac.il/I2I-SiteEngine.

3   Classification of Protein-Protein Interfaces

We applied I2I-SiteEngine to classify a *pilot* dataset consisting of 64 common protein-protein interfaces. As a result, these were clustered into 22 different groups, which are detailed on our website. Protein complexes with interfaces that belong to the same cluster are considered to share similar spatial and chemical organizations of their interacting regions.

Similarity Ranking and Clustering. The first stage was to rank the dataset proteins according to their similarity to each other. To achieve this, I2I-SiteEngine was applied to all pairs of dataset interfaces (except for where surface area differs by more than 50% of the smaller interface). Specifically, each time a certain interface is used as a query. The query is compared to all other dataset interfaces, ranking them in a decreasing order according to their similarity to this query. Table 1 provides three examples of such comparisons. The query interfaces in these examples were: (a) Clip bound to class II Mhc Hla-Dr3 (1a6a), (b) Trypsin(ogen) complexed with Pancreatic Trypsin inhibitor (1bzx) and (c) Glycine N-Methyltransferase (1d2h). The top ranking solutions are the interfaces that are most similar to the query. As can be seen, many of them are comprised of proteins that share very low sequence similarity with the query. In addition, most of these proteins have totally different overall protein folds. Therefore, the similarity between them can not be recognized by sequence alignment methods or by structural alignment methods that align the overall backbones of the proteins. Yet, the similarity of these interfaces is successfully recognized by I2I-SiteEngine. In the next stage, a clustering algorithm uses the rank lists of all pairwise comparisons to cluster the interfaces into different groups. In this preliminary implementation we applied a greedy clustering procedure that performs several iterations with decreasing cut off values defined by the *Match Score*.

The Obtained Clusters. Following the classification procedures, two types of interface clusters were obtained. The full clusters list is detailed on our website: `http://bioinfo3d.cs.tau.ac.il/I2I-SiteEngine`. Here we present few examples of relatively large clusters that are divided to two types: (1) similar interfaces comprised by proteins with similar overall folds, (2) similar interfaces comprised by proteins with different overall folds. While type 1 is more straightforward, type 2 can suggest either preferred binding organizations or similar functions shared by unrelated proteins.

Consider the rank lists of three clusters presented in bold in Table 1. Table 1(a) presents an example of a type 1 cluster composed of MHC-antigen interfaces. Although in this cluster the MHC molecules are bound to different peptides the similarity between their interfaces is successfully recognized and thus they are clustered together. A more interesting type 2 cluster is presented in Table 1(b). This well-studied [4] functional class of 'serine protease with inhibitor' is comprised of two different folds: the Trypsin-like serine proteases with inhibitor (ranked 1-4 in the Table) and the Subtilisin-like with inhibitor (ranked 5-6). In spite of the different overall folds of the chains, these proteins exhibit a similar

Table 1. The ten top ranking solutions obtained when three interface queries (ranked 1) were searched against the *pilot* dataset: (a) Clip bound to class II Mhc Hla-Dr3 (1a6a), (b) Trypsin with inhibitor (1bzx), and (c) Glycine N-Methyltransferase (1d2h). For each comparison, the first column presents the PDB codes followed by the corresponding chains, in the order determined by the correspondence identified by the method. Column two presents the score that indicates the extent of similarity. Column three presents the overall sequence similarity to the query chains, e.g. 1sbnEI - 15/13, means that sequence similarity of chains E of 1sbn and of 1bzx is 15%, and of chains I is 13%. Column four presents the algorithm running times. In bold are the interfaces that later on were classified to belong to the same cluster as the query.

Rank	(a) Rank List of 1a6a				(b) Rank List of 1bzx				(c) Rank List of 1d2h			
	PDB	Match Score	Seq. Sim.(%)	Time (sec.)	PDB	Match Score	Seq. Sim.(%)	Time (sec.)	PDB	Match Score	Seq. Sim.(%)	Time (sec.)
1	**1a6aBC**	100	100/100	8	**1bzxEI**	100	100/100	14	**1d2hAB**	100	100/100	5
2	1aqdBC	52	88/20	8	1tgsZI	56	64/14	21	1axcCA	38	14/14	7
3	1dlhBC	44	87/20	8	1gl1AI	51	41/17	15	1kbaBA	34	5/5	5
4	1d9kDP	38	60/6	10	1acbEI	48	40/13	16	1cdtBA	34	8/8	7
5	1jk8BC	38	61/6	9	3tecEI	37	13/13	15	1clyBA	34	6/11	6
6	1f3jBP	35	60/6	10	1sbnEI	36	15/13	18	1kklAH	33	12/9	6
7	1ydtEI	27	13/5	13	1h28BE	28	15/8	11	1czvAB	32	12/12	4
8	1axcAC	27	13/1	11	1cxzAB	27	17/12	13	1b77AB	32	15/15	4
9	1gl1AI	26	15/5	13	1clyAB	27	18/10	12	1ao7AD	30	15/11	3
10	1clyBA	25	12/1	10	1d9kDP	27	15/3	12	3tecIE	30	7/16	10

function that is related to the interface properties. As expected, our method grouped together all interfaces of serine proteases with inhibitors bound to their catalytic region.

Another interesting example of a type 2 cluster can be seen in Table 1(c). The cluster contains complexes that are comprised of a total of 7 different folds: (1) Methyltransferases (1d2h), (2) Snake toxin-like (1cdt,1kba), (3) DNA clamp (1axc,1b77), (4) PEP carboxykinase-like bound to Hpr-like (1kkl), (5) Galactose-binding domain-like (1czv), (6) P-loop containing nucleoside triphosphate hydrolases bound to Beta-Grasp (1cly), (7) Defensin-like (1dfn, which was added to this cluster at the last iteration of the clustering algorithm and therefore is not presented in the Table). Figure 5 presents a superimposition of members of this cluster according to the transformation obtained by the alignment of corresponding interfaces. Figure 5(a) shows the alignment Coagulation factor V and Cardiotoxin V4II. Figure 5(b) shows the alignment between Glycine N-Methyltransferase and Proliferating Cell Nuclear Antigen. In both cases, methods for sequence comparison [3] failed to recognize any significant pattern. All of the proteins in this cluster perform different functions, but are recognized to have similar interfaces. This may suggest that similar interface binding organizations can be shared by different protein families.

4   Additional Applications

Besides its usefulness for classification of interfaces, our method can be also applied to search databases of protein-protein interfaces. Searches of this type have two major biological applications: (1) fast classification of newly determined complexes. (2) prediction of binding partners and binding modes by recognition of similarity to known complexes. In the first application, an interface of a 'newly' determined complex is extracted, compared to known interfaces and classified.

This may provide valuable knowledge regarding the interface and its biological function. In the second, a complete protein structure is compared to a database of known interfaces to predict its potential binding sites.

These database searches can be performed more efficiently with the help of a dataset of classified interfaces. Instead of comparing to each interface, we compare only to cluster *representatives*, thus gaining efficiency and speed. The *representative* of each cluster is selected as the largest interface of that cluster. Below, we provide examples of the two search procedures applied on the cluster representatives of our *pilot* dataset.

Interface Type Recognition. In this section we describe classification of 'new' interfaces that were not used in the *pilot* dataset. A new interface (query) is compared to all the cluster representatives, except for those whose surface area differs by more than 50% from the query. These examples show the correctness of the obtained clusters and of their representation by representative interfaces.

Beta-Defensin BD. In the following example, an interface of Beta-Defensin BD (1fd4) was compared to all cluster representatives. The top ranking interface was that of G-protein Rap1A in complex with the Ras-Binding-Domain (RBD) of C-Raf1 Kinase (1c1y). Examination of the cluster members represented by this complex revealed an interface comprised by Defensin HNP-3 protein (1dfn), which belongs to the same Defensin family as the query. Although the comparison was only between Beta-Defensin BD and the cluster representatives (none of which is from Defensin family), the resulting classification was correct.

Subtilisin Carlsberg with Eglin C. When the interface of Subtilisin Carlsberg in complex with Eglin C (1cse) was compared to all of the cluster representatives, the interface formed by Trypsin complexed with Animal Kazal-type inhibitor (1tgs) received the highest rank. In spite of the fact that this representative interface belongs to a fold different from that of the query, the 'new' interface was correctly classified to belong to the cluster of serine proteases in complex with an inhibitor.

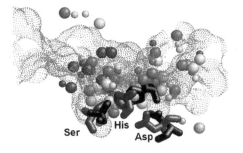

Fig. 4. The interfaces of subtilisin Carlsberg with Eglin C (1cse) and Trypsin with Animal Kazal-type inhibitor (1tgs). The surfaces are depicted as green dots and the pseudocenters as balls (1cse smaller, 1tgs larger, colored as in Figure 1. The catalytic residues of 1tgs (S195-H57-D102) are black and of 1cse (S221-H64-D32) are brown.

SH3 Domain with a Peptide. Another example is classification of an interface of Abl tyrosine kinase SH3 domain complexed with a peptide. Once again, this interface was compared to all the representatives. The top

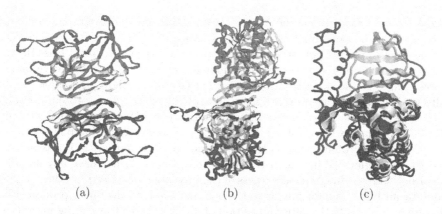

(a)   (b)   (c)

Fig. 5. (a) Coagulation factor V (PDB:1czv, chain A - blue, chain B - red) and Cardiotoxin V4II(PDB:1cdt, chain A - green, chain B - yellow) **(b)** Alignment between Glycine N-Methyltransferase (PDB:1d2h, chain A - blue, chain B - red) and Proliferating Cell Nuclear Antigen (PDB:1axc, chain A - green, chain C - yellow). **(c)** Superimposition of the three top ranking results obtained in searching with a complete structure of G-protein cH-p21 Ras (yellow). In blue are the G-proteins of the recognized solutions. In green (1cly), purple (1cxz) and red (1cee) are the binding partners of these interfaces that can potentially bind to the query.

ranking interface was the catalytic domain of Prommp-2 E404Q mutant complexed with inhibitor (1eak). The cluster represented by this interface contains another interface created by c-Src tyrosine kinase SH3 domain complexed with a peptide. As previously, this classification shows the consistency of the method. It is interesting to note that all of the peptides that create the interfaces of this cluster contain at least three Proline residues. Spatial similarity of two of these was recognized by the method. The matching of these relatively rigid residues may be an explanation for the obtained classification.

Prediction of Binding Partners and Binding Modes. Here we show how the presented method and classification can be used to predict the potential binding partners and modes of a cH-p21 Ras (G-protein). In this application we search the complete surface of this protein for the presence of a binding site, similar to those that constitute known interfaces. Such recognition can provide information regarding its potential binding partners and their binding modes.

From the algorithmic standpoint, in this application we combine the approaches of I2I-SiteEngine and SiteEngine. In the matching stage all I-triangles of an interface are compared to all triangles of the complete protein. The two binding sites of the interface are scored separately. A region of a protein that achieves the highest score in a comparison to one of the binding sites is selected. The binding partners of that site can potentially bind to the recognized region. The superimposition of the complex of the interface on the protein provides prediction of its binding mode. When ranking the results of searching for different interfaces on the surface of a protein it is important to prevent the automatic selection of large interfaces, which have more features, thus receiving a higher score. Therefore, we divide the score of each pairwise comparison by the normalized score of the same binding site when it is searched in its native protein [8].

In order to predict potential binding sites of the G-protein cH-p21 Ras (1he8), its complete protein structure was compared to all of the representative interfaces. Three top ranking solutions were: (1) G-protein RhoA in complex with effector domain of the protein kinase pkn/prk1 (1cxz), (2) G-protein CDC42 in complex with the Gtpase binding domain of Wasp (1cee) and (3) G-protein Rap1A in complex with c-Raf1 RBD (1c1y). Therefore, the binding partners of these three proteins may potentially bind to G-protein cH-p21 Ras at the recognized regions. Whereas in the first two cases we do not have any specific information regarding the correctness of the prediction, the third interface is created by an RBD domain similar to the one that binds to the query protein [15]. Figure 5(c) presents the superimposition of all these three top scoring interfaces on the query by the transformation recognized by the comparison of the corresponding interfaces. Here we have shown the correctness of the method by examples that can be verified by methods of sequence and structural alignment. However, the presented method considers only the surfaces and the functional groups of the interfaces and thus can recognize similarities shared by proteins with different overall sequences or folds. Thus it can recognize similarities that can not be detected by other methods.

5   Conclusions and Future Work

Recognition of similar patterns of interactions between evolutionary unrelated proteins is important for various biological applications. Here we presented a method that can recognize such similarities without any assumption regarding the similarity of the sequences or the folds. Our method considers physicochemical and geometrical considerations of the side-chains as well as the backbone. It is efficient and can be applied to large scale database searches and classifications. However, it has several weaknesses. First, it addresses protein molecules as rigid bodies and considers flexibility only through a set of thresholds that allow a certain variability in the locations. Second, there is no implicit treatment of electrostatic potentials that are known to have an impact on the protein interactions. Such issues will be addressed in future research.

We have applied our method to classify a *pilot* dataset of protein interfaces and have shown its usefulness for searching applications. Although the constructed dataset is limited in size, it is sufficient to already show a clear representation of the interfaces clusters. Motivated by this experience we intend to apply I2I-SiteEngine to all PDB complexes for the complete classification of known interfaces. We hope that this will suggest preferred chemical organizations shared by similar interfaces. The insight we have gained from the pairwise interface alignment will facilitate the development of a tool for multiple interfaces alignment based on functional groups that is currently under development.

Acknowledgments

We would like to thank Maxim Shatsky and Dina Schneidman for useful discussions and for their contribution of software to this project. We would like to thank Dr. Shuo Liang Lin for valuable suggestions. This research has been supported in part by the "Center of Excellence in Geometric Computing and its Applications" funded by the Israel Science Foundation. The research of H.J.W. and A.S-P. is partially supported by the H. Minkowski-Minerva Center for Geometry at TAU. The research of R.N. has been funded in whole or in part with Federal funds from the NCI, NIH, under contract number NO1-CO-12400. The content of this publication does not necessarily reflect the view or policies of the Department of Health and Human Services, nor does mention of trade names, commercial products, or organization imply endorsement by the U.S. Government.

References

1. Valdar, W.S., Thornton, J.M.: Protein-protein interfaces: analysis of amino acid conservation in homodimers. Proteins **42** (2001) 108–124
2. Lo Conte, L., Chothia, C., Janin, J.: The atomic structure of protein-protein recognition sites. J. Mol. Biol. **285** (1999) 2177–2198
3. Falquet, L., Pagni, M., Bucher, P., Hulo, N., Sigrist, C., Hofmann, K., Bairoch, A.: The PROSITE database, its status in 2002. Nucleic Acids Res. **30** (2002) 235–238
4. Wallace, A.C., Laskowski, R.A., Thornton, J.M.: Derivation of 3D coordinate templates for searching structural databases: application to Ser-His-Asp catalytic triads in the serine proteinases and lipases. Protein Sci. **5** (1996) 1001–1013
5. Keskin, A., Tsai, C.H., Wolfson, H.J., Nussinov, R.: A new, structurally nonredundant, diverse dataset of protein-protein interfaces and its implications. Protein Sci. (2004) in press.
6. Schmitt, S., Kuhn, D., Klebe, G.: A new method to detect related function among proteins independent of sequence or fold homology. J. Mol. Biol. **323** (2002) 387–406
7. Kinoshita, K., Nakamura, H.: Identification of protein biochemical functions by similarity search using the molecular surface database ef-site. Protein Sci. **12** (2003) 1589–1595
8. Shulman-Peleg, A., Nussinov, R., Wolfson, H.J.: Recognition of functional sites in protein structures. J. Mol. Biol. **339(3)** (2004) 607–633
9. Connolly, M.: Analytical molecular surface calculation. J. Appl. Cryst. **16** (1983) 548–558
10. Connolly, M.L.: Measurement of protein surfaces shape by solid angles. J. Mol. Graph. **4** (1986) 3–6
11. Duhovny, D., Nussinov, R., Wolfson, H.J.: Efficient unbound docking of rigid molecules. In Guigo, R., Gusfield, D., eds.: Workshop on Algorithms in Bioinformatics. Volume 2452. LNCS, Springer Verlag (2002) 185–200
12. Cormen, T.H., Leiserson, C.E., Rivest, R.L.: Introduction to Algorithms. The MIT Press (1990)
13. Mehlhorn, K.: The LEDA platform of combinatorial and geometric computing. Cambridge University Press (1999)
14. Kabsch, W.: A discussion of the solution for the best rotation to relate two sets of vectors. Acta Crystallogr. **A 34** (1978) 827–828
15. Paduch, M., Jelen, F., Otlewski, J.: Structure of small G proteins and their regulators. Acta Biochim Pol. **48** (2001) 829–50

ATDD: An Algorithmic Tool
for Domain Discovery in Protein Sequences

Stanislav Angelov[1,*], Sanjeev Khanna[1,**], Li Li[2,***], and Fernando Pereira[1,†]

[1] Department of Computer and Information Science, School of Engineering
University of Pennsylvania, PA 19104, USA
{angelov,sanjeev,pereira}@cis.upenn.edu
[2] Department of Biology, School of Arts and Sciences
University of Pennsylvania, Philadelphia, PA 19104, USA
lili4@sas.upenn.edu

Abstract. The problem of identifying sequence domains is essential for understanding protein function. Most current methods for protein domain identification rely on prior knowledge of homologous domains and construction of high quality multiple sequence alignments. With rapid accumulation of enormous data from genome sequencing, it is important to be able to automatically determine domain regions from a set of proteins solely based on sequence information.
We describe a new algorithm for automatic protein domain detection that does not require multiple sequence alignment and differs from alignment based methods by allowing arbitrary rearrangements (both in relative ordering and distance) of the domains within the set of proteins under study. Moreover, our algorithm extracts domains by simply performing a comparative analysis of a given set of sequences, and no auxiliary information is required. The method views protein sequences as collections of overlapping fixed length blocks. A pair of blocks within a sequence gets a "vote of confidence" to be part of a domain if several other sequences have similar pairs of blocks at roughly the same distance from each other. Candidate domains are then identified by discovering regions in each protein sequence where most block pairs get strong votes of confidence. We applied our method on several test data sets with a fixed choice of parameters. To evaluate the results we computed sensitivity and specificity measures using SMART-derived domain annotations as a reference.

1   Introduction

With the rapidly increasing amount of sequence data coming from various genome projects, one of the major goals of comparative genomics is to determine molecular and cellular functions of proteins encoded by these genomes.

* Supported in part by NSF ITR 0205456 and NIGMS 1-P20-GM-6912-1.
** Supported in part by an Alfred P. Sloan Research Fellowship, NSF Career Award CCR-0093117, and NSF ITR 0205456.
*** Supported in part by Burroughs Wellcome Fund.
† Supported in part by NSF ITR 0205456.

I. Jonassen and J. Kim (Eds.): WABI 2004, LNBI 3240, pp. 206–217, 2004.
© Springer-Verlag Berlin Heidelberg 2004

A general approach for functional characterization of unknown proteins is to infer protein functions based on sequence similarity to annotated proteins in sequence databases using database search tools such as BLAST [1] and FASTA [2]. Such automated procedures are fast and powerful, but also limited [3]. In particular they do not take into consideration the modular organization of proteins. Many proteins consist of multiple independently evolving domains [4]. A protein domain is generally defined as a region within a protein that either forms a specific function, such as substrate binding, or constitutes a stable, compact structural unit. Ignoring the modularity of protein sequences may lead to false transfer of annotation between proteins that have distinct functions but share some highly conserved domains. Therefore, automated techniques for the identification of protein domains can greatly contribute to the understanding of protein function and provide insights into physiology of an organism.

Many methods have been developed to classify proteins into distinct families based on identification of a shared domain. Most of them start with a seed multiple sequence alignment (MSA) of proteins that are known to be functionally related or represent conserved domains and then use the alignment to characterize the domain family. For example, Pfam [5, 6] and SMART [7] use Hidden Markov Models (HMM) built from the seed alignments as profiles for each domain family, BLOCKS [8] constructs position specific matrices, while Prosite [9] defines motif signatures of the functional sites. All of those methods depend crucially on carefully chosen seed alignments. Manual selection of sequences to be included in the seed alignments and hand editing of MSAs by experts is often required to ensure the quality of the seed alignments. Furthermore, alignment tools will often not handle remotely related sequences. These problems led us to investigate alternative methods in which protein domains or conserved regions are identified automatically in a set of proteins from sequence information alone, allowing for domain reordering and avoiding the need for seed alignments.

Our algorithm predicts candidate domain regions by comparative sequence analysis of protein sequences. Each protein sequence is viewed as a collection of overlapping fixed length blocks. Similarities between pairs of blocks are calculated from un-gapped sequence alignments using a specific substitution matrix. A pair of blocks within a sequence gets a "vote of confidence" if several other sequences have similar pairs of blocks separated from each other at roughly similar distance. Candidate domains are then identified by discovering regions in each protein sequence where most block pairs get strong votes of confidence. By appropriately adjusting various parameters such as what constitutes block similarity and when does a pair of blocks get a vote from another sequence, we get a non-local alignment technique that allows us to detect shared domains among sequences that may be remotely related. We evaluated our algorithm using the InterPro [10] annotation. InterPro is a comprehensive documentation resource of protein families, domains and functional sites, where the major protein signature databases, PROSITE [9], Pfam [5, 6], PRINTS [11], ProDom [12] and SMART [13] have been manually integrated and curated. Because of inconsistencies among annotations from these different databases, and possible omissions of putative domains, accuracy metrics with respect to manually annotated data can be misleading. Therefore, we also evaluated our algorithm on synthetic data sets constructed by embedding annotated domains into protein sequences generated by random sampling of regions without domain annotations.

2   Algorithm

Our algorithm can be viewed as composed of three phases. In the first phase (Witness Matrix), we compare each protein sequence σ with all other sequences to extract the pairs of amino acid positions in σ that are potentially correlated i.e. are part of the same domain. In the second phase (Segment Matrix), we translate this pairwise correlation information into a score for each amino acid position in σ, indicating the likelihood of that position being a domain. In the third and the final phase (Domain Extraction), we use a relative thresholding scheme to label each amino acid position as a domain or a non-domain position. We next describe these phases in detail.

The Witness Matrix. The first phase of the algorithm is to create a matrix W for each protein sequence σ, in a given set of proteins P, which collects evidence on which pairs of amino acids in σ are likely to be in a domain. The entry $W[i,j]$ contains a subset $P' \subseteq P$ such that each sequence in P' is a vote for the pair of locations i, j in σ being contained in some domain.

Definition 1 (Block). *A block in a sequence σ is a substring of length b for some fixed parameter b. We will denote by $\sigma(i)$ the block in σ that starts at location i.*

Definition 2 (Block Similarity Function). *A block similarity function f is a Boolean function that is true of a pair of blocks iff the two blocks are "similar" to each other.*

Fix any sequence $\sigma \in P$. We will describe the construction of the witness matrix W for σ. The construction scheme is identical for all sequences. For each pair of blocks $\sigma(i)$, $\sigma(j)$ (possibly overlapping), we perform the following computation. For each sequence $\sigma' \in P$, we check if there is a pair of blocks $\sigma'(i)$, $\sigma'(j')$ such that:

(i) $f(\sigma(i), \sigma'(i'))$ and $f(\sigma(j), \sigma'(j'))$, and
(ii) $(j' - i')/s \leq j - i \leq s(j' - i')$, where s is the *stretch* parameter.

If so, we add σ' as a witness for all pairs of amino acid positions x, y in σ where x is in $\sigma(i)$ and y is in $\sigma(j)$.

In this phase of the algorithm we determine which block pairs of the sequence σ are *preserved* across the protein set P, and thus likely to be together in a domain. The intuition is that a pair of blocks contained within a domain will also appear in other sequences sharing the same domain. In addition, we want to filter block pairs that appear often but in different relative order, or with significantly different distances, as they are likely to be part of related but different domains.

The block similarity function should absorb local amino acid substitutions and a limited number of insertions and deletions between homologous blocks without introducing too many false positives. A possible choice for similarity function is gapped alignment with a substitution matrix and an alignment score cutoff value. As a reasonable approximation for relatively small block sizes, as well as for efficiency reasons, we have used un-gapped alignment. The stretch parameter allows for more significant mutation events such as larger gaps, inserts, or deletes.

The Segment Matrix. The second phase of the algorithm refines the witness information gathered above to classify each amino acid position in σ as a domain or a non-domain position. To do so, we define the notion of a *segment* in a sequence: a segment $[s, e]$ is a substring of σ that starts at location s and ends at location e in σ. For each segment $[s, e]$, we compute a score $M[s, e]$ as follows:

(a) For each sequence $\sigma' \in P$, we compute total number of pairs (i, j) where $s \leq i, j \leq e$ such that $\sigma' \in W[i, j]$. Let $n(\sigma')$ denote this count.
(b) A sequence $\sigma' \in P$ is said to be *relevant* to a segment $[s, e]$ if $n(\sigma')$ is a significant fraction of the total number of pairs in $[s, e]$ (say, $(e - s + 1)^2/2$). The score $M[s, e]$ is defined to be the sum of the counts $n(\sigma')$ of the k relevant sequences with the highest count, normalized by $(e - s + 1)^2$, the total number of pairs in the segment $[s, e]$.

For a segment, we are interested in only relevant sequences $\sigma' \in P$ and filter the others to penalize for gaps, and to prevent accumulation of votes due to *noise* that may potentially outscore conserved but infrequent domains. We further select the top k of the highest voters to hedge against over-represented domains and more conserved stretches within a domain (normalize the score). An example is given in Fig. 1 where the domain with accession number IPR000008 (for clarity denoted by 8) is found to appear more often (it is shared by large fraction of the sequences) than the rest of the domains in the studied data set.

Domain Extraction. Once we have computed the matrix M, we next define an array A of length $|\sigma|$ such that the entry $A[i]$ is the score $M[s, e]$ corresponding to the highest scoring segment of length at least B that contains location i (Through the parameter B we filter noise from small frequently occurring regions as can be seen in Fig. 1). Let $a_{\max}$ be the highest entry in the array A. We now pick a threshold $\theta \in (0, 1]$ and classify each location i in σ as domain or non-domain by the following simple rule: if $A[i] \geq \theta a_{\max}$, then location is part of a domain and it is a non-domain location otherwise.

2.1 Complexity

We now analyze the worst case running time of the algorithm in terms of the number of proteins sequences, say m, the length n of each protein sequence, and

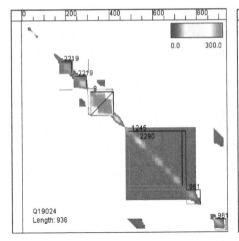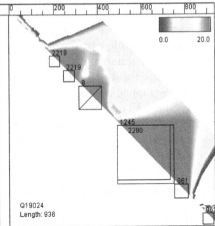

Fig. 1. The 'heatmap' in the left panel shows the list sizes in the witness matrix W, for each pair of positions (i, j). Here, W is computed for the example sequence Q19024. Sizes of less than 30 are not shown for clarity. The corresponding segment matrix M (in the right panel) shows the computed segment scores ($k = 20$) where a point (s, e) represents a segment that starts at position s and ends at e. The rectangular boxes outline the actual domain boundaries (SMART) for the example sequence.

the maximum domain size d (typically between 300 and 400 amino acids). For sequences σ and σ', we can compute for all blocks in σ the respective sorted lists of similar blocks, identified by their starting positions, in σ' in $O(n^2)$ time via dynamic programming by reusing computations for overlapping blocks. Similarly, once the witness matrix for σ is constructed, we compute the segment scores in $O(mn^2)$ time reusing information from overlapping segments.

The dominant term in the algorithm complexity is the computation of the witness lists for the $O(nd)$ ordered block pairs in σ. Recall that blocks that are further than d positions apart are not in a domain by definition. Given a pair of blocks in σ of distance $d' \leq d$ positions and their corresponding similar blocks in σ', we determine if σ' is a witness for that pair by essentially merging the two sorted lists of similar blocks. More precisely, we add d' to each entry in the first list and do the merge in time proportional to the sizes of the lists. This allows us to compute the corresponding block pair in σ' that is of distance closest to d' in $O(n)$ time (each list is at most n in size). In practice, however, lists of similar blocks are significantly shorter than n depending on the similarity threshold we use and the sequence complexity. The experiments we performed suggest that the algorithm in fact spends most time in computing block similarity.

To summarize, the running time per sequence against the whole protein data set is proportional to mn^2d.

3   Implementation and Evaluation

A platform-independent, preliminary version of the method was implemented in Java. It was used and tested on various Windows and Linux machines running

Sun's Java 2 platform. Several parameters control the behavior of the algorithm. We used blocks of size 25 each, and computed block similarity using un-gapped alignment with amino acid substitution matrix BLOSUM62 [14], using a cutoff level of 30. The stretch factor we allowed between pairs of block was set to 1.5 and the maximum allowed distance between blocks to 400 amino acids (aa). For each position in the sequence, we computed its likelihood of being a domain by considering the top 20 voters for each of the regions that contain the amino acid of length at least 50, and normalizing by the highest score. We used these choices of parameters for all the data sets we studied.

3.1 Data Sets

To evaluate the performance of the algorithm, we studied four different data sets and their domain annotations that are available from InterPro database at the EBI[1]. From the corresponding protein families we have chosen random subsets that are typically annotated with both SMART and Pfam. Table 1 summarizes the selected data sets.

Table 1. Test data sets.

Accession No	Description	Number of Sequences	Average Length	Coverage by SMART	Masked Positions
1. IPR001192	Phosphoinositide-specific phospholipase C (PLC), and	704	708aa	44%	8%
IPR001565	synaptotagmin				
2. IPR008936	Rho GTPase activation protein, and	1916	795aa	43%	9%
IPR001849	Pleckstrin-like				
3. IPR001356	Homeobox	1011	348aa	28%	15%
4. IPR001245	Tyrosine protein kinase	1701	706aa	57%	7%

The low complexity regions in the input sequences were masked using the software **seg** [15] with default parameters '12 2.2 2.5 -x'. The 'Masked Position' column in Table 1 shows the average percentage of low complexity positions in a sequence for each data set.

3.2 Evaluating Performance

We assessed the quality of the algorithm's output based on the specificity and sensitivity measures using SMART-derived domain annotations as a reference. Specificity is the ratio of accurately predicted domain positions to the total number of predicted domain positions by our algorithm. Sensitivity is the proportion of accurately predicted domain positions over total number of actual

---

[1] http://www.ebi.ac.uk/interpro

domain positions. Given an accurate reference annotation, the ideal domain discovery method will have both specificity and sensitivity equal to 1. On the other hand, a naive algorithm that marks the entire protein sequence as a domain will have perfect sensitivity, but specificity equal to the percentage of total positions covered by the reference annotation (see Table 1).

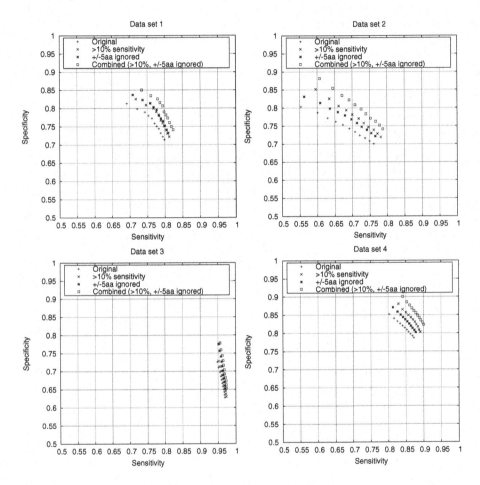

Fig. 2. Trade-off between specificity and sensitivity for threshold θ ranging from 0.9 to 1. High threshold implies lower sensitivity but higher specificity.

Figure 2 shows the trade-off between average per sequence sensitivity and specificity when the likelihood threshold θ is varied from 0.9 to 1 (step 0.01). In addition we removed sequences with less than 10% sensitivity (between 5 and 10% of the input data). We also computed an adjusted specificity-sensitivity curve to eliminate bias introduced around the domain boundaries (summary of actual values is given in Table 2). More precisely, we have ignored ±5aa around the SMART annotated domain boundaries when computing specificity

Table 2. Average specificity and sensitivity for threshold $\theta = 0.98$ using SMART annotation as a reference.

Data Set	Sensitivity (ignored ±5aa)	Specificity	Data Set	Sensitivity (ref. extended by 10aa)	Specificity
1.	0.78	0.83	1.	0.74	0.88
2.	0.67	0.83	2.	0.64	0.88
3.	0.96	0.74	3.	0.94	0.88
4.	0.86	0.88	4.	0.84	0.90

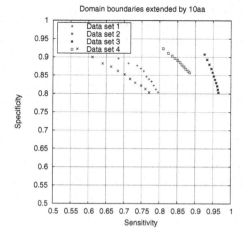

Fig. 3. Trade-off between specificity and sensitivity for extended annotated domain boundaries by 10aa.

and sensitivity. The significant improvement in both sensitivity and specificity when ignoring 5aa around the domain boundaries is evidence that our method find the rough location of domains even when it fails to identify the domains' exact boundaries.

To test if our algorithm consistently overmatches the annotated domains due to its methodology we have also performed the following experiment. We extended the SMART-derived domain annotations by 10aa at each of the domain boundaries and computed the specificity and sensitivity measures removing outliers with less than 10% sensitivity and varying the threshold as above (summary of results shown in Fig. 3). We again observe a significant increase in specificity. The result implies, especially in the case of data set 3 (where we also observe the highest sensitivity) that one weakness of the current approach is a tendency to extend the domain boundaries by several amino acids. One explanation is that blocks partially overlapping with a domain are classified as being part of it, if their common portion is sufficiently well preserved (depending on the similarity cutoff level and the size of the overlap) in the remaining (witness) instances of the same domain. Also, small extensions of conserved regions may still get a high enough votes of confidence for lower values of the threshold θ and thus help incorrectly classify the extensions as domain positions.

Synthetic Data Sets. One of the challenges in evaluating the performance is lack of perfect agreement between known domain annotations. In order to overcome this difficulty we created a synthetic data set with near "perfect" annotation and yet containing realistic domain sequences. We embedded actual SMART-derived sequence domains from data set 1 in random proteins preserving their relative order and distance. We generated the random protein sequences by first building a dictionary with all words of size 3 that appear in non-domain regions of the original sequences, and then sampling from these words at random based on their frequencies. The computed results are shown in Fig. 4. We obtained more than 97% specificity when we extended the domain boundaries of the SMART annotation as outlined above. This suggests that we can further improve the method in terms of specificity by taking into a special consideration blocks at tentative domain boundaries without increasing the top k parameter (Fig. 5).

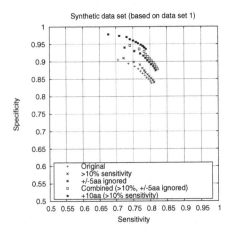

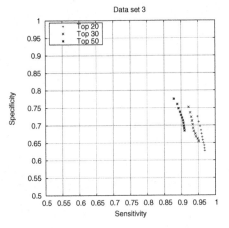

Fig. 4. SMART-derived domains embedded in random sequences.

Fig. 5. Change in specificity and sensitivity for data set 3 for parameter top $k = 20, 30$, and 50.

4   Discussion and Future Directions

The experimental results of our method are encouraging. We were able to obtain simultaneously good sensitivity and specificity for the studied data sets based exclusively on the available sequence information with a fixed, predetermined, set of parameters. The method can further be improved, and at each step of the algorithm several variations are possible. To determine block correspondence between two sequences, we can apply (un)gapped alignment or use the longest

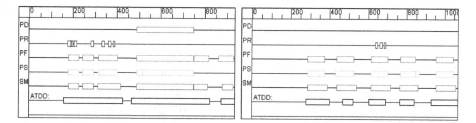

Fig. 6. Alternative domain annotations for two example sequences Q19024 (left) and O94848 (right). The annotations include ProDom (PD), PRINTS (PR), Pfam (PF), PROSITE, Profile (PS), SMART (SM), and the extracted domain positions by our algorithm (ATDD) for $\theta = 0.98$ and $k = 20$.

common subsequence (LCS) as a similarity measure. The segment evaluation function may be more sophisticated taking into account the distribution of the correlated amino acids within a region. Also, we feel that the domain (boundaries) extraction procedure may be further improved taking into account all of the collected information by the method.

Block Similarity. We have performed limited number of experiments using gapped alignment [16] to determine block similarity and found that results are similar to that of using un-gapped alignment. We have also evaluated data set 1 using the LCS measure with a 0/1 substitution matrix as our block similarity function. The substitution matrix was obtained from the equivalency set based on chemical nature of amino acids[2]. For block size of 25 we considered two blocks to be similar if their LCS is more than 17. We determined that in such a setting specificity increases but sensitivity is poor.

Segment Evaluation. We have described and assessed a segment evaluation function based on the density of votes (number of votes divided by total number of possible votes for a segment) and number of relevant sequences for that segment. Another possibility is to weight votes based on the distance of the corresponding block pairs. Such scheme decreases the penalty of small gaps but potentially merges near-by domains. We have computed the sensitivity and specificity of such method using the existing domain extraction procedure for data set 1. We determined that its performance is comparable to the current method modulo a small shift in the threshold values used since the method favors longer domains within a sequence. This bias leads to an increase in specificity but a decrease in sensitivity.

Domain Extraction. The current method to compute the domain positions is quite simplistic and leads to occasional fusion of consecutive proposed domains (see left panel of Fig. 6). Another problem is that uniform thresholding

---

[2] http://cbcsrv.watson.ibm.com/Tspd.html

(θ) potentially misses some of the domains – in multi-domain proteins, all domains to be discovered, either there should be comparable evidence (votes) in the considered data set for all the different domains, or the threshold should be sufficiently low without hurting the specificity of the method. This problem is partially solved by considering only the top voters for all regions in the input sequence. To improve the method we can use non-uniform threshold for different regions in the sequence under consideration, as well as, the gathered information of which positions are part of the same domain to delineate domain boundaries more accurately.

References

1. Altschul, S., Gish, W., Miller, W., Myers, E., Lipman, D.: Basic local alignment search tool. J Mol Biol **215** (1990) 403–410
2. Pearson, W., Lipman, D.: Improved tools for biological sequence comparison. Proc Natl Acad Sci USA **85** (1988) 2444–2448
3. Bork, P., Koonin, E.: Predicting functions from protein sequences–where are the bottlenecks. Nat Genet **18** (1998) 313–318
4. Hegyi, H., Bork, P.: On the classification and evolution of protein modules. J Protein Chem **16** (1997) 545–551
5. Bateman, A., Coin, L., Durbin, R., Finn, R.D., Hollich, V., Griffiths-Jones, S., Khanna, A., Marshall, M., Moxon, S., Sonnhammer, E.L.L., Studholme, D.J., Yeats, C., Eddy, S.R.: The Pfam protein families database. **32** (2004) D138–141
6. Sonnhammer, E., Eddy, S., Birney, E., Bateman, A., Durbin, R.: Pfam: multiple sequence alignments and hmm-profiles of protein domains. Nucl. Acids. Res. **26** (1998) 320–322
7. Letunic, I., Goodstadt, L., Dickens, N.J., Doerks, T., Schultz, J., Mott, R., Ciccarelli, F., Copley, R.R., Ponting, C.P., Bork, P.: Recent improvements to the SMART domain-based sequence annotation resource. Nucl. Acids. Res. **30** (2002) 242–244
8. Henikoff, J., Pietrokovski, S., McCallum, C., Henikoff, S.: Blocks-based methods for detecting protein homology. Electrophoresis **21** (2000) 1700–1706
9. Falquet, L., Pagni, M., Bucher, P., Hulo, N., Sigrist, C.J.A., Hofmann, K., Bairoch, A.: The PROSITE database, its status in 2002. Nucl. Acids. Res. **30** (2002) 235–238
10. Mulder, N., Apweiler, R., Attwood, T., Bairoch, A., Bateman, A., Binns, D., Biswas, M., Bradley, P., Bork, P., Bucher, P., Copley, R., Courcelle, E., Durbin, R., Falquet, L., Fleischmann, W., Gouzy, J., Griffith-Jones, S., Haft, D., Hermjakob, H., Hulo, N., Kahn, D., Kanapin, A., Krestyaninova, M., Lopez, R., Letunic, I., Orchard, S., Pagni, M., Peyruc, D., Ponting, C., Servant, F., Sigrist, C.: Interpro: an integrated documentation resource for protein families, domains and functional sites. Brief Bioinform **3** (2002) 225–235
11. Attwood, T., Beck, M., Bleasby, A., Parry-Smith, D.: PRINTS–a database of protein motif fingerprints. Nucl. Acids. Res. **22** (1994) 3590–3596
12. Corpet, F., Servant, F., Gouzy, J., Kahn, D.: ProDom and ProDom-CG: tools for protein domain analysis and whole genome comparisons. Nucl. Acids. Res. **28** (2000) 267–269

13. Letunic, I., Copley, R.R., Schmidt, S., Ciccarelli, F.D., Doerks, T., Schultz, J., Ponting, C.P., Bork, P.: SMART 4.0: towards genomic data integration. Nucl. Acids. Res. **32** (2004) D142–144

14. Henikoff, S., Henikoff, J.: Amino acid substitution matrices from protein blocks. Proc Natl Acad Sci U S A **89** (1992) 10915–10919

15. Wootton, J.C., Federhen, S.: Statistics of local complexity in amino acid sequences and sequence databases. Computers in Chemistry **17** (1993) 149–163

16. Smith, T., Waterman, M.: Identification of common molecular subsequences. J Mol Biol **147** (1981) 195–197

Local Search Heuristic for Rigid Protein Docking[*]

Vicky Choi[1], Pankaj K. Agarwal[2], Herbert Edelsbrunner[3], and Johannes Rudolph[4]

[1] Departments of Computer Science and Biochemistry
Duke University, Durham, North Carolina
[2] Departments of Computer Science and Mathematics
Duke University, Durham, North Carolina
[3] Departments of Computer Science and Mathematics, and Raindrop Geomagic
Duke University, Durham, North Carolina
[4] Departments of Biochemistry and Chemistry
Duke University, Durham, North Carolina

Abstract. We give an algorithm that locally improves the fit between two proteins modeled as space-filling diagrams. The algorithm defines the fit in purely geometric terms and improves by applying a rigid motion to one of the two proteins. Our implementation of the algorithm takes between three and ten seconds and converges with high likelihood to the correct docked configuration, provided it starts at a position away from the correct one by at most 18 degrees of rotation and at most 3.0Å of translation. The speed and convergence radius make this an attractive algorithm to use in combination with a coarse sampling of the six-dimensional space of rigid motions.

1 Introduction

Protein interactions are the molecular basis for many essential components of life. In this paper we contribute to the growing body of work on *protein docking*, which is the computational approach to predicting protein-protein interactions.

Field of Protein Docking. The reliable prediction of protein interactions from three-dimensional structures alone is one of the grand challenges in computational biology. There is ample experimental evidence from X-ray crystallography and other structure determination methods that interactions require proteins to exhibit extensive local shape complementarity. Nevertheless, the precise mechanism that brings about interactions is poorly understood. The observed shape complementarity of docked proteins suggests we start with the geometric structures of individual proteins and search for a good local fit. This raises intriguing but hard questions about the relative importance of physical forces (e.g. van der Waals interactions, hydrogen bonds, ion pairs, etc.) and shape, particularly as to the precise meaning of shape when objects are not rigid.

The known structures of protein complexes form a benchmark for computational tools and the attempt to reassemble proteins to their observed, native configurations is

[*] All authors are supported by NSF under grant CCR-00-86013. VC, JR, and HE are also supported by a BGT Postdoc Program from Duke University. JR and HE are also supported by NIH under grant R01 GM61822-01. PA is also supported by NSF under grants EIA-01-31905 and CCR-02-04118 and by the U.S.-Israel Binational Science Foundation.

I. Jonassen and J. Kim (Eds.): WABI 2004, LNBI 3240, pp. 218–229, 2004.
© Springer-Verlag Berlin Heidelberg 2004

referred to as *bound protein docking*. Even if we ignore physical forces and focus exclusively on shape, the high dimension of the search space makes this a difficult problem. Following the policy of small steps, it thus makes sense to simplify the problem by assuming rigidity. The task in *rigid protein docking* is to find a motion that positions one rigid protein relative to the other into the correct docked configuration. The dimension of the search space is still high, three for translations plus three for rotations. Implementation of a fast and accurate rigid docking algorithm would allow for future methods that add the higher dimensionality of conformational changes seen in real docking problems. Possible approaches to flexibility include tolerance to collisions [8, 14] and pre-calculation of multiple residue conformations [16].

Prior Work. Many different approaches have been taken to solving the rigid docking problem and we refer to several survey articles in the general area [7, 10, 13]. All of these methods consist of essentially two parts. First, one creates a scoring function that discriminates correctly docked conformations from incorrect ones. The scoring is based primarily on shape recognition but often includes electrostatics or hydrogen bonds. Because of the size of the search space and the number of atoms for each protein, simplifications or data reduction methods are often employed. Second, one creates a search algorithm that finds the correct solution using the scoring function. Many rigid protein docking algorithms based primarily on shape have been implemented using diverse approaches to search the space of rigid motions, including cube coverings, fast Fourier transforms, spherical harmonics, and geometric hashing. One major limitations in these methods is that they can yield anywhere from a few to thousands of false positives, incorrect configurations that have a higher score than the native configuration. Thus, a re-ranking of the docked configurations is usually implemented based on a wide variety of methods including solvation potentials, empirical atom-atom or residue-residue contact energies, optimal positioning of hydrogen bonds, etc. Bespamyatnikh et al. developed a shape-based docking algorithm and demonstrated that it correctly docks a diverse set of 25 protein complexes without any false positives [3]. This result was achieved using a scoring function that approximates the van der Waals interactions by counting pairs of atoms and by high-resolution sampling of the space of rigid motions. The major limitation of this method is the amount of computation time needed, with even a modest size problem taking a day on a cluster of 100 processors. This does not allow for facile experimentation or implementation of flexibility.

Local Search. We envision a more efficient algorithm that first uses a coarse sampling in the space of rotations to generate a set of possible solutions with at least one not too far from the correct docking configuration. The second step is a rapid search for the solution using a local improvement method. Based on our previous results [3], the correct docking configuration will yield the highest score following this second step. A multistage local search method for rigid protein docking has been recently reported that mimics the physical events of protein binding [5]. Starting from complexes as much as 10Å root-mean-square-distance from the native configuration, the method guides protein docking, first with desolvation and electrostatics, then adding partial van der Waals potentials as the proteins come closer together. This energy funnel method was shown to work well with a set of eight different complexes, but appears computationally expensive.

Our Results. As in [5], we do not explicitly address the generation of initial configurations in this paper. Instead the main question we pursue is the convergence radius of our local search heuristic. In other words, how far away from the native configuration can we start and still have a good chance to recover (a good approximation of) the native configuration? For the time being, we do not have any theoretical results and our approach to finding answers is purely experimental and restricted to the case of known structures of docked proteins.

We performed computational experiments using the barnase-barstar complex whose structure can be found in the protein databank [2]. Our findings show that the chances of recovering the correct, native configuration are about 80% provided we start with a configuration generated by a local perturbation with rotation angle at most 18° and translation distance at most 3.0Å. We extended the experiments to nineteen additional protein complexes and found that the bounds on the local perturbation are about the same and perhaps universal for protein complexes.

Outline. Section 2 introduces the geometric and algorithmic background used in this paper. Section 3 presents the local search heuristic. Section 4 describes the results of the computational experiments that test the performance of the heuristic. Section 5 summarizes our findings and points toward future directions.

2   Background and Definitions

In this section, we introduce the notation and the main geometric and algorithmic concepts used in the design of our local search heuristic.

Notation and Assumptions. We use solid spheres to represent atoms and space-filling diagrams to model proteins as unions of such spheres. Writing a_i for the center and r_i for the radius of the i-th sphere, we let $\mathcal{A} = \{A_i = (a_i, r_i) \mid 1 \le i \le m\}$ be the set of spheres defining the first protein. Similarly, we let $\mathcal{B} = \{B_j = (b_j, s_j) \mid 1 \le j \le n\}$ be the set of spheres defining the second protein. Following the work of Bespamyatnikh et al., we say A_i and B_j *collide* if the two spheres overlap, and they *score* if they are within a pre-specified distance but do not overlap. Formally,

$$\text{score}(i,j), \text{collision}(i,j) = \begin{cases} 0,1 & \text{if} & \|a_i - b_j\| < r_i + s_j, \\ 1,0 & \text{if } r_i + s_j & \le \|a_i - b_j\| \le r_i + s_j + \lambda, \\ 0,0 & \text{if } r_i + s_j + \lambda < \|a_i - b_j\|, \end{cases}$$

where the constant is experimentally set to $\lambda = 1.5$Å [3]. The *total score* and the *total collision number* are $\text{Score}(\mathcal{A}, \mathcal{B}) = \sum_{i,j} \text{score}(i,j)$ and $\text{Collision}(\mathcal{A}, \mathcal{B}) = \sum_{i,j} \text{collision}(i,j)$. Using a second constant, χ, we can now formally define the rigid docking problem as finding a rigid motion μ that maximizes the total score between $\mathcal{A}$ and $\mu(\mathcal{B})$ while keeping the number of collisions at or below χ. The second threshold is experimentally set to $\chi = 5$ [3]. Using van der Waals radii for the spheres, physics dictates that there are no collisions at all, but in order to compensate for measurement errors and other modeling inaccuracies, we allow for a small number of violations of that dictum. We make two assumptions on the geometric input data motivated by the application to organic molecules. To state them, let δ be the minimum distance between centers of any two spheres in $\mathcal{A}$ or in $\mathcal{B}$, and let $R_{\min}$ and $R_{\max}$ be the minimum and maximum radii of the spheres in these sets.

I. There are constants $c \leq C$ such that $R_{\max}/C \leq \delta \leq R_{\min}/c$.
II. The difference between the extreme radii satisfies $R_{\max} - R_{\min} < \lambda$.

We note that Assumption II is implied by Assumption I and $C - c < \lambda/\delta$, requiring that the two constants in I are not too different. Our algorithm crucially depends on Assumption I, and it makes use of Assumption II, but that dependence could be avoided. In the data retrieved from the protein databank [2], we observe $\delta = 1.18$Å and get $c = 1.14$ and $C = 1.60$. In our experiments, we use only five different radii, between 1.348Å and 1.880Å, which clearly satisfy Assumption II.

Preprocessing. Consider a sphere B with radius $s = (R_{\min} + R_{\max})/2$. It scores with a sphere A_i iff its center b lies in the shell centered at a_i whose inner and outer radii are $r_i + s$ and $r_i + s + \lambda$, respectively. Consistent with the terminology introduced above, we define the *score* of B equal to the number of shells that contain b. Each sphere in $\mathcal{A}$ defines a shell, giving an arrangement of $2m$ (non-solid) spheres that decompose $\mathbb{R}^3$ into cells of constant score, as considered in [6]. The arrangement is useful for distinguishing desirable from undesirable positions for the sphere B but it has two drawbacks, namely it suggests regions and not specific positions, and not all spheres in $\mathcal{B}$ have radius s. We remedy both by replacing the shell around a_i by its mid-sphere,

$$S_i \;\; = \;\; \{x \in \mathbb{R}^3 \mid \|x - a_i\| = r_i + s + \frac{\lambda}{2}\}.$$

By Assumption II, S_i lies within the shell around a_i defined for each radius r in $[R_{\min}, R_{\max}]$. The mid-spheres intersect pairwise in circles and triplewise in points, the latter being the vertices of the arrangement. We use some of these vertices as target positions for the spheres B_j in $\mathcal{B}$, as illustrated in Figure 1. We compute and evaluate these vertices in a preprocessing step, which we now describe.

Step 1. Compute the set of vertices of the arrangement of mid-spheres.
Step 2. For each vertex u, compute the score of $B_u = (u, s)$ and the number of collisions between $B'_u = (u, s + \tau)$ and spheres in $\mathcal{A}$.
Step 3. Let $\mathcal{U}$ be the set of vertices u for which B'_u has zero collisions and there is no vertex $v \in \mathcal{U}$ nearby that dominates u in terms of scoring.

The constant τ used in Step 2 will be discussed shortly. Step 1 is greatly helped by Assumption I, which implies that each mid-sphere intersects only a constant number of other mid-spheres. It follows that the number of vertices is only $O(m)$, and using the grid data structure of Halperin and Overmars [9] we find them in time $O(m \log m)$. Using the same data structure, we compute the scores and collision numbers of the spheres B_u and B'_u in time $O(m \log m)$.

The purpose of the vertices is to act as target locations for the spheres in $\mathcal{B}$. It thus makes sense to eliminate vertices u for which the enlarged sphere B'_u has non-zero collisions with spheres in $\mathcal{A}$. We use the experimentally determined constant $\tau = 0.2$Å for the enlargement. Of the remaining vertices, we keep only the ones with locally maximum score. More specifically, we remove a vertex u for which there is a vertex v

Fig. 1. The dotted circles represent mid-spheres. The vertices formed by the mid-spheres are black or white depending on whether a sphere centered at that vertex forms a near collision or not. The region of scoring, non-colliding positions is shaded.

at distance at most $\lambda/2$ such that the set of spheres A_i scoring with B_u is a proper subset of the set scoring with B_v. Although the vertices do not observe a constant separation bound, it is easy to prove from Assumption I that there are only a constant number of vertices within a constant distance from any point in space. We can therefore use the same grid data structure to implement Step 3 within the same time bound as the first two steps. It follows that all three preprocessing steps together take time $O(m \log m)$. We note that instead of one we may use several arrangements, each catering to a small range of radii of spheres in $\mathcal{B}$. In our implementation, we use five arrangements, one each for the five different radii in our data sets. As long as the number of arrangements is a constant, the running time is not affected by more than a constant factor.

Least Square Rigid Motion. In the local search heuristic, we will repeatedly compute local rigid motions by solving a least-square optimization problem. An instance is given by a subset $G = \{g_1, g_2, \ldots, g_\ell\}$ of the vertices in $\mathcal{U}$, a subset $Z = \{z_1, z_2, \ldots, z_\ell\}$ of the centers of spheres in $\mathcal{B}$, and a bijection between G and Z specified by shared indices. The objective is to find a rigid motion μ that minimizes the sum of square distances, $\sum_{k=1}^{\ell} \|g_k - \mu(z_k)\|^2$. The problem of computing μ is known as the absolute orientation problem in computer vision. Every rigid motion can be written as a translation followed by a rotation about the origin. Assuming the centroid of the vertices is the origin, $\sum_{k=1}^{\ell} g_k = 0$, the translational component of the optimal motion necessarily moves the centroid $\bar{z} = \frac{1}{\ell} \sum_{k=1}^{\ell} z_k$ to the origin. It remains to compute the optimal rotation for the points $z_k - \bar{z}$, which reduces to solving a small eigenvalue problem. The matrix for this problem can be computed in time $O(\ell)$ from the sets G and Z, using either the formalism of rotation matrices [15] or that of quaternions [11]. We remark that the reduction to an eigenvalue problem allows for more general correspondences between G and Z than bijections, and it can be modified to use a set of weights $W = \{w_1, w_2, \ldots, w_\ell\}$. In other words, we can compute in time $O(\ell)$ the rigid motion $\mu = \text{OptRM}(G, Z, W)$ that minimizes $\sum_{k=1}^{\ell} w_k \|g_k - \mu(z_k)\|^2$.

3 Local Search Heuristic

In this section, we describe the algorithm that locally improves the fit between the space-filling representations of two proteins. We begin by explaining the overall structure of the algorithm and follow up by detailing its loops.

High-Level Structure. Given sets of spheres $\mathcal{A}$ and $\mathcal{B}$, we aim at finding a local rigid motion that we can apply to $\mathcal{B}$ to improve the fit. By a local rigid motion we mean a rigid motion that is small, and we will be specific in Section 4 about how small. Here, we focus on the structure of the algorithm, which repeatedly solves one of two types of weighted least-square problems. The types are distinguished by the intended effect on the fit:

- *score-improving* instances are prepared and solved in the outer loop, and
- *collision-reducing* instances are prepared and solved in the inner loop.

The success of the algorithm crucially depends on how we define these instances. We follow two intuitions:

1. worthwhile target positions for spheres in $\mathcal{B}$ are collision-free and locally maximize the score;
2. an effective collection of target positions is approximately congruent to the configuration of corresponding sphere centers.

We satisfy the first intuition by using the vertices in $\mathcal{U}$ and the second intuition by limiting our attention to spheres and vertices that are near each other. Letting Z be the set of centers of spheres in $\mathcal{B}$ that have a vertex in their neighborhood, we define a bijection between Z and a subset G of the vertices; we refer to G as the set of *tentative goals*. Assuming a set of weights, W, and a bijection between Z and G, we can now describe the algorithm.

> for $\#_{\text{outer}}$ times do
> > prepare a score-improving instance of the least-square problem,
> > compute $\mu = \text{OptRM}(Z, G, W)$, and let $\mathcal{B} = \mu(\mathcal{B})$
> > while Collision$(\mathcal{A}, \mathcal{B}) > \chi$ and μ is not negligible do
> > > prepare a collision-reducing instance of the least-square
> > > problem, compute $\mu = \text{OptRM}(Z, G, W)$, and let $\mathcal{B} = \mu(\mathcal{B})$
> > endwhile
> endfor; return best fit encountered during the iteration.

We cannot prove, and indeed do not expect, that the algorithm always successfully finds a fit with sufficiently high score and small collision number. We therefore implement the algorithm with a constant limit, $\#_{\text{inner}}$, on how often the inner loop can be repeated. the outer loop $\#_{\text{outer}}$ times and return the best fit encountered during any of the iterations.

Score-Improving Outer Loop. We describe how to prepare an instance of the weighted least-square problem that aims at improving the score of the fit. We define a distance threshold, D_j, for each sphere B_j in $\mathcal{B}$, and we let the corresponding tentative goal of B_j be a vertex u_j in $\mathcal{U}$ within distance D_j from b_j that maximizes the score as computed in the preprocessing step. If there is no vertex within distance D_j from b_j, the tentative goal of B_j is undefined. We let Z be the set of centers of spheres with tentative goals, and G the corresponding set of tentative goals. Finally, we set all weights to one.

It remains to describe how we choose the distance threshold $D_j = \min\{D, d_j\}$, where D is a general and d_j is an individual threshold. The general threshold depends on the current fit between $\mathcal{A}$ and $\mathcal{B}$ and gets smaller as the fit gets better. The individual threshold depends on whether or not B_j collides with a sphere in $\mathcal{A}$. Let $A_i \in \mathcal{A}$ be the sphere that minimizes $\|b_j - a_i\| - s_j - r_i$. If A_i and B_j are disjoint then we choose d_j small enough so that moving b_j within this limit avoids a collision. Otherwise, we choose d_j large enough to give B_j a chance to undo the collision:

$$
d_j = \begin{cases} \|b_j - a_i\| - s_j - r_i & \text{if } \|b_j - a_i\| \geq s_j + r_i, \\ \|b_j - a_i\| + s_j + r_i + \lambda & \text{if } \|b_j - a_i\| < s_j + r_i. \end{cases}
$$

When we compute the sphere A_i of B_j, we do not have to look farther than distance $D + s_j + R_{\max}$ from b_j. Since this is a constant, we can again use the grid data structure and explore the neighborhood of b_j in constant time. After collecting Z and G, we compute the optimal rigid motion, μ, and apply it to $\mathcal{B}$. We expect that the total score increases but there is no guarantee that this really happens. Simultaneously, the total collision number may also increase.

Collision-Reducing Inner Loop. This brings us to the preparation of instances of the weighted least-square problem that aim at reducing the number of collisions. We distinguish spheres B_j with and without collisions. If B_j collides with at least one sphere in $\mathcal{A}$ then we find the closest collision-free position in $\mathcal{U}$ within distance $D + s_j$ from b_j and let it be the tentative goal of b_j. If B_j scores and has no collision then we encourage it to stay put by setting its tentative goal equal to its own center, b_j. Let Z contain the centers of all spheres B_j that receive tentative goals, and let G be the corresponding set of tentative goals. Finally, we use weights to counterbalance the usual relative abundance of collision-free spheres. Letting $F \subseteq Z$ be the subset of centers of collision-free spheres, we set the weight of a sphere center $z_k \in Z$ equal to

$$
w_k = \begin{cases} 1 & \text{if } z_k \in F, \\ \omega/\|z_k - g_k\|^2 & \text{if } z_k \in Z - F, \end{cases}
$$

with *weight factor* ω. Initially, we set $\omega = |F|/|(Z - F)|$, but if this leads to an increase in the number of collisions we adjust the weight factor as explained shortly. After computing Z, G, and W, we determine the optimal rigid motion, μ, and apply it to $\mathcal{B}$. We expect that the total number of collisions decreases but there is no guarantee that this really happens. Simultaneously, the total score may decrease although we counteract that tendency by including collision-free scoring spheres in the weighted least-square problem.

Adjusting the Weight Factor. If the rigid motion $\mu = \mathrm{OptRM}(Z, G, W)$ leads to an increase in the number of collisions, then we adjust the weight factor, ω, and redo the step in the inner loop. Recall that the rigid motion $\mu = \mathrm{OptRM}(Z, G, W)$ minimizes

$$
\mathrm{Error}(\mu') = \sum_{z_k \in F} \|\mu'(z_k) - z_k\|^2 + \sum_{z_k \in Z - F} w_k \|\mu'(z_k) - g_k\|^2,
$$

where the minimum is taken over all rigid motions μ'. We set

$$\omega_{new} \;=\; \sum_{z_k \in F} \|\mu(z_k) - z_k\|^2 \;\Big/\; \sum_{z_k \in Z-F} \left(1 - \frac{\|\mu(z_k) - g_k\|^2}{\|z_k - g_k\|^2}\right).$$

We then set $\omega = \omega_{new}$ and repeat the computations unless $\mathrm{Collision}(\mathcal{A}, \mu(\mathcal{B})) \leq \mathrm{Collision}(\mathcal{A}, \mathcal{B})$. Experimentally, the adjustment is not always needed, and if necessary it seems to take at most four iterations. In our implementation, we limit the number of iterations to at most a constant $\#_{wf}$. If the iteration ends with higher collision number for $\mu(\mathcal{B})$ than for $\mathcal{B}$ then we set $\mu = \mathrm{id}$.

Implementation and Running Time. We complete the description of the algorithm by defining the constants we used but have left unspecified. Most importantly, we use a distance threshold D in the preparation of various least-square problems. In the inner loop, we use $D = 2\text{Å}$ for the collision reducing instances. The situation is more complicated in the outer loop, where D depends on the current fit. Assuming $\mathrm{Score}(\mathcal{A}, \mathcal{B}) > 180$, we set

$$D \;=\; \begin{cases} 2\text{Å} & \text{if} \quad \mathrm{Collision}(\mathcal{A}, \mathcal{B}) \leq 5, \\ 3\text{Å} & \text{if} \quad 5 < \mathrm{Collision}(\mathcal{A}, \mathcal{B}) \leq 15, \\ 4\text{Å} & \text{if } 15 < \mathrm{Collision}(\mathcal{A}, \mathcal{B}) \leq 30. \end{cases}$$

We set $D = 4.5\text{Å}$ if $\mathrm{Collision}(\mathcal{A}, \mathcal{B}) > 30$ or $\mathrm{Score}(\mathcal{A}, \mathcal{B}) \leq 180$. The other constants we use are the upper bounds on the number of iterations, $\#_{inner} = 20$, $\#_{outer} = 13$, and $\#_{wf} = 5$. In each case, the search for the tentative goal of a sphere B_j is limited to a constant size neighborhood. Assumption I implies that there are only a constant number of spheres and vertices to consider, which takes only constant time. The total amount of time needed for an iteration of either loop is therefore bounded by $O(n)$. However, the constant number of spheres and vertices searched for each B_j can be rather large, warranting the implementation of a hierarchical search structure, e.g., the kd-tree, assisting the grid data structure. The total number of iterations is again at most a constant. It follows that the running time of the entire algorithm is $O(n)$, not including the time for preprocessing, which has been accounted for in Section 2.

We have implemented the algorithm in C++ and refer to the software as LILAC. Depending on the protein complex, LILAC takes between half a minute and five minutes for the preprocessing step, and between three and ten seconds for a local search, on a PC with a Pentium III processor with clock speed of 929 MHz and memory of 600 MB. Our main focus is on accelerating the search since we are interested in applications in which the preprocessing time can be amortized over a potentially large number of local searches.

4 Experimental Results

We applied the software to a number of known protein-protein complexes. In this section, we describe these experiments and analyze the results we obtain.

Statement of Questions. We test the effectiveness of the local search heuristic by running LILAC on a number of protein-protein complexes with known structure. For each such complex, we generate two sets of spheres, $\mathcal{A}_{nat}$ and $\mathcal{B}_{nat}$, one for each protein. We then perturb the native configuration by applying a local rigid motion π to the second set and use the local search heuristic on $\mathcal{A} = \mathcal{A}_{nat}$ and $\mathcal{B} = \pi(\mathcal{B}_{nat})$, obtaining another rigid motion, μ. Define $\mathcal{C} = \mu(\mathcal{B}) = \mu(\pi(\mathcal{B}_{nat}))$. In the ideal case, μ is the inverse of π and $\mathcal{C} = \mathcal{B}_{nat}$. In general, we measure how different $\mathcal{C}$ is from $\mathcal{B}_{nat}$ and use this information to distinguish successful from unsuccessful applications of the search heuristic. Letting $\mathcal{B}_{nat}^* \subseteq \mathcal{B}_{nat}$ be the set of scoring spheres in the native configuration, we define $\mathcal{C}^* = \mu(\pi(\mathcal{B}_{nat}^*))$ and compute the root-mean-square-distance between corresponding centers,

$$\mathrm{RMSD}^* = \mathrm{RMSD}(\mathcal{B}_{nat}^*, \mathcal{C}^*) = \sqrt{\frac{1}{\ell} \sum_{b_j \in \mathcal{B}_{nat}^*} \| b_j - \mu(\pi(b_j)) \|^2},$$

where $\ell = |\mathcal{B}_{nat}^*|$. Recall the main question formulated in Section 1, which we can now rephrase to asking how much the native configuration can be perturbed so we still have a good chance to recover the complex using our local search heuristic. To approach this question, we let $\bar{b}^*$ be the centroid of the centers of the spheres in $\mathcal{B}_{nat}^*$ and we write each rigid motion, π, as a rotation about $\bar{b}^*$ followed by a translation. We measure the rotation and the translation separately, letting $\theta(\pi)$ be the angle of the rotation and $t(\pi)$ be the distance of the translation. Using oriented rotation axes, we may assume that both the angle and the distance are non-negative. A *local rigid motion* is one for which θ and t are small. We can now rephrase our main question again.

QUESTION A. What are the largest thresholds Θ and T such that for a perturbation with $\theta(\pi) \leq \Theta$ and $t(\pi) \leq T$ we have a good chance the local search heuristic recovers a good approximation of the native configuration?

We also address several related, more detailed questions, such as whether or not the success rate of the search heuristic is influenced by the angle between the rotation axis and the translation vector, or by the number of collisions in the perturbed starting configuration. The second main question addresses the variation over different complexes.

QUESTION B. Does the behavior or the search heuristic depend significantly on the protein complexes or are there universal thresholds T and Θ that apply to all or most complexes?

Statistical results of the experiments aimed at answering the two questions are now presented.

Convergence Thresholds. Our first experiment explores the dependence of the performance of the local search heuristic on the size of the initial perturbation. We use the experimentally well-studied barnase-barstar complex (1BRS) as a test case. In this complex, the ribonuclease (barnase) exhibits extensive geometric surface complementarity docked to its natural protein inhibitor (barstar). The interaction surface is large, measuring about 800Å^2, and contains many of the features typically seen at protein interfaces, including a few hot-spot residues that contribute the majority of the interaction energy, electrostatic interactions, buried water molecules and the lack of a deep binding groove

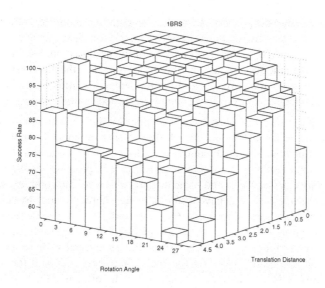

Fig. 2. The success rate of the local search heuristic as a function of the rotation angle and translation distance. Each prism corresponds to a pair θ, t and represents 1,024 trials. Of course for $\theta = t = 0.0$ all rigid motions are the same. The height of a prism gives the percentage of successes within the corresponding collection of trials.

that characterizes small molecule binding. To define a perturbation, we select two directions $\mathbf{u}, \mathbf{v} \in \mathbb{S}^2$, an angle θ and a distance t. The thus specified perturbation first rotates by θ around the oriented axis passing through $\bar{b}^*$ in the direction $\mathbf{u}$ and second translates by adding $t\mathbf{v}$. We note that a uniform sampling of directions, angles, and distances favors small over large rigid motions. While generally undesirable, this bias is acceptable in our application, which focuses on local rigid motions. We sample the space of perturbations using 32 directions (defined by the vertices and face normals of the regular icosahedron), 10 angles $(0°, 3°, \ldots, 27°)$, and 10 lengths $(0\text{Å}, 0.5\text{Å}, \ldots, 4.5\text{Å})$. This gives a total of about one hundred thousand perturbations or trials. For each trial, we compute the score, collision number, RMSD^* of the native, perturbed, and computed configurations. We declare a trial successful if the computed configuration has score at least some fraction of the native score and small RMSD^*. To be specific, we consider the computation of $\mathcal{C}$ from $\pi(\mathcal{B})$ a *success* if

$$\text{Score}(\mathcal{A}, \mathcal{C}) \geq p \cdot \text{Score}(\mathcal{A}, \mathcal{B}_{\text{nat}}) \text{ and}$$
$$\text{RMSD}(\mathcal{B}^*_{\text{nat}}, \mathcal{C}^*) \leq \varepsilon,$$

where we use $p = 0.9$ and $\varepsilon = 1.5\text{Å}$. We note that $\text{Collision}(\mathcal{A}, \mathcal{C}) \leq \chi$ is understood since this is the threshold used in the search heuristic. The results of this experiment are displayed in Figure 2. Using these results, we set $\Theta = 18°$ and $T = 3.0\text{Å}$. For convenience, we introduce a norm to measure the size of a perturbation, $\|\pi\| = [(\theta(\pi)/\Theta)^2 + (t/T)^2]^{1/2}$. We use it to display the experimental results with graphs of one-dimensional functions, such as in Figure 3. We looked into the question of whether or not the search heuristic performs better for some perturbations than others. Comparing the RMSD* caused by rotations versus by translations it seems that the algorithm is about twice as sensitive to the rotational part of the perturbation. We did not find any correlation between the success rate and the angle formed by the directions defining the rotation and the translation. We also did not find any correlation between the success rate and the number of collisions in the initial, perturbed configuration.

Universality of Convergence Thresholds. We repeated the same computational experiment for nineteen additional protein complexes to test the applicability of our algorithm to a diverse set of protein-protein complexes. The types of interactions tested include some that fall into commonly observed classes such as protease-inhibitor complexes or antibody-antigen complexes, and some that are one-of-a-kind complexes, often with broad featureless interfaces that have historically been harder to dock by computational methods. We used the same thresholds, $\Theta = 18°$ and $T = 3.0$Å to define the norm of a perturbation and $p = 0.9$ and $\varepsilon = 1.5$Å to distinguish successful from unsuccessful trials. As shown in Figure 3, the results exhibit the same trend as those for 1BRS. The search heuristic was more successful for twelve and less successful for seven of the additional complexes. The answer to Question B is therefore that the thresholds Θ and T appear to be universal for protein complexes. Of course, this applies only to bound, rigid structures.

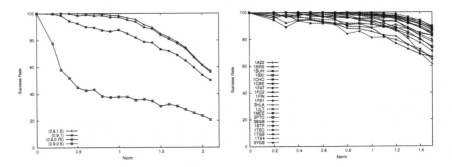

Fig. 3. Left: comparison of success rates obtained for different root-mean-square thresholds $\varepsilon = 0.5, 0.75, 1.0, 1.5$Å. Right: comparison of success rates for different complexes, while setting $p = 0.9$ and $\varepsilon = 1.5$Å.

5   Discussion

The main contribution of this paper is a local improvement algorithm for rigid protein docking and an experimental study of its performance. Comparison of our algorithm to other docking methods is beyond the scope of this paper as we have presented here only a local improvement step, not an overall docking method. The results are encouraging but we hope to improve them through additional tests and fine-tuning of the various steps in the algorithm. We plan to use this algorithm to accelerate the exhaustive search method of Bespamyatnikh et al. [3] using a coarse sampling of rigid motions that relates to the observed convergence radius. We also plan to use the algorithm in combination with more sophisticated strategies for coarse sampling that are based on protein shape measurements [1].

The most important new research direction is the inclusion of flexibility in protein docking. This can be approached in a variety of ways, including the use of rigidity analysis of protein structures [12], normal mode analysis [4], and ensembles of alternative configurations [16]. The experimental results reported in [3] suggest that a successful prediction of protein interaction is possible based on geometric criteria only but requires a fine sampling of the space of rigid motions and careful accounting of details. Any method incorporating flexibility will sacrifice some of the specificity we find in rigid docking and will need to balance speed and specificity.

References

1. P. K. AGARWAL, H. EDELSBRUNNER, J. HARER AND Y. WANG. Extreme elevation on a 2-manifold. *In* "Proc. 20th Ann. Sympos. Comput. Geom., 2004", 357–365.
2. H. M. BERMAN, J. WESTBROOK, Z. FENG, G. GILLILAND, T. N. BHAT, H. WEISSIG, I. N. SHINDYALOV AND P. E. BOURNE. The protein data bank. *Nucleic Acid Res.* **28** (2000), 235–242.
3. S. BESPAMYATNIKH, V. CHOI, H. EDELSBRUNNER AND J. RUDOLPH. Accurate bound protein docking by shape complementarity. Manuscript, Dept. Comput. Sci., Duke Univ., Durham, North Carolina, 2003.
4. B. R. BROOKS AND M. KARPLUS. Harmonic dynamics of proteins: normal modes and fluctuations in bovine pancreatic trypsin inhibitor. *Proc. Natl. Acad. Sci.* **80** (1983), 3696–3700.
5. C. J. CAMACHO AND S. VAJDA. Protein docking along smooth association pathways. *Proc. Natl. Acad. Sci.* **98** (2001), 10636–10641.
6. V. CHOI AND N. GOYAL. A combinatorial shape matching algorithm for rigid protein docking. *To appear in* "Proc. 15th Ann. Sympos. Combin. Pattern Matching, 2004".
7. A. H. ELCOCK, D. SEPT AND J. A. MCCAMMON. Computer simulation of protein-protein interactions. *J. Phys. Chem.* **105** (2001), 1504–1518.
8. J. FERNANDEZ-RECIO, M. TOTROV AND R. ABAGYAN. Soft protein-protein docking in internal coordinates. *Protein Sci.* **11** (2002), 280–291.
9. D. HALPERIN AND M. H. OVERMARS. Spheres, molecules, and hidden surface removal. *Comput. Geom. Theory Appl.* **11** (1998), 83–102.
10. I. HALPERIN, B. MA, H. WOLFSON AND R. NUSSINOV. Principles of docking: an overview of search algorithms and a guide to scoring functions. *Proteins* **47** (2002), 409–443.
11. B. K. P. HORN. Closed-form solution of absolute orientation using unit quaternions. *J. Opt. Soc. Amer. A* **4** (1987), 629–642.
12. D. J. JACOBS, A. J. RADER, L. A. KUHN AND M. F. THORPE. Protein flexibility predictions using graph theory. *Proteins* **44** (2001), 150–165.
13. J. JANIN AND S. J. WODAK. The structural basis of macromolecular recognition. *Adv. Protein Chem.* **61** (2002), 9–73.
14. F. JIANG AND S.-H. KIM. "Soft-docking": matching of molecular surface cubes. *J. Mol. Biol.* **219** (1991), 79–102.
15. W. KABSCH. A discussion of the solution for the best rotation to relate two sets of vectors. *Acta Crystallogr. Sect. A* **34** (1978), 827–828.
16. D. M. LORBER, M. K. UDO AND B. K. SHOICHET. Protein-protein docking with multiple residue conformations and residue substitutions. *Protein Sci.* **11** (2002), 1393–1408.

Sequence Database Compression for Peptide Identification from Tandem Mass Spectra

Nathan Edwards* and Ross Lippert**

Informatics Research, Advanced Research and Technology, Applied Biosystems
45 W. Gude Drive, Rockville MD, 20850
{EdwardNJ,LipperRA}@AppliedBiosystems.com

Abstract. The identification of peptides from tandem mass spectra is an important part of many high-throughput proteomics pipelines. In the high-throughput setting, the spectra are typically identified using software that matches tandem mass spectra with putative peptides from amino-acid sequence databases. The effectiveness of these search engines depends heavily on the completeness of the amino-acid sequence database used, but suitably complete amino-acid sequence databases are large, and the sequence database search engines typically have search times that are proportional to the size of the sequence database.

We demonstrate that the peptide content of an amino-acid sequence database can be represented by a reformulated amino-acid sequence database containing fewer amino-acid symbols than the original. In some cases, where the original amino-acid sequence database contains many redundant peptides, we have been able to reduce the size of the amino-acid sequence to almost half of its original size. We develop a lower bound for achievable compression and demonstrate empirically that regardless of the peptide redundancy of the original amino-acid sequence database, we can compress the sequence to within 15-25% of this lower bound. We believe this may provide a principled way to combine amino-acid sequence data from many sources without unduly bloating the resulting sequence database with redundant peptide sequences.

1   Introduction

The identification of peptides from tandem mass spectra is an important part of many high-throughput proteomics pipelines. In the high-throughput setting, the spectra are typically identified using software that matches tandem mass spectra with putative peptides from amino-acid sequence databases. The effectiveness of these search engines depends heavily on the completeness of the amino-acid sequence database used, but suitably complete amino-acid sequence databases are large, and the search engines typically have search times that are proportional

---

* Corresponding Author.

** Current address: Department of Mathematics, Massachusetts Institute of Technology. Email: lippert@math.mit.edu

I. Jonassen and J. Kim (Eds.): WABI 2004, LNBI 3240, pp. 230–241, 2004.

to the size of the sequence database. See [1, 2] for an extensive discussion of the running time cost of these search engines.

An inherent weakness of sequence database search engines such as SEQUEST [3], Mascot [4], and SCOPE [5], is that they struggle to identify the tandem mass spectra of peptides whose sequence is missing from the sequence database. The obvious solution, then, is to construct amino-acid sequence databases containing more of the peptide sequences that we might need in order to identify tandem mass spectra. For example, the sequences of all protein isoforms listed as variant annotations in Swiss-Prot [6, 7] records can be enumerated using the program varsplic [8, 9]. The resulting amino-acid sequence database is more than 1.5 times the size of Swiss-Prot, but at most, it contains about 2% additional peptides candidates. Since sequence database search engines typically have search times proportional to the size of the input sequence database, this increase in search time is a significant cost. Similarly, when we form the union of a number of sequence databases in order to create a comprehensive search database, merely ensuring non-redundancy at the protein sequence level leaves significant peptide level redundancy. In this paper, we strive to rewrite amino-acid sequence databases in such a way that most of the peptide redundancy is eliminated without losing any peptide content.

Current mass spectrometers are capable of reliably acquiring tandem mass spectra from ions with mass of up to about 3000 Daltons. However, the tandem mass spectra from peptides with more than 20 amino-acids or charge state 4 or more are rarely successfully interpreted by current sequence database search engines. Further, most peptide identification workflows digest proteins with trypsin, which cuts at either lysine or arginine (unless followed by proline). These amino-acids occur with sufficient frequency that peptides longer than 25 amino-acids are rare. Bearing all this in mind, we can conservatively upper-bound the length of peptides that sequence database search engines need to consider at about 30 amino-acids. We adopt the terminology of DNA sequence analysis and refer to a sequence of k amino-acids as a k-mer. The peptide content of an amino-acid sequence database is therefore represented by the set of 30-mers it contains.

We will refer to a number of commonly used amino-acid sequence databases to demonstrate our approach. We will refer to the Swiss-Prot section of the UniProt Knowledgebase [6, 7] as Swiss-Prot. The union of the Swiss-Prot, TrEMBL, and TrEMBL-New sections of the UniProt Knowledgebase will be referred to as UniProt. Many Swiss-Prot and UniProt entries contain protein isoform annotations, but only a single sequence is provided per entry. To construct a sequence database containing all of these isoforms, we use varsplic [8, 9] with the command line options

```
-which full -uniqids -fasta -varsplic -variant -conflict
```

to enumerate all original sequences plus all splice forms, variants and conflicts. These sequence databases will be referred to as Swiss-Prot-VS and UniProt-VS respectively. MSDB [10] (Mass Spectrometry protein sequence DataBase)

Table 1. Sequence statistics of some sequence databases used for peptide identification via tandem mass spectrometry.

Sequence Database	Sequence Length	Distinct 30-mers	Overhead
IPI-HUMAN	20358846	12115520	68%
IPI	54145883	29769766	81%
Swiss-Prot	56454588	44374286	27%
Swiss-Prot-VS	89541275	45307827	97%
UniProt	472581860	274510105	72%
UniProt-VS	506796094	275391669	84%
MSDB	481919777	276523755	74%
NRP	495502241	283160529	75%
NCBI-nr	619132252	378721915	63%
UnionNR	674700840	385369671	75%
Union	2157353500	385369671	460%

is a composite non-identical protein sequence database built from a number of primary source databases. MSDB is often used in conjunction with the Mascot sequence database search engine for protein identification. NCBI-nr [11] is a composite, non-redundant protein database from NCBI constructed from various sources. NRP [12] is a composite, non-redundant protein sequence database from the Advanced Biomedical Computing Center at NCI in Frederick, MD. The international protein index [13] (IPI) sequence databases, from EBI, are human, mouse and rat sections from Swiss-Prot, TrEMBL, RefSeq and Ensembl. We denote the human protein index by IPI-HUMAN, and the union of the three pieces as IPI. Finally, we form the concatenation of each of these comprehensive sequence databases to form a new comprehensive sequence database, Union, comprising UniProt-VS, MSDB, NCBI-nr, NRP, and IPI. For completeness, we also do the standard exact protein sequence redundancy elimination for Union, since its constituent sequence databases contain common protein entries. The protein level non-redundant version of Union is called UnionNR. Table 1 shows the size, or sequence length, of these sequence databases, as well as the number of distinct 30-mers they contain.

Suppose that we could rewrite each of these amino-acid sequence databases as a new amino-acid sequence database that is:

Complete
Every 30-mer from the original sequence database is present,
Correct
Only 30-mers from the original sequence database are present, and
Compact
No 30-mer is present more than once.

Suppose further that such a complete, correct and compact sequence database consists of a single sequence. Since the sequence is complete, a sliding 30-mer window on this sequence will generate all the 30-mers of our original sequence database. Therefore, this sequence has at least as many amino-acids as our original sequence database has distinct 30-mers. Hence the number of distinct 30-mers of our original sequence database is a lower bound on the size of any complete sequence database. The column **Overhead** of Table 1 shows the amount of additional sequence contained in various sequence databases over and above this lower bound. We will demonstrate how to construct complete, correct, compact sequence databases that come as close as possible to this lower bound.

We note that the well known shortest (common) superstring (SCS) problem [14, 15] cannot be applied as it does not guarantee its output to be correct or compact. Further, since the input sequences must be represented intact, the SCS cannot take advantage of redundancy in the interior of the protein sequences.

2   Sequence Databases and SBH-Graphs

In order to compress the amino-acid sequences of our input sequence database, we must first build a representation of its 30-mers.

Sequencing-by-Hybridization, proposed by Bains and Smith [16], Lysov *et. al.* [17], and Drmanac *et. al.* [18], is a technique by which DNA is interrogated by hybridization to determine the presence or absence of all possible length k DNA sequences. The information from these experiments can be represented in a graph, which we call the SBH-graph, first proposed by Pevzner [19]. The graph contains a directed edge for every observed k-mer probe, from a node representing the first $(k-1)$-mer of the probe to a node representing the last $(k-1)$-mer of the probe. Determining the original DNA sequence is then a matter of finding a path through the SBH-graph that uses every edge, representing an observed k-mer, at least once. See Figure 1 for a small SBH-graph example. As shown in Figure 1, each node has its $(k-1)$-mer sequence associated with it, while each edge holds the nucleotide that is appended to the sequence at the tail of the edge to form the k-mer it represents. We will use a SBH-graph to represent all the k-mers of our input sequence database.

We point out the connection here to de Bruijn graphs [20]. The SBH-graph is really just a subgraph of a de Bruijn digraph. A de Bruijn graph has edges for all k-mers from the alphabet, and hence nodes for all $(k-1)$-mers from the alphabet. A de Bruijn sequence is a circular sequence from the alphabet that represents all k-mers of the alphabet exactly once. A de Bruijn sequence can be enumerated by constructing an Eulerian tour on the corresponding de Bruijn graph, which happens to be an Eulerian graph. We call our de Bruijn subgraphs SBH-graphs to emphasize that we are representing some subset of all the k-mers, those k-mers found in the input sequence database. Despite this distinction, the de Bruijn sequence represents the motivation for the results to follow.

Given G, the SBH-graph representation of the k-mers of the sequence database S, we can identify the sequences of S with paths of G. In fact, any sequence s containing only k-mers of S can be represented by a path on G. Given s, we first locate the node representing the first $(k-1)$-mer of s. The edge

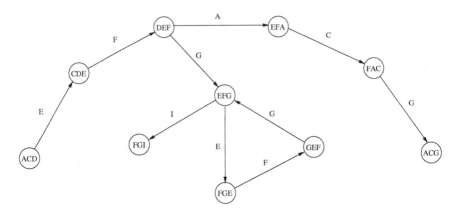

Fig. 1. SBH-graph for 4-mers from the sequences: ACDEFGI, ACDEFACG, DEFGE-FGI.

representing the first k-mer of s must be present since s contains only k-mers from S. We follow this edge, and look for an edge representing the next k-mer of s. The series of edges traced by this procedure corresponds to a path on G. Conversely, we construct the sequence represented by some path by outputting the $(k-1)$-mer of the initial node of the path and then the sequence on each edge of the path.

With this understanding of the equivalence of sequences on the k-mers of S and paths on the SBH-graph, we must now determine what the complete, correct, compact constraints imply for paths on our SBH-graph. Since each edge of the SBH-graph represents a distinct k-mer from our input sequence database, we obtain a complete, correct, compact sequence database by finding a path set that uses each edge exactly once and by generating the sequences represented by these paths.

Next, we must quantify the size of our new amino-acid sequence database. Let $S' = \{s'_1, \ldots, s'_l\}$ be our complete, correct, compact sequence database, and let N_k be the number of distinct k-mers in our original sequence database. Since S' is complete, correct, and compact, we know that all k-mers are observed, no extra k-mers are observed, and that each k-mer window generates a distinct k-mer. Therefore

$$N_k = \sum_{i=1}^{l} \text{windows}(s'_i) = \sum_{i=1}^{l} \text{length}(s'_i) - (k-1)$$

$$\implies \quad l(k-1) + N_k = \sum_{i=1}^{l} \text{length}(s'_i)$$

Consequently, the size of S' is

$$|S'| = \text{length}(s'_1) + 1 + \cdots + 1 + \text{length}(s'_l) = N_k + lk - 1$$

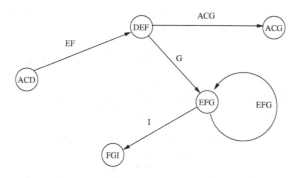

Fig. 2. Compressed SBH-graph for 4-mers from the sequences: ACDEFGI, ACDE-FACG, DEFGEFGI.

The extra character we output with each sequence is an end-of-sequence marker, necessary to ensure we don't consider k-mers that straddle the end of one sequence and the start of the next. We can, of course, suppress the last end-of-sequence marker.

The important thing to notice here is that the only parameter we are free to change in order to reduce the size of S' is the number of sequences it contains. Therefore, constructing a minimum size complete, correct, compact sequence database is equivalent to minimizing the number of paths in a SBH-graph path set such that each edge is used exactly once.

The best possible scenario, then, is that our SBH-graph admits an Eulerian path. In this case, one path is sufficient. Note that this lemma effectively restates our lower bound on the minimum size of any complete sequence database.

Lemma 1. *The set of k-mers of S can be represented by a complete, correct, and compact sequence database of size $(k-1) + N_k$ if and only if the SBH-graph of S admits an Eulerian path.*

In this work, we will use a *compressed* SBH-graph to represent the k-mers of our sequence database. The compressed SBH-graph (CSBH-graph) represents the same information as the SBH-graph, except that paths through trivial nodes, those with in and out degree one, are turned into a single edge. Where an edge of the CSBH-graph replaces a path through trivial nodes, we associate the concatenation of the character on the edges of the path with the new edge. Figure 2 shows the CSBH-graph corresponding to the SBH-graph of Figure 1. It should be clear that all of the preceding discussion, including Lemma 1, holds just as well for CSBH-graphs as for SBH-graphs.

Lemma 2. *The set of k-mers of S can be represented by a complete, correct, and compact sequence database of size $(k-1) + N_k$ if and only if the CSBH-graph of S admits an Eulerian path.*

We use the CSBH-graph instead of the SBH-graph because it is much smaller, with many fewer nodes and edges. However, building the CSBH-graph for a

Table 2. SBH-graph and CSBH-graph sizes for some sequence databases used for peptide identification via tandem mass spectrometry.

Sequence Database	SBH-graph Nodes	SBH-graph Edges	CSBH-graph Nodes	CSBH-graph Edges
IPI-HUMAN	12119290	12115520	115246	111476
IPI	29645471	29769766	545356	669651
Swiss-Prot	44352317	44374286	550060	572029
Swiss-Prot-VS	45259553	45307827	609679	657953
UniProt	274510105	274510105	4075920	4445958
UniProt-VS	274995795	275391669	4132200	4528074
MSDB	276094660	276523755	4313501	4742596
NRP	282706577	283160529	4443077	4897029
NCBI-nr	384256196	378721915	5534281	5978394
Union(NR)	384866007	385369671	5758507	6262171

given amino-acid sequence database is non-trivial, particularly for $k = 30$ and an alphabet of size 20. Clearly we could first build the SBH-graph and remove trivial nodes, but the initial cost to build the SBH-graph quickly becomes prohibitive. We avoid these issues by constructing the CSBH-graph directly, using a suffix-tree data-structure. Table 2 shows the sizes of the SBH and CSBH-graphs for each of the sequence databases of Table 1.

3   Optimal Complete, Correct, Compact (C^3) Enumeration

For a directed graph $G = (V, E)$, we define $\delta_i(v)$ and $\delta_o(v)$ to be the in and out degree of node $v \in V$. Further, we define $b(v) = \delta_i(v) - \delta_o(v)$. $b(v)$ is called the degree deficit if $b(v) < 0$ and the degree surplus if $b(v) > 0$. A node v with $b(v) = 0$ is called balanced, otherwise it is called unbalanced. A graph or connected component is called balanced if all of its nodes are balanced and unbalanced otherwise. We define V_+ to be the set of surplus degree nodes and n_+ to be number of these nodes. Similarly, we define V_- to be the set of deficit degree nodes and n_- to be the number of these nodes. The total degree surplus is defined to be $B_+ = \sum_{v \in V_+} b(v)$ while the total degree deficit is defined as $B_- = \sum_{v \in V_-} b(v)$. The degree surplus of a connected component $C \subseteq V$ is defined to be $B_+(C) = \sum_{v \in V_+ \cap C} b(v)$, with the $B_-(C)$ defined analogously.

In practice, CSBH-graphs built from amino-acid sequence databases with $k = 30$ fail to be Eulerian on two counts. First, very few nodes of the graph are balanced, and second, the graphs usually have more than one connected component. Table 3 shows the extent to which the CSBH-graphs of 30-mers built for our test set of amino-acid sequence databases fail to be Eulerian.

We must, of course, have at least one path in our path set per component. For each balanced component, we require exactly one path, an Eulerian tour, as

Table 3. CSBH-graph statistics of some sequence databases used for peptide identification via tandem mass spectrometry.

Sequence Database	Degree Surplus Nodes	Total Surplus Degree	Degree Deficit Nodes	Total Deficit Degree	Components (Balanced)
IPI-HUMAN	57275	57971	56975	57971	23076 (1)
IPI	267896	273052	267329	273052	35728 (2)
Swiss-Prot	270279	276262	270228	276262	93611 (0)
Swiss-Prot-VS	299410	307551	299154	307551	93624 (0)
UniProt	1992448	2086977	1988855	2086977	626503 (5)
UniProt-VS	2019828	2116632	2015947	2116632	626470 (5)
MSDB	2112761	2213341	2101795	2213341	629636 (6)
NRP	2175883	2281329	2164551	2281329	643496 (6)
NCBI-nr	2712544	2826497	2701270	2826497	850325 (7)
Union(NR)	2822070	2943180	2810160	2943180	863078 (8)

per Lemma 2. We denote the number of balanced components by m. What then, for unbalanced components? Lemma 3 prescribes a lower bound on the number of paths required.

Lemma 3. *If C is an unbalanced component of a CSBH-graph, then we require at least $B_+(C)$ paths, in order to use each edge exactly once.*

Proof. Given a path set, suppose we consider each path, in turn, and delete its edges from C. The deletion of the edges of a path from s to t increases $b(s)$ by one, decreases $b(t)$ by 1, and leaves the remaining $b(v), v \neq s, t$ unchanged. Since the path set uses every edge exactly once, when the process is complete we must have $b(v) = 0$ for all nodes $v \in V$. Therefore, there must have been $-b(v)$ paths starting at nodes $v \in V_- \cap C$ and $b(v)$ paths ending at nodes $v \in V_+ \cap C$. Therefore, we require at least $\max\{B_+(C), -B_-(C)\}$ paths in the path set, each of which starts at a node of $V_- \cap C$ and ends at a node of $V_+ \cap C$. By induction on the edges of C, it is straightforward to show that $B_+(C) = -B_-(C)$. □

What we have shown then is that our path set must contain at least $B_+ + m$ paths. All that remains is to demonstrate how this bound can be achieved.

Lemma 4. *Given an unbalanced component C of a CSBH-graph G, there exists a path set of size $B_+(C)$ that uses each edge exactly once.*

Proof. We add $B_+(C) - 1$ artificial *restart* edges from nodes of $V_+ \cap C$ to nodes of $V_- \cap C$. The edges are added in such a way that $v \in V_+ \cap C$ has at most $b(v)$ outgoing restart edges, while $v \in V_- \cap C$ has at most $-b(v)$ incoming restart edges. The resulting CSBH-graph must contain exactly two unbalanced nodes, s with $b(s) = -1$ and t with $b(t) = 1$. Therefore, we can construct an Eulerian path between s and t that uses every edge exactly once. This Eulerian path can then broken into $B_+(C)$ paths at the restart edges and the lemma is shown. □

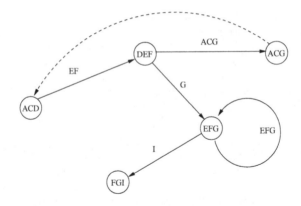

Fig. 3. Compressed SBH-graph with artificial edge (dashed) for 4-mers from the sequences: ACDEFGI, ACDEFACG, DEFGEFGI.

Thus, we have demonstrated how to construct a complete, correct, compact sequence database of minimum size.

Theorem 1. *Let G be the CSBH-graph of an input sequence S for some mer size k with m balanced components and total degree surplus B_+. Then the optimal complete, compact, correct sequence database with respect to S and k has size*

$$N_k + k(m + B_+) - 1.$$

Figure 3 shows the CSBH-graph of Figure 2 with its required $B_+ - 1 = 2 - 1 = 1$ artificial restart edge(s) inserted. The Eulerian path of this graph leads to the minimum length 4-mer enumeration: DEFACG, ACDEFGEFGI; which uses 17 characters, instead of the original 25 characters, to represent 10 distinct 4-mers. The lower bound implied by Lemma 2 is 13 characters.

4   Computational Experiments

Table 4 shows the performance of the C^3 30-mer enumeration strategy as a compression technique for the amino-acid sequence databases of Table 1.

We notice that despite the widely varying degree of overhead of the original sequence databases, the overhead of the compressed version is typically in the 15-25% range. For those sequence databases, such as Swiss-Prot, which have little overhead, the compression is not particularly impressive, but for those sequence databases containing lots of peptide level redundancy, such as Swiss-Prot-VS, UniProt-VS, or Union, the compression is significant. Even for sequence databases with moderate overhead, the compression is substantial. This compression technique has reduced each of the sequence databases to approximately the same level of peptide redundancy.

Most encouraging is the performance of the compression on the Swiss-Prot-VS, UniProt-VS and Union sequence databases. These sequence databases are

Table 4. Size of optimal complete, correct, compact 30-mer enumerations for sequence databases used for peptide identification via tandem mass spectrometry.

Sequence Database	C^3 30-mer Enumeration	Overhead	Compression	Compression Bound
IPI-HUMAN	13854679	14.35%	68.05%	59.51%
IPI	37961385	27.52%	70.11%	54.98%
Swiss-Prot	52662145	18.68%	93.28%	78.60%
Swiss-Prot-VS	54534356	20.36%	60.90%	50.60%
UniProt	337119564	22.81%	71.34%	58.09%
UniProt-VS	338890778	23.06%	66.87%	54.34%
MSDB	342924164	24.01%	71.16%	57.38%
NRP	351600578	24.17%	70.96%	57.15%
NCBI-nr	463517034	22.39%	74.87%	61.17%
UnionNR	473665310	22.91%	70.20%	57.12%
Union	473665310	22.91%	21.96%	17.86%

constructed specifically to create richer peptide candidate lists, but they necessarily contain many redundant peptide candidates. As discussed in the introduction, the Swiss-Prot-VS sequence database contains about 2% additional distinct 30-mers, despite being more than 1.5 times the size of Swiss-Prot. After compression, Swiss-Prot-VS is smaller than the original Swiss-Prot sequence database. With no modification of existing search engines, we can search all the isoforms of Swiss-Prot in less time than it would take to search the original Swiss-Prot sequence. The Union sequence database is compressed to about three quarters of the size of one of its constituents, NCBI-nr. Again, without any modification to our existing search engines, we can search all isoforms of UniProt, all of MSDB, all of NRP, all of IPI, and all of NCBI-nr in less time than it takes to search the original NCBI-nr sequence.

5   Conclusion

We have shown that the peptide content of an amino-acid sequence database can be represented in a reformulated amino-acid sequence database containing fewer amino-acid symbols than the original. We have demonstrated how to construct an enumeration of all k-mers of a sequence database that is complete, correct and compact, and shown that this representation is as small as possible. We believe that this technique will be most useful for sequence databases that contain lots of peptide level redundancy, such as those constructed to enumerate protein isoforms or merge sequence databases from different sources. In some cases, this technique makes it possible to search a richer set of peptide candidates using sequence databases that are no bigger than their less expressive counterparts.

There remains much work to be done. We have not addressed a number of issues that must be resolved before we can use such a sequence database with Mascot or SEQUEST. In the process of our k-mer enumeration, our amino-acid sequences lose all explicit connection to their original protein sequence and annotation. We have effectively decoupled the problem of associating peptide sequences with spectra and the problem of determining which proteins the peptides represent. Fortunately, searching for exact peptide sequences in large sequence databases can be done very quickly using well established string matching techniques, particularly since we only need to do this for 10-20 peptides per spectrum. The key piece of infrastructure required to accomplish this is a tool that can take a Mascot or SEQUEST output file and a sequence database as input, and output a new Mascot or SEQUEST output file with the protein information inserted appropriately.

For some use cases, our requirement that we represent all 30-mers is too strong. In particular, for identification of tandem mass spectra, we often search only for those peptides that have trypsin digest motifs at each end. A complete, correct, compact tryptic peptide enumeration would be a natural specialization of this work. However, in order to implement this, we would need either to be able to build a CSBH-graph like data-structure for variable length mers, or to consider tryptic fragment "characters" as input to the problem.

For those use cases, such as peptide identification, in which 30-mers are merely an upper bound on the length of relevant sequences, we note that there is no particular advantage in insisting that all 30-mers are distinct, since this guarantees nothing about the number of occurrences of shorter sequences. In fact, we believe that it should be possible to compress our input sequences databases still further by relaxing the compactness constraint, permitting k-mers to appear more than once in the output. Instead of adding restart edges, we may, perhaps, reuse edges to get from the end of one path to the start of another. Given the similarity of this variant of the problem to the well known "Chinese Postman Problem" [21], we believe that this variant may also be solved to optimality in polynomial time using some sort of matching formulation.

A less obvious application of this k-mer enumeration is in suffix tree compression. For applications such as the peptide candidate generation [1], in which we only traverse the suffix tree nodes to a bounded depth k, the suffix tree of the k-mer enumeration contains the same suffixes (to depth k) as our original sequence database's suffix tree. However, since the memory requirements of suffix trees are linear in size of the underlying sequence, the suffix tree built on the k-mer enumeration will require a smaller memory footprint.

Acknowledgments

The authors thank Bjarni Halldórsson and Clark Mobarry for helpful comments and suggestions on the manuscript.

References

1. Edwards, N., Lippert, R.: Generating peptide candidates from amino-acid sequence databases for protein identification via mass spectrometry. In: Proceedings of the Second International Workshop on Algorithms in Bioinformatics, Springer-Verlag (2002) 68–81

2. Cieliebak, M., Erlebach, T., Lipták, Z., Stoye, J., Welzl, E.: Algorithmic complexity of protein identification: Combinatorics of weighted strings. Discrete Applied Mathematics **137** (2004) 27–46
3. Eng, J., McCormack, A., Yates, J.: An approach to correlate tandem mass spectral data of peptides with amino acid sequences in a protein database. Journal of American Society of Mass Spectrometry **5** (1994) 976–989
4. Perkins, D., Pappin, D., Creasy, D., Cottrell, J.: Probability-based protein identification by searching sequence databases using mass spectrometry data. Electrophoresis **20** (1997) 3551–3567
5. Bafna, V., Edwards, N.: SCOPE: A probabilistic model for scoring tandem mass spectra against a peptide database. Bioinformatics **17** (2001) S13–S21
6. Apweiler, R., Bairoch, A., Wu, C.H., Barker, W.C., Boeckmann, B., Ferro, S., Gasteiger, E., Huang, H., Lopez, R., Magrane, M., Martin, M.J., Natale, D.A., O'Donovan, C., Redaschi, N., Yeh, L.S.L.: UniProt: the Universal Protein knowledgebase. Nucl. Acids. Res. **32** (2004) D115–119
7. Welcome to UniProt — UniProt [the Universal Protein Resource] [online, cited June 24, 2004]. Available from: http://www.uniprot.org/
8. Kersey, P., Hermjakob, H., Apweiler, R.: VARSPLIC: Alternatively-spliced protein sequences derived from SWISS-PROT and TrEMBL. Bioinformatics **16** (2000) 1048–1049
9. UniProt/Swiss-Prot Tools [online, cited June 24, 2004]. Available from: http://www.ebi.ac.uk/swissprot/tools.html
10. CSC/ICSM Proteomics Section Home Page [online, cited June 24, 2004]. Available from: http://csc-fserve.hh.med.ic.ac.uk/msdb.html
11. The BLAST Databases [online, cited June 24, 2004]. Available from: ftp://ftp.ncbi.nlm.nih.gov/blast/db/
12. NRP (Non-Redundant Protein) Database [online, cited June 24, 2004]. Available from: ftp://ftp.ncifcrf.gov/pub/nonredun/
13. EBI Databases — International Protein Index [online, cited June 24, 2004]. Available from: http://www.ebi.ac.uk/IPI/IPIhelp.html
14. Garey, R., Johnson, D.: Computers and Intractability: A guide to the theory of NP-completeness. W. H. Freeman and Company, San Francisco (1979)
15. Gusfield, D.: Algorithms on Strings, Trees, and Sequences: Computer Science and Computational Biology. Cambridge University Press (1997)
16. Bains, W., Smith, G.: A novel method for nucleic acid sequence determination. Journal of Theoretical Biology **135** (1988) 303–307
17. Lysov, Y., Floretiev, V., Khorlyn, A., Khrapko, K., Shick, V., Mirzabekov, A.: DNA sequencing by hybridization with oligonucleotides. Dokl. Acad. Sci. USSR **303** (1988) 1508–1511
18. Drmanac, R., Labat, I., Bruckner, I., Crkvenjakov, R.: Sequencing of megabase plus DNA by hybridization. Genomics **4** (1989) 114–128
19. Pevzner, P.A.: l-tuple DNA sequencing: Computer analysis. J. Biomol. Struct. Dyn. **7** (1989) 63–73
20. de Bruijn, N.: A combinatorial problem. In: Proc. Kon. Ned. Akad. Wetensch. Volume 49. (1946) 758–764
21. Kwan, M.K.: Graphic programming using odd or even points. Chinese Mathematics **1** (1962) 273–277

Linear Reduction for Haplotype Inference*

Jingwu He and Alex Zelikovsky

Department of Computer Science, Georgia State University, Atlanta, GA 30303
jimhe@alla.cs.gsu.edu, alexz@cs.gsu.edu

Abstract. Haplotype inference problem asks for a set of haplotypes explaining a given set of genotypes. Popular software tools for haplotype inference (e.g., PHASE, HAPLOTYPER) as well as new algorithms recently proposed for perfect phylogeny inference (DPPH) are often not well scalable. When the number of sites (SNP's) comes to thousands these tools often cannot deliver answer in reasonable time even if the number of haplotypes is small. In this paper we propose a new linear algebra based method which drastically reduces the number of sites in the original data. After solving a reduced instance, linear decoding allows to recover haplotypes of full length for given genotypes. Experiments show that our method significantly speeds up popular haplotype inference tools while finding almost the same solution practically in all cases thus not compromising the quality of the known haplotype inference methods.

Keywords: Haplotype inference, linear independence, perfect phylogeny

1 Introduction

In diploid organisms each chromosome has two "copies" which are not completely identical. Each of two single copies is called a haplotype, while a description of the data consisting of mixture of the two haplotypes is called a genotype. For complex diseases caused by more than a single gene it is important to obtain haplotype data which identify a set of gene alleles inherited together. In haplotype description it is important only positions where the two copies are different which are called single nucleotide polymorphisms (SNP's). A SNP is a single nucleotide site where exactly two (of four) different nucleotides occur in a large percentage of the population. The SNP-based approach is the dominant one, and high density SNP maps have been constructed across the human genome with a density of about one SNP per thousand nucleotides.

In general, it is costly and time consuming to examine the two copies of a chromosome separately, and genotype data rather than haplotype data are only available, even though it is the haplotype data that will be of greatest use. Data from m sites (SNP's) in n individual genotype are collected, where each site can have one of two states (alleles), which we denote by 0 and 1. For each individual, we would ideally like to describe the states of the m sites on each of the two chromosome copies separately, i.e., the haplotype. However, experimentally determining the haplotype pair is technically difficult or expensive. Instead,

---

* Partially supported by NIH Award 1 P20 GM065762-01A1.

I. Jonassen and J. Kim (Eds.): WABI 2004, LNBI 3240, pp. 242–253, 2004.

the screen will learn the 2m states (the genotype) possessed by the individual, without learning the two desired haplotypes for that individual. One then uses computation to extract haplotype information from the given genotype information. Several methods have been explored and some are intensively used for this task [5, 6, 9, 11, 17, 15, 16]. None of these methods are presently fully satisfactory, although many give impressively accurate results.

The rest of the paper is organized as follows. In the next section we give a formal definition of the Haplotype Inference problem. In Section 3 we describe intuition behind suggested linear reduction of haplotype inference methods. In Section 4 we describe the linear reduction in terms of matrix multiplication and in Section 5 we describe a graph based decoding algorithm which fixes errors introduced by matrix multiplication. In Section 6 we show how the proposed liner reduction drastically reduces the runtime of DPPH [1, 2] without compromising quality, compare original and linearly reduced HAPLOTYPER [16] and PHASE [19].

2   The Haplotype Inference Problem

The input and the output of the Haplotype Inference problem admits the following traditional combinatorial description (see e.g., [3]).

The input population is given in the form of an $n \times m$ *genotype matrix* $G = \{g_{ij}\}$ with all values $g_{ij} \in \{0, 1, 2\}$. Each row g_i, $i = 1, \ldots, n$, of the matrix G corresponds to a genotype and each column s_j, $j = 1, \ldots, m$, corresponds to a site of interest on the chromosome, namely, a SNP. When the site s_j is homozygous for the genotype g_i, then $g_{ij} = 0$ if the associated chromosome site has that state 0 on both copies and, respectively, $g_{ij} = 1$ if the site has state 1 on both copies. When the site s_j is heterogenous for the genotype g_i, i.e., the site has different state on the two copies, then $g_{ij} = 2$.

The output of Haplotype Inference problem is a $2n \times m$ *haplotype matrix* $H = \{h_{ij}\}$, with all values $h_{ij} \in \{0, 1\}$. A consecutive pair of rows (h_{2i-1}, h_{2i}) corresponds to a pair of haplotypes which is a feasible "explanation" of the genotype vector g_i, $i = 1, \ldots, n$. For any homozygous site s_j of the genotype g_i, i.e., the site with value 0 (respectively, 1), the corresponding haplotypes should both have value 0 (respectively, 1) in its j-th position, i.e., if $g_{ij} = 0$, then $h_{2i-1,j} = h_{2i,j} = 0$ and if $g_{ij} = 1$, then $h_{2i-1,j} = h_{2i,j} = 1$. For any heterogenous site s_j of the genotype g_i, i.e., the site with value 2, the corresponding haplotypes should have different values in its j-th position, i.e., if $g_{ij} = 2$, then $h_{2i-1,j} = 1 - h_{2i,j}$.

Thus, the Haplotype Inference problem asks for a haplotype matrix H which is a feasible "explanation" of a given genotype matrix G. Although the input and the output of the Haplotype Inference problem are very well formalized, it is still ill-formulated since, in general as well as in common biological setting, there is exponential number of possible haplotype matrices for the same input matrix. Indeed, an individual genotype with k heterozygous sites can have 2^{k-1} haplotype pairs that could appear in H. Without additional biological insight, one cannot deduce which of the exponential number of solutions is the best, i.e., the most biologically meaningful.

In this paper, we try to avoid discussing the genetic models referring reader to the rich literature (see, e.g., [5]). Nevertheless, in order to measure the quality of the proposed linear reduction we compare the original methods with the linearly reduced methods as well as with the original simulated and real data.

3   Linear Dependence of Sites, Haplotypes and Genotypes

In this section we give motivation and informal description of ideas behind suggested linear reduction of haplotype inference methods.

Usually, in genetic sequences derived from human haplotypes (see [10, 18]), the number of sites is much larger than the number of individuals. Because of such disproportion many columns corresponding to SNP sites are similar. Indeed, as noted in [18], the number of *synonymous* sites in real data is considerably large, here two sites are synonymous (or equivalent) if the corresponding 0-1-columns either the same or the complimentary (i.e., the same after each entry x is replaced with $1 - x$). It is common to keep only one site out of several synonymous sites since they are assumed not to carry any additional information [18]. Thus if the site column s_i is equal or complementary to the site column s_j, then one of them can be dropped. From haplotype inference point of view, we infer the haplotypes in one of the synonymous sites the same way as in another.

In this paper we make the next inductive step: if k columns are "dependent", or k-th site can be "expressed" in terms of $k - 1$ others, then we suggest to drop the k-th site. Indeed, the k-th site arguably does not carry any information additional to one which we can derive from the first $k - 1$ sites. Inductively, if we decide how to infer haplotypes in the first $k - 1$ site, then we should consistently infer haplotypes in the k-th site.

In order to make this idea work, we need to formalize the notion of "dependent" or "expressed" in a such way that it should be easy and fast to derive and manipulate. The most suitable approach is to rely on the standard linear dependence. Unfortunately, two synonymous $0 - 1$-columns are not linearly dependent in a standard arithmetic. It is not difficult to see that replacing 0's with -1's will resolve that issue. Indeed, in the new notations, multiplication by (-1) corresponds to complementing the column in the traditional notations. Thus

Remark 1. In $(-1, 1)$-notations, two sites are synonymous if and only if they are collinear (i.e., linearly dependent).

We also need to change notations for genotypes. Ideally, a genotype obtained from two haplotypes should be linear dependent from these haplotypes, then we can hope that linear dependency between columns of the genotype matrix will correspond to linear dependency between columns of the haplotype matrix. It is easy to see that replacing 0's with -1's (as for haplotypes) and replacing 2's with 0's makes this idea work. In the new notations,

Remark 2. In $(-1, 1, 0)$-notations, a genotype vector g is obtained from haplotype vectors h and h' if and only if $g = (h + h')/2$.

In the rest of the paper we apply linear algebra to haplotype inference using new notations. The proposed linear reduction consists of the following three steps:

1. (*encoding*) reduce the genotype matrix by keeping only linearly independent sites and dropping all linearly dependent sites;
2. apply an arbitrary haplotype inference method to the resulted site-reduced genotype matrix obtained;
3. (*decoding*) complement the inferred site-reduced haplotype matrix with linearly dependent column-sites which are obtained using original linear combinations of inferred haplotype columns.

Obviously, the number of linearly independent columns r cannot be more than the size of population, i.e., the number of rows. Also, Remark 2 implies that r is at most h, where h is the number of haplotypes. In this paper we explore how the linear reduction can reduce the runtime for all known haplotype inference methods.

4   Implementation of Linear Reduction Based on Matrix Multiplication

In this section we describe linear algebra behind the suggested implementation of our linear reduction. Everywhere further we will only use new $(-1, 1, 0)$-notations for genotypes and haplotypes.

Let G be a $(-1, 1, 0)$-genotype matrix consisting of n rows corresponding to genotypes and m columns corresponding to SNP sites. We will modify the $(-1, 1)$-haplotype matrix H by removing all duplicate rows, i.e., if a haplotype is used for different genotypes, then only a single its copy remains in H. Let the modified matrix H' has h rows. The dependency between G and H' can be expressed as a graph $X = (H, G)$ with h vertices corresponding to haplotypes and n edges corresponding to genotypes – an edge connects two vertices if the corresponding genotype row is a sum of the corresponding two haplotype rows. Let I_X be an $n \times h$ incidence matrix of the graph X, i.e., each of row e_i of I_X corresponds to a genotype g_i and consists of all 0's except exactly two 1's in two columns corresponding to the two vertices-haplotypes connected by e_i. Thus, using matrix multiplication we can express this dependency as follows

$$G = \frac{1}{2} I_X \times H' \qquad (1)$$

One can reformulate the Haplotype Inference problem as follows: given a $(-1, 1, 0)$-matrix G, find a $(-1, 1)$-matrix H' and a graph X, such that the equality (1) holds. In other words, the Haplotype Inference problem becomes equivalent to a matrix factorization problem (1).

Let $rh = rank(H')$ be the rank of the matrix H'. Note that the number of sites is often larger than the number of haplotypes, $m >> h$, therefore $rank(H') << m$ often coincide with the number of rows h. The matrix H' can be represented as follows

$$H' = H_{rh} \times (E_{rh}|C) \qquad (2)$$

where the matrix H_{rh} consists of rh linearly independent columns of H' and $(E_{rh}|C)$ is a $(rh \times m)$ matrix with the first rh columns and rows forming the identity matrix E_{rh} (1's on the main diagonal and 0's elsewhere) and C is a $(rh \times (m - rh))$ matrix. Substituting (2) into (1), we obtain

$$G = \frac{1}{2} I_X \times H_{rh} \times (E_{rh}|C) \qquad (3)$$

On the other hand, using $O(n^2m)$ Gaussian elimination, we can extract $r = rank(G)$ linearly independent columns from the matrix G such that

$$G = G_r \times (E_r|C') \qquad (4)$$

where the matrix G_r consists of r linearly independent columns of G and $(E_r|C')$ is a $(r \times m)$ matrix with the first r columns forming the identity matrix E_r and C' is a $(r \times (m - r))$ matrix.

If $rank(I_X) = rh$ (note that $rank(I_X) \leq rh$), then $r = rh$. If we can choose the same linearly independent sites for G and H, then (3) and (4) implies that $C = C'$ and

$$G_r = \frac{1}{2} I_X \times H_{rh} \qquad (5)$$

Thus, we have reduced the Haplotype Inference problem (1) to the *linearly reduced* Haplotype Inference problem (5). Indeed, in time $O(n^2m)$ we find representation (4), then after solving factorization (5), we can find H' using (2) in time $O(h^2m)$. For example, the PPH problem can be solved in time $O(n^2m)$ rather than in time $O(nm^2)$.

5   Fixing Caveats in Linear Reduction Approach

Unfortunately, the plan described in the previous section, may fail due to the following two caveats. First, the factorization problem (5) can have more solutions than the original problem (1). Secondly, it is possible that the matrix H' obtained from (2) contains entries not equal to -1 and 1 or, even worse, there is no feasible matrix H' which can be obtained from H_{rh}. In this section we show how to enhance the original linear reduction idea to deal with the above two caveats.

In our experiments we have found that sometimes the matrix multiplication

$$H_{rh} \times (E_r|C') \qquad (6)$$

which presumably gives the decoded complete haplotype matrix H', results in non-feasible product with entries unequal to -1 and 1. This happens when the matrix I_X does not have full column rank, i.e., when $rank(I_X)$ is less than the number of columns of I_X. The following theorem characterizes *nontrivial* graphs, i.e., graphs which have full column rank incidence matrix I_X.

Theorem 1. *The graph X is nontrivial if and only if each connected component of X is not bipartite.*

Proof. Let the graph $X = (V, E)$ have a bipartite connected component with vertex set C, i.e., $C = C_A \cup C_B$ and all edges with at least one endpoint in C should connect a vertex from C_A with a vertex from C_B. Then the sum of all columns corresponding to C_A equals to the sum of all columns corresponding to C_B since whenever C_A-column has 1 in a certain row, the 1 corresponding to the opposite endpoint of the edge will be in C_B columns in the same row.

It is easy to see that X is nontrivial if and only if each connected component is nontrivial. Wlog assume that X is a connected graph with an odd cycle B. By removing edges we can assume that X is a connected graph with a single odd cycle. If X does not have leaves, then $X = B$ and by sorting vertices of X in order of the traversal of B, we obtain that I_X is a square matrix with $det(I_X) = 2$. If X has a leaf l, then by induction I_{X-l} has full column rank and $r(I_X) = r(I_{X-l}) + 1$ since the column l has a single 1 in a new row.

The following remark shows why one would prefer a nontrivial graph X.

Remark 3. Let X be a trivial graph. Then there exists such a genotype matrix G and two different haplotype matrix H_1 and H_2 such that $G = \frac{1}{2}I_X \times H_1$ and $G = \frac{1}{2}I_X \times H_2$.

Proof. If X is a trivial graph, then there exists a connected component X' which can be colored into 2 colors. It is easy to see that we can find a matrix H with a column C having all 0's in the rows corresponding to the genotypes-edges from X'. We can color all vertices of X' into two colors (-1) and 1 such that no two adjacent vertices will have the same color. Obviously, such phase assignment corresponding to H_1 is feasible as well as "opposite" phase assignment $H_2 = -H_1$ obtained from H_1 by multiplying each phase assignment in X_1 by (-1).

Note that the probability that X is nontrivial is very high when the number of edges (genotypes) is large enough with respect to the number of haplotypes. Nevertheless, we propose a simple decoding algorithm which reconstructs H' from H_{rh} which relies on the reconstruction of the graph X rather than on the matrix multiplication.

As soon as the haplotype graph X_r for the reduced set of sites of size r is obtained by any haplotype inferring algorithm, we need to extend this graph to the graph X_m for all m sites. Very often the graphs X_r and X_m are isomorphic, i.e., they are the same graphs but with the different labels – X_r-vertices (resp. edges) are labeled by haplotypes (resp. genotypes) with the reduced set of r sites while X_m-vertices (resp. edges) are labeled by haplotypes (resp. genotypes) with

---

Input: The reduced genotype matrix $G = \{s_{ij}\}$, the graph X_r and the haplotype matrix H_r

Output: The graph X_m and the haplotype matrix H'

For each site s_j and each connected component C of X_r do
(1) Find an edge e_i with with non-zero label s_{ij}
(2) Label one endpoint u of e_i with s_{ij}, i.e., $s_{u,j} = s_{ij}$
(3) In breadth-first-search manner, propagate the labels over C:
    for each edge $e_i = (u, v)$, $s_{v,j} = s_{u,j} - s_{ij}$
Output haplotype matrix H' which are the labels of the vertices of the graph X_m.

Fig. 1. The Decoding Algorithm.

the required site-length of m. The following Decoding Algorithm (see Figure 1) will always restore X_m from X_r if X_m and X_r are isomorphic.

Note that the Decoding Algorithm would stuck at the step (1) if all edges of a connected components have a zero label for a certain site. But in this case, as we know from Remark 3, the solution is not unique if the connected component is bipartite and no feasible labels are possible otherwise. Therefore we have proved the following

Theorem 2. *The Decoding Algorithm correctly reconstructs the graph X_m for the original Haplotype Inference problem from the graph X_r for the site-reduced Haplotype Inference problem if X_m and X_r are isomorphic and such reconstruction unique.*

Unfortunately, it is possible that the graphs X_r and X_m are not isomorphic. Indeed, consider the following two genotypes with three sites:

$$g_1 = (1, 0, 1) \text{ and } g_2 = (0, -1, -1) \tag{7}$$

The reduced site set includes the first two sites since the third column-site just equals the sum of the first two column-sites. The corresponding reduced haplotype graph X_r has 3 vertices labeled $h_1 = (1, 1)$, $h_2 = (1, -1)$ and $h_3 = (-1, -1)$ and edges $g_1 = (h_1, h_2)$ and $g_2 = (h_2, h_3)$ (see Figure 2(a)), while X_m has 4 vertices (see Figure 2(b)).

Thus, if the graphs X_m and X_r are not isomorphic, then we should apply splitting of vertices. Fortunately, in our extensive experimental study we never got any instance where splitting was necessary.

6   Experimental Results

In generating the test data, following [3] we have used the haplotype generator ms [14]. This generator is a well-known standard based on the coalescent

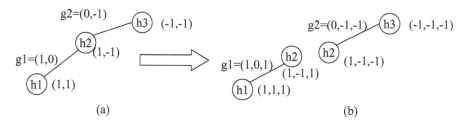

Fig. 2. (a) The reduced haplotype graph with 3 vertices. (b) Result of splitting of the vertex h_2 into two vertices).

Table 1. The comparison of the running times of DPPH and Linearly Reduced DPPH. Each value is averaged over 100 datasets. E and D is the CPU time for encoding and decoding and RD is DPPH runtime for the reduced instance.

Datasets		Average Running Time (seconds)			
		DPPH	Linearly Reduced DPPH		
sites	pop	Total	Total	E and D	RD
30	50	0.0297	0.0141	0.0042	0.0099
100	100	0.2168	0.0793	0.0446	0.0374
300	150	1.9243	0.2144	0.1499	0.0645
500	250	7.9553	0.5825	0.4548	0.1277
1000	500	56.4723	2.6373	2.2311	0.4062
2000	1000	549.415	9.4355	8.0125	1.4230

model of SNP sequence evolution. The ms generator has capability to generate a given number of haplotypes with the prescribed number of sites and recombination rate. In our tests, we have generated $2n$ haplotypes for n sites, and then randomly paired them to obtain n genotypes. For the Table 1 we have set the recombination rate to 0. For all other testcases the recombination rate is specified in the corresponding tables. Although in principle, the linearly reduced haplotype inference methods may need to split vertices (see Section 2), in *all* our testcases it is never happened.

Three different methods of perfect phylogeny reconstruction (DPPH [1,2], GPPH [12], and HPPH [8]) have been compared in [3]. The experimental data show that the fastest method DPPH is slowly increases advantage in runtime over the second best HPPH achieving factor of 3 for the largest instances. In Table 1, using similar testcases we compare DPPH with the suggested in this paper Linearly Reduced DPPH. For the Linearly Reduced DPPH we report separately the total CPU time as well as time taken by linear encoding and decoding and time taken by DPPH to solve the reduced instance. Our results show that the advantage in runtime of Linearly Reduced DPPH grows fast with testcase size and reaches factor of 60 for the largest instances. In *all* testcases, if DPPH finds a unique solution for the Haplotype Inference problem, then so does the Linearly Reduced DPPH and the solution is identical.

In Tables 2 and 3 we compare the runtime and quality of haplotype inference of the standard PHASE [19] and proposed Linearly Reduced PHASE. As we can see, the runtime is drastically reduced while the quality (measured in the

Table 2. The comparison of the running times of PHASE and Linearly Reduced PHASE. Each value is averaged over 25 datasets.

Datasets		Average Running Time (seconds)			
		PHASE	Linearly Reduced PHASE		
sites	pop	Total	Total	E and D	RP
60	30	201.9131	42.2205	0.1093	42.2124
60	60	501.2166	148.2856	0.0262	148.2594
100	50	1019.1806	162.4863	0.0392	162.4471
100	100	2103.3299	274.7445	0.0722	274.6722
140	70	2367.4557	141.2005	0.0606	141.1399
140	140	4460.5307	179.8020	0.1012	179.7008

Table 3. The comparison of the quality of haplotyping of Linearly Reduced PHASE (LRP) and PHASE (P) vs the original haplotypes (O). Here the difference in haplotype data sets, Hapset1/Hapset2 is the arithmetic mean of numbers of false-positive and false-negative haplotypes over the number of haplotypes Hapset2 times 100%. Each value is averaged over 25 datasets.

Datasets		Recombination Rate											
		0			4			16			40		
sites	pop	LRP/P	LRP/O	P/O	LRP/P	LRP/O	P/O	LRP/P	LRP/O	P/O	LRP/P	LRP/O	P/O
60	30	6.25	9.91	8.84	10.53	14.25	11.98	19.42	20.07	12.80	20.08	17.55	15.92
60	60	6.47	9.28	7.99	10.76	8.88	10.15	8.05	9.86	6.62	10.72	12.68	9.19
100	50	9.02	12.92	11.46	9.00	10.24	8.88	12.84	14.81	11.28	22.38	22.61	10.36
100	100	5.88	8.39	8.46	6.01	8.14	7.63	5.66	8.78	5.07	7.34	9.38	6.64
140	70	8.39	9.38	10.02	8.59	8.91	9.84	9.32	6.25	8.48	12.92	11.60	10.49
140	140	6.29	9.66	11.11	7.97	10.18	9.82	9.28	11.17	10.54	12.65	14.37	11.56

Table 4. The comparison of the quality of haplotyping of Linearly Reduced PHASE (LRP) and PHASE (P) vs the original haplotypes (O) over the instances with nontrivial graphs.

Datasets		Recombination Rate					
		0			4		
sites	pop	LRP/P	LRP/O	P/O	LRP/P	LRP/O	P/O
60	30	1.28	1.91	2.23	1.18	4.71	4.71
60	60	2.68	4.46	3.57	7.10	5.41	7.88
100	50	3.71	3.71	5.56	4.34	6.08	5.21
100	100	4.17	6.70	7.29	4.57	5.32	6.01
140	70	5.92	7.77	8.51	6.12	5.33	7.45
140	140	5.05	8.19	10.50	4.78	8.08	8.28

average number of errors versus original haplotypes) is practically the same. Table 4 compares the quality of the standard and linearly reduced PHASE's for the cases when genotypes form nontrivial graphs. One can easily see that the quality of the both methods is much better than in the general case. Moreover, the Linearly Reduced PHASE consistently beats in quality the original PHASE.

Similarly, the Table 5 and 6 compare the runtime and quality of haplotype inference of the standard HAPLOTYPER [16] and proposed Linearly Reduced HAPLOTYPER. Table 7 shows that the quality considerably improves for the nontrivial graphs for HAPLOTYPER similar to what happend with PHASE.

The last two tables are devoted to biological datasets. The first three testcases are derived from the drosophila haplotypes and last two testcases are derived from the 616 kilobase region from [7].

Table 5. The comparison of the running times of HAPLOTYPER and Linearly Reduced HAPLOTYPER. Each value is averaged over 25 datasets.

Datasets		Average Running Time (seconds)			
		HAPLOTYPER	Linearly Reduced HAPLOTYPER		
sites	pop	Total	Total	E and D	RH
40	20	0.0943	0.0291	0.0019	0.0272
40	40	0.1578	0.0801	0.0072	0.0729
80	40	0.3516	0.1335	0.0242	0.1093
80	80	0.6382	0.2580	0.0415	0.2165
100	50	0.6631	0.1747	0.0562	0.1185
100	100	1.2606	0.6035	0.0646	0.5389

Table 6. The comparison of the quality of haplotyping of Linearly Reduced HAPLOTYPER (LRH) and HAPLOTYPER (H) vs the original haplotypes (O). Here the difference in haplotype data sets, Hapset1/Hapset2 is the arithmetic mean of numbers of false-positive and false-negative haplotypes over the number of haplotypes Hapset2 times 100%. Each value is averaged over 25 datasets.

Datasets		Recombination Rate										
		0			4			16			40		
sites	pop	LRH/H	LRH/O	H/O	LRH/H	LRH/O	H/O	LRH/H	LRH/O	H/O	LRH/H	LRH/O	H/O
40	20	11.33	16.31	13.44	18.44	21.28	11.48	21.32	22.42	16.86	24.86	28.97	18.85
40	40	9.84	15.72	12.40	9.21	12.21	8.71	9.43	14.10	11.46	14.93	18.58	16.18
80	40	12.83	17.84	16.92	13.62	18.74	15.11	17.47	20.74	15.46	27.11	30.99	20.94
80	80	15.30	26.85	22.53	10.96	15.87	11.43	11.51	15.38	12.13	16.77	18.38	10.32
100	50	17.76	25.03	17.25	11.98	16.69	11.87	14.21	17.76	13.64	20.42	21.87	13.07
100	100	8.67	15.28	11.83	6.69	11.46	9.90	11.51	13.87	9.67	20.58	21.91	11.82

Table 7. The comparison of the quality of haplotyping of Linearly Reduced HAPLOTYPER (LRH) and HAPLOTYPER (H) vs the original haplotypes (O) over the instances with nontrivial graphs.

Datasets		Recombination Rate					
		0			4		
sites	pop	LRH/H	LRH/O	H/O	LRH/H	LRH/O	H/O
40	20	4.12	10.01	7.05	8.82	8.82	5.88
40	40	6.19	8.48	5.35	2.61	5.71	5.71
80	40	2.38	11.36	9.09	4.11	11.80	11.10
80	80	4.16	6.69	7.29	0.76	6.25	7.03
100	50	4.71	7.41	9.26	4.16	9.45	5.41
100	100	3.48	10.46	9.30	6.96	9.58	7.50

Table 8. The comparison of the running times on real data.

sites	pop	H	LRH	PHASE	LRP
43	11	0.0903	0.00846	37.5904	2.8928
43	22	0.1276	0.0192	59.8368	5.5232
43	33	0.1712	0.0318	105.0596	8.6668
21	95	0.3250	0.2310	381.6423	321.4914
50	48	0.5510	0.0720	617.3057	262.6013

We also tried to randomly permute columns before running our algorithms. This allows to choose a random basis. Unfortunately, such method doesn't show significant improvement in quality.

Table 9. The comparison of Linearly Reduced HAPLOTYPER (LRH), HAPLO-
TYPER(H), Linearly Reduced PHASE (LRP), PHASE (P), and original haplotypes
(O) on biological data.

sites	pop	LRH/H	LRH/O	H/O	LRP/P	LRP/O	P/O
43	11	8.62	11.33	8.89	7.84	11.17	8.22
43	22	1.99	7.11	6.22	1.79	7.89	6.88
43	33	1.48	6.01	6.89	2.23	7.33	6.22
21	95	6.52			2.53		
50	48	2.54			0.29		

7   Conclusions and Future Work

We have suggested a linear reduction method which drastically reduces the runtime of haplotype inference while not significantly compromising the quality of
inference. For the perfect phylogeny reconstruction we reduce the runtime by
factor of 60.

In our future work, we will explore possibility of finding a basis of genotype
matrix without 4-gamete-rule-violations even though the given instance contain
such violations. Then haplotype inference can be done based on perfect phylogeny model.

Also we are going to investigate an application of the suggested linear reduction to finding a small number of representative sites sufficient to distinguish all
haplotypes. The known methods still find such a set of too large size while linear
reduction offers comparatively small number of sites.

Acknowledgments

We thank Dan Gusfield for making available the DPPH software and encouraging
us to write the paper.

References

1. V. Bafna, D. Gusfield, G. Lancia, and S. Yooseph. Haplotyping as perfect phylogeny: A direct approach. Technical report, UC Davis, Department of Computer
Science, 2002.
2. V. Bafna, D.Gusfield, G. Lancia, and S. Yooseph. Haplotyping as perfect phylogeny: A direct approach. J. Computational Biology, 10:323–340, 2003.
3. R.H. Chung and D. Gusfield. Empirical exploration of perfect phylogeny haplotyping and haplotypers. In Proceedings of COCOON 03 The 9'th International
Conference on Computing and Combinatorics, volume 2697 of LNCS, pages 5–19,
2003.
4. R.H. Chung and D. Gusfield. Perfect phylogeny haplotyper: Haplotype inferral
using a tree model. Bioinformatics, 19(6):780–781, 2003.
5. A. Clark. Inference of haplotypes from PCR-amplified samples of diploid populations. Mol. Biol. Evol, 7:111–122, 1990.

6. A. Clark, K. Weiss, and D. Nickerson et. al. Haplotype structure and population genetic inferences from nucleotide-sequence variation in human lipoprotein lipase. Am. J. Human Genetics, 63:595–612, 1998.

7. M. Daly, J. Rioux, S. Schaffner, T. Hudson, and E. Lander. High resolution haplotype structure in the human genome. Nature Genetics, 29:229–232, 2001.

8. E. Eskin, E. Halperin, and R. Karp. Efficient reconstruction of haplotype structure via perfect phylogeny. Technical report, UC Berkeley, Computer Science Division (EECS), 2002.

9. M. Fullerton, A. Clark, Charles Sing, and et. al. Apolipoprotein E variation at the sequence haplotype level: implications for the origin and maintenance of a major human polymorphism. Am. J. of Human Genetics, pages 881–900, 2000.

10. Gabriel, G., Schaffner, S., Nguyen, H., Moore, J., Roy, J., Blumenstiel, B., Higgins, J., DeFelice, M., Lochner, A., Faggart, M., Liu-Cordero, S., Rotimi, C., Adeyemo, A., Cooper, R., Ward, R., Lander, E., Daly, M. and Altshuler, D. (2002). The structure of haplotype blocks in the human genome. Science, 296:2225–2229.

11. D. Gusfield. Inference of haplotypes from samples of diploid populations: complexity and algorithms. Journal of computational biology, 8(3), 2001.

12. D. Gusfield. Haplotyping as Perfect Phylogeny: Conceptual Framework and Efficient Solutions (Extended Abstract). In Proceedings of RECOMB 2002: The Sixth Annual International Conference on Computational Biology, pages 166–175, 2002.

13. E. Halperin and E.Eskin. Haplotype reconstruction from genotype data using imperfect phylogeny. Bioinformatics. Advance Access published on February 26, 2004.

14. R. Hudson. Gene genealogies and the coalescent process. Oxford Survey of Evolutionary Biology, 7:1–44, 1990.

15. S. Lin, D. Cutler, M. Zwick, and A. Cahkravarti. Haplotype inference in random population samples. Am. J. of Hum. Genet., 71:1129–1137, 2003.

16. T. Niu, Z. Qin, X. Xu, and J.S. Liu. Bayesian haplotype inference for multiple linked single-nucleotide polymorphisms. Am. J. Hum. Genet, 70:157–169, 2002.

17. S. Orzack, D. Gusfield, and V. Stanton. The absolute and relative accuracy of haplotype inferral methods and a consensus approach to haplotype inferral. Abstract Nr 115 in Am. Society of Human Genetics, Supplement 2001.

18. Patil, N., Berno, A., Hinds, D., Barrett, W., Doshi, J., Hacker, C., Kautzer, C., Lee, D., Marjoribanks, C., McDonough, D., Nguyen, B., Norris, M., Sheehan, J., Shen, N., Stern, D., Stokowski, R., Thomas, D., Trulson, M., Vyas, K., Frazer, K., Fodor, S. and Cox, D. (2001). Blocks of limited haplotype diversity revealed by high-resolution scanning of human chromosome 21. Science, 294, 1719–23.

19. M. Stephens, N. Smith, and P. Donnelly. A new statistical method for haplotype reconstruction from population data. Am. J. Human Genetics, 68:978–989, 2001.

A New Integer Programming Formulation for the Pure Parsimony Problem in Haplotype Analysis

Daniel G. Brown and Ian M. Harrower

School of Computer Science, University of Waterloo
Waterloo ON N2L 3G1 Canada
{browndg,imharrow}@cs.uwaterloo.ca

Abstract. We present a new integer programming formulation for the haplotype inference by pure parsimony (HIPP) problem. Unlike a previous approach to this problem [2], we create an integer program whose size is polynomial in the size of the input. This IP is substantially smaller for moderate-sized instances of the HIPP problem. We also show several additional constraints, based on the input, that can be added to the IP to aid in finding a solution, and show how to find which of these constraints is active for a given instance in efficient time. We present experimental results that show our IP has comparable success to the formulation of Gusfield [2] on moderate-sized problems, though it is is much slower. However, our formulation can sometimes solve substantially larger problems than are practical with Gusfield's formulation.

1 Introduction

Single Nucleotide Polymorphisms, or SNPs, constitute a large portion of genomic variation in human populations. Diploid organisms contain pairs or chromosomes, one inherited from each parent. The DNA sequence of each chromosome is its *haplotype*, and knowing this sequence is useful in identifying the origin of genetic disease. However, current sequencing technology can only determine the *genotype*: the conflation of the pair of haplotypes. This limitation gives the *haplotype inference problem*: determining haplotype sequences of a number of chromosomes, given genotypes derived from them.

Several statistical and inference methods have been proposed for this problem. In 1990, Clark [1] proposed an inference method to attempt to solve this problem. In 2001, Stephens *et al.* presented a statistical approach to the problem [9]. In 2002, Niu *et al.* [8] presented a Bayesian approach to the problem.

Recent work by Gusfield in 2003 [2], and by Wang and Xu in 2003 [10] has explored the effectiveness of the parsimony condition for haplotype inference. This optimization condition seeks the minimum number of haplotypes that explain a given set of genotypes. The algorithms in these papers expand all sets of possible explaining parents for each genotype, giving exponential-sized formulations. Gusfield [2] formulated the problem as an integer program, and showed that it

I. Jonassen and J. Kim (Eds.): WABI 2004, LNBI 3240, pp. 254–265, 2004.

can be used to optimize the pure-parsimony objective for problems of moderate size. However, the exponential size of Gusfield's IP makes it impractical for some problems. Specifically, Halldórsson *et al.* in 2002 noted that for haplotype inference, exponential-time algorithms are only practical for SNPs around a gene, and that polynomial-time algorithms are needed for SNPs across regions spanning multiple genes [4]. Here we present a polynomial-sized integer programming formulation for the haplotype inference by pure parsimony problem (Section 3), and discuss its usefulness.

Unlike previous approaches, our polynomial-sized formulation does not expand possible explaining parents, but has decision variables representing the haplotype sequences themselves. In addition to the polynomial-sized formulation, we present constraints that can be added to the formulation to assist in solving it (Section 4). One of these classes of cuts is a class of data-driven constraints based on patterns in the input genotype data. These constraints have interesting property that as the size of the pattern increases, the constraints become stronger, and have great usefulness in solving our instances.

Finally, we present experiments that show that our formulation can solve problems of the same size as previous formulations, and that it can solve much larger problems that are outside of the range of exponential formulations.

We note that very recently, Dan Gusfield [3] has alerted us to the existence of another paper with a different polynomial-sized integer program for the same problem [7]. This program is similar to ours, but its authors do not present cuts for the problem, and as such, experienced poor performance when they tried to solve it.

2   Models and Notation

We first formally define the haplotype inference by pure parsimony problem (HIPP). Our input will be genotype vectors. Formally, a *genotype* g is a string $g = (g[1], g[2], \ldots, g[m])$ of length m over the alphabet $A = \{0, 1, 2\}$, so $g \in \{0, 1, 2\}^m$. Let $G = (g_1, g_2, \ldots, g_n)$ be a collection of genotypes.

We wish to explain the genotypes using haplotypes. A *haplotype* $h = (h[1], h[2], \ldots, h[m])$ is a binary string of length m, so $h \in \{0, 1\}^m$. A pair of haplotypes h_1 and h_2 is an *explaining pair* for genotype g if the following hold:

- For every position i where $g[i] = 0$, $h_1[i] = 0$ and $h_2[i] = 0$
- For every position i where $g[i] = 1$, $h_1[i] = 1$ and $h_2[i] = 1$
- For every position i where $g[i] = 2$, either $h_1[i] = 0$ and $h_2[i] = 1$, or $h_1[i] = 1$ and $h_2[i] = 0$

Here, genotype positions with letter 0 and 1 correspond to homozygous positions on both chromosomes. Genotype positions with letter 2 correspond to heterozygous positions on the chromosomes. If h_1 and h_2 are an explaining pair for genotype g, then h_1 and h_2 are possible *parents* of g. We order the parents of a genotype lexicographically, to resolve one ambiguous site in the genotype.

Given n genotypes $G = (g_1, g_2, \ldots, g_n)$, each of length m, the *haplotype inference by pure parsimony problem* is to find a minimum cardinality set of haplotypes H, such that each genotype in G has an explaining pair in H. This problem was shown to be NP-Hard by Lancia et $al.$ in 2002 [6].

3 Integer Linear Programming Formulation

In this section, we present an integer programming formulation for the HIPP problem. Previous approaches to this problem [2] have focused on enumerating all possible explaining pairs for each genotype. This approach gives rise to exponential-sized problem formulations. If a genotype has k positions with letter 2, then that genotype has 2^{k-1} possible explaining pairs. As the length of the genotypes and the number of heterozygous positions in a problem instance increases, the exponential size of these formulations makes them impractical. In contrast, our integer program (IP) is of polynomial size.

An IP is a linear objective function we wish to optimize, subject to a system of linear constraints. Our polynomial-sized IP for the HIPP problem encodes the solution without expanding possible explaining parents of a given genotype. Instead, we represent the actual haplotypes that explain each genotype. For each genotype g_i, we create binary variables representing each of the m characters of the explaining haplotypes. We say that g_i is explained by h_{2i-1} and h_{2i}. For genotype g_i, the variable that represents the $h_{2i-1}[k]$ is $y_{2i-1,k}$ and the variable that represents $h_{2i}[k]$ is $y_{2i,k}$. There are $2nm$ of these variables.

Our IP includes a set of constraints ensuring that solutions to the IP properly explain every genotype. For all $1 \leq i \leq n$ and $1 \leq k \leq m$:

$$y_{2i-1,k} = 0 \text{ and } y_{2i,k} = 0, \quad \text{if } g_i[k] = 0 , \tag{1}$$

$$y_{2i-1,k} = 1 \text{ and } y_{2i,k} = 1, \quad \text{if } g_i[k] = 1 , \text{ and} \tag{2}$$

$$y_{2i-1,k} + y_{2i,k} = 1, \quad \text{if } g_i[k] = 2 . \tag{3}$$

There are $O(nm)$ constraints of this form, each with $O(1)$ non-zeros.

The parsimony condition requires the program to count the number of unique haplotypes explaining the input genotypes. We use variables to mark pairwise differences between the haplotypes. For each pair of haplotypes $1 \leq i < j \leq 2n$, we create a variable $d_{i,j}$. We will force $d_{i,j}$ to 1 if $h_i \neq h_j$. There are $O(n^2)$ of these variables. If $h_i \neq h_j$, then there will be some position k in the haplotypes where $h_i[k] = 1$ and $h_j[k] = 0$ or vice versa. To force $d_{i,j}$ to 1 when $h_i \neq h_j$, we add these constraints for all $1 \leq i < j \leq 2n$ and $1 \leq k \leq m$:

$$y_{i,k} - y_{j,k} \leq d_{i,j} , \tag{4}$$

$$y_{j,k} - y_{i,k} \leq d_{i,j} . \tag{5}$$

Recall that $y_{i,k}$ and $y_{j,k}$ are binary variables. Thus, if they are not equal, one of the two constraints (4) or (5) will be equivalent to $1 \leq d_{i,j}$, forcing $d_{i,j}$ to 1. There are $O(n^2m)$ constraints of this form, each with 3 non-zeros. Recall that if $g_i[k]$ is 0 or 1, then the value of $y_{2i-1,k}$ and $y_{2i,k}$ are known based on the input. Thus, some of the $d_{i,j}$ may be forced to 1 based directly on the input.

Finally, we count unique haplotypes. For each haplotype h_i, we introduce a binary variable x_i which is 1 if h_i is unique in $(h_1, \ldots, h_i)$. There are $2n$ such variables. We set $x_1 = 1$ since h_1 is the only haplotype in (h_1). If haplotype h_i is unique in $(h_1, \ldots, h_i)$, then $\sum_{j<i} d_{j,i} = i - 1$, while if it is found more than once, this sum is strictly less than $i - 1$. We thus add the following constraint for each i between 2 and $2n$, inclusively:

$$x_i \geq 2 - i + \sum_{j=1}^{i-1} d_{j,i} \ . \tag{6}$$

There are $O(n)$ constraints of this form, each with $O(n)$ non-zeros.

This gives us the final formulation for the IP:

$$\text{minimize} \sum_{i=1}^{2n} x_i, \ \text{such that}$$

$$y_{2i-1,k} = 0 \ \text{and} \ y_{2i,k} = 0, \quad \text{for all } i, k \text{ such that } g_i[k] = 0 \ ,$$

$$y_{2i-1,k} = 1 \ \text{and} \ y_{2i,k} = 1, \quad \text{for all } i, k \text{ such that } g_i[k] = 1 \ ,$$

$$y_{2i-1,k} + y_{2i,k} = 1, \quad \text{for all } i, k \text{ such that } g_i[k] = 2 \ ,$$

$$d_{i,j} \geq y_{i,k} - y_{j,k}, \quad \text{for all } 1 \leq i < j \leq 2n \text{ and } 1 \leq k \leq m \ ,$$

$$d_{i,j} \geq y_{j,k} - y_{i,k}, \quad \text{for all } 1 \leq i < j \leq 2n \text{ and } 1 \leq k \leq m \ ,$$

$$x_i \geq 2 - i + \sum_{j=1}^{i-1} d_{j,i}, \quad \text{for all } 1 \leq i \leq 2n \ ,$$

$$x_i, y_{i,k}, d_{i,j} \in \{0,1\}, \quad \text{for all } i, j \text{ and } k \ .$$

This IP has $O(n^2 m)$ constraints, $\Theta(nm + n^2)$ variables and $O(n^2 m)$ non-zero entries in the constraint matrix. All non-zero entries are $+1$ or -1, with the exception of the $2n$ constraints on the x_i, which include numbers only of size at most $2n$. The IP can be stored in $O(n^2 m)$ space.

4   Cuts and Extensions

Solving an IP is NP-Hard. Typically, one solves an IP by relaxing the program, replacing integrality constraints with corresponding range constraints, producing the linear program (LP) relaxation. Feasible solutions to the IP are also feasible for the LP relaxation, so an optimal solution to the LP relaxation gives a lower bound on the optimal solution to the IP.

The LP relaxation can be solved efficiently. If its solution is integral, it optimizes the IP. Otherwise, a common approach is to find new constraints, called *valid cuts*, violated by the solution to the LP relaxation, but satisfied by all feasible integer solutions. Or, one can *branch* by picking a single variable and creating a series of new programs where it is set to each of its possible integer values. The best solution found is the optimal solution to the original IP. Adding valid cuts during branching is called *branch-and-cut*.

Unfortunately, the IP formulation in the previous section does not behave well under LP relaxation. In heterozygous positions in genotypes, the $y_{i,k}$ variables will often have value 0.5 in the LP solution. This does not resolve the haplotypes, and also allows the $d_{i,j}$ variables to have value zero, giving no information. Thus, we need to strengthen the formulation. In this section, we describe a series of enhancements to tighten the gap between the optimal solution to the LP relaxation and the optimal solution to the IP. We both augment the objective function and show several new valid constraints.

Augmented Objective Function. A weakness of our IP (and of the LP relaxation) is that optimal solutions may allow $d_{i,j} = 1$ even when $h_i = h_j$. This is largely harmless for the IP, but in the case of the LP, it allows the $y_{i,k}$ variables to take fractional values. Fundamentally, it also violates our intuition in designing the program itself. The $d_{i,j}$ variables should only equal 1 when $h_i \neq h_j$. To bias the $d_{i,j}$ variables towards 0, we extend the objective function to take into account a secondary goal of minimizing the sum of the $d_{i,j}$ variables. To ensure that an optimal solution still minimizes the number of unique explaining parents, the coefficients of the $d_{i,j}$ variables must be such that their total sum is strictly less than 1. In addition, we ensure that the coefficients of the $d_{i,j}$ variables are of the same scale. If all of the $d_{i,j}$ variables have the same coefficient, there may be many possible solutions with the same objective value, which is inconvenient for a branch-and-cut algorithm. Hence, we add a small random number $r_{i,j}$ to each coefficient, which gives the following extended objective function:

$$\text{minimize} \sum_{i=1}^{2n}(x_i) + \sum_{i=2}^{2n}\sum_{j=1}^{i-1}(\epsilon + r_{i,j}) \cdot d_{i,j} \ . \tag{7}$$

By choosing ϵ so that $8\epsilon n^2 < 1$, and enforcing that each $0 \leq r_{i,j} < \frac{\epsilon}{2}$ we ensure that the coefficients are all in the same scale and that sum of all the coefficients is strictly less than 1.

Transitivity Constraints. The $d_{i,j}$ variables are to equal 1 if h_i and h_j are different and 0 otherwise. The expanded objective function partially obtains this behavior, but adding new constraints can help further. We add the following transitivity constraints for all $1 \leq i < j < k \leq 2n$:

$$d_{i,j} + d_{i,k} \geq d_{j,k} \ , \tag{8}$$
$$d_{i,j} + d_{j,k} \geq d_{i,k} \ , \tag{9}$$
$$d_{i,k} + d_{j,k} \geq d_{i,j} \ . \tag{10}$$

Consider three haplotypes h_i, h_j and h_k. If $h_i = h_j$ and $h_i = h_k$ then we can conclude that $h_j = h_k$. Therefore, if $d_{i,j}$ and $d_{i,k}$ are both 0, constraint (8) forces $d_{j,k}$ to also be 0. The constraint also enforces the contrapositive: If $h_j \neq h_k$ then $d_{j,k} = 1$ and constraint (8) ensures that $d_{i,j} + d_{i,k} \geq 1$. Since these variables are binary, this constrains at least one of h_j and h_k to be different from h_i.

While it is possible that integer solutions to the IP will violate this new constraint, there are always equivalent integer solution that do not. We show that these constraints are still valid for an optimal solution to the HIPP problem by considering any optimal solution to the problem. If we modify the solution such that $d_{i,j} = 1$ if h_i and h_j are different, and $d_{i,j} = 0$ when h_i and h_j are the same, leaving all other variable the same, this is an optimal solution for the HIPP instance that satisfies all the transitivity constraints.

Data-Driven Cuts. We next consider cuts that depend on input data patterns. These constraints are violated by many fractional solutions. They thus close the gap between the optimal solutions of the LP relaxation and the IP.

Consider input genotypes g_1 and g_2, and suppose there exists a k such that $g_1[k] = 2$ and $g_2[k] = 0$. Then $d_{1,3} + d_{2,3} \geq 1$, since both parents of g_1 can not be 0 at position k, while both parents of g_2 must be 0 at position k. This can be concluded directly from the transitivity constraint (10) $d_{1,3} + d_{2,3} \geq d_{1,2}$, since the parents of g_1 must be different, and $d_{1,2} = 1$. In general, if

$$g_i[k] = 2$$
$$g_j[k] = 0$$

then the following are valid cuts:

$$d_{2i-1,2j-1} + d_{2i,2j-1} \geq 1 \,, \tag{11}$$
$$d_{2i-1,2j} + d_{2i,2j} \geq 1 \,. \tag{12}$$

This first pair of constraints is implied by the transitivity constraints, but these data-driven cuts can be extended to stronger cuts based on larger patterns. Consider once again two genotypes, but consider two characters in each, say k_1 and k_2. Then an occurrence of

$$g_i[k_1, k_2] = 20$$
$$g_j[k_1, k_2] = 02$$

gives the valid constraint:

$$d_{2-1i,2j-1} + d_{2i,2j-1} + d_{2i-1,2j} + d_{2i,2j} \geq 3 \,. \tag{13}$$

This constraint requires that at most one parent of g_i is the same as a parent of g_j. Assume without loss of generality that the first parent of g_i equals the first of g_j. Then both parents have value 0 in positions k_1 and k_2. The second haplotype for g_i has $h_{2i}[k_1, k_2] = 10$, while the second haplotype for g_j has $h_{2j}[k_1, k_2] = 01$, so they match none of these haplotypes. This shows that the constraint $d_{2-1i,2j-1} + d_{2i,2j-1} + d_{2i-1,2j} + d_{2i,2j} \geq 3$ is valid for all feasible integer solutions to the IP.

This new constraint is strictly stronger than the transitivity constraints ((11) and (12)). Consider the fractional solution $d_{2i-1,2j-1} = d_{2i,2j-1} = d_{2i-1,2j} = d_{2i,2j} = \frac{1}{2}$. These variable assignments satisfy $d_{2i-1,2j-1} + d_{2i,2j-1} \geq 1$ and

$d_{2i-1,2j} + d_{2i,2j} \geq 1$, but violate the new constraint $d_{2-1i,2j-1} + d_{2i,2j-1} + d_{2i-1,2j} + d_{2i,2j} \geq 3$, which forces the average of the four pairwise difference variables to be greater than $\frac{3}{4}$.

This pattern extends extends generally to ℓ length patterns in ℓ strings. As before, we generate a constraint pattern that allows at most one parent of the first genotype g_i to be the same as a parent of the remaining $\ell - 1$ genotypes $g_{j_1}, g_{j_2}, \ldots, g_{j_{\ell-1}}$ in the pattern. The pattern enforces that the genotypes $g_{j_1}, g_{j_2}, \ldots, g_{j_{\ell-1}}$ all have different parents. This is established by having for each pair, a character position k_i where one string has a 1 and the other a 0. There will be one column k_1 in which $g_i[k_1] = 2$ and all other characters will be the same non-2 value. This column k_1 guarantees that if h_{2i-1} equals one of the other parents, then h_{2i} is different from all the other parents. Finally, we will have a column k_ℓ in which $g_i[k_\ell]$ is not 2, and all other genotypes are 2. This column, as in the previous constraint, forces the parents of each g_j to be different from each other and thus equal to only one of g_i's parents. When such a pattern occurs, we obtain the valid cut:

$$\sum_{x=1}^{\ell-1} (d_{2i-1,2j_x-1} + d_{2i-1,2j_x} + d_{2i,2j_x-1} + d_{2i,2j_x}) \geq 4\ell - 5 .$$

This constraint allows at most one parent of the first genotype g_i to be a parent of at most one of the $\ell - 1$ remaining genotypes. It forces the average of the pairwise difference variables to be at least $\frac{4\ell-5}{4\ell-4}$.

For example, here is a case for $\ell = 4$:

$$g_i[k_1, k_2, k_3, k_4] = 2220$$
$$g_{j_1}[k_1, k_2, k_3, k_4] = 0002$$
$$g_{j_2}[k_1, k_2, k_3, k_4] = 0012$$
$$g_{j_3}[k_1, k_2, k_3, k_4] = 0112$$

From the argument above, this pattern gives the valid cut:

$$\sum_{x=1}^{3} (d_{2i-1,2j_x-1} + d_{2i-1,2j_x} + d_{2i,2j_x-1} + d_{2i,2j_x}) \geq 11 . \qquad (14)$$

This constraint is stronger than the constraints for the sub-patterns of length 3. By assigning all pairwise difference variables the fractional value $\frac{7}{8}$, all of the length-3 constraints will be satisfied. However, the length-4 constraint is violated by this fractional solution, since $12 \cdot \frac{7}{8} < 11$.

These patterns do not strictly need to occur as described. In particular, the final column k_ℓ is establishing that for each g_j, there exists a column where g_i is 0 or 1 and g_j is 2. This condition need not occur in the same column for all g_j. Also, the forced differences between genotypes $g_{j_1}, g_{j_2}, \ldots, g_{j_{\ell-1}}$ need not all occur in a block, but could occur at differing positions.

We can compute all forced pairwise differences between the genotypes in $O(n^2)$ time. Moreover, this work is done while constructing the IP. Also, for each genotype we can mark if it contains a 2 or not in $O(nm)$ time. For a fixed ℓ and genotypes $g_i, g_{j_1}, \ldots, g_{j_{\ell-1}}$, we can test the above conditions by first examining fixed $d_{i,j}$ variables, and then searching for a column k_1 with $g_i[k_1] = 2$ and all other characters equal to the same non-2 value. For a fixed ℓ, we can find all such constraints by brute force in $O(n\binom{n}{\ell-1}\binom{\ell}{2}\ell m)$ time.

In practice, cuts for $\ell \geq 3$ are not practical to find with this simple brute-force algorithm. However, they form a general class of valid cuts, getting stronger as ℓ increases, which can be found in polynomial time.

Permutation Constraints. One drawback our IP is that its feasible region depends on the order of the genotypes. Under the LP relaxation, the $d_{i,j}$ variables may take fractional values. If two such variables in the same constraint take value 0.5, the x_i variable can be set to 0. Under a different ordering, it might be possible that more x_i variables are assigned the value 1 under the relaxation.

This order dependency introduces an interesting class of permutation constraints. In a second instance of the problem with a different genotype order, the underlying problem would still be the same, so the objective function value would be, too. Conceptually, the permutation constraints combine these two programs such that each has its own x_i variables, but all other variables are shared. We then add the constraint that the two objective function values are equal.

Let π be a permutation of $\{1, \ldots, 2n\}$. Define new variables $x_1^\pi, \ldots, x_{2n}^\pi$, and add constraints analogous to those defined for the x_i's. Let $d_{i,j} = d_{j,i}$ if $i > j$, then for $1 \leq i \leq 2n$ we have:

$$x_i^\pi \geq 2 - i + \sum_{j=1}^{j<i} d_{\pi(j),\pi(i)} \ . \tag{15}$$

These constraints are equivalent to constraints (6), only with a different ordering. For integer points, these x_i^π count the number of unique haplotypes used, just as before. Therefore the additional valid cut can be added:

$$\sum_{i=1}^{2n} x_i = \sum_{i=1}^{2n} x_i^\pi \ , \tag{16}$$

which forces the number of unique haplotypes for both orders to be equal.

If the sum constraint (16) is violated, we arbitrarily increase the needed x variables. This will not change any of the other types of variables in the formulation. We would like to have the property that if $x_i = 1$ then, for all j less than i, $d_{i,j} = 1$, and similarly for x_i^π. This can be forced by adding this constraint for all $j < i$:

$$x_i^\pi \leq d_{\pi(j),\pi(i)} \ . \tag{17}$$

Some permutation constraints will strengthen the LP relaxation. However, there are $(2n)!$ possible permutations. Given a fractional solution, it is not clear how to identify permutations with constraints violated by that solution. In addition, these constraints add additional variables to the problem, which may increase the runtime of solving the LP relaxation.

5   Experiments

We have given a polynomial-sized IP formulation for the HIPP problem. However, since in general solving an IP is NP-Hard, we need to explore if we can solve our IP in reasonable time. Previous papers [2, 10] have addressed whether the parsimony condition is effective for haplotype inference. Thus, we restrict our attention to the problem of solving maximum parsimony.

Here, we present the model used for our experiments and show our findings on both problems of size comparable to those solved by Gusfield [2] and on significantly larger problems. Our results show that on small problems, there is no significant difference in the fraction of instances solved. We also show that we can solve problems that are impractical for exponential formulations in one third of the cases.

Implementation. We created a branch-and-cut implementation to solve instances of the HIPP problem. We use CPLEX 8.1.1 to solve the LP relaxations, and then add cuts or set branched variables directly. Commercial LP solvers provide powerful IP solvers which may have some success solving this problem. However, implementing the simple branch-and-cut algorithm allows us to explore different branching rules and cuts.

We branch only on the pairwise difference variables $d_{i,j}$, because they have strong influence on the IP. If all $d_{i,j}$ variables are integral and all transitivity constraints are satisfied, there is an integral solution to the problem with the same objective value. In practice, when the $d_{i,j}$ variables solved to integer values, all other variables were assigned integer variables by the solver. Setting a $d_{i,j}$ to 0 is very strong: it forces the corresponding parents to be equal. However, there is still flexibility in the actual values of those parents.

We added data-driven constraints based on 2 column patterns and transitivity constraints when violated by the solution to the LP relaxation. We use the number calls to the LP solver, or equivalently the number of branches, as a measure of the ease in solving the problem.

We also implemented Gusfield's RTIP formulation [2] to compare the effectiveness of our IP with his.

Data Generation. We experimented with simulated data. We created parent haplotypes using Hudson's program ms [5]. It simulates neutral evolution and recombination. We then paired haplotypes randomly to produce offspring populations. The model in ms allows multiple identical haplotypes to be created. We eliminated all duplicates, creating harder instances of the problem, because completely resolved genotypes are less likely. Thus, our input sets are harder than those created in previous experiments with HIPP data.

We generated two classes of data sets. First, we ran experiments on inputs of sizes comparable to those used by Gusfield [2]. The data consist of collections of $n = 50$ offspring with haplotypes of length either $m = 10$ or 30. The second class of inputs are designed to be problem instances that are too large to formulate in an exponential formulation. In these cases, we use collections of 30 offspring with haplotypes of length 50, 75 and 100.

Problem Results. We wished to show that our formulation can solve problem instances comparable to those solved by Gusfield. We generated four data sets of inputs with sizes equal to those tested in his work. Each set consists of 15 problem instances, each with 50 population members. We tested data sets with both 10 sites, and 30 sites. In the 10-site experiments, we evaluated recombination levels of 0, 4 and 16. The results are shown in Table 1.

Table 1. Results for populations of size 50. Minimum and maximum number of branches are reported over the solved instances. For small recombination, the success rate is comparable to that reported by Gusfield [2].

# of Sites	Recombination Level	Max # of 2s in a genotype	Fraction Solved Poly	Fraction Solved Exp	Branch Limit	Min # Branches	Max # Branches
10	0	8	15/15	15/15	2000	1	1064
10	4	9	15/15	15/15	2500	1	2254
10	16	9	12/15	15/15	3000	48	2716
30	0	23	11/15	7/15	2000	44	1808

With 10 sites and no recombination, our program was able to find the optimal solution to our IP formulation in every instance. The time to solve these instances was slower than other HIPP programs. The reason for this is that our problem sizes are cubic in the number of offspring, leading to larger formulations on these simple problems. Also, as recombination increased, our program performed worse. This is interesting, since in Gusfield's formulation, runtimes improved as recombination increased.

With 30 sites and no recombination, our program found optimal solutions to the IP in 11 of 15 instances. There is no significant difference between this result and the reported 13 out of 15 instances solved by Gusfield's formulation. However, our implementation of Gusfield's RTIP formulation was only able to solve 7 out of 15 instances within a two hour time limit. For the RTIP formulation, 6 of the instances where solved less than 5 seconds, with the final instance taking just under 1 hour. In each case, the RTIP was solved with only one call to the LP solver. Of the 11 solution to our IP, 7 were solved within 6 minutes, and only 1 took more than 35 minutes.

These experiments show that we can solve problem instances comparable to those of previous formulations.

Larger Problem Instances. The advantage of a polynomial-sized IP formulation is that one can formulate the IP for larger problems. We generated problem instances that would not be solvable with exponential-sized IP formulations.

Each group of tests were of populations with 30 offspring. The number of sites evaluated were 50, 75 and 100 sites. Even with only 50 sites, our input data

Table 2. Results for populations of size 30. Minimum and maximum number of branches are reported over the solved instances. Even on large inputs, the formulation can often be solved.

# of Sites	Max No. of 2s in a genotype	Fraction Solved	Branch Limit	Min # Branches	Max # Branches
50	42	27/50	2000	34	1698
75	52	4/10	2500	11	1265
100	68	3/10	3000	388	2616

contained genotypes with approximately 30 heterozygous sites. At these levels, exponential approaches are not practical, as a given genotype can be explained by billions of haplotype pairs. The results are summarized in Table 2.

Of the problems with 50 sites, more than half of the instances could be solved, often with very few branches. With 75 and 100 sites, approximately one third of the instances solved.

The IP formulations remained very small, even for large inputs. For the 50 site problems, we reached a maximum of 2614 variables, much smaller than an exponential IP formulation would achieve. The largest problem instance had 80326 constraints when the algorithm terminated, of which 3031 of the constraints were added as valid cuts. The number of variables has very little dependency on the number of sites. With 100 sites, only 3598 variables were needed in the largest formulation. By contrast, an exponential-sized IP would require billions of constraints. (Note that some of the variables are preset to 0 or 1 because of the data; we only have to find values for y variables for heterozygous positions. The small increase in the number of variables is due to the increase in the number of 2s in the input.) We needed 144847 constraints in the largest formulation. At this size, the LP solver we were using became significantly slower.

For moderate-sized instances of the problem, we still have a reasonable likelihood of successful solution, even when exponential-sized solvers would be creating hopelessly large IPs. Moreover, adding data-directed cuts helps substantially in solving the problems.

6 Conclusions

Pure parsimony has been shown to be a reasonable objective function for the haplotype inference problem [2, 10]. All previous exact approaches to the HIPP problem use a formulation that involves either explicitly or implicitly expanding out possible parents, leading to exponential problem size.

Instead, we have given a polynomial-sized IP formulation of the HIPP problem. Our approach is different from previous IP approaches to this problem in that we do not expand out all explaining parents. In addition, we have shown how to strengthen the problem using valid cuts derived from properties of the input genotypes. Our experiments show that although our formulation takes longer to solve, we can solve problems the same size as previous formulations. Our formulation is also capable of solving problem instances significantly larger than previous formulations.

Acknowledgements

We thank Joseph Cheriyan and Prabhakar Ragde for useful discussions. This work was supported by the Natural Science and Engineering Research Council of Canada, and by the Human Frontier Science Program.

References

1. A. G. Clark. Inference of haplotypes from PCR-amplified samples of diploid populations. *Molecular Biology and Evolution*, 7(2):111–112, 1990.
2. D. Gusfield. Haplotype inference by pure parsimony. In *Proceedings of the 14th Annual Symposium on Combinatorial Pattern Matching (CPM 2003)*, volume 2676 of *LNCS*, pages 144–155. Springer-Verlag, 2003.
3. D. Gusfield. Personal communication, June 2004.
4. B. V. Halldórsson, V. Bafna, N. Edwards, R. Lippert, S. Yooseph, and Sorin Istrail. A survey of computational methods for determining haplotypes. In *Computational Methods for SNPs and Haplotype Inference: DIMACS/RECOMB Satellite Workshop*, volume 2983 of *LNCS*, pages 26–47. Springer-Verlag, 2002.
5. R. R. Hudson. Generating samples under a Wright-Fisher neutral model of genetic variation. *Bioinformatics*, 18(2):337–338, 2002.
6. G. Lancia, C. M. Pinotti, and R. Rizzi. Haplotyping populations: Complexity and approximations. Techical report DIT-02-0080, University of Ternto, October 2002.
7. G. Lancia, C. M. Pinotti, and R. Rizzi. Haplotyping populations by pure parsimony: Complexity, exact, and approximation algorithms. *INFORMS Journal of Computing*, 2004. To appear.
8. T. Niu, Z. S. Qin, X. Xu, and J. S. Liu. Bayesian haplotype inference for multiple linked single-nucleotide polymorphisms. *American Journal of Human Genetics*, 70(1):157–159, 2002.
9. M. Stephens, N. J. Smith, and P. Donnelly. A new statistical method for haplotype reconstruction from population data. *American Journal of Human Genetics*, 68(4):978–979, 2001.
10. L. Wang and Y. Xu. Haplotype inference by maximum parsimony. *Bioinformatics*, 19(14):1773–1780, 2003.

Fast Hare: A Fast Heuristic for Single Individual SNP Haplotype Reconstruction

Alessandro Panconesi and Mauro Sozio

Informatica, La Sapienza
via Salaria 113, 00198, Roma, Italy
{ale,sozio}@di.uniroma1.it

Abstract. We study the single individual SNP haplotype reconstruction problem. We introduce a simple heuristic and prove experimentally that is very fast and accurate. In particular, when compared with a dynamic programming of [8] it is much faster and also more accurate. We expect Fast Hare to be very useful in practical applications. We also introduce a combinatorial problem related to the SNP haplotype reconstruction problem that we call Min Element Removal. We prove its NP-hardness in the gapless case and its $O(\log n)$-approximability in the general case.

1   Introduction

A Single Nucleotide Polymorphism, or SNP (pronounced "snip"), is a small genetic change, involving a single base, that can occur within a person's DNA sequence. In other words, a SNP is a single position (base) in the human DNA where variability across a set of individuals is observed (see Figure 1). SNPs are the most frequent form of human genetic variability and it is hoped that their understanding will increase our ability to treat humane diseases, among other things [6, 9].

In this paper we present a fast and accurate heuristic for the single individual SNP haplotyping reconstruction. This problems was introduced in [5] and can be described in the following way. Roughly, we have a set of small fragments of two original strings, called haplotypes. The problem asks for reconstructing the haplotypes from the fragments, in spite of the fact that the latter may contain reading errors. Contrary to what happens in the case of fragment assembly here we know the span of a fragment, i.e. the set of positions it covers. More precisely, we have two strings of A's and B's called the *red* and the *blue haplotypes*. (The following is not meant to be an accurate description of the way the biological data are generated, but as a way to describe the computational problem.) The two haplotypes correspond to the projection of the two DNA strands of a certain individual to those positions that are known to be SNPs. That is, we just consider those positions along the DNA known to be SNPs (see Figure 2).

Without loss of generality we can assume that the two haplotypes are strings of A's and B's of the same length, rather than strings over the four letter alphabet A, C, G, T [5]. Consider the following process. The blue and red haplotypes are

I. Jonassen and J. Kim (Eds.): WABI 2004, LNBI 3240, pp. 266–277, 2004.

SNP   SNP   SNP
↓   ↓   ↓

DNA strand 1 → acgta **G** agaatacag **A** ttacagatacacag **T** tcacaggctaa
DNA strand 2 → acgta **C** agaatacag **A** ttacagatacacag **A** tcacaggctaa
DNA strand 3 → acgta **G** agaatacag **A** ttacagatacacag **A** tcacaggctaa
DNA strand 4 → acgta **G** agaatacag **T** ttacagatacacag **A** tcacaggctaa

Fig. 1. SNP's in the DNA.

chromosome c, paternal: acgta**G**agaatacag**A**ttacagatacacag**T**tcacaggctaa
chromosome c, maternal: acgta**C**agaatacag**A**ttacagatacacag**A**tcacaggctaa

Haplotype 1 →   **G**   **A**   **T**
Haplotype 2 →   **C**   **A**   **A**

Fig. 2. The two haplotypes of a single individual.

copied many times, and each copy is then broken into fragments. Let $F :=$ $\{f_1, \ldots, f_n\}$ be the set of fragments. An unknown subset of these fragments are read and arranged in a matrix M whose columns are the SNPs (i.e. column i correspond to position i of the haplotype) and whose rows are the fragments. If fragment f_k goes from position i to position j, the entries $M[k, \ell]$, for $\ell \notin \{i, i+1, \ldots, j\}$, will be set to $-$, denoting lack of information, while the entries $M[k, \ell]$ for $i \leq \ell \leq j$ will be either A or B. An entry whose value is $-$ will be called a *null* entry. This process corresponds to the shotgun sequencing process used to generate and read the fragments out of DNA strands. The crucial point is that in building the matrix M we lose important information in two different ways. First, we lose the information concerning the strand to which the fragment belongs, i.e., we do not know if a fragment comes from the red or the blue haplotype. The second loss of information is that a certain number of non null matrix entries are corrupt, i.e. we have an A while we should have B, and viceversa. This corresponds to the fact that the hardware that reads DNA fragments commits reading errors. The biologist estimate of the error rate is between 3% and 5%.

We can now define our computational problem,

PLOTTER RECONSTRUCTION: Given the matrix M, reconstruct the red and blue haplotypes as accurately as possible.

Whether and with what accuracy the problem can be solved depends on several factors. One is the percentage of reading errors. Another important aspect is the number of fragments that cover a given position, or *coverage*. Clearly, if the coverage for a certain position is very low, it is unlikely that we can reconstruct the two strands in that position with good accuracy. In practice, the coverage can be assumed to be around 10 (with roughly half of the fragments coming from each strand). In this paper, we present some heuristic algorithms that, given M, produce two haplotypes. The accuracy of the solution is the number of positions of the (hidden) blue and red haplotypes that are guessed correctly. For instance, assume that the true strands are

$$\text{red} = \texttt{AAAAAAAAAAAAAAAAA}$$
$$\text{blue} = \texttt{BBBBBBBBBBBBBBBBB}$$

and that the output is

$$s_1 = \texttt{BBBBBBBABBBBBBABB}$$
$$s_2 = \texttt{AAABAAAAA--AAABAA.}$$

If we identify s_1 with the blue haplotype we have that 6 positions are wrong, two of which due to $-$ (when we identify the output strings with red and blue we can chose the best out of the two possibilities). The main result of this paper is a thorough empirical study demonstrating that our heuristic Fast Hare is both accurate and fast for realistic scenarios. In particular, it is much faster than the dynamic programming algorithm of [5], and also somewhat more accurate. We remark that, given the regrettable unavailability of real data in the public domain, we had to rely on realistic synthetic data, including SNP sequences generated with the well-known generator CELSIM [7].

As observed in [5] it is the presence of reading errors that makes the problem challenging. Consider the following graph $G(M)$ in which rows (fragments) correspond to vertices and where ij is an edge if the ith row and the jth have a *conflict*. Denoting with r_i the ith row, we say that r_i and r_j have a conflict if there exists a column k such that $A = M[i, k] \neq M[j, k] = B$ or $B = M[i, k] \neq M[j, k] = A$ ($-$ does not give raise to conflicts). In words, two fragment conflict if they share a common position but have differing values for that position. In [5] it is observed that if there are no reading errors then $G(M)$ is bipartite. Hence the two original haplotypes can be recovered simply by computing the two sides of the bipartition. In particular, the reconstruction is unique if $G(M)$ is connected. The converse is not strictly true, but essentially true in most cases. When $G(M)$ is bipartite we say that M is *bipartisan*. The presence of reading errors can induce odd cycles and it makes this approach unfeasible. For this reason the authors of [5] introduce the following computational problem:

MIN FRAGMENT REMOVAL: Given M, remove the smallest number of rows so that the remaining matrix is bipartisan.

In [5, 8] the complexity of this and related problems is analyzed, and a polynomial time dynamic programming algorithm is given for the *gapless* case of MIN FRAGMENT REMOVAL. A fragment matrix M is *gapless* if every row consists of a (possibly empty) sequence of $-$, followed by a sequence of A's and B's, followed by a (possibly empty) sequence of of $-$. That is, a matrix is gapless if every fragment is a contiguous substring of one of the two haplotypes. The gapless case is perhaps the most relevant from the point of view of the applications and it is the one considered in this paper. It is also the most challenging from the algorithmic point of view (both for upper and lower bounds).

Once an algorithmic solution $\mathcal{A}$ for MIN FRAGMENT REMOVAL is developed there are two possible approaches. On the one hand, one can study, empirically or otherwise, how well $\mathcal{A}$ solves MIN FRAGMENT REMOVAL. However, given that the original motivation is to try to solve PLOTTER RECONSTRUCTION, a second, equally sensible approach would be to study how well $\mathcal{A}$ solves PLOTTER RECONSTRUCTION (this would also show how good an approach is to use MIN FRAGMENT REMOVAL in order to solve PLOTTER RECONSTRUCTION). This is what we do in this paper. We perform an empirical study of how accurately and how quickly the dynamic programming solution developed in [8] solves PLOTTER RECONSTRUCTION. We shall refer to this algorithm as DYNPROG. As already mentioned, a comparison between DYNPROG and our heuristic Fast Hare shows the latter to be not only much faster but also more accurate. This shows that a more direct approach to PLOTTER RECONSTRUCTION is to prefer to a more round-about approach in which an auxiliary combinatorial problem such as MIN FRAGMENT REMOVAL is introduced. This general conclusion is further confirmed by the discussion below.

In the same spirit of [5, 8] we introduce the following problem,

MIN ELEMENT REMOVAL: Given a gapless M, change the smallest number of non null entries in order to make M bipartisan.

This problem was independently introduced by Lancia before us, but in an unpublished manuscript (pers. comm.). This problem formulation is a different approach to solving PLOTTER RECONSTRUCTION satisfactorily. One devises an algorithmic solution for MIN ELEMENT REMOVAL and then sees how well it solves PLOTTER RECONSTRUCTION.

In this paper, we also study the complexity of MIN ELEMENT REMOVAL (in the gapless case, the most difficult and interesting case). We prove that the problem is NP-hard, and that it is $O(\log n)$-approximable in polynomial-time. From the practical point of view, we have also developed an exact dynamic programming algorithm. But since its performance is quite comparable to DYNPROG, we do not discuss it. This further confirms that the more direct approach embodied by Fast Hare to solve PLOTTER RECONSTRUCTION pays off. As a final remark, we note that Fast Hare is able to process also fragments with gaps, but since our emphasis here is on the comparative analysis with DYNPROG we have limited ourselves to the gapless case.

2 Complexity Results for Min Element Removal

In this section we briefly present some complexity results concerning MIN ELEMENT REMOVAL.

Theorem 1. MIN ELEMENT REMOVAL *is NP-hard even in the gapless case.*

Proof. (Sketch) MIN ELEMENT REMOVAL is a special case of Hypercube Segmentation [3].

Theorem 2. MIN ELEMENT REMOVAL *is $O(\log n)$-approximable in polynomial-time in the general (not necessarily gapless) case.*

Proof. (Sketch) MIN ELEMENT REMOVAL can be reduced to the problem of removing the smallest number of vertices from a graph in order to make it bipartite [2].

We have also developed a dynamic programming algorithm of complexity $O(3^{2k}n)$ that solves MIN ELEMENT REMOVAL exactly, where n is the number of rows of the input matrix and k is the maximum length of a fragment. In realistic applications the value of k is between 3 and 8, so that the algorithm is expensive but practical. In practice we have found it to have a performance similar to DYNPROG. For this reason it is not further discussed in this paper.

3   The Heuristic Fast Hare

In this section we describe our heuristic Fast Hare. The input to Fast Hare is a SNP-fragment matrix M with n rows and m columns, together with a parameter t that will be introduced in the sequel. The output consists of three objects:

– Two haplotypes of length n, i.e. two n-bit strings h_1 and h_2 over the alphabet $\{-, A, B\}$. These are the algorithm's guesses of the blue and red haplotype.
– A SNP-fragment matrix M' with m columns and n rows. In this matrix the non null entries of each row of M can be modified.
– A partition of the rows of M (fragments) into two groups. Each group corresponds to one of the two haplotypes in output.

We need some notation. Let n_x be the number of entries of column c that are equal to x. Let $f_A := n_A/(n_A + n_B)$ and $f_B := n_B/(n_A + n_B)$. The column under consideration will be clear from the context.

For $x, y \in \{-, A, B\}$ let

$$d(x,y) := \begin{cases} +1 \text{ if } (x = y = A) \text{ or } (x = y = B) \\ -1 \text{ if } (x = A \text{ and } y = B) \text{ or } (x = B \text{ and } y = A) \\ 0 \quad \text{otherwise.} \end{cases} \quad (1)$$

Given two strings $S := s_1, \ldots, s_n$ and $T := t_1, \ldots, t_n$ over the alphabet $\{-, A, B\}$, let

$$D(S,T) := \sum_{i=1}^{n} d(s_i, t_i). \quad (2)$$

Basically $D(X,Y)$ counts the number of matches minus the number of mismatches between the two strings, without $-$ contributing to it. As we shall see below, Fast Hare builds the two haplotypes from left to right, inserting the fragments one at a time. In order to decide which haplotype the current fragment should be assigned to one could be tempted to take only the mismatches into account. Figure 3 shows why this is not a good idea.

Haplotype h_1: ABAAAAAAAAAA
Haplotype h_2: ABBABB - - - - - -

Fragment f_i: - - - AABAAAAAA

Fig. 3. By considering the number of mismatches only the fragment could be assigned to what clearly is the wrong haplotype. If one counts in the number of matches the problem is avoided.

Let $S := \{M[i_1, -], \ldots, M[i_k, -]\}$ be a set of rows of a SNP-fragment matrix M. Let $A^c(S, M)$ be the number of A's in the set $\{M[i_1, c], \ldots, M[i_k, c]\}$, and similarly for $B^c(S, M)$. That is, $A^c(S)$ $(B^c(S))$ is the number of A's (B's) in column c when restricted to the rows in S. Given S and M, the *consensus haplotype* of S and M, denoted as $h(S, M)$ is the string $h(S, M) := h_1(S, M), \ldots, h_m(S, M)$ where,

$$h_c(S, M) := \begin{cases} A \text{ if } A^c(S, M) > B^c(S, M) \\ B \text{ if } B^c(S, M) > A^c(S, M) \\ - \text{ otherwise} \end{cases} \quad (3)$$

Let us now describe the heuristic Fast Hare. Fast Hare divides the rows of the input M (i.e. the fragments) into two groups; then each group is transformed into a haplotype. Finally the two haplotypes are used to compute the output matrix M'.

The first step of Fast Hare is the elimination of all columns of M in which $f_A \leq t$ or $f_B \leq t$. In our experiments we set $t := 0.2$. Let $\hat{M}$ denote the new matrix. The intuition is that if (the non null part of) a column consists for more than 80% of the same character, say, A, then the remaining B entries are very likely to be reading errors. These position do not help us to decide in which of the two groups the fragments should be placed and are therefore discarded. In the final output matrix M' these discarded columns are inserted by converting all non null entries to the majority value, i.e. if there are more A's then all B's are converted into A's, and viceversa. For future reference, we say that if the non null entries of a discarded column k become all A's the column is *A-fied*, while if they become all B's then it is *B-fied*.

The second step is to sort the rows of $\hat{M}$ by the starting positions of the non null stretch. That is, if r_1 is non null in the interval $[i_1, \ldots, j_1]$ and r_2 is non null in the interval $[i_2, \ldots, j_2]$, then $r_1 \leq r_2$ if and only if $i_1 \leq i_2$. Let $r_1, \ldots, r_n$ denote the row ordering so obtained.

The third step is to process the rows of $\hat{M}$ (i.e. the fragments) according to the row ordering with the aim of partitioning them into two groups S_1 and S_2. We start by placing r_1 into S_1. To place r_i we compute the consensus haplotypes $h(S_1, \hat{M})$ and $h(S_2, \hat{M})$ and place r_i in the group corresponding to the largest value of $D(\cdot, \cdot)$. That is, if $D(r_i, h(S_1, \hat{M})) \geq D(r_i, h(S_2, \hat{M}))$ we place r_i in S_1, otherwise we place r_i in S_2.

The haplotypes produced by Fast Hare are $h(S_1, \hat{M})$ and $h(S_2, \hat{M})$ augmented by the bits of the columns discarded by step 1. If column k is A-fied then position k of both haplotypes will also be A, while if column k is B-fied it will be B. Let us denote the final haplotypes as h_1 and h_2.

The third and final object output by Fast Hare is the output matrix M', computed in the following way. Let r be the kth row (fragment) of the input matrix M, and let S_i the group r belongs to. Then r is matched to h_i and all non null positions of r that differ from the haplotype are changed. The resulting row becomes the kth row of M'.

The running time of Fast Hare is $O(n \log n + nm)$. A contribution of $O(n \log n)$ is due to the sorting of the rows. The contribution of $O(nm)$ is due to the inductive procedure used to compute the two groups of fragments, and to the initial preprocessing used to eliminate the columns.

4   The Instances

As remarked in the introduction, to the best of our knowledge real data in the public domain are not available. We have therefore generated artificial data under some realistic assumptions.

The haplotypes we used were strings of about 100 characters, generated according to two approaches. In the first approach, a first haplotype h_1 of length $n \approx 100$ is generated at random. The second haplotype h_2 has the same length and is generated by flipping each bit of h_1 at random in such a way that the hamming distance (i.e. the number of positions in which the two differ) between h_1 and h_2 is equal to a parameter d. Then, we make c copies of each haplotype and each copy is broken uniformly at random in order to generate k fragments for each copy. Finally, each bit of every fragment is flipped with probability p. This corresponds to introducing reading errors. Realistic values for the parameters are as follows: $d \approx 0.2$; $c = 5, \ldots, 10$; $k \approx 20$ and, $p \in [0.02, 0.05]$. The value of c defines the *coverage*, i.e. the number of fragments that cover the same position of the haplotype. The value of k is chosen so that the expected length of a fragment is between 3 and 7. We shall refer to this group of instances as the *home made* instances.

This set of instances is quite realistic except perhaps for the fact that the procedures used to break the DNA do produce random fragments, but the breaks do not occur with uniform probability at all positions.

To obviate this problem we have made use of CELSIM, a well-known simulator of the shotgun assembly process [7]. In this case two haplotypes are first generated according to the same procedure described above. These two strings have hamming distance equal to d. Subsequently, CELSIM is invoked to produce f fragments, each of which of length between *lMin* and *lMax*. The output fragments are then processed to plant reading errors with probability p, as above. To summarize, the parameters of this procedure are d, f, *lMin*, *lMax*, and p. We shall refer to these instances as the *CelSim* instances. Since CELSIM generates the fragments by extracting a fragment at random from the haplotype (as opposed to breaking a haplotype into fragments) we have computed the value of f in order to have on average the same coverage specified for the home made instances.

For each input we have run each algorithm once. When the variance was low we have run 100 tests (each point of our data plot corresponds to the average over 100 runs). Low variance was observed for example when testing the running times. For tests with higher variance, typically those measuring the accuracy of the algorithms under various metrics, each point of the data plots is the average over 500 runs. We thus expect our observations to be quite robust.

5 Implementation Issues

We ran our experiments on an Intel Pentium III 1266 MHZ with a 1 GByte of RAM. The operating system was Linux RedHat 8.0 and the compiler was g++ version 3.2.

6 Performance Evaluation

In this section we describe our evaluation of Fast Hare and of DYNPROG. The input for both algorithms is a SNP-fragment matrix M with n rows and m columns generated from two haplotypes *red* and *blue* according to the procedures described in § 4. In the following discussion we shall refer to the set of fragments *before* reading errors are planted as the *perfect* fragments.

We are interested in evaluating the following parameters:

1. Percentage of haplotype positions that are guessed wrongly. That is, denoting with H the hamming distance, with h_i a true haplotype and with s_i the output of the algorithm, $\min\{H(h_1, s_1) + H(h_2, s_2), H(h_1, s_2) + H(h_2, s_1)\}$.
2. Percentage of errors in the output fragments. This metric is described in detail below.
3. Running time.

Not surprisingly, Fast Hare is much faster than DYNPROG. The running time is reported in Figure 4, for which home made input instances have been used.

Concerning (1), recall that DYNPROG removes rows in order to produce a bipartisan output matrix M'. If the conflict graph $G(M')$ is not connected however the bipartition is not uniquely determined and thus there is no unique way two reconstruct the two haplotypes. In order to make a significant comparison we therefore limited ourselves to using only those input matrices for which $G(M')$ is connected, so that the two haplotypes are unambiguously determined. The results are shown in Figure 5. As it can be seen, Fast Hare is slightly more accurate for this set of instances. Here too we used home made instances.

We also performed some experiments whose aim is to assess the tendency of the algorithms to create disconnected outputs. Note that Fast Hare too can create an output matrix whose conflict graph is disconnected. The results are shown in Figure 6. DYNPROG creates disconnected outputs more frequently than Fast Hare. In our opinion this is another very useful property of Fast Hare. The instances used were home made.

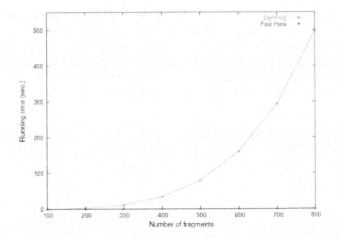

Fig. 4. Running Time. Each point is the average over 100 runs, using home made instances. For each run a new input was generated. The haplotypes were 100 bits long, differing on 20% of the bits. Each haplotype was broken into 25 fragments. In order to increase the number of input fragments (x-axis) the number of copies of each haplotype was varied between 2 and 16.

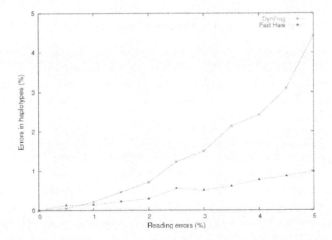

Fig. 5. Errors in the haplotypes. Every dot is the average over 500 runs (one run, one input), using home made instances that are not disconnected by DYNPROG. Haplotypes were 100 bit long, differing on 20% of the bits. Each haplotype was copied 10 times and broken into 20 fragments.

The above discussion motivates the introduction of a different criteria to evaluate the two algorithms on all inputs, as opposed to limiting the comparison to instances that will not be disconnected. This is done by using the percentage of errors in the fragments. The "life" of a fragment can be described as follows.

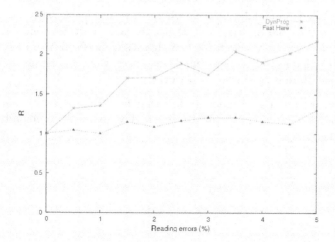

Fig. 6. Creating a disconnected output. The value of R (y-axis) is defined as follows. Consider the input fragments *before* reading errors are planted. Let U be the SNP-fragment matrix defined by these fragments, and let $G(U)$ be the conflict graph. And let M' be the matrix output by the algorithm. R is the ratio between the number of connected components of $G(M')$ and that of $G(U)$. Here haplotypes were 100 bits long, differing on 20% of the positions. The instances were home made. Each haplotype was copied 10 times and broken into 20 fragments. Each point is the average over 500 runs. The x-axis is the percentage of reading errors planted in the fragments.

Initially the fragment is *perfect*, i.e. without reading errors. Then reading errors might be planted in it. Third, the algorithm processes the fragments. In the case of DYNPROG a fragment can only be eliminated, while Fast Hare does not eliminate any fragment but it can modify it (this happens when the fragment is compared with the output haplotype). We want to measure the percentage of non null entries of the fragments that are wrong. Let P denote the matrix defined by the perfect fragments, i.e. the input matrix before reading errors are planted. Let M' denote as usual the output matrix, and define

$$f(x,y) := \begin{cases} 1 \text{ if } - \neq x \neq y \neq - \\ 0 \qquad \text{otherwise} \end{cases} \qquad (4)$$

and,

$$g(x) := \begin{cases} 1 \text{ if } x \neq - \\ 0 \text{ otherwise} \end{cases}. \qquad (5)$$

Then, the percentage of errors in the fragments is given by,

$$\frac{\sum_{ij} f(P[i,j], M'[i,j])}{\sum_{ij} g(M'[i,j])}. \qquad (6)$$

where i and j ran over the rows and columns of M'. DYNPROG can make two kinds of mistakes. A false negative occurs when a row (fragment) without reading error is discarded, and a false positive when a row (fragment) with reading errors is retained in the final matrix. Note that only false positives contribute to this metric. Figure 7 shows how well the two algorithms perform with respect to this metric, for CelSim as well as home made instances. Both algorithms behave very well, but Fast Hare is somewhat superior.

To summarize, our experiments show that Fast Hare is a fast and accurate heuristic for SNP haplotype reconstruction and we expect it to be quite effective in practical scenarios.

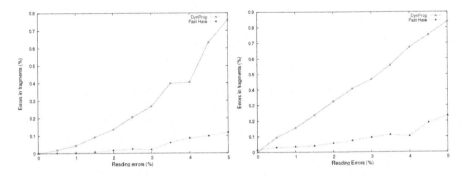

Fig. 7. Errors in fragments. The left hand side refers to home made instances, while the right hand side it refers to CelSim instances. Here each haplotype has 100 bits, and the two haplotypes differ on 20% of the positions. The coverage for each positions is 10. For home made instances each haplotype is broken into 20 random pieces, while for CelSim instances the min and max fragment length is between 3 and 6. The error (y-axis) is plotted against the percentage of reading errors planted in the fragments (x-axis). Each point is the average over 100 runs.

Acknowledgments

We thank Gene Myers for kindly providing us with CELSIM and Russell Schwartz and Giuseppe Lancia for useful advice.

References

1. V.Bafna, B.Halldórsson, R.Schwartz, A.Clark, and S.Istrail, *Haplotypes and Informative SNP Selection Algorithms: Don't Block Out Information*, Proceedings of ACM RECOMB 2003
2. N.Garg, V.Vazirani, and M.Yannakakis, *Multiway Cuts in Directed and Node Weighted Graphs*. In Proceedings of the 21st Int. Colloquium on Automata, Languages and Programming (ICALP), LCNS 820, 487-498, Springer-Verlag, 1994.
3. J.Kleinberg, C.Papadimitriou, and P.Raghavan, *Segmentation Problems*. Proceedings of the ACM STOC, 473-482, 1998.

4. G.Lancia, *Mathematical Programming Approaches for Computational Biology Problems*, In Modelli e Algoritmi per l'Ottimizzazione di Sistemi Complessi, Agnetis and Di Pillo Eds., Pitagora Editrice, 265-310

5. G.Lancia, V.Bafna, S.Istrail, R.Lippert, and R.Schwartz, *SNPs Problems, Complexity and Algorithms*, European Symposium on Algorithms (ESA 01), Lecture Notes in Computer Science, 2161, 182-193, Springer-Verlag eds., 2001

6. The SNP Consortium, http://snp.cshl.org

7. G.Myers, *A Dataset Generator for Whole Genome Shotgun Sequencing.* Proc Int Conf Intell Syst Mol Biol 202-210 (1999)

8. R.Rizzi, V.Bafna, S.Istrail, and G.Lancia, *Practical Algorithms and Fixed-Parameter Tractability for the Single Individual SNP Haplotyping Problem*,Proceedings of 2nd Workshop on Algorithms in Bioinformatics (WABI), Lecture Notes in Computer Science, 2002

9. *Just the Facts: A Basic Introduction to the Science Underlying NCBI Resources. SNPs: VARIATIONS ON A THEME*, available at http://www.ncbi.nlm.nih.gov/About/primer/snps.html

Approximation Algorithms for the Selection of Robust Tag SNPs

Yao-Ting Huang[1], Kui Zhang[2], Ting Chen[3], and Kun-Mao Chao[1]

[1] Department of Computer Science and Information Engineering
National Taiwan University, Taiwan
{d92023,kmchao}@csie.ntu.edu.tw
[2] Section on Statistical Genetics
Department of Biostatistics
University of Alabama at Birmingham, USA
kzhang@ms.soph.uab.edu
[3] Department of Biological Sciences
University of Southern California, USA
tingchen@usc.edu

Abstract. Recent studies have shown that the chromosomal recombination only takes places at some narrow hotspots. Within the chromosomal region between these hotspots (called haplotype block), little or even no recombination occurs, and a small subset of SNPs (called tag SNPs) is sufficient to capture the haplotype pattern of the block. In reality, the tag SNPs may be genotyped as missing data, and we may fail to distinguish two distinct haplotypes due to the ambiguity caused by missing data. In this paper, we formulate this problem as finding a set of SNPs (called robust tag SNPs) which is able to tolerate missing data. To find robust tag SNPs, we propose two greedy and one LP-relaxation algorithms which give solutions of $(m + 1)\ln\frac{K(K-1)}{2}$, $\ln((m + 1)\frac{K(K-1)}{2})$, and $O(m\ln K)$ approximation respectively, where m is the number of SNPs allowed for missing data and K is the number of patterns in the block.

Keywords: approximation algorithm, haplotype block, missing data, SNP

1 Introduction

In recent years, *Single Nucleotide Polymorphisms* (SNPs) [9] have become more and more popular for association studies of genetic diseases or traits. Although the cost of genotyping SNPs is gradually reducing, it is still uneconomical to genotype all SNPs for association study [1]. However, recent findings [3,5,6] showed that the chromosomal recombination only occurs at some narrow hotspots. The chromosomal region between these hotspots is called a "haplotype block." Within a haplotype block, there is little or even no recombination occurred, and the SNPs in the block tend to be inherited together. Due to the low haplotype diversity within a block, the information carried by these SNPs is highly redundant. Thus, a small subset of SNPs (called "tag SNPs") is sufficient to capture the haplotype pattern of the block. Haplotype blocks with

I. Jonassen and J. Kim (Eds.): WABI 2004, LNBI 3240, pp. 278–289, 2004.

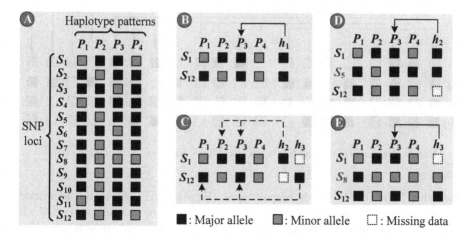

Fig. 1. The influence of missing data and corresponding auxiliary tag SNPs.

corresponding tag SNPs are quite useful and cost-effective in association studies as it does not require genotyping all SNPs. Many studies have tried to minimize the number of tag SNPs required in each block. In a large-scale study of chromosome 21, Patil et al. [5] developed a greedy algorithm to partition the haplotypes into 4,135 blocks with 4,563 tag SNPs. Zhang et al. [6, 7] used a dynamic programming approach to reduce the numbers of blocks and tag SNPs to 2,575 and 3,562, respectively. Bafna et al. [1] showed that the problem of minimizing tag SNPs is NP-hard and gave efficient algorithms for special cases of this problem.

In reality, a tag SNP may be genotyped as missing data if it does not pass the threshold of data quality [5, 8]. However, missing data may cause ambiguity in identification of an unknown haplotype sample. Figure 1 illustrates the influence of missing data when identifying haplotype samples. In this figure, a haplotype block[1] (Figure 1 (A)) defined by 12 SNPs is presented. Each column represents a haplotype pattern (P_1, P_2, P_3, and P_4) and each row represents a SNP locus (S_1, S_2, ..., and S_{12}). The black and grey boxes stand for the major and minor alleles at each SNP locus, respectively. Suppose we select SNPs S_1 and S_{12} as tag SNPs. The haplotype sample h_1 is identified as haplotype pattern P_3 unambiguously (Figure 1 (B)). Consider haplotype samples h_2 and h_3 with one tag SNP genotyped as missing data (Figure 1 (C)). h_2 can be identified as haplotype patterns P_2 or P_3, and h_3 can be identified as P_1 or P_3. As a result, these missing data result in ambiguity when identifying haplotype samples.

Although we can not avoid the occurrence of missing data, the remaining SNPs within the haplotype block may provide abundant information to resolve the ambiguity. For example, if we re-genotype an additional SNP S_5 for h_2 (Figure 1 (D)), h_2 is identified as haplotype pattern P_3 unambiguously. On

---

[1] This haplotype block is redrawn from the haplotype database of chromosome 21 published by Patil et al. [5, 10]. We follow the same assumption as Patil, Zhang, and Bafna et al. that all SNPs are biallelic (i.e., taking on only two values).

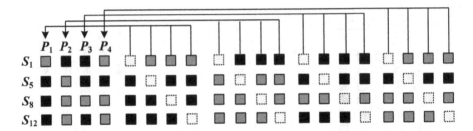

Fig. 2. A set of robust tag SNPs and haplotype samples with missing data.

the other hand, if SNP S_8 is re-genotyped (Figure 1 (E)), h_3 is also identified unambiguously. These additional SNPs are referred to as "auxiliary tag SNPs," which can be found from the remaining SNPs in the block and are able to resolve the ambiguity caused by missing data.

Alternatively, instead of re-genotyping auxiliary tag SNPs whenever encountering missing data, we work on a set of SNPs which is not affected by the the occurrence of missing data. For example (see Figure 2), suppose we select SNPs S_1, S_5, S_8, and S_{12} to be genotyped. Note that no matter which SNP is genotyped as missing data, the remaining three SNPs can still identify haplotype samples unambiguously. We refer to these SNPs as "robust tag SNPs," which correctly identify haplotype samples regardless of missing data occurred at any SNP locus. When the occurrence of missing data is frequently, the cost of re-genotyping processes can be reduced by robust tag SNPs.

This paper studies the problems of finding robust and auxiliary tag SNPs. Our study indicates that auxiliary tag SNPs can be found efficiently if robust tag SNPs have been computed in advance. This paper is organized as follows. In Section 2, we formulate the problem of finding robust tag SNPs and prove its NP-hardness. Section 3 gives a greedy algorithm to find robust tag SNPs with $(m+1)\ln(\frac{K(K-1)}{2})$ approximation, where m is the number of SNPs allowed for missing data and K is the number of patterns in the block. Section 4 illustrates the second greedy algorithm which gives a better solution of $\ln((m+1)\frac{K(K-1)}{2})$ approximation. Section 5 describes an iterative LP-relaxation algorithm which returns a solution of $O(m \ln K)$ approximation. Section 6 presents an efficient algorithm to find auxiliary tag SNPs when robust tag SNPs have been computed in advance. In Section 7, we show the experimental results of our algorithms applied to the public haplotype database. Finally, concluding remarks are given in Section 8.

2   Finding Robust Tag SNPs

Assume we are given a haplotype block consisting of N SNPs and K haplotype patterns. This block is denoted by an $N \times K$ matrix M_h. Let $M_h[i,j] \in \{1,2\}$ for each $i \in [1, N]$ and $j \in [1, K]$, where 1 and 2 represent the major and minor alleles, respectively. Define C as the set of all SNPs in M_h. The problem of finding a set of robust tag SNPs $C' \subseteq C$ which allows at most m SNPs genotyped as missing data is defined as follows.

Problem: Minimum Robust Tag SNPs (MRTS)
Input: An $N \times K$ matrix M_h and an integer m.
Output: The minimum subset of SNPs $C' \subseteq C$ which satisfies:
(1) for each pair of patterns P_i and P_j, these is a SNP $S_k \in C'$ such that $M_h[k, i] \neq M_h[k, j]$;
(2) when at most m SNPs are discarded from C' arbitrarily, (1) still holds.

We reformulate MRTS to a variant of the *set covering problem* [4]. Each SNP $S_k \in C$ (the k-th row in M_h) is reformulated to a set $S'_k = \{(i, j) \mid M[k, i] \neq M[k, j]$ and $i < j\}$. For example, suppose the k-th row in M_h is $\{1,1,1,2\}$. The corresponding set $S'_k = \{(1, 4), (2, 4), (3, 4)\}$. In other words, S'_k stores the pair of patterns distinguished by the SNP S_k. Let P be the set that contains each pair of these K patterns (i.e., $P = \{(i, j) \mid 1 \leq i < j \leq K\} = \{(1, 2), (1, 3), ..., (K - 1, K)\}$). For each pair of patterns, C' must contain more than m SNPs which are able to distinguish it. Otherwise, if C' contains m (or less) SNPs for some pair of patterns and they are all genotyped as missing data, it may cause ambiguity since no SNPs can distinguish this pair of patterns.

Lemma 1. $C' \subseteq C$ *is the set of robust tag SNPs which allows at most m SNPs genotyped as missing data iff each element in P is covered by the reformulated sets of C' for at least $(m + 1)$ times.*

Now we show the NP-hardness of the MRTS problem, which implies there is no polynomial time algorithm to solve MRTS unless P=NP.

Theorem 1. *The MRTS problem is NP-hard.*

Proof. When $m = 0$, MRTS is the same as the original problem of finding minimum number of tag SNPs, which is known as the *minimum test set* problem [4, 7]. Since the minimum test set problem is NP-hard and can be reduced to a special case of MRTS, MRTS is NP-hard.   □

3   The First Greedy Approximation Algorithm

In this section, we describe an approximation algorithm to solve MRTS by a greedy approach. By Lemma 1, we need to select a subset of SNPs $C' \subseteq C$ such that each element in P is covered for at least $(m + 1)$ times. Assume that the SNPs selected by this algorithm are stored in a $(m + 1) \times |P|$ table (see Figure 3 (A)) and each grid is empty initially. At each iteration of the algorithm, a SNP is selected to cover the grids of the column (i, j), where (i, j) stands for the patterns P_i and P_j distinguished by this SNP. Let R_i be the set of uncovered grids at the i-th row. This algorithm works by covering the grids from the first row to the $(m + 1)$-th row, and greedily selects a SNP which covers most uncovered grids in the i-th row at each iteration. In other words, while working on the i-th row, a

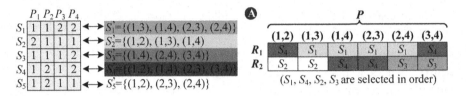

Fig. 3. The SNPs selected by the first greedy algorithm.

SNP is selected if its reformulated set S maximizes $|S \cap R_i|$. Figure 3 illustrates an example for this algorithm with $m = 1$, where SNPs S_1, S_4, S_2, and S_3 are selected in order. Then, the uncovered grids in other rows are updated according to the remaining elements in S (i.e., $S - R_i$). When all grids in this table are covered, each pair of patterns is covered by $(m + 1)$ SNPs. Thus, the set of SNPs in the table is the set of robust tag SNPs which allows at most m SNPs genotyped as missing data.

Algorithm: GREEDY-ROBUST-TAG-SNPs-1(C, P, m)
1    $R_i \leftarrow P, \forall i \in [1, m + 1]$
2    $C' \leftarrow \phi$
3   **for** $i = 1$ to $m + 1$ **do**
4   **while** $R_i \neq \phi$ **do**
5   select and remove an S from C that maximizes $|S \cap R_i|$
6            $C' \leftarrow C' \cup S$
7            $j \leftarrow i$
8   **while** $S \neq \phi$ **and** $j \leq m + 1$ **do**
9                $S_{tmp} \leftarrow S \cap R_j$
10               $R_j \leftarrow R_j - S_{tmp}$
11               $S \leftarrow S - S_{tmp}$
12               $j \leftarrow j + 1$
13   **endfor**
14   **return** C'

The time complexity of this algorithm is analyzed as follows. At Line 4, the number of iterations of the intermediate loop is bounded by $|R_i| \leq |P|$. Within the loop body (Lines 5-12), Line 5 takes $O(|C||P|)$ because we need to check all SNPs in C and examine the uncovered grids of R_i. The inner loop (Lines 8-12) takes only $O(|S|)$. Thus, the entire program runs in $O(m|C||P|^2)$.

We now show the solution C' returned by the greedy algorithm is not too larger than the optimal solution C^*. Suppose the algorithm selects the k-th SNP when working on the i-th row. Let S_k^c be the set of grids in the i-th row covered by the k-th selected SNP (i.e., $S_k^c = S \cap R_i$ at Line 5 in GREEDY-ROBUST-TAG-SNPs-1). We incur 1 unit of cost to each selected SNP [2], and spread this cost among the grids in S_k^c. In other words, each grid at the i-th row and j-th column is assigned a cost C_j^i (see Figure 4), where

$$C' \left\{ \begin{array}{l} S_1=\{(1,3),\ (1,4),\ (2,3),\ (2,4)\} \\ S_4=\{(1,2),\ (1,4),\ (2,3),\ (3,4)\} \\ S_2=\{(1,2),\ (1,3),\ (1,4)\} \\ S_3=\{(1,4),\ (2,4),\ (3,4)\} \end{array} \right.$$

	P					
	(1,2)	(1,3)	(1,4)	(2,3)	(2,4)	(3,4)
	1/2	1/4	1/4	1/4	1/4	1/2
	1/2	1/2	0	0	1/2	1/2

Fig. 4. The cost C_j^i of each grid for the first greedy algorithm.

$$C_j^i = \begin{cases} \frac{1}{|S_k^c|} & \text{if the algorithm selects the } k\text{-th SNP} \\ & \text{when working on the } i\text{-th row;} \\ 0 & \text{otherwise.} \end{cases}$$

Since each selected SNP is assigned 1 unit of cost, the sum of C_j^i for each grid in the table is equal to $|C'|$, i.e.,

$$|C'| = \sum_{i=1}^{m+1} \sum_{j=1}^{\frac{K(K-1)}{2}} C_j^i \ . \tag{1}$$

Let R_k^i be the number of uncovered grids in the i-th row before the k-th iteration (i.e., $(k-1)$ SNPs have been selected by the algorithm), and C_i' be the set of SNPs selected by the algorithm to cover the i-th row. We can rewrite (1) as

$$\sum_{i=1}^{m+1} \sum_{j=1}^{\frac{K(K-1)}{2}} C_j^i = \sum_{i=1}^{m+1} \sum_{k \in C_i'} (R_{k-1}^i - R_k^i) \frac{1}{|S_k^c|} \ . \tag{2}$$

Lemma 2. *The k-th selected SNP has $|S_k^c| \geq \frac{R_{k-1}^i}{|C^*|}$.*

Proof. Suppose the algorithm is working on the i-th row at the beginning of the k-th iteration. Let C_k^* be the set of SNPs in C^* (the optimal solution) that has been selected by the algorithm before the k-th iteration, and the set of remaining SNPs in C^* be $C_{\bar{k}}^*$. We claim that there exists a SNP in $C_{\bar{k}}^*$ which can cover at least $\frac{R_k^i}{|C_{\bar{k}}^*|}$ grids in the i-th row. Otherwise (i.e., each SNP in $C_{\bar{k}}^*$ covers less than $\frac{R_k^i}{|C_{\bar{k}}^*|}$ grids), all SNPs in $C_{\bar{k}}^*$ will cover less than ($\frac{R_k^i}{|C_{\bar{k}}^*|} \times |C_{\bar{k}}^*| = R_k^i$) grids in the i-th row. But since $C_k^* \cup C_{\bar{k}}^* = C^*$, this implies that C^* can not cover all grids in R_k^i, which is a contradiction. Because all SNPs in $C_{\bar{k}}^*$ are candidates to the greedy algorithm, the k-th selected SNP must cover at least $\frac{R_k^i}{|C_{\bar{k}}^*|}$ grids in the i-th row, which implies $|S_k^c| \geq \frac{R_{k-1}^i}{|C^*|}$ since $|C^*| \geq |C_{\bar{k}}^*|$ and $|R_k^i| \leq |R_{k-1}^i|$.   □

Theorem 2. *The first greedy algorithm gives a solution of $(m+1) \ln \frac{K(K-1)}{2}$ approximation.*

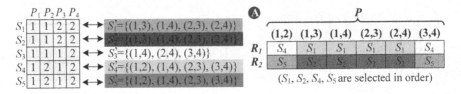

Fig. 5. The SNPs selected by the second greedy algorithm.

Proof. Define the d-th harmonic number as $H(d) = \sum_{i=1}^{d} \frac{1}{i}$ and $H(0) = 0$. By (2) and Lemma 2,

$$\sum_{i=1}^{m+1} \sum_{j=1}^{\frac{K(K-1)}{2}} C_j^i = \sum_{i=1}^{m+1} \sum_{k \in C_i'} (R_{k-1}^i - R_k^i) \frac{1}{|S_k^c|} \leq \sum_{i=1}^{m+1} \sum_{k \in C_i'} (R_{k-1}^i - R_k^i) \frac{|C^*|}{R_{k-1}^i}$$

$$= \sum_{i=1}^{m+1} \sum_{k \in C_i'} \left(\sum_{l=R_k^i+1}^{R_{k-1}^i} \frac{|C^*|}{R_{k-1}^i} \right)$$

$$\leq |C^*| \sum_{i=1}^{m+1} \sum_{k \in C_i'} \sum_{l=R_k^i+1}^{R_{k-1}^i} \frac{1}{l} \qquad (l \leq R_{k-1}^i)$$

$$= |C^*| \sum_{i=1}^{m+1} \sum_{k \in C_i'} \left(\sum_{l=1}^{R_{k-1}^i} \frac{1}{l} - \sum_{l=1}^{R_k^i} \frac{1}{l} \right)$$

$$\leq |C^*| \sum_{i=1}^{m+1} \sum_{k \in C_i'} (H(R_{k-1}^i) - H(R_k^i))$$

$$\leq |C^*| \sum_{i=1}^{m+1} (H(R_0^i) - H(R_{|C_i'|}^i))$$

$$\leq |C^*|(m+1)\max\{H(R_0^i)\} \qquad (R_{|C_i'|}^i = 0 \text{ and } H(0) = 0)$$

$$\leq |C^*|(m+1)\ln|P| . \qquad (H(R_0^i) \leq H(|P|)) \qquad (3)$$

By (1) and (3), we get

$$\frac{|C'|}{|C^*|} \leq (m+1)\ln|P| = (m+1)\ln\frac{K(K-1)}{2} .$$

$\square$

4   The Second Greedy Approximation Algorithm

This section gives the second greedy algorithm which returns a solution of better approximation than that in Section 3. Let R_i be the set of uncovered grids at the i-th row. Unlike the row-by-row manner of the first algorithm, this algorithm greedily selects a SNP that covers most uncovered grids in the table (i.e., its reformulated set S maximizing $|S \cap (R_1 \cup \cdots \cup R_{m+1})|$). Figure 5 illustrates an example for this algorithm with $m = 1$, where the SNPs S_1, S_2, S_4, and S_5 are selected in order. Let T be the collection of R_i (i.e., the set of all uncovered grids in the table). If all grids in the i-th row are covered (i.e., $R_i = \phi$), we remove R_i from T. This algorithm runs until $T = \phi$ (i.e., all grids in the table are covered).

Algorithm: GREEDY-ROBUST-TAG-SNPS-2(C, P, m)
1    $R_i \leftarrow P, \forall i \in [1, m+1]$
2    $T \leftarrow \{R_1, R_2, \cdots, R_{m+1}\}$
3    $C' \leftarrow \phi$
4   **while** $T \neq \phi$ **do**
5   select and remove an S from C that maximizes $|S \cap (R_1 \cup \cdots \cup R_{m+1})|$
6          $C' \leftarrow C' \cup S$
7   **for each** $R_i \in T$ **and** $S \neq \phi$ **do**
8                      $S_{tmp} \leftarrow S \cap R_i$
9                      $R_i \leftarrow R_i - S_{tmp}$
10                     $S \leftarrow S - S_{tmp}$
11   **if** $R_i = \phi$ **then** $T \leftarrow T - R_i$
12   **endfor**
13   **return** C'

The time complexity of this algorithm is analyzed as follows. At Line 4, the number of iterations of the loop is bounded by $|T| = (m+1)|P|$. Within the loop, Line 5 takes $O(|C||P|)$ time because we need to check each SNP in C and examine if it can cover any uncovered grid in each column. The inner loop (Lines 7-12) is bounded by $O(|S|) < O(|P|)$. Thus, the running time of this program is $O(m|C||P|^2)$.

We now evaluate the solution returned by the second greedy algorithm. Let C' and C^* be the set of SNPs selected by this algorithm and the optimal solution, respectively. Define $|S_k^c|$ as the number of grids in the table covered by the k-th selected SNP, and T_k as the number of uncovered grids in the table before the k-th iteration. We have the following lemma similar to Lemma 2 and the proof is omitted.

Lemma 3. *The k-th selected SNP has $|S_k^c| \geq \frac{T_{k-1}}{|C^*|}$.*

Theorem 3. *The second greedy algorithm gives a solution of $\ln((m+1)\frac{K(K-1)}{2})$ approximation.*

Proof. Each grid at the i-th row and j-th column is assigned a cost $C_j^i = \frac{1}{|S_k^c|}$ if it is covered by the k-th selected SNP. The sum of C_j^i for each grid is

$$|C'| = \sum_{i=1}^{m+1} \sum_{j=1}^{\frac{K(K-1)}{2}} C_j^i = \sum_{k=1}^{|C'|} (T_{k-1} - T_k) \frac{1}{|S_k^c|} \qquad \text{(see (1) and (2))}$$

$$\leq \sum_{k=1}^{|C'|}(T_{k-1} - T_k)\frac{|C^*|}{T_{k-1}} \qquad \text{(by Lemma 3)}$$
$$\leq |C^*|(H(T_0) - H(T_{|C'|})) \quad \text{(see the proof in Theorem 2)}$$
$$\leq |C^*|\ln((m+1)|P|). \tag{4}$$

By (4), we have

$$\frac{|C'|}{|C^*|} \leq \ln((m+1)|P|) = \ln((m+1)\frac{K(K-1)}{2}) \ .$$

□

5   The Iterative LP-Relaxation Algorithm

In practice, a probabilistic approach is sometimes more useful since the randomization can explore different solutions. In this section, we describe an alternative method to find robust tag SNPs by an iterative LP-relaxation algorithm. The MRTS problem is reformulated to an *Integer Programming* (IP) [4] problem, and an algorithm for this reformulation is described as follows.

Step 1. Given a haplotype block of N SNPs and K haplotype patterns. Let $X = \{x_1, x_2, ..., x_N\}$ be integer variables for the N SNPs, where $x_k = 1$ if the SNP S_k is selected and $x_k = 0$ otherwise. Define $D(P_i, P_j)$ as the set of SNPs which are able to distinguish P_i and P_j patterns. To allow at most m SNPs genotyped as missing data, each pair of patterns must be distinguished by at least $(m+1)$ SNPs (see Section 2). In other words, for each set $D(P_i, P_j)$, at least $(m+1)$ SNPs have to be selected to distinguish P_i and P_j patterns. Thus, this problem can also be defined as follows:

$$\textbf{Minimize } \sum_{k=1}^{N} x_k$$
$$\textbf{Subject to } \sum_{k\in D(P_i,P_j)} x_k \geq m+1, \quad \text{for all } 1 \leq i < j \leq K, \tag{5}$$
$$x_k = 0 \text{ or } 1.$$

This reformulated problem is the IP problem, which is known to be NP-hard.
Step 2. We relax the integer constraint of x_k, and the IP problem becomes a *Linear Programming* (LP) problem defined as follows:

$$\textbf{Minimize } \sum_{k=1}^{N} y_k$$
$$\textbf{Subject to } \sum_{k\in D(P_i,P_j)} y_k \geq m+1, \quad \text{for all } 1 \leq i < j \leq K, \tag{6}$$
$$0 \leq y_k \leq 1.$$

Define OPT(LP) and OPT(IP) as the optimal solution of above LP and IP problems, respectively. Since the solution space of LP includes that of IP, OPT(LP) $\leq$ OPT(IP)=OPT(MRTS).

Step 3. Let y_1, y_2, ..., y_N be the values of the optimal solution in (6), where $0 \leq y_k \leq 1$. We assign 0 or 1 to x_k by the following randomized rounding method:

$$\text{Assign} \begin{cases} x_k = 1 \text{ with probability } y_k, \\ x_k = 0 \text{ with probability } 1 - y_k. \end{cases}$$

Step 4. The randomized rounding may invalidate some of the inequalities in (5). We repeat Steps 1, 2, and 3 for those unsatisfied inequalities until all of them are satisfied. Then we construct a final solution by the following rule:

$$\text{Assign} \begin{cases} x_k = 1 \text{ if } x_k \text{ is assigned to 1 in any one of the iterations;} \\ x_k = 0 \text{ otherwise.} \end{cases}$$

The following theorem gives the approximation bound of this algorithm, and the proof is omitted.

Theorem 4. *The iterative LP-relaxation algorithm gives a solution of $O(m \ln K)$ approximation.*

6   Finding Auxiliary Tag SNPs

This section describes the problem of finding auxiliary tag SNPs. Define M_h as in Section 2 and $C_{tag} \subseteq C$ as the set of tag SNPs genotyped from a haplotype sample (with missing data). A formal definition for this problem is given below.

Problem: Minimum Auxiliary Tag SNPs (MATS)
Input: An $N \times K$ matrix M_h, and a set of tag SNPs C_{tag}.
Output: The minimum subset of SNPs $C_{aux} \subseteq C - C_{tag}$ such that each pair of ambiguous patterns can be distinguished by SNPs in C_{aux}.

Theorem 5. *MATS is NP-hard.*

Proof. Consider that all tag SNPs are genotyped as missing data. This problem is just like finding another set of tag SNPs to distinguish those K patterns, which is already known as NP-hard [1].   □

Although the MATS problem is NP-hard, we show that auxiliary tag SNPs can be found efficiently when robust tag SNPs have been computed in advance, which is described as follows. Without loss of generality, assume that these robust tag SNPs are stored in an $(m + 1) \times |P|$ table T_r (see Section 3).

Step 1. The patterns that match the haplotype sample are stored into a set A. For example (see Figure 6), if we genotype SNPs S_1, S_2, and S_3 for the sample h_2 and S_1 is missing data, patterns P_1 and P_3 both match h_2.

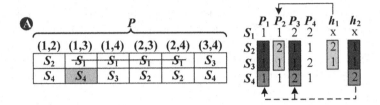

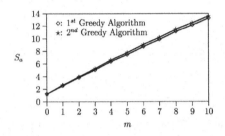

Fig. 6. An example to find the auxiliary tag SNPs.

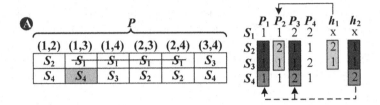

Fig. 7. The comparison of two greedy algorithms.

Table 1. The total number of robust tag SNPs found by the three algorithms.

Algorithm	$m = 0$	$m = 1$	$m = 2$
First Greedy	13	20	29
Second Greedy	13	19	27
LP-relaxation	13	21	28

(80 haplotypes, 160 SNPs)

Step 2. If $|A|=1$, the sample is identified unambiguously and we are done (e.g., h_1 in Figure 6). If $|A| > 1$ (e.g., h_2), for each pair of ambiguous patterns in A (e.g., P_1 and P_3), traverse the corresponding column in T_r, find the next unused SNP (e.g., S_4), and add the SNP to C_{aux}. As a result, the SNPs in C_{aux} can distinguish each pair of ambiguous patterns and thus are the auxiliary tag SNPs for the haplotype sample.

The worst case of this algorithm is that all SNPs in C_{tag} are genotyped as missing data, and we need to traverse each column in T_r. Thus, the running time of this algorithm is $O(|T_r|) = O(m|P|)$.

7   Experimental Results

We apply two approximation algorithms mentioned in Sections 3 and 4 to the public haplotype database of Human Chromosome 21 [5, 10]. This data set includes 20 haplotypes of 24,047 SNPs spanning over about 32.4MB. We evaluate these two algorithms with respect to m (i.e., the number of SNPs allowed to be genotyped as missing data). Let S_a be the average number of robust tag SNPs for each block. Figure 7 plots S_a with respect to m for both algorithms. Note that since m SNPs are allowed to be genotyped as missing data, the set of robust tag SNPs must contain more than m SNPs. We observe that S_a grows linearly with respect to m for both algorithms. Therefore, this phenomenon indicates that the result of both algorithms is still close to the optimal solution even when m increases.

We also test the greedy and iterative LP-relaxation algorithms on a simulated large block composed of 80 haplotypes samples and 160 SNPs. The iterative

LP-relaxation algorithm is executed for 10 runs and we output the minimum solution among them. Table 1 lists the number of robust tag SNPs found by each algorithm with respect to m. The solution returned by the iterative LP-relaxation algorithm is close to those by greedy approaches.

8 Conclusion

In this paper, we study the problems of finding robust and auxiliary tag SNPs. One future direction is to assign weights to different types of SNPs (e.g., SNPs in coding or non-coding regions), and design algorithms for the selection of weighted tag SNPs.

Acknowledgements

We thank the reviewers for their helpful comments. We are grateful for valuable suggestions from Chia-Yu Su. Yao-Ting Huang and Kun-Mao Chao were supported in part by an NSC grant 92-2213-E-002-059. Ting Chen was sponsored in part by NIH NIGMS PAR-02-021, Center of Excellent in Genome Sciences.

References

1. Bafna, V., Halldorsson, B.V., Schwartz, R., Clark, A.G., and Istrail, S. Haplotypes and Informative SNP Selection Algorithms: Don't Block Out Information. *Proceedings of the Seventh Annual International Conference on Research in Computational Molecular Biology*, pages 19–27, 2003.
2. Cormen T.H., Leiserson, C.E., Rivest, R.L., and Stein, C. *Introduction to Algorithms*. The MIT Press, 2001.
3. Daly, M.J., Rioux, J.D., Schaffner, S.F., Hudson, T.J., and Lander, E.S. High-Resolution Haplotype Structure in the Human Genome. *Nature Genetics*, 29:229–232, 2001.
4. Garey, M.R. and Johnson, D.S. *Computers and Intractability*. Freeman, New York, 1979.
5. Patil, N., Berno, A.J., Hinds, D.A., Barrett, W.A., Doshi, J.M., Hacker C.R., Kautzer, C.R., Lee, D.H., Marjoribanks, C., McDonough, D.P., *et al.* Blocks of Limited Haplotype Diversity Revealed by High-Resolution Scanning of Human Chromosome 21. *Science*, 294:1719–1723, 2001.
6. Zhang, K., Deng, M., Chen, T., Waterman, M.S., and Sun, F. A Dynamic Programming Algorithm for Haplotype Block Partitioning. *Proceedings of the National Academy of Sciences of the United States of America*, 99:7335–7339, 2002.
7. Zhang, K., Sun, F., Waterman, M.S., and Chen, T. Dynamic Programming Algorithms for Haplotype Block Partitioning: Applications to Human Chromosome 21 Haplotype Data. *Proceedings of the Seventh Annual International Conference on Research in Computational Molecular Biology*, pages 332–340, 2003.
8. Zhao, J.H., Lissarrague, S., Essioux, L., and Sham, P.C. GENECOUNTING: Haplotype Analysis with Missing Genotypes. *Bioinformatics*, 18:1694–1695, 2002.
9. http://www.ncbi.gov/dbsnp.
10. http://www.perlegen.com/haplotype/.

The Minisatellite Transformation Problem Revisited: A Run Length Encoded Approach

Behshad Behzadi and Jean-Marc Steyaert

LIX, Ecole Polytechnique, Palaiseau cedex 91128, France
{Behzadi,Steyaert}@lix.polytechnique.fr

Abstract. In this paper we present a more efficient algorithm for comparison of minisatellites which has complexity $O(n'^3 + m'^3 + mn'^2 + nm'^2 + mn)$ where n and m are the lengths of the maps and n' and m' are the sizes of run-length encoded maps. We show that this algorithm makes a significant improvement for the real biological data, dividing the computing time by a factor 30 on a significant set of data.

1 Introduction

Comparing sequences is a long-addressed problem in computer science as well as in biology. Numerous algorithms have been designed starting from `diff` in Unix and ending (for the moment) at the subquadratic algorithm of Crochemore et al. (see [13]). Our interest in this paper is devoted to a structured comparison of sequences when complex operations can be used to transform strings. These notions intervene naturally in the algorithmic study and treatment of minisatellites – a very important concept in biology. These genomic subsequences are commonly used to understand the dynamic of mutations in particular for interallelic gene conversion-like processes at autosomal loci [8,9]. Jobling et al. [7] have characterized the Y-specific locus MSY1, a haploid minisatellite, which is composed of 48 to 114 copies of a repeat unit of length 25, rich in AT and predicted to form stable hairpin structures. These sequences are of great interest since they constitute markers for Y chromosome diversity: therefore they allow to trace male descendence proximity in populations.

Modelling minisatellite evolution is therefore necessary in order to provide biologists with a rigorous tool for comparing sequences and establishing likely conclusions as to their proximity. Bérard and Rivals [6] have proposed a combinatorial algorithm to solve the edit distance problem for minisatellites: they considered the five operations – amplification, contraction, mutation, insertion, deletion – with symmetric costs for each type of operation and designed an $O(n^4)$ algorithm. We showed in [4] that it is possible to take into account the generalized cost model, and we designed an algorithm which runs in time $O(n^3)$, thus being more efficient even in a more involved context. In this paper we propose a new enhancement of the method, by making use of a renewed vision based on the run-length encoding.

A string s is called *run length encoded* if it is described as an ordered sequence of pairs (x, i), often denoted x^i, where x is an alphabet symbol and i is an integer.

I. Jonassen and J. Kim (Eds.): WABI 2004, LNBI 3240, pp. 290–301, 2004.

Each pair corresponds to a *run* in s consisting of i consecutive occurrences of x. For example, the string $aaaabbbbbcccabbbbcc$ is encoded as $a^4b^4c^3a^1b^4c^2$. Letters of two consecutive runs are different. Such a run-length encoded string can be significantly shorter than the standard string representation. Run-length encoding is a usual image compression technique, since many images typically contain large runs of identically-valued pixels. Among biological sequences, minisatellites are ideal ones for using the run-length encoded technique, because basically they consist of a large number of tandem repeats.

Different algorithms have been developed for comparing run-length encoded (RLE) strings. Bunke and Csirik [3] as well as Apostolico, Landau, and Skiena [1] present algorithms for the LCS problem on RLE strings. Mäkinen, Navarro and Ukkonen in [5], Arbell, Landau and Mitchell in [2] and Crochemore, Landau and Ziv-Ukelson in [13] presented algorithms for edit distance of RLE strings.

In this paper, we extend these algorithms and propose an algorithm for computing the transformation distance between two RLE minisatellite maps. The framework we propose has its full generality; operation costs can be almost arbitrary with the only feature that amplifications and contractions are of low cost.

In Section 2, we describe the mathematical model for the minisatellite evolution, and we state the problem in its general form.

In Section 3, we state different lemmas which are essential to prove the correctness of our algorithm.

In Section 4, we show how our method can be adapted for the simplest transformation distance using the arch concept developed by Bérard and Rivals [6].

Section 5 is dedicated to the algorithm. It consists of two parts: Preprocessing and the Core algorithm both of which use the dynamic programming method.

In section 6, we discuss about the performance of the new algorithm compared with the previous ones on randomly generated data and real biological data. We show that our new algorithm works much faster on the real minisatellite data.

2   Model Description

The symbols are elements from a finite alphabet Σ. We will use the letters x, y, $z,...$ for the symbols in Σ and $s, t, u, v,...$ for strings[1] over Σ. The empty string is denoted by ϵ. We will denote by $s[i]$ the symbol in position i of the string s (the first symbol of a string s is $s[1]$). The substring of s starting at position i and ending at position j is denoted by $s[i..j] = s[i]s[i+1]\ldots s[j]$. A substring $s[i..j]$ is called a *run* if $s[i-1] \neq s[i] = s[i+1] = s[i+2]...s[j-1] = s[j] \neq s[j+1]$. A string obtained by replacing each of the runs of string s by a letter of that run is called the *compact* representation of s and is denoted by s'.

---

[1] Throughout the paper we use the word string to designate what biologists call sequences or maps [6]. The word sequence will refer to a sequence of operations on a string.

In the evolutionary model, five elementary operations are considered on strings. These operations are mutation (replacement), insertion, deletion, amplification and contraction. The first three are the well-known string edit distance operations (see for example [12]). The last two are new operations which are significant in the study of the evolution of minisatellite strings and more generally whenever large substrings in the genome are multiply repeated or deleted. Amplification of a symbol x in a string s amounts to repeating this symbol after one of its occurrences in s. A p-plication of a symbol x which is an amplification of *order* p amounts to $p - 1$ times repeat symbol x *after* the initial symbol x. Conversely, the p-contraction of a symbol x means to delete $p - 1$ consecutive symbols x provided that the symbol just *before* them is also an x. Given two strings s and t, there are infinitely many sequences of elementary operations which transform the string s into the string t. Among this infinity, some evolution sequences are more likely; in order to identify them, we introduce a cost function for each elementary operation depending on the symbols involved in the operation: $I(x)$ and $D(x)$ are the costs of an insertion or a deletion of symbol x. $M(x, y)$ is the cost of the replacement of symbol x by symbol y in the string. For $p > 1$, $A_p(x)$ is the cost of a p-plication of symbol x in the string and finally $C_p(x)$ is the cost of a p-contraction of a symbol x. In this paper we consider only the amplifications (and contractions) of order 2. Whenever we use the terms amplification and contraction, we mean duplication and 2-contraction. Note that the costs can be non symmetric ($I(x)$ may be different from $I(y)$, etc.). We suppose that the mutation cost function satisfies the triangle inequality property: $M(x, y) + M(y, z) \geq M(x, z)$ for all different x, y, z in $\Sigma \cup \{\epsilon\}$. In addition, $M(x, x) = 0$ for any symbol x and all other values of all of the cost functions are strictly greater than zero. These hypotheses do not reduce the generality of our statements. The main hypothesis to consider in comparison of minisatellites is the following:

Hypothesis 1 *The cost of duplications (and contractions) is less than the cost of all other operations.*

A *transformation* of s into t amounts to applying a sequence of operations on s transforming it into t. When s is transformed into t by a sequence of operations we write by $s \xrightarrow{*} t$ and when s is transformed into t in one elementary operation we use the notation $s \rightarrow t$. The cost of a transformation is the sum of the costs of its operations. The *transformation distance* from s into t is the minimum cost for a possible transformation from s into t. The transformation which gives this minimum is called *optimal transformation* (it is not necessarily unique). Our objective in this paper is to find this distance between two strings and one of their optimal transformations. In the next section we will study the optimal transformation properties.

It will be convenient to add an extra special symbol \$ to the alphabet and to consider that the value of all the functions with \$ as one of their variables is ∞ (with exception of $M(\$, \$) = 0$). Whenever we are asked to find the transformation distance between strings s and t, we will compute the optimal transformation of \$$s$ into \$$t$. By our assumption these two values are equal. This is a way to forbid any insertion (and deletion) at the beginning of strings. (So from now without loss of generality we suppose that the insertions (and deletions) are allowed *only after* symbols.)

3   Optimal Transformation Properties

A transformation applies a sequence of operations on a string s and the result will be a string t. This sequence is called *transformation sequence* from s into t. In a transformation of s into t, each symbol of s, generates a substring of t: this substring can be the empty string. The symbols of s which generate a non-empty substring of t are called *generating symbols* and the other symbols are called *vanishing symbols*.

Now consider the transformation of one symbol x to a non-empty string s: This transformation is called *generation*. The generations which use only mutations, amplifications and insertions are called non-decreasing generations. A non-decreasing generation can be represented by a tree. The tree construction rules are the following:

1) The root of the tree has label x.
2) For any duplication of a symbol y, add two new nodes with label y as children of that node.
3) The insertion of a letter z after a symbol y is shown by adding two children to the corresponding node y which have labels y and z from left to right.
4) The mutation of a symbol y into z is represented by a single child with label z for the node with label y.

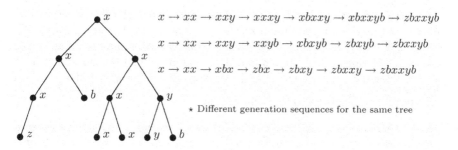

$$x \rightarrow xx \rightarrow xxy \rightarrow xxxy \rightarrow xbxxy \rightarrow xbxxyb \rightarrow zbxxyb$$

$$x \rightarrow xx \rightarrow xxy \rightarrow xxyb \rightarrow xbxyb \rightarrow zbxyb \rightarrow zbxxyb$$

$$x \rightarrow xx \rightarrow xbx \rightarrow zbx \rightarrow zbxy \rightarrow zbxxy \rightarrow zbxxyb$$

⋆ Different generation sequences for the same tree

Generation Cost $= 2A_2(x) + 2I(b) + I(y) + M(x, z)$

Fig. 1. The tree representation of a non-decreasing generation.

Each internal node in the tree corresponds to an operation. As shown in the Figure 1, different generation sequences can have the same tree representation. They differ by the order of operations but their costs are the same. A distinguished generation sequence that one can construct for a given tree is the sequence which is obtained by a *left depth first search* of the tree (visit the children of a node from left to right). This sequence is called *left-to-right* generation sequence. We recall two lemmas about the optimal transformation properties from [4]. The proofs are given in [4].

Lemma 1. *(The generation lemma):*
The optimal generation of a non-empty string s from a symbol x can be achieved by a non-decreasing generation.

Lemma 2. *(The independency of contractions)*
There exists an optimal transformation of a string s into string t in which all the contractions are done before all the amplifications.

Symmetrically, the optimal *reduction* of a non-empty string s to a single symbol x can be obtained by using only mutations, deletions and contractions. Now we study the properties of runs in an optimal transformation. The next lemma considers a transformation of a single run string into another single run string.

Lemma 3. *(Transformation of x^k into y^l)*
The cost of the optimal transformation of the string $s = x^k$ into $t = y^l$ is:
For $k \leq l$:
(1) $(k-1) \times C_2(x) + M(x,y) + (l-1) \times A_2(y)$ *if $M(x,y) \geq A_2(y) + C_2(x)$*
(2) $k \times M(x,y) + (l-k) \times A_2(y)$ *if $M(x,y) < A_2(y) + C_2(x)$*
For $k > l$:
(3) $(k-1) \times C_2(x) + M(x,y) + (l-1) \times A_2(y)$ *if $M(x,y) \geq A_2(y) + C_2(x)$*
(4) $k \times M(x,y) + (k-l) \times C_2(x)$ *if $M(x,y) < A_2(y) + C_2(x)$*

Proof: Let u be the number of x's which are mutated into y. Then $k - u$ contractions on $s = x^k$ are necessary to delete the extra symbols. If $l \geq k$ then $l - u$ duplications should be applied after the mutations. The total cost can be expressed as a function of u: $f(u) = u \times M(x,y) + (l-u) \times A_2(y) + (k-u) \times C_2(x)$. This linear function of u is minimized at $u = 1$ or $u = k$ depending on the sign of $M(x,y) - A_2(y) - C_2(x)$. These two possible cases for the minimum are the expressions (1) and (2) of the lemma. A similar proof can be considered in the case $l < k$ for the expressions (3) and (4). Note that $M(x,y) = 0$ if $x = y$. $\square$

Hypothesis 1 leads us to the following fact:

Fact 1 *There exists an optimal generation of a non-empty string t from a single symbol x in which for every run of size $k > 1$ in t, the $k-1$ right symbols of the run are generated by duplications of the leftmost symbol of the run.*

Lemma 4. *(First run lemma)*
There exists an optimal transformation of string s into string t in which for any generating symbol x the following is true: If x generates $t[i..j]$ and $t[i]$ is not the first symbol of a run in t then the first run in $t[i..j]$ has length one.

Proof: Consider an optimal transformation of s into t. Let x be the rightmost symbol which violates the statement of the lemma. This means that the first run in $t[i..j]$ generated by x is a run of length at least two and $t[i-1] = t[i]$. If we move $t[i]$ from this generated substring to the substring immediately generated at its left, the total cost remains the same (Fact 1). By iterating this operation the first run generated by x will have length one. By iterating the whole procedure all the symbols will satisfy the lemma statement.   □

Lemma 4 is important for the design of our algorithm. In fact in an optimal transformation of string s into string t, if $s[n]$ is a generating symbol it will generate a suffix of t and there are $O(m')$ candidate suffixes in t (and not $O(m)$) for this generation.

Lemma 5. *(Generating and vanishing symbols of a run)*
There exists an optimal transformation of s to t such that for any run in s, all the vanishing symbols are at the right of all the generating symbols.

Proof: Consider a run of size $k = k_v + k_g$ such that k_v symbols are vanishing and k_g symbols are generating. If $k_g = 0$ the statement of the lemma is correct. If $k_g > 0$, the reduction of k_v symbols costs minimum when they have an identical symbol at their left because all these symbols can use contractions for their deletions.   □

We call a substring of s a *vanishing substring* if all the symbols of the substring are vanishing symbols. A vanishing substring is maximal if there is no other vanishing symbol just before or just after it. By lemma 5, there exists an optimal transformation of s into t in which the last symbol of all maximal vanishing substrings of s is the last (rightmost) symbol of some run in s. This is an important fact for the design of our algorithm. In a transformation of string s (with compact form s') into string t, the maximum number of maximal vanishing substrings is not more than $n' = |s'|$ and the rightmost symbol of any of these vanishing substring is one of the n' special positions in s.

Consider a maximal vanishing substring $s[i..j]$ and let i' be the smallest number in the interval $[i, j]$ which is a first symbol of a run in s if such a number exists. If i' exists the evolution can be considered as two reductions: First $s[(i'-1)..j]$ is reduced into $s[i'-1]$ and then $s[(i-1)..(i'-1)]$ is reduced into $s[i-1]$. The rightmost symbol of both of these reductions is a rightmost symbol of some run in s, so the whole number of these reductions in a transformation is $O(m')$. As a conclusion we have the following lemma.

Lemma 6. *(Reduction types)*
There exists an optimal transformation of s into t, in which the last symbol of any reduced substring (reduction) is the last symbol of a run in s and the reduction is one of the two following types:
(a) Some complete runs of s vanish.
(b) A suffix of one of the runs of s vanishes.

4   Arch Representation

Bérard and Rivals [6] use the notion of arches in order to represent the duplications (and contractions). For a given string s an arch is identified by a pair of integers (i, j) such that $i < j$ and $s[i] = s[j]$. Two arches are called *compatible* if both of the duplications (or contractions) can happened together in an evolution history. Formally, two arches (i, j) and (i', j') where $i \leq i'$ are *incompatible* if $i < i' < j$ and $j' \geq j$. Two arches are compatible if they are not incompatible (Figure 2).

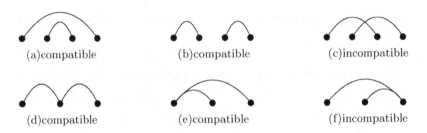

(a)compatible   (b)compatible   (c)incompatible

(d)compatible   (e)compatible   (f)incompatible

Fig. 2. Different cases for compatible and incompatible arches.

In their simple model, where the cost of operations does not depend on the symbols, they show that the minimum generation cost of a substring can be computed from the maximum number of compatible arches in the substring. The following proposition shows how one can compute the maximum number of the compatible arches faster using the runs idea:

Proposition 1 *Let s' be the string obtained by replacing each run of the string s by a single letter. The maximum number of compatible arches in s equals the maximum number of compatible arches in s' plus the difference of lengths of the strings s and s'.*

Proof: Let n and n' be the lengths of the strings s and s' respectively. The proof consists of two parts.

Let k' be the maximum number of compatible arches in s'. Let $\mathcal{A}'$ be a set of k' compatible arches in s'. We construct a set $\mathcal{A}$ of arches in s in the following way: For each pair (i', j') in $\mathcal{A}'$, add an arch connecting the last symbol of the i'-th run to the first symbol of the j'-th run (*inter-run arches*). Add all arches of the form $(i, i+1)$ into $\mathcal{A}$ (*in-run arches*). The number of in-run arches is $n - n'$. All of the arches in $\mathcal{A}$ are compatible; hence we have at least $k = k' + (n - n')$ compatible arches in s.

Let k be the maximum number of compatible arches in s and $\mathcal{A}$ be a set of k compatible arches in s. If the number of in-run arches of form $(i, i+1)$ in $\mathcal{A}$ is less than $n - n'$, then there exists an in-run arch of form $(i, i+1)$ which is not included in $\mathcal{A}$. Let $(i, i+1)$ be the leftmost such arch. By definition of the incompatible arches, in $\mathcal{A}$ there is at most one arch with $i+1$ as the right endpoint (Figure 2(f)). So there is at most one arch in $\mathcal{A}$ incompatible with $(i, i+1)$. By maximality of $\mathcal{A}$, there is exactly one arch in $\mathcal{A}$ incompatible with $(i, i+1)$. We replace this arch by $(i, i+1)$ in $\mathcal{A}$. By iterating the same procedure, we arrive to a set of k compatible arches such that $n - n'$ are in-run arches of form $(i, i+1)$. Since all the possible in-run arches of form $(i, i+1)$ are present in this set, there is at most one inter-run arch between two given runs. If we reduce every run to a single symbol and keep only the inter-run arches, we obtain a set of $k' = k - (n - n')$ compatible arches for s'. This completes the proof.$\square$

As a result of this proposition, instead of computing the maximum number of compatible arches in a string s, one can compute the maximum number of compatible arches in s' (which is eventually much smaller than s) and add $n - n'$ to the computed value. Let $MA[i, j]$ be the maximum number of compatible arches in substring $s'[i..j]$. The simple dynamic programming algorithm given in Figure 3 computes the maximum number of compatible arches for all substrings of s'. The proof is easy and is omitted. The time complexity is $O(n'^3)$.

Algorithm 1 Compatible Arches

If $i \geq j$ **Then** $MA[i, j] = 0$

Else $MA[i, j] = \max \begin{cases} MA[i+1, j] \\ MA[i, k-1] + MA[k, j] + 1 \quad \textit{if } s'[i] = s'[k], \forall i < k \leq j \end{cases}$

Fig. 3. Recurrence relations for Maximum Compatible arches calculation.

5 The Algorithm

In this section we describe an algorithm to compute the transformation distance from a string s into a string t. Firstly, in the preprocessing part we determine the cost of generation of substrings of the target string t from any given symbol in the alphabet. Then we will compute the transformation distance from s to t by applying a dynamic programming algorithm using the preprocessing results.

5.1 Pre-processing

Let $t[i..j]$ be the string generated by a symbol x. The compact representation of this substring is $(t[i..j])'$. By Fact 1 and the proof of Proposition 1, the optimal cost of $x \xrightarrow{*} t[i..j]$ can be obtained by adding all duplication costs of two identical neighbor symbols in $t[i..j]$ into the optimal cost of the generation $x \xrightarrow{*} (t[i..j])'$. Let $b(i)$ denotes the number of the runs including $t[i]$. $NDup[i]$ is defined as below:

$$NDup[1] = 0 \text{ and } \forall 1 < i \ NDup[i] = NDup[i-1] + \begin{cases} 0 & \text{if } t[i] \neq t[i-1] \\ A_2(t[i]) & \text{if } t[i] = t[i-1] \end{cases}$$

If t' is the compact representation of t, the minimum generation cost of $t[i..j]$ from a symbol x denoted by $G_t(x, i, j)$ is equal to $(NDup[j] - NDup[i]) + G_{t'}[x, b[i], b[j]]$. The table $G_{t'}$ is computed by dynamic programming algorithm given in Figure 4. This algorithm is a simplified version of the pre-processing algorithm given in [4]. The proof is identical to the proof in [4].

Symmetrically to $G_t(x, i, j)$, we compute $K_s(x, i, j)$ which is the minimum cost of reduction of $s[i..j]$ into the symbol x. The complexity of the preprocessing part for a fixed alphabet is $O(n + m + n'^3 + m'^3)$ in time and $O(m + n + n'^2 + m'^2)$ in space.

Algorithm 2 Generation_Costs

Initialization:
$$\forall 0 < i \leq m', \ \forall x \in \Sigma \ : \ G_{t'}[x, i, i] = M(x, t'[i])$$

Recurrence:
$\forall i < j \ \forall x \in \Sigma :$

1) $T[x, i, j] = \min \begin{cases} A_2(x) + \min_{i < k \leq j} \{G_{t'}[x, i, k-1] + G_{t'}[x, k, j]\} \\ G_{t'}[y, i+1, j] + I(y) \qquad\qquad \text{if } t'[i] = x, \ \forall y \in \Sigma \end{cases}$

2) $G_{t'}[x, i, j] = \min_{y \in \Sigma} \{T[y, i, j] + M(x, y)\}$

Fig. 4. Recurrence relations for generation costs.

5.2 Core Algorithm: Dynamic Programming Algorithm

An optimal transformation which satisfies the lemmas 4, 5 and 6 is called a *good optimal transformation*. Dynamic programming is the tool to find a good optimal transformation of string s into string t. Let $TD[i..j]$ be the minimum cost for a transformation of $s[1..i]$ into $t[1..j]$ which can be extended into a good transformation of $s[1..n]$ into $t[1..m]$; n' and m' denote the number of runs in strings s and t respectively. Let $e_s(k)$ be the position of the last element of the k-th run in s for any $1 \leq k \leq n'$. E_s is the set of all values of $e_s(k)$ for $1 \leq k \leq n'$. For each $i \leq n$, $e_s^*(i)$ is defined as the largest $e_s(k) < i$ for $k \geq 1$ if such a k exists and 0 otherwise; $e_t(l)$, E_t and $e_t^*(j)$ are defined similarly for $1 \leq l \leq m'$ and $1 \leq j \leq m$. All these functions can be computed in linear time, before the execution of the core algorithm. The core algorithm is given in Figure 5.

Proposition 2 *The Algorithm given in Figure 5 determines correctly the transformation distance of s into t.*

Proof Sketch: Let us explain how the recurrence relations given in Figure 5 determine the cost of a good optimal transformation. The first lines correspond to the fact there is a special symbol $ in the head of the strings. We have four different cases for the recurrence relation. Each case corresponds to different positions of the current indices w.r.t a run in s and t; then we apply repeatedly Lemmas 4, 5 and 6 to the situation and evaluate systematically the costs of all possible analyses that can be done. □

Algorithm 3 Transformation_Distance(S,T)

$TD[0,0] = 0$

For $i \leftarrow 1$ **to** n **do** $TD[i,0] = \infty$

For $j \leftarrow 1$ **to** m **do** $TD[0,j] = \infty$

For $i \leftarrow 1$ **to** n **do**

For $j \leftarrow 1$ **to** m **do**

$$TD[i,j] = \begin{cases}
\text{A) if } i \notin E_s \text{ and } j \in E_t \\
\quad \min\{TD[i\text{-}1,l] + G_t(s[i], l+1, j)\}_{l \leq j:\ l \in E_t \text{ or } l+1 \in E_t \text{ or } l = j-1} \\[1ex]
\text{B) if } (i \notin E_s \text{ and } i\text{-}1 \notin E_s) \text{ and } j \notin E_t \\
\quad \min\{TD[i\text{-}1,j\text{-}1] + G_t(s[i], j, j),\ TD[i\text{-}1, e_t^*(j)] + G_t(s[i], e_t^*(j)+1, j)\} \\[1ex]
\text{C) if } (i \in E_s \text{ or } i\text{-}1 \in E_s) \text{ and } j \notin E_t \\
\quad \min \begin{cases}
TD[i\text{-}1, j\text{-}1] + G_t(s[i], j, j) \\
TD[i\text{-}1, e_t^*(j)] + G_t(s[i], e_t^*(j)+1, j) \\
\{TD[i\text{-}1,l] + G_t(s[i], l+1, j)\}_{l \leq j:\ l \in E_t \text{ or } l+1 \in E_t} \quad \text{if } i\text{-}1 \in E_s \\
\{TD[k,j] + K_s(s[k], k, i)\}_{k < i:\ k \in E_s \text{ or } k > e_s^*(i)}
\end{cases} \\[1ex]
\text{D) if } i \in E_s \text{ and } j \in E_t \\
\quad \min \begin{cases}
\{TD[i\text{-}1,l] + G_t(s[i], l+1, j)\}_{l \leq j:\ l \in E_t \text{ or } l+1 \in E_t \text{ or } l = j-1} \\
\{TD[k,j] + K_s(s[k], k, i)\}_{k < i:\ k \in E_s \text{ or } k > e_s^*(i)}
\end{cases}
\end{cases}$$

Fig. 5. Recurrence relations for Transformation Distance.

The complexity of the algorithm is $O(mn + mn'^2 + nm'^2)$ in time and $O(mn)$ in space. Analysis of the complexity can be done by computing the total complexity of each single line in the recurrence relation separately and then adding the results.

6 Discussion

In this section, we compare the running time of the algorithms presented in this paper with the algorithms presented in [4] and [6]. For this aim we firstly compare the running time on randomly generated strings on an alphabet Σ. In a second part we run the three algorithms on a real biological database of minisatallites.

Let us just remark that the number of runs R in a random sequence of length n with equally distributed letters follows a Bernoulli law: $P\{R = k\} = \binom{n-1}{k-1}\left(\frac{|\Sigma|-1}{|\Sigma|}\right)^{k-1}\left(\frac{1}{|\Sigma|}\right)^{n-k}$. The mathematical expectation of the number of runs for a string of size n is $1 + (n-1)\frac{|\Sigma|-1}{|\Sigma|}$.

The number of runs of a string which is generated randomly on an alphabet Σ grows linearly with n. The compact representations of randomly generated strings have size proportional to n on the average. We observe on biological minisattelite samples that the compact representation is much shorter than the original string.

Type 1: CACAATATACATGATGTATATTATA

Type 2: CATAATATACATGATGTATATTATA

Type 3: CACAATATACATCATGTATATTATA

Type 4: CATAATATACATCATGTATATTATA

Type 5: CATAATATACATGATGTATAATATA

0	1	1	2	2
1	0	2	1	1
1	2	0	1	3
2	1	1	0	2
2	1	3	2	0

Fig. 6. Five variant MSY1 repeats identified and a simple mutation table on them based on the number of different nucleotides.

Table 1. Running times of different algorithms on random and biological datasets.

Algorithm	Random sequences		minisatellites data	
	PreProcessing	Core	PreProcessing	Core
Bérard & Rivals(2002)	15.90 sec	1058.44 sec	16.37 sec	1062.23 sec
Behzadi & Steyaert(2003)	2.51 sec	1014.32 sec	2.49 sec	1012.38 sec
This paper	2.14 sec	810.93 sec	0.03 sec	32.54 sec

The MSY1 repeats are AT rich (75%-80%) sequences. Five variant repeats designated were identified. Each of these repeats contains 25 bp. Repeat types 1-4 differ at two sites, a C/T transition at position 3, and a C/G transition at position 13. The single type 5 repeat, differs from the type 2 repeat by a transition at position 21.

We used a dataset provided by M. Jobling in which minisattelite maps for 690 Y chromosome from worldwide population samples were determined. The length of each of these 690 sequences is between 48 and 118. We compute the transformation distances between each pair of these strings by the three mentioned algorithms. The running times are given in table 1. The PreProcessing part is executed once for each of these strings and the core algorithm is considered for any pairs of these strings (690×690 pairs). The given times correspond to the total time needed for all these computations. Note that the random sequences have the same lengths as the sequences in the database and the alphabet is the same.

7   Conclusion

We have also considered a model in which we have amplifications (and contractions) of order greater than two. The results can be easily generalized.

As a final remark, let us just point out again, that the minisatellite problem is an instance of the general problem of transforming a chain into another and the framework is now at its maximum generality.

References

1. Apostolico, A., Landau, G.M. and Skiena, S.: Matching for Run Length Encoded Strings. Journal of Complexity, 15, 1, 4-16 (1999).
2. Arbell, O., Landau, G.M., Mitchell, J.S.B: Edit Distance of Run-Length Encoded Strings. Information Processing Letter, 83(6), 307-314, 2002.
3. Bunke, H. and Csirik, J.: An Improved Algorithm for Computing the Edit Distance of Run Length Coded Strings. Information Processing Letters, 54, 93-96 (1995).
4. Behzadi B. and Steyaert J.-M.: An Improved Algorithm for Generalized Comparison of Minisatellites. Proc. of 14th CPM. Lecture Notes in Computer Science (2003).
5. Mäkinen, V., Navarro, G., Ukkonen, E.: Approximate Matching of Run-Length Compressed Strings. Proc. of 12th CPM, Lecutre Notes in Computer Science 2089, Springer-Verlag, 31-49 (2001).
6. Bérard, S., Rivals, E.: Comparison of Minisatellites. Proceedings of the 6th Annual International Conference on Research in Computational Molecular Biology. ACM Press, (2002).
7. Jobling, M.A., Bouzekri, N., Taylor, P.G.: Hypervariable digital DNA codes for human paternal lineages: MVR-PCR at the Y-specific minisatellite, MSY1(DYF155S1). Human Molecular Genetics, Vol. 7,No. 4. (1998)643–653.
8. Bouzekri, N., Taylor, P.G., Hammer M.F, Jobling, M.A.: Novel mutation processes in the evolution of haploid minisatellites, MSY1: array homogenization without homogenization. Human Molecular Genetics,Vol. 7, No. 4. (1998)655–659
9. Jeffreys, A.J., Tamaki, K., Macleod, A., Monckton, D.G., Neil, D.L and Armour, J.A.L: Complex gene conversion events in germline mutation at human minisatellites. Nature Genetics, 6. (1994)136–145.
10. Brión, M., Cao, R., Salas, A., Lareu M.V., Carracedo A.: New Method to Measure Minisatellite Variant Repeat Variation in Population Genetic Studies. American Journal of Human Biology, Vol. 14.(2002) 421–428.
11. Elemento, O., Gascuel, O., Lefranc, M.-P.: Reconstructing the duplication history of tandemly repeated genes. Molecular Biology and Evolution,vol 19(3). (2002) 278–288.
12. Sankoff, D. and Kruskal, J.B: Time Warps, String Edits and Macromolecules: The Theory and Practice of Sequence Comparison. Addison-Wesley. (1983).
13. Crochemore, M., Landau, G. M., Ziv-Ukelson, M.: A sub-quadratic sequence alignment algorithm for unrestricted cost matrices. SODA'2002. ACM-SIAM. (2002)679–688.

A Faster and More Space-Efficient Algorithm for Inferring Arc-Annotations of RNA Sequences Through Alignment

Jesper Jansson[1], See-Kiong Ng[2], Wing-Kin Sung[1], and Hugo Willy[1]

[1] Department of Computer Science, National University of Singapore
3 Science Drive 2, Singapore 117543
{jansson,ksung,hugowill}@comp.nus.edu.sg
[2] Institute for Infocomm Research, 21 Heng Mui Keng Terrace, Singapore
skng@i2r.a-star.edu.sg

Abstract. This paper considers the problem of inferring the optimal nested arc-annotation of a sequence given another nested arc-annotated sequence by maximizing the weighted alignment between the bases and arcs in the two sequences. The problem has a direct application in predicting the secondary structure of an RNA sequence given a closely related sequence whose secondary structure is already known. The currently most efficient algorithm for this problem requires $O(nm^3)$ time and $O(nm^2)$ space where n is the length of the sequence with known arc-annotation while m is the length of the sequence to be inferred. We present an improved algorithm which runs in $\min\{O(nm^2 \log n), O(nm^3)\}$ time and $\min\{O(m^2 + mn), O(m^2 \log n)\}$ space. The time improvement is achieved by applying sparsification to the dynamic programming algorithm, while the space is reduced to a more practical quadratic complexity by using a Hirschberg-like traceback technique together with a simple compression.

1 Introduction

Recent research shows that RNA functions as catalysts and regulators in nucleic acid processing and gene expression in addition to its commonly known intermediary role in DNA transcription and translation process. It is generally known that much of RNA's functionalities depend on its structural features. Unfortunately, although massive amount of sequence data are continuously generated, the number of known RNA structures is still very limited since experimental methods, such as NMR and Crystallography, require expertise and long experimental time. Therefore, computational methods for predicting RNA structure are very useful.

There exist a number of computational approaches to predict the structure of RNA in the literature. Basically, they can be classified into three categories: Energy Minimization, Comparative, and Structure Inferring methods. The first approach tries to compute the structure of an RNA molecule which has the lowest free energy. Representatives of this approach are the methods of Nussinov et al [16] and Zuker et al [15, 20, 21]. Since the current energy model is not

I. Jonassen and J. Kim (Eds.): WABI 2004, LNBI 3240, pp. 302–313, 2004.

accurate enough and RNA may not fold into the lowest energy structure, the prediction accuracy of this method is usually not high. For the Comparative method, we are given a number of RNA sequences which are expected to have similar structure called the homologous sequences. By aligning those RNA sequences, we compute the consensus structure. Representatives of this approach include Maximum Weighted Matching (MWM) [3, 18] and Stochastic Context Free Grammars (SCFGs) [8, 7, 17]. The Comparative approach is currently the best way to predict RNA structures [9, 12]. However, when the number of homologous sequences is not large enough, the accuracy can be low. If we only have a few homologous RNA sequences where the structure of one of the sequences is known, the RNA structure can be predicted using the Structure Inferring method [2, 19]. Consider two sequences S_1 and S_2 of length n and m. Assuming that the secondary structure of S_1 is known, this method infers the secondary structure of S_2 by aligning S_1 and S_2. Bafna et al [2] propose a dynamic programming solution to this problem and solve it using $O(n^2m^2 + nm^3)$ time and $O(n^2m^2)$ space. Zhang [19] improves their result and gives an algorithm which runs in $O(nm^3)$ time and $O(nm^2)$ space. In this paper, we further improve the running time of the inference algorithm to $\min\{O(nm^2 \log n), O(nm^3)\}$ and at the same time bring down the space requirement to $\min\{O(m^2 + mn), O(m^2 \log n)\}$.

Our improvement in the running time stems from sparsification. We observe that the entries in every row and every column in the dynamic programming tables are monotonically increasing, enabling us to calculate less entries in the tables without losing any information. We also designed a new recursive dynamic programming algorithm that gives a better worst-case space requirement in the case of computing only the score of the alignment of S_1 and S_2. Finally, by incorporating the latter into an algorithm similar to Hirschberg's traceback [10] together with a simple compression method, we can recover the optimal inferred structure from the table within the stated reduced space complexity. Note that the space improvement is critical in our application since currently the length of a typical RNA sequence used in lab experiments is around 3K to 5K bases. Assuming that $n \approx m$, the memory requirement of an $O(nm^2)$ space algorithm could easily reach over tens of gigabytes. This memory requirement is not impossible to meet but it is highly impractical.

This paper is organized as follows. Section 2 contains the formal statement of the problem with some basic definitions. Section 3 presents the algorithm and is divided into three parts. The first part presents the original algorithm given in [19], noting the bottleneck of the computation. The following two parts present our techniques to improve the running time of the algorithm. The Hirschberg-like traceback algorithm is described in Section 4. Finally, Section 5 concludes this paper with some possible extensions of the problem.

2   Preliminaries

We use a slightly different notation from the one in [19] where the secondary structure of the first sequence is represented as a tree. Each internal node in the tree represents a base pair and the bases in the loop created by the base pair are the children of the node.

In our algorithm, we represent an RNA sequence and its secondary structure information using the *arc-annotated* sequence [4]. Consider a sequence S over a fixed alphabet $\Sigma = \{A, C, G, U\}$. We define $S[i]$ to be the ith character in S and $S[i..j]$ to be the substring of S in positions between i and j (inclusive). For any $x \in \Sigma$, let $Complement(x)$ be the complementary base of x based on the Watson-Crick base pairing. For example $Complement(A)$ is U and $Complement(G)$ is C. An unordered pair of positions (i, j), where $i < j$, indicates that $S[i]$ and $S[j]$ form a base pair in the RNA structure. Such pair is called an *arc*. For RNA sequences, it is required that $S[j] = Complement(S[i])$ and vice versa. A set P of arcs is called an *arc-annotation*, and the pair (S, P) is called an *arc-annotated* sequence. Arc-annotated sequences are well-studied [1, 4, 6, 11, 13, 14, 19] and are commonly used in computational biology to represent the structure of RNA and protein sequences. Since we are considering RNA secondary structures, we assume that the RNA sequences we are dealing with do not have any pseudoknots. The corresponding arc-annotation construct for such RNA structures is the *nested* arc-annotation [1, 11, 13, 14] where, given two arcs, either one is within the other, or they are completely disjoint ($\forall (i_1, j_1), (i_2, j_2) \in P$, $i_1 \in [i_2, j_2] \Leftrightarrow j_1 \in [i_2, j_2]$). For any arc $u \in P$, we denote u_l and u_r to be the left and the right endpoints of u, respectively. The *size* of an arc u is denoted by $|u| = u_r - u_l + 1$. We say that position i is *free* if i is not an endpoint of any arc in P. A position i is *covered* by an arc u if $u_l < i < u_r$ and there exist no other arc u' such that $u'_l < i < u'_r$. The set of all positions covered by u is called the *arc cover* of u, denoted by $C(u)$.

Given two arc-annotated sequences (S_1, P_1) and (S_2, P_2), we can define the similarity of the sequences by aligning the bases and the arcs in them. We need to define a scoring function for each type of alignment. Let χ be the function to score the alignment of unpaired bases in the two sequences where, for $a, b \in \{A, C, G, U, \sqcup\}$, $\chi(a, b) = \beta$ if $a = b$ and 0 otherwise ('$\sqcup$' denotes a blank character). For any arc u, which represents paired bases in the RNA structure, let δ be a scoring function for arcs alignment whose value is defined as:

$$\delta((S_1[u_l], S_1[u_r]), (S_2[j], S_2[j'])) = \begin{cases} \alpha_1 & \text{if } S_1[u_l] = S_2[j] \text{ and } S_1[u_r] = S_2[j'] \\ \alpha_2 & \text{if } S_1[u_l] \neq S_2[j] \text{ and } S_1[u_r] \neq S_2[j'] \\ & \text{but } S_2[j] = Complement(S_2[j']) \\ -\infty & \text{otherwise} \end{cases}$$

β, α_1, and α_2 are positive integer constants. Usually the parameters are set such that $\beta \leq \alpha_2 \leq \alpha_1$ which reflects that an arc-alignment(α_1 or α_2) takes precedence over single base alignment(β). Moreover, an arc alignment with exactly the same base pairs should score higher (α_1) since both bases and their arc are aligned. One can also have constraints on the arc width. For example, when $|j - j'|$ is less than some minimum arc width parameter, we can define $\delta = -\infty$. Now given the definition of the arc annotation and the scoring functions, we formally state our problem (slightly altered from the one in [19]) as follows.

The Weighted Largest Common Substructure(WLCS) of two arc annotated sequences (S_1, P_1) and (S_2, P_2) is defined as the maximum weighted alignment between S_1 and S_2 where free bases are aligned to free bases and arcs are aligned to arcs. The WLCS score is then defined as the sum of all bases and arcs alignment scores. The problem we address in this paper is: Given a nested arc-annotated sequence (S_1, P_1) and a plain sequence S_2, infer the nested arc-annotation P_2 for S_2 that maximizes their WLCS score.

3   Algorithm Description

This section reviews Zhang's algorithm (presented in [19]) for inferring the RNA secondary structure P_2 for S_2 that maximizes the WLCS score between (S_1, P_1) and (S_2, P_2). Let $|S_1| = n$ and $|S_2| = m$. Let $DP_{(i,i')}[j, j']$, where $1 \leq i \leq i' \leq n$ and $1 \leq j \leq j' \leq m$, denotes the score of the weighted largest common substructure between $(S_1[i..i'], P_1)$ and $S_2[j..j']$. Note that $DP_{(i,i')}[j, j'] = 0$ whenever $i > i'$ or $j > j'$. Zhang presented an algorithm which runs in $O(nm^3)$ time and uses $O(nm^2)$ space based on a two-step dynamic programming. Given an arc u, the first step computes the value of DP for the arc-cover of u i.e. computes $DP_{(u_l+1, u_r-1)}[j, j']$ for all $1 \leq j \leq j' \leq m$. Then, the next step computes the value of DP for the whole arc u, that is $DP_{(u_l, u_r)}[j, j']$ for all $1 \leq j \leq j' \leq m$. Below are the three equations in [19] to compute the two steps in the algorithm. Please refer to the paper for the proofs.

Lemma 1. *(Lemma 4 in [19]) If either i' is free or i' is an endpoint of an arc whose other endpoint is not in $[i..i' - 1]$,*

$$DP_{(i,i')}[j, j'] = \max \begin{cases} DP_{(i,i'-1)}[j, j' - 1] + \chi(S_1[i'], S_2[j']), \\ DP_{(i,i'-1)}[j, j'] + \chi(S_1[i'], \sqcup), \\ DP_{(i,i')}[j, j' - 1] + \chi(\sqcup, S_2[j']) \end{cases}$$

Lemma 2. *(Lemma 5 in [19]) For any arc $u \in P_1$ and $i < u_l$,*

$$DP_{(i,u_r)}[j, j'] = \max_{j-1 \leq j'' \leq j'} \{DP_{(i,u_l-1)}[j, j''] + DP_{(u_l,u_r)}[j'' + 1, j']\}$$

Lemma 3. *(Lemma 3 in [19]) For any arc $u \in P_1$,*

$$DP_{(u_l,u_r)}[j, j'] = \max \begin{cases} DP_{(u_l+1,u_r-1)}[j+1, j' - 1] + \\ \qquad \delta((S_1[u_l], S_1[u_r]), (S_2[j], S_2[j'])), \\ DP_{(u_l+1,u_r-1)}[j, j'], \\ DP_{(u_l,u_r)}[j + 1, j'], \\ DP_{(u_l,u_r)}[j, j' - 1] \end{cases}$$

Definition 1. *If i' is free or i' is a right endpoint of an arc whose left endpoint is not in $[i..i']$, then given the table $DP_{(i,i'-1)}$, $DP_{(i,i')}$ can be computed by using Lemma 1. We define the computation of $DP_{(i,i')}$ from $DP_{(i,i'-1)}$ as the operation $EXTEND(DP_{(i,i'-1)})$.*

$WLCS(S_1, P_1, P_2)$

For every arc $u \in P_1$ from the innermost to the outermost, left to right,

Step 1 : Compute $DP_{(u_l+1, u_r-1)}$ as follows.

 For every $i \in C(u)$ in increasing order,

  – if i is free, compute $DP_{(u_l+1, i)}$ by EXTEND($DP_{(u_l+1, i-1)}$).

  – if $i = v_r$ for some arc v, compute $DP_{(u_l+1, i)}$ by
MERGE($DP_{(u_l+1, v_l-1)}, DP_{(v_l, v_r)}$).

  – if $i = v_l$, do nothing.

Step 2 : Compute $DP_{(u_l, u_r)}$ by ARC-MATCH($DP_{(u_l+1, u_r-1)}$).

Fig. 1. The algorithm from [19] described in terms of EXTEND, MERGE and ARC-MATCH operations.

Definition 2. *Consider any arc s. The operation* $MERGE(DP_{(i,s_l-1)}, DP_{(s_l,s_r)})$ *is defined to be the computation of the table* $DP_{(i,s_r)}$ *given* $DP_{(i,s_l-1)}$ *and* $DP_{(s_l,s_r)}$ *using Lemma 2.*

Definition 3. *Consider any arc s. The operation* $ARC\text{-}MATCH(DP_{(s_l+1,s_r-1)})$ *is defined to be the computation of the table* $DP_{(s_l,s_r)}$ *given* $DP_{(s_l+1,s_r-1)}$ *using Lemma 3.*

Fig. 1 describes the procedure $WLCS(S_1, P_1, S_2)$ which computes $DP_{(1,n)}[j, j']$ for all $1 \le j \le j' \le m$ based on the algorithm in [19]. As analyzed in the latter, EXTEND takes $O(m^2)$ time. There are $O(n)$ free bases in S_1; thus, all calls to EXTEND require a total of $O(nm^2)$ time. The procedure MERGE will need to fill $O(m^2)$ entries in the combined table, each requires $O(m)$ time to compute because we need to find the maximum over $O(m)$ sums, in the worst case. Since MERGE is only invoked on arcs and the number of arcs in P_1 could reach $O(n)$; in total, all calls to MERGE require $O(nm^3)$ time. ARC-MATCH computes the term in Lemma 3 over $O(m^2)$ (j, j') pairs for each arc in P_1. Based on a similar argument on the number of arcs in P_1, ARC-MATCH requires $O(nm^2)$ time. As for the space requirement, assuming the standard traceback for inferring the secondary structure of the sequence S_2, we must store all intermediary DP tables computed by $WLCS(S_1, P_1, S_2)$. The cardinality of the latter is bounded by $O(n)$ as the number of free bases and arcs are both bounded by $O(n)$. In conclusion, the time and space complexity of the whole algorithm is $O(nm^3)$ and $O(nm^2)$, respectively.

3.1 The Sparsification Technique –
Monotonically Increasing Property of DP

The previous section shows that the bottleneck of the computation of the WLCS score is in the procedure MERGE. Here, we describe how to speed up the computation of MERGE by taking advantage of the properties of $DP_{(i,i')}$.

Observation 1 *For any $i \le i'$, $DP_{(i,i')}$ satisfies the following properties.*

1. In every row j of $DP_{(i,i')}$, the entries are monotonically increasing, i.e., $DP_{(i,i')}[j,j'] \leq DP_{(i,i')}[j,j'+1]$.
2. In every column j' of $DP_{(i,i')}$, the entries are monotonically decreasing, i.e., $DP_{(i,i')}[j,j'] \geq DP_{(i,i')}[j+1,j']$.

The observations above motivate the following definitions.

Definition 4. [5] For every row j of $DP_{(i,i')}$, a position j^* satisfying $j \leq j^* \leq m$ is defined to be a row interval point if $DP_{(i,i')}[j,j^*-1] < DP_{(i,i')}[j,j^*]$. The set of row interval points j^* in the jth row of $DP_{(i,i')}$ is denoted by $\mathrm{RowIP}_j(DP_{(i,i')})$.

Definition 5. [5] For every column j of $DP_{(i,i')}$, a position j^* satisfying $1 \leq j^* \leq j$ is defined to be a column interval point if $DP_{(i,i')}[j^*,j] > DP_{(i,i')}[j^*+1,j]$. The set of column interval points j^* in the jth column of $DP_{(i,i')}$ is denoted by $\mathrm{ColIP}_j(DP_{(i,i')})$.

Lemma 4. Let $\alpha = \max\{\beta,\alpha_1,\alpha_2\}$. Then there are at most $(\min\{\alpha(i'-i+1),(m-j+1)\})$ row interval points in any row j of $DP_{(i,i')}$.

Proof. (Sketch) The total number of row interval points (which are all distinct) in any row j of $DP_{(i,i')}$ is bounded by the minimum of the maximum (integer) score and the number of columns in the row. □

Corollary 1. There are at most $(\min\{\alpha(i'-i+1),j'\})$ column interval points in any column j' of $DP_{(i,i')}$.

In [19], for every (j,j') pair where $j \leq j'$, the procedure $\mathrm{MERGE}(DP_{(i,i')}, DP_{(i'+1,i'')})$ tries every possible $j'' \in [j-1..j']$ to compute the one that maximizes

$$DP_{(i,i')}[j,j''] + DP_{(i'+1,i'')}[j''+1,j'] \tag{1}$$

The following lemma states that it is unnecessary to consider all $j'' \in [j-1..j']$ to find the maximum of (1).

Lemma 5. The equation from Lemma 2 can be computed by

$$DP_{(i,u_r)}[j,j'] = \max_{\substack{j^* \in \mathrm{RowIP}_j(DP_{(i,u_l-1)}) \cup \{j-1\} \\ j^* \leq j'}} \{DP_{(i,u_l-1)}[j,j^*] + DP_{(u_l,u_r)}[j^*+1,j']\}$$

which checks at most $(\min\{\alpha(u_l-i)+1,(j'-j+1)\})$ candidates of j^*.

Proof. (Sketch) Let $F(j'') = DP_{(i,i')}[j,j''] + DP_{(i'+1,i'')}[j''+1,j']$. By Lemma 2, $DP_{(i,u_r)}[j,j'] = \max\{F(j-1),\max_{j'' \in [j..j']}F[j'']\}$. For each $j'' \in [j..j']$, we observe that there exists a $j^* \in \mathrm{RowIP}_j(DP_{(i,u_l-1)})$ such that $j^* \leq j''$ and $DP_{(i,u_l-1)}[j,j^*] = DP_{(i,u_l-1)}[j,j'']$. Furthermore, since $j^* \leq j''$, we have $DP_{(u_l,u_r)}[j^*+1,j'] \geq DP_{(u_l,u_r)}[j''+1,j']$. Hence, for such j^* we have $F[j^*] \geq F[j'']$ resulting in $\max_{j'' \in [j..j']}F[j''] = \max_{j^* \in \mathrm{RowIP}_j(DP_{(i,u_l-1)})}F[j^*]$. □

Corollary 2. *The equation from Lemma 2 can be computed by*

$$DP_{(i,u_r)}[j,j'] =$$

$$\max_{\substack{j^*+1\in\text{ColIP}_{j'(DP_{(u_l,u_r)})\cup\{j'+1\}} \\ j^*+1\geq j}} \{DP_{(i,u_l-1)}[j,j^*] + DP_{(u_l,u_r)}[j^*+1,j']\}$$

which checks at most $(\min\{\alpha|u|+1,(j'-j+1)\})$ *candidates of* j^*.

By Lemma 5 and Corollary 2, the time complexity of MERGE($DP_{(i,i'-1)}$, $DP_{(i',i'')}$) is improved to $O(\min\{\alpha(i''-i')m^2, \alpha(i'-i)m^2, m^3\})$.

3.2   The Recursive Dynamic Programming Algorithm

We now introduce a new algorithm $WLCS_r(S_1, P_1, S_2)$ which computes the table $DP_{(1,n)}$ using a carefully designed recursive dynamic programming algorithm. This improved algorithm guarantees that each MERGE operation is applied only on arcs whose size is at most half of its parent's[1].

Let us start with some definitions. The followings are with respect to a *nested* annotated structure. An arc u is a *parent* of an arc v (denoted by $Parent(v)$) if $u_l < v_l < v_r < u_r$ and there is no arc w such that $u_l < w_l < v_l < v_r < w_r < u_r$. Conversely, v is referred as the *child* of the arc u. The set of children of an arc u is denoted by $Child(u)$. A *core-arc*, with respect to an arc u, is a child of u which has the biggest size (denoted as core-arc(u)). All other children of u are named *side-arcs* and form the set *side-arcs*(u). A *terminal-arc* is defined to be an arc which has no child. For any arc $u \in P_1$, the *core-path* $CP(u)$ is an ordered set of core-arcs $\{c_1, c_2, \cdots, c_\ell\}$, where $c_1 = u$ and for any c_i, c_{i+1} is core-arc(c_i)

$WLCS_r(S_1, S_2)$ first finds the largest arc u in $[1..n]$ and processes every core-arc $c \in CP(u)$ from the innermost to the outermost. For terminal arcs t, $DP_{(t_l,t_r)}$ can be computed by using EXTEND operations only. For the remaining arcs c, $DP_{(c_l,c_r)}$ is obtained using a *two-part computation*. Let c' be core-arc(c). Due to the bottom-up ordering, $DP_{(c'_l,c'_r)}$ will have been computed at this point of time. We first compute the value of $DP_{(c_l+1,c'_l-1)}$ (the *LEFT Part* phase) using EXTEND and MERGE operations. Given $DP_{(c_l+1,c'_l-1)}$, we proceed using EXTEND and MERGE to compute $DP_{(c'_l,c_r-1)}$(the *RIGHT Part* phase). In both phases, whenever we encounter a side-arc s, we first compute $DP_{(s_l,s_r)}$ by recursively calling $WLCS_r(S_1[s_l..s_r], S_2)$. Next, we apply MERGE on $DP_{(c_l+1,c'_l-1)}$ and $DP_{(c'_l,c_r-1)}$ to compute $DP_{(c_l+1,c_r-1)}$. Finally, $DP_{(c_l,c_r)}$ is obtained by applying ARC-MATCH($DP_{(c_l+1,c_r-1)}$). If $(1,n) \in P_1$, then $u = (1,n)$ and we are done. Otherwise, we need to compute $DP_{(1,n)}$ using the same two-part

---

[1] The routine $WLCS(S_1, P_1, P_2)$ given in [19] computes the DP tables according to the postorder of the nodes in their tree representation. The problem of this approach is that we may need to perform MERGE on arcs with large sizes causing an $\Omega(nm^2)$ space requirement even if we only wish to compute the WLCS score of (S_1, P_1) and S_2. We shall prove this claim in the full version of this paper.

WLCS$_r(S_1, P_1, S_2)$ /* $|S_1| = n, |S_2| = m$ */

- Let u be biggest arc in P_1 and $CoreArcs = CP(u)$.
- Let $DP_{(i,i')}$ be the $m \times m$ score matrix of $DP_{(i,i')}[j, j']$, $1 \leq j \leq j' \leq m$.
- For the terminal-arc $t \in CoreArcs$,
    - For $k = t_l + 1$ to $t_r - 1$
        - ⋄ Compute $DP_{(t_l+1,k)}$ using EXTEND($DP_{(t_l+1,k-1)}$).
    - Compute $DP_{(t_l,t_r)}$ by ARC-MATCH($DP_{(t_l+1,t_r-1)}$).
- For every core-arc $c \in CoreArcs$ in bottom-up order, $c' = $ core-arc(c)
    - **LEFT Part**
      For $k = c_l + 1$ to $c'_l - 1$ where $k \in C(c)$,
        - ○ If k is free,
            - ⋄ Compute $DP_{(c_l+1,k)}$ using EXTEND($DP_{(c_l+1,k-1)}$).
        - ○ If $k = s_r$ for some $s \in$ side-arcs(c),
            - ⋄ Compute $DP_{(s_l,s_r)}$ recursively by $WLCS_r(S_1[s_l..s_r], P_1, S_2)$.
            - ⋄ Compute $DP_{(c_l+1,k)}$ using MERGE($DP_{(c_l+1,s_l-1)}, DP_{(s_l,s_r)}$).
    - **RIGHT Part**
      For $k = c'_r + 1$ to $c_r - 1$ where $k \in C(c)$,
        - ○ If k is free,
            - ⋄ Compute $DP_{(c'_l,k)}$ using EXTEND($DP_{(c'_l,k-1)}$).
        - ○ If $k = s_r$ for some $s \in$ side-arcs(c),
            - ⋄ Compute $DP_{(s_l,s_r)}$ recursively by $WLCS_r(S_1[s_l..s_r], P_1, S_2)$.
            - ⋄ Compute $DP_{(c'_l,k)}$ using MERGE($DP_{(c'_l,s_l-1)}, DP_{(s_l,s_r)}$).
    - Compute $DP_{(c_l+1,c_r-1)}$ by MERGE($DP_{(c_l+1,c'_l-1)}, DP_{(c'_l,c_r-1)}$).
    - Compute $DP_{(c_l,c_r)}$ using ARC-MATCH($DP_{(c_l+1,c_r-1)}$).
- If $u \neq (1, n)$
    - Compute $DP_{(1,u_l-1)}$ by the **LEFT Part** computation.
    - Compute $DP_{(u_l,n)}$ by the **RIGHT Part** computation.
    - Compute $DP_{(1,n)}$ by MERGE($DP_{(1,u_l-1)}, DP_{(u_l,n)}$).

Fig. 2. The algorithm $WLCS_r(S_1, P_1, S_2)$.

computation technique: first compute $DP_{(1,u_l-1)}$, followed by $DP_{(u_l,n)}$, and then obtain $DP_{(1,n)}$ by MERGE($DP_{(1,u_l-1)}, DP_{(u_l,n)}$). Our complete algorithm $WLCS_r(S_1, P_1, S_2)$ is listed in Fig. 2.

Lemma 6. $WLCS_r(S_1, P_1, S_2)$ *runs in* $\min\{O(\alpha nm^2 \log n), O(nm^3)\}$ *time.*

Proof. To obtain the execution time of $WLCS_r(S_1, P_1, S_2)$, we analyze the total execution time of the EXTEND, MERGE, and ARC-MATCH operations separately. The time required by EXTEND and ARC-MATCH operations is still the same as in [19], namely $O(nm^2)$, as they are still applied at most once on every free bases and arcs, respectively. Note that MERGE is now invoked on all arcs which belong to the set side-arc(u) for some arc $u \in P_1$ and on the merging of the LEFT part and the RIGHT part of all non-terminal arcs. For any side-arc s, merging the table $DP_{(s_l,s_r)}$ into some table $DP_{(i,s_l-1)}$ takes at most $O(\min\{\alpha|s|m^2, \alpha(s_l - i)m^2, m^3\})$ which is at most $O(\min\{\alpha|s|m^2, m^3\})$ time. In the second type of MERGE invocations, we execute MERGE($DP_{(c_l+1,c'_l-1)}$,

$DP_{(c'_l, c_r - 1)}$) for all non-terminal arcs c where c' =core-arc(c). The latter requires $O(\min\{\alpha(c'_l - c_l)m^2, \alpha(c_r - c'_l)m^2, m^3\}) \leq O(\min\{\alpha(c'_l - c_l)m^2, m^3\})$. Let r be an imaginary arc where $r = (0, n+1)$ and $T(S_1)$ be the total execution time for all MERGE operations in $WLCS_r(S_1, P_1, S_2)$.

$$T(S_1) = \sum_{c \in CP(r)} \left(\sum_{s \in \text{side-arcs}(c)} T(S_1[s_l..s_r]) + O(\min\{\alpha|s|m^2, m^3\}) \right) +$$

$$\sum_{c \in CP(r)} O(\min\{\alpha(c'_l - c_l)m^2, m^3\}) \tag{2}$$

$$= \sum_{\substack{s \in \text{side-arcs}(c) \\ c \in CP(r)}} T(S_1[s_l..s_r]) + \sum_{\substack{s \in \text{side-arcs}(c) \\ c \in CP(r)}} O(\min\{\alpha|s|m^2, m^3\}) +$$

$$\sum_{c \in CP(r)} O(\min\{\alpha(c'_l - c_l)m^2, m^3\}) \tag{3}$$

$$= \sum_{\substack{s \in \text{side-arcs}(c) \\ c \in CP(r)}} T(S_1[s_l..s_r]) + O(\min\{\alpha n m^2, m^3\}) \tag{4}$$

Both the second and the third summation terms in (3) sum up to $O(\min\{\alpha n m^2, m^3\})$ since all side-arcs $s \in \text{side-arcs}(c)$ as well as the ranges $[c'_l..c_l]$ for $c \in CP(r)$ are non overlapping. Based on the fact that $\sum |s| \leq |c|$ and $|s| \leq \frac{|c|}{2}$, by inspection, the solution of the recurrence is $O(\alpha n m^2 \log n)$ if $\min\{\alpha n m^2, m^3\} = \alpha n m^2$ or $O(n m^3)$ otherwise. Combining the running time of the three operations, the lemma follows. $\qquad \square$

4   Traceback Using a Hirschberg-Based Technique

Using the standard traceback algorithm, one is required to store all DP tables corresponding to any arc $u \in P_1$. Alternatively, we can make use of the recursive technique introduced by Hirschberg in [10] and use $WLCS_r(S_1, P_1, S_2)$ only to compute the WLCS score. We shall refer to the latter as the *score-only* $WLCS_r(S_1, P_1, S_2)$.

Lemma 7. *Computing the score-only* $WLCS_r(S_1, P_1, S_2)$ *requires* $\min\{O(m^2 \log n), O(m^2 + \alpha m n)\}$ *space.*

Proof. To compute the score-only $WLCS_r(S_1, P_1, S_2)$, since we do not have to traceback, we can just store the information needed to compute the alignment score. This corresponds to $O(m^2)$ space for the EXTEND and ARC-MATCH operations. As for the MERGE operations, when there is no recursive call involved (the second type of MERGE), the space requirement is also in $O(m^2)$. Otherwise, referring back to Fig. 2, when we invoke the recursive call $WLCS_r(S_1[s_l..s_r], P_1, S_2)$, we observe that we need to store $DP_{(i, s_l - 1)}$ for some fixed i. Storing only the row interval points takes $O(\min\{\alpha(s_l - i)m, m^2\})$ space (by Lemma 4). Since recursive calls are only applied on side-arcs, we have at

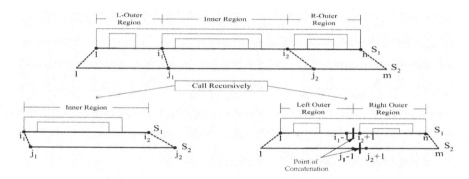

Fig. 3. The recursion on the partitioned continuous region by Lemma 8.

most $O(\log n)$ recursion levels giving an upper bound of $O(m^2 \log n)$ on the space complexity. We further claim that the space required is smaller than $O(\alpha nm)$ since, in each recursion level x, we only store $DP_{(i_x, s_{l_x}-1)}$ where all of the intervals $[i_x..s_{l_x}-1]$ are disjoint. Hence, $\sum_x O(\alpha(s_{l_x} - i_x)m) \leq O(\alpha nm)$. Combining the three terms, the lemma follows. $\square$

Following the idea of Hirschberg in [10], we compute the WLCS alignment between (S_1, P_1) and S_2 as follows,

1. Divide S_1 into a constant number of non-overlapping regions $S_{11}, S_{12}, ..S_{1c}$.
2. For each region S_{1i}, find the region S_{2i} in S_2 such that the optimal WLCS alignment will align S_{1i} to S_{2i}.
3. Recursively compute the optimal WLCS alignments between S_{1i} and S_{2i} for $i = 1, 2, .., c$.

To do the first step, since S_1 is arc-annotated, we must divide S_1 in such a way that we do not break any arc in P_1. The solution is to divide S_1 into *inner* and *outer* regions. Given two points i_1 and i_2, $1 \leq i_1 \leq i_2 \leq n$, the *inner* region with respect to i_1 and i_2 is $S_1[i_1..i_2]$ and the *outer* region is the concatenation of $S_1[1..(i_1 - 1)]$ and $S_1[(i_2 + 1)..n]$ (see Fig. 3). The latter is also referred as a *gapped* region since it has a discontinuous interval ($S_1[i_1..i_2]$ is removed). Let $\star$ be a special character that denotes the gap in the sequence such that the gapped region can be written as $S_1[1..(i_1 - 1)] \star S_1[(i_2 + 1)..n]$. If a region has no gap in it, we say it is *continuous*. We shall show that we can bound the size of each region by ϕn for some constant ϕ, $0 < \phi < 1$. Due to space constraints, the proofs of the following lemmas will appear in the journal version of this paper.

Lemma 8. *We can always partition a continuous region into 2 non-overlapping subregions, where one of them is continuous and the other is gapped. Every subregion's size is at most $\frac{2}{3}$ of the original region.*

Lemma 9. *We can always partition a gapped region into at most 4 non-overlapping subregions, where at most one of them is continuous. Every subregion's size is at most $\frac{2}{3}$ of the original region.*

After dividing S_1 into at most 4 subregions, where each is denoted by S_{1i} for $i \leq 4$, we now need to compute the regions S_{2i} in S_2 to which the subregions S_{1i} is aligned by the optimal WLCS alignment.

Lemma 10. *For any $1 \leq i_1 \leq i_2 \leq n$, we can compute $1 \leq j_1 \leq j_2 \leq m$, such that the optimal alignment between (S_1, P_1) and S_2 aligns $S_1[i_1..i_2]$ to $S_2[j_1..j_2]$, within the same time and space complexity of the score-only $WLCS_r(S_1, P_1, S_2)$.*

Lemma 11. *For any $1 \leq i_1 \leq i_2 \leq i_3 \leq i_4 \leq n$, we can compute $1 \leq j_1 \leq j_2 \leq j_3 \leq j_4 \leq m$, such that the optimal alignment between (S_1, P_1) and S_2 aligns $S_1[i_1..i_2] \star S_1[i_3..i_4]$ to $S_2[j_1..j_2] \star S_2[j_3..j_4]$, within the same time and space complexity of the score-only $WLCS_r(S_1, P_1, S_2)$.*

By Lemmas 10 and 11, the second step of this algorithm can be executed within the same time and space complexity of the score-only $WLCS_r(S_1, P_1, S_2)$ since the number of the subregions is constant. Next, we proceed by applying the algorithm recursively on each pair (S_{1i}, S_{2i}). While applying the algorithm on the continuous region is straightforward, the gapped region needs a bit of extra care. In this case, $\star$ in S_{1i} must be aligned to $\star$ in S_{2i} because they represent the subregion pair(s) computed in the other recursive call(s). To implement such constraint, we add into the base scoring function the following cases: $\chi(\star, \star) = 0$ and $\chi(\star, x) = \chi(x, \star) = -\infty$ for $x \in \{A, C, G, U, \sqcup\}$. This way, the optimal alignment between the two sequences is forced to align $\star$ in the first sequence to $\star$ in the second in order to have a non-negative score.

Lemma 12. *Our new algorithm can recover the optimal WLCS alignment in $\min\{O(\alpha nm^2 \log n), O(nm^3)\}$ time and $\min\{O(m^2 \log n), O(m^2 + \alpha mn)\}$ space.*

5   Concluding Remarks

Consider two homologous RNA sequences S_1 and S_2 where S_1 has a known structure. This paper presents an improved algorithm to solve the problem of inferring the structure of S_2 such that the WLCS score between the two structures are maximized. The same algorithm can easily be applied to *the longest arc-preserving common subsequence problem* (LAPCS) (see, e.g., [4, 11]). In particular, we improve the time and space complexity of LAPCS (nested, plain) problem from $O(nm^3)$ and $O(nm^2)$[11] to $\min\{O(nm^2 \log n), O(nm^3)\}$ and $\min\{O(m^2 + mn), O(m^2 \log n)\}$.

One interesting extension of the problem discussed in this paper is to incorporate a more realistic, non-linear scoring function into the base and arc matching function. Another possible direction is to attempt some special cases of *crossed* arc-annotation structures, which can represent pseudoknotted structures in RNA sequences, by applying the algorithm iteratively.

References

1. J. Alber, J. Gramm, J. Guo, and R. Niedermeier. Towards optimally solving the longest common subsequence problem for sequences with nested arc annotations in linear time. In *CPM*, pages 99–114, 2002.

2. V. Bafna, S. Muthukrishnan, and R. Ravi. Computing similarity between RNA strings. In *CPM*, volume 937, pages 1–16, 1995.
3. R.B. Carey and G.D. Stormo. Graph-theoretic approach to RNA modeling using comparative data. In *ISMB*, pages 75–80, 1995.
4. P. A. Evans. *Algorithms and Complexity for Annotated Sequence Analysis*. PhD Thesis, University of Victoria, 1999.
5. W. Fu, W. K. Hon, and W. K. Sung. On all-substrings alignment problems. In *COCOON*, volume 2697, pages 80–89, 2003.
6. J. Gramm, J. Guo, and R. Niedermeier. Pattern matching for arc-annotated sequences. In *FSTTCS*, volume 2556, pages 182–193, 2002.
7. L. Grate, M. Herbster, R. Hughey, I. S. Mian, H. Noller, and D. Haussler. RNA modeling using Gibbs sampling and stochastic context free grammars. In *ISMB*, pages 138–146, 1994.
8. L. Grate. Automatic RNA secondary structure determination with stochastic context-free grammars. In *ISMB*, pages 136–144, 1995.
9. R.R. Gutell, N. Larsen, and C.R. Woese. Lessons from an evolving rRNA: 16S and 23S rRNA structures from a comparative perspective. *Microbiological Reviews*, 58(1):10–26, 1994.
10. D. S. Hirschberg. Algorithms for the longest common subsequence problem. *J. Association of Computing Machinery*, 24(4):664–675, 1977.
11. T. Jiang, G. H. Lin, B. Ma, and K. Zhang. The longest common subsequence problem for arc-annotated sequences. In *CPM*, volume 1848, pages 154–165, 2000. To appear in Journal of Discrete Algorithms.
12. D.A.M. Konings and R.R. Gutell. A comparison of thermodynamic foldings with comparatively derived structures of 16s and 16s-like rRNAs. *RNA*, 1:559–574, 1995.
13. G. H. Lin, Z. Z. Chen, T. Jiang, and J. Wen. The longest common subsequence problem for sequences with nested arc annotation. *Journal of Computer and System Sciences*, 65:465–480, 2002.
14. G. H. Lin, B. Ma, and K. Zhang. Edit distance between two RNA structures. In *RECOMB*, pages 211–200, 2001.
15. R. B. Lyngsø, M. Zuker, and C.N.S. Pedersen. Internal loops in RNA secondary structure prediction. In *RECOMB*, pages 260–267, 1999.
16. R. Nussinov and A.B. Jacobson. Fast algorithm for predicting the secondary structure of single stranded RNA. In *PNAS*, volume 77(11), pages 6309–6313, 1980.
17. Y. Sakakibara, M. Brown, R. Hughey, I.S. Mian, K. Sjölander, R.C. Underwood, and D. Haussler. Recent methods for RNA modeling using stochastic context-free grammars. In *Proc. of the Asilomar Conference on Combinatorial Pattern Matching*, 1994.
18. J.E. Tabaska, H.N. Gabow R.B. Cary, and G.D. Stormo. An RNA folding method capable of identifying pseudoknots and base triples. *Bioinformatics*, 14(8):691–699, 1998.
19. K. Zhang. Computing similarity between RNA secondary structures. In *IEEE International Joint Symposia on Intelligence and Systems*, pages 126–132, 1998.
20. M. Zuker. Prediction of RNA secondary structure by energy minimization. In *Methods in Molecular Biology*, volume 25, pages 267–94, 1994.
21. M. Zuker and P. Stiegler. Optimal computer folding of large RNA sequences using thermodynamics and auxiliary information. *Nucleic Acid Res. 9*, pages 133–148, 1981.

New Algorithms for Multiple DNA Sequence Alignment

Daniel G. Brown and Alexander K. Hudek*

School of Computer Science, University of Waterloo
Waterloo ON N2L 3G1 Canada
{browndg,akhudek}@uwaterloo.ca

Abstract. We present a mathematical framework for anchoring in global multiple alignment. Our framework uses anchors that are hits to spaced seeds and identifies anchors progressively, using a phylogenetic tree. We compute anchors in the tree starting at the root and going to the leaves, and from the leaves going up. In both cases, we compute thresholds for anchors to minimize errors. One innovative aspect of our approach is the approximate inference of ancestral sequences with accomodation for ambiguity. This, combined with proper scoring techniques and seeding, lets us pick many anchors in homologous positions as we align up a phylogenetic tree, minimizing total work. Our algorithm is reasonably successful in simulations, is comparable to existing software in terms of accuracy and substantially more efficient.

1 Introduction

We present new techniques for anchoring global multiple alignments. A novel idea we discuss is the use of a mathematical model to adjust the parameters of anchor finding algorithms. The goal is to obtain as many correct anchors as possible while avoiding choosing any incorrect anchors.

We apply this idea to two different anchor choosing strategies. The first is picking anchors for multiple global alignment at each node in a guiding phylogenetic tree, starting at the root, and working to the leaves. Using this, we anchor the alignment at biologically relevant positions found in some sequences but not all, guiding us towards a good and efficient multiple alignment.

We support this by changing the requirements on anchors at nodes of the guide tree, using the tree structure and lengths of the regions being aligned. We give a mathematical formulation to produce good choices for the size and rarity of these anchors so few false ones are chosen, and we still obtain efficient run time. To increase the prevalence of good anchors, we use spaced seeds [17]. These have advantages in many contexts [17, 16, 4, 5], but we focus on their most important: they find more, and more independent anchors. Thus, we divide the initial sequences into smaller bits to be aligned in lower nodes of the tree. This strategy reduces run time while keeping high quality alignment.

---

* Work funded by NSERC and the Human Frontier Science Program.

I. Jonassen and J. Kim (Eds.): WABI 2004, LNBI 3240, pp. 314–325, 2004.
© Springer-Verlag Berlin Heidelberg 2004

Second, we emphasize progressive anchoring of alignments based on existing alignments. Our goal is to use strategies for pairwise alignment in this process, approximately inferring the ancestral sequences of alignments being aligned. Rather than trying to do this exactly, we allow uncertainty in the ancestral sequence, using a broader alphabet allowing ambiguity. We show how to pick good anchors in these sequences, with a log-odds framework for scoring sequence positions. Then, we give a seeding strategy to pick anchors in these alignments. The regions in between these anchors can be aligned using any algorithm that aligns alignments; we use a recently developed heuristic of Ma, Wang and Zhang [18].

Finally, we present experiments that show our method effectively aligns simulated sequences subject to much mutation and insertion. Our algorithms rarely err in choosing anchors to align these sequences, which confirms our approach. The heuristic for aligning alignments that we use makes occasional errors.

Our goal throughout is to give a more mathematical framework in which more robust models of both sequence and its evolution can be incorporated into designing effective anchoring strategies. Our current work primarily discusses simple evolutionary models, but can be expanded to this more robust domain.

2   Related Work

Historically, multiple alignment algorithms have been applied to protein sequences. Typical algorithms have slow run time that grows rapidly with sequence length. As such they are inappropriate for large DNA sequence comparison.

Instead, faster heuristics are needed. This problem has become a recent focus of research. New aligners take advantage of two things: the evolutionary tree for the sequences, and conserved regions that can anchor the alignment. For example, MLAGAN [7] anchors alignments with local alignments discovered by the program CHAOS [8] and builds alignments of alignments as it works its way up the tree. MGA [13] anchors alignments based on maximum exact matches found in all sequences. DIALIGN [19] builds multiple alignments by assembling exact matches found between the sequences. It has been recently modified to use CHAOS alignments as anchors [6]. A recent approach [21] adapts the Eulerian path problem to this domain, but requires closely related sequences.

A common feature of many programs is their reliance on anchors in alignment. This shrinks the sequences to align, reducing run times. The types of anchors they use vary quite a bit. Several programs [13, 19, 21] anchor with consecutive sequence matches, while others use local alignments [6, 7].

Choosing anchors is a basic task of many global multiple aligners. Another is aligning two existing alignments, which is NP-hard [18] with affine gap penalties and the standard sum-of-pairs scoring function. Several heuristics for this problem exist [7, 18, 14]. We use a recent heuristic due to Ma, Wang, and Zhang[18] which appears promising, but we note that this is an area needing substantial improvement. Our work focuses on the anchoring process.

Finally, we note that independent of our work, two recent programs have appeared [1, 2] which also address this problem.

3   Notation and Definitions

Our sequences are over the the 4-letter DNA alphabet $\Sigma = \{A, C, G, T\}$. We denote the jth character of sequence S_i by $S_i[j]$.

Alignments. We represent alignments by inserting a special character $-$ into the sequences. The resulting alignment $A = (A_1, \ldots, A_m)$ has all strings of equal length. The jth column of the alignment A is all symbols found in position j of the strings A_i; these symbols are aligned to each other.

To score an alignment, we use the sum-of-pairs score. Given an alignment A, its score is the sum of the scores of the pairwise alignments induced by A on all pairs of sequences, where we remove columns in which both sequences have a gap. This scoring is used extensively in practice [9] and can be augmented by placing weights on sequences [20], which we discuss in Section 7.

Seeds, anchors and fragments. Our anchors are hits to long spaced seeds [17]. These require matches in specific positions in the sequences. A seed is a binary vector q; its number of 1s is its *weight*. Informally, positions in q with value 1 must match in all sequences; the others are "don't care" positions. Formally, we say that q *hits* at $((S_1, p_1) \ldots, (S_k, p_k))$, where the S_i are sequences and p_i are position offsets, when $S_1[p_1 + c - 1] = S_2[p_2 + c - 1] = \ldots = S_k[p_k + c - 1]$ for all positions c where $q[c] = 1$.

An *anchor* is a restriction on the chosen alignment. It requires the alignment of certain positions in certain sequences. An anchor is a set of k triples (S_i, p_i, ℓ_i) of sequences, start positions and anchor lengths. Anchors restrict the alignment, requiring that regions between anchors should be homologous. This shrinks the problems to solve, and gives anchor-based alignment its usefulness.

Our algorithms for finding anchors generate many we will not use. A *fragment* is a possible anchor, discovered by a *fragment detection* algorithm; an *anchor choosing algorithm* chooses a set of anchors from a set of fragments. The anchors chosen must not conflict; that is, if $i < j$, and anchors α_i and α_j both refer to positions in sequence S_k, then if $(k, p_i, \ell_i) \in \alpha_i$ and $(k, p_j, \ell_j) \in \alpha_j$, then $p_i + \ell_i$ must be less than or equal to p_j. (The beginning of the second anchor in sequence k must follow the end of the first.) The most common anchor choosing algorithms choose the largest or highest-scoring set of non-conflicting fragments (*e.g.*, [13]); we use this strategy here as well.

4   Anchors Down the Tree

Our anchor-based strategy uses a single phylogenetic tree. With a simple model of evolution, we compute match thresholds for good anchors as we go down the tree. Few anchors are chosen at the upper levels of most trees, as there is enough mutation to prevent them. However, as we go down the tree, more anchors are found. This *progressive anchoring* reduces the workload in the next phase, as we build alignments up the tree.

4.1   Choosing Anchors at the Nodes Going Down

The anchors we produce going down the tree are hits to a good spaced seed, 1110010110111. This seed has the highest probability of all weight-9 seeds with

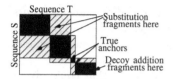

Fig. 1. Two types of bad fragments: Decoy addition fragments are fragments found between two consecutive good fragments, while substitution fragments can be replace a single good fragment.

length at most 13 of hitting local alignments across a wide range of sequence identities and lengths. We cut this seed off if we need a seed with weight below 9, or repeat it if we need higher weight.

Our primary focus in this section is on choosing this weight. If we choose a low weight seed, many chosen anchors will be false. If, instead, we choose too high weight, so few anchors will be found that it becomes useless to anchor at all. We first introduce good fragments, then a notion of alignment work, and then introduce two kinds of bad anchors, and attempt to minimize them.

Good fragments. At any node in the tree, a good fragment is a set of homologous positions in all sequences below that node. Any fragment detection algorithm will find a set of good fragments; let the longest such set of good fragments be that fragment detection algorithm's good anchor list. Our fragment detection algorithm finds all hits to a spaced seed.

The distance between homologous hits in all sequences is random and can be estimated by extending the algorithm of Keich *et al.*[15] to the context of estimating the distance between seed hits. We compute the probability p that all leaves below the current node have the same base at a homologous position. The distribution of the length x between good fragments is the same as for seed hits in a sequence of independent Bernoulli variables with hit probability p.

We approximate work remaining after anchor choice by the total m-dimensional volume of blocks between anchors. Assuming only good fragments are chosen, for sequences of length n, this is approximately $np^w E[x^m]$. Since x is approximately geometrically distributed with mean $1/p^w$; this volume is approximated [11]: it is $np^w[m!(p^{-w})^m] = nm!p^{w(1-m)}$. For a given weight w, we can approximate the extent to which anchoring with seeds of that weight will help.

Bad fragments. Our fragment-detection algorithm will also generate bad fragments. Given the set of possible fragments, and an anchor choosing strategy, we must also control the number of bad anchors.

Bad fragments are among not entirely homologous positions. We consider two types of bad fragments. A *decoy addition fragment* is a fragment which could be inserted between two consecutive good fragments. *Substitution fragments* are bad fragments that can replace a good fragment. An anchor choosing algorithm that simply picks the longest sequence of non-conflicting fragments as anchors, may choose either of these types of bad fragments, thus ruining the alignment in their local area. Figure 1 shows these kinds of bad fragments. We choose a seed whose weight will restrict the number of such bad fragments to be chosen as anchors. If the expected number is close to zero, the expected number of bad anchors of all types is, as well.

Numbers of bad anchors. Suppose that two good fragments with one good fragment between them occur distance ℓ apart. Suppose that the good fragments occur at positions 0 and ℓ in all m sequences. Then any possible fragment occurs at a position in $[0, \ell]^m$. For simplicity, we assume a gapless alignment, so positions with equal coordinate in two sequences are homologous. (This assumption can be removed; we choose this simpler case here.)

There are $\ell^m - \ell$ possible choices of coordinates in $[0, \ell]^m$ that could give a decoy addition or substitution fragment. In the $\ell(\ell - 1) \ldots (\ell - m + 1)$ positions where all coordinates are different, it is easy to compute the probability of a hit to a spaced seed of weight w when the sequences are random: it is $4^{-w(m-1)}$.

However, for many possible bad fragments, some coordinates are equal. These possible bad fragments are much more likely because a match between homologous positions is much more likely than between unrelated ones. For example, the possible bad fragment at positions $(120, 120, 180)$ is much more likely than at $(100, 28, 126)$, since the probability of a match in the homologous positions is greater than 4^{-w}. Thus, we must account for the number of equal coordinates in positions that could give bad fragments.

Let the number of positions with k distinct coordinates in $[0, \ell]^m$ be $\alpha_{k,\ell,m}$. We approximate the probability of a bad fragment in those positions as $4^{-w(k-1)}$, by assuming that homologous positions do match, and bound the expected number of bad fragments by $\sum_{i=1\ldots m} \alpha_{i,\ell,m} \cdot 4^{-w(k-1)}$.

The value of $\alpha_{i,\ell,m}$ is $\alpha_{i,\ell,m} = \binom{\ell}{i} [i^m - \sum_{j<i} \binom{i}{j} \alpha_{j,j,w}]$. Since ℓ is the sum of two geometrically distributed variables, we can use the approximation for $E[\ell^i]$ to approximate the number of bad anchors found between good fragments two apart. Any such bad fragment has a high chance of being chosen by our anchor-choosing algorithm.

Thus, for a given alignment length n, and a given expected distance ℓ between consecutive anchors, we can bound above the expected number of positions that give rise to potential decoy or substitution anchors.

Each weight w gives rise to a different value of ℓ, and to a different expected number of bad fragments. Using our approximation, we choose a seed weight w, differently each time we generate fragments, so that the expected number of bad anchors chosen will be well below 1. In many cases, we cannot choose such an anchor weight. Then, we choose no anchors, and recurse to lower levels of the tree. This happens often in large trees, or in trees with much mutation.

5   Anchors Up the Tree

Our strategy of dividing sequences into homologous segments going down the tree will reach the leaves. We have identified putatively homologous regions, and now align these sequences, working from the leaves to the root.

The basic operation is alignment of alignments, multiple or pairwise. As we go up the tree, the multiple alignments that we are aligning may be long. Optimal alignment of multiple alignments is NP-hard [18], so we want to align short segments. We again anchor: we identify homologous points in the alignments and align the alignments between these anchors. To compute the fragments, we estimate a consensus sequence for each alignment, and then use vector seeds [3] to identify possible fragments.

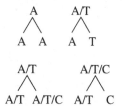

Fig. 2. Consensus symbols. We approximately infer the ancestral sequence at the nodes whose alignments are being aligned. For identical letters, the ancestral inference is that letter. When two ambiguous symbols have base[s] in common, their consensus is that base[s], while if the two symbols disagree entirely, their consensus is their union. This process is repeated going up the phylogenetic tree to the root.

Consensus sequences to obtain good anchors. We approximate the ancestral sequences at the nodes whose alignments are being aligned. We thus find anchors in only *two* sequences. This accommodates our biological intuition of evolution and its representation in a mulitple alignment.

We infer consensus sequences of the alignments being aligned. Our approach allows each position in the sequence to be a non-empty subset of the DNA alphabet, rather than a unique character. Our approach is essentially identical to that of Fitch [12] for computing most parsimonious sequences in a phylogenetic tree, except that we allow the constructed ancestral sequence to keep its ambiguity, and that some positions may be gaps, which are ignored. The ancestral sequence is thus over the 15-letter IUPAC extended DNA alphabet. See Figure 2 for examples of this process.

Probabilities in consensus sequences. For a given tree, we now compute the distribution of consensus symbols found at each node of the tree by our algorithm. We then compute the distribution of pairs of consensus symbols found at homologous positions in the two children of any internal node. This allows us to score matches with a log-odds framework. We assume simple Jukes/Cantor-style evolution, where every edge e has probability p_e of mutation, with probability $p_e/3$ of having mutated to any other base.

Let $C_z(c, b)$ be the probability that a position in the consensus sequence at a given node z in the tree is the potentially ambiguous symbol c, given that the true ancestral sequence at that site is the DNA base b. We seek these values for all c and b at all nodes z of the tree.

Theorem 1. *Given a tree on m taxa, the values of $C_z(c, b)$ for all choices of z, c and b can be computed in $O(m)$ time.*

Proof. The algorithm is tree-based dynamic programming. At the leaves of the tree, the sequence is known. The consensus matches the true base, so $C_z(b, b) = 1$ for all DNA bases b, while for all symbols $c \neq b$, $C_z(c, b) = 0$. Let $T_x(b)$ be the event that the true ancestral base at x is b.

Now, consider nodes x and y that are both children of node z. Suppose we know $C_x(c_x, b_x)$ and $C_y(c_y, b_y)$ for all values of c_x, c_y, b_x and b_y. We are computing $C_z(c, b)$. Let $\mathcal{C}_c$ be all pairs (c_x, c_y) whose consensus is the symbol c. We want to add up the probabilities of all entries in $\mathcal{C}_c$ for each choice of b_x and b_y, multiplied by the conditional probabilities of b_x and b_y given that the true value at z is b. This gives the following formula:

$$C_z(c,b) = \sum_{\substack{b_x,b_y, \\ (c_x,c_y)\in \mathcal{C}_c}} [\Pr[T_x = b_x | T_z = b] \Pr[T_y = b_y | T_z = b] C_x(c_x,b_x) C_y(c_y,b_y)].$$

The probabilities can be found in the tree. These sums are over a constant number of terms and can be computed in constant time at each node.   □

With these probabilities, we can compute the probability that at homologous positions in sister taxa x and y, the consensus symbols are A and A/T. This is $\sum_{b_x,b_y,b_z} [Pr[T_x(b_x), T_y(b_y) | T_z(b_z)] C_x(\text{A/T}, b_x) C_y(\text{A}, b_y)]$. We can also compute the probability that at *unrelated* positions, the consensus symbols are A and A/T. If the true sequences have the same frequency for each nucleotide, this is $\frac{1}{16} \sum_{b_x \in \Sigma} C_x(\text{A/T}, b_x) \sum_{b_y \in \Sigma} C_y(\text{A}, b_y)$. We score this match in a log-odds framework. For two consensus symbols c_x and c_y, let $S_z(c_x, c_y)$ be the log-odds ratio of seeing the two symbols in related and unrelated positions.

Finding consensus fragments. With consensus sequences, and a way to score matches in them, we now discuss fragment finding. We locate positions that match well using a spaced seed, and filter to accept fragments in positions with a local alignment of high quality according to the scoring matrix.

Given two consensus sequences C_x and C_y, and a spaced seed v, we identify a fragment at positions i in C_x and j in C_y when for all k such that $v_k = 1$, $C_x[i + k - 1] \subseteq C_y[j + k - 1]]$ or vice versa. With these potential fragments, we then add the scores at all positions in the seed. When this is above a threshold T, we identify a fragment. (This is a slight modification of the original definition of vector seeds [3].)

We use the same spaced seed as before (and repeats of it when we need a support greater than 9) for this process, and pick the value of T as before: over the set of possible thresholds and seed weights, we choose the one that minimizes the expected total area between good anchors, setting the probability of bad anchors as close to zero as possible.

Finally, we choose a set of anchors. We score each fragment according to the score of all positions in the fragment. Then, we identify the set of non-conflicting anchors of maximum total score. This can be done in $O(k \log k)$ time [10] with k fragments. We divide the multiple alignments at our chosen anchors, and align the alignments in the regions between the anchors. Then, we group them together into a full alignment. We continue this process at other nodes of the tree, working our way up to the root, until we produce a full multiple alignment of all sequences.

6   Experiments

We have implemented a multiple aligner using our ideas. We present an overview of its implementation, and discuss its performance in simulation experiments.

Our program is successful at producing good multiple alignments, and validates our ideas about choosing thresholds for anchors. Some problems do arise due to errors in aligning alignments between anchors. With this in mind, we note that our algorithm produces alignments slightly worse than MLAGAN [7], while not suffering severe slowdowns when aligning a large number of sequences.

Implementation details. Our aligner is implemented in C++, using the standard template library and the Boost C++ library.

The implementation differs from what we have described in that we have added several filtering steps in both anchoring phases. The consensus fragment detection algorithm does not allow hits at consensus sequence positions with large amounts of ambiguity, since these would require a much larger hash table, and usually give rise to fragments with score below the threshold. Also, when a position hits too many other positions with the spaced seed, we ignore that hash table entry: repetitive sequence gives useless anchors.

Experimental hypotheses. Our experiments test several hypotheses. We propose these hypotheses first, and then discuss results.

The first hypothesis is that on closely related sequences, the anchoring strategy as we go down the tree produces good anchors, and, ultimately, multiple alignments of very high quality. Our second hypothesis is that consensus anchors going up the tree, by themselves, also guide us to good alignments. Our third hypothesis is that as the amount of mutation grows, our algorithm properly anchors the sequences, and produces good results, just with fewer anchors. Next, we predict that allowing a moderate amount of insertion does not affect the success of our anchor choosing. Finally, we predict that as the number of sequences being aligned grows, the consensus anchoring will allow us to identify proper anchors, and give faster run times than for programs like MLAGAN [7] which require alignment of all pairs of sequences.

All of our experiments are on simple simulated data. Our focus is on anchoring, rather than full alignment, and we do not believe that our test data are severely limited for this focus. However, we are in the process of revising our mathematical models for more realistic settings.

6.1   Experimental Results

Our experiments have validated our hypotheses, though they also show some surprises. We implemented our consensus anchor choosing strategy with a quadratic-time chaining algorithm for simplicity, and because we are interested in suboptimal chains. This is a serious time bottleneck in long, very similar regions. Also, while our anchor-choosing algorithms are successful, the algorithm we use to align between anchors inserts too many gaps, especially when insertion mutations are common.

Simple sequences for root anchors. We chose two simple 4-taxon trees, shown in Figure 3. We used low mutation rates to keep anchors common, and simulated

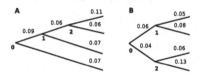

Fig. 3. Trees used in our experiments. We show the point mutation rates for each edge in both trees. Subsequent experiments substantially increase this mutation rate.

Table 1. (a) Anchors going down the tree help in closely related sequences. For 100 kB sequences generated by the trees in Figure 3, we show how many anchors are chosen at nodes of the tree. In 20 experiments, very few false anchors were chosen, and we dramatically reduced the size of aligned regions. **(b)** Consensus anchors going up the tree. Given 100 kB sequences generated by the trees in Figure 3, we show the mean number of consensus anchors chosen at the nodes of the tree. Very few false anchors were chosen, and these anchors allow substantial segmenting of the alignments into homologous regions.

Tree	Node	Mean number of anchors chosen	bad	Seed weight
A	0	18	0	18
A	1	679	0	14
A	2	2988	0.1	12
B	0	50	0	17
B	1	4318	0.2	11
B	2	2613	0.25	12

(a)

Tree	Node	Mean number of anchors chosen	bad
A	0	799	0.15
A	1	1217	0
B	0	1140	0.1

(b)

100 kB sequences with very few insertions. We show results in Table 1(a). Almost no mistakes are made in the anchors as we move down the tree and the anchors divided the sequence up into substantially smaller pieces. (When we examined non-root nodes, we considered the entire sequence at the node.)

Simple sequences for consensus anchors. Here, we allowed much more mutation to show that consensus anchors are useful in regions of lower conservation. We used the same trees as before with 1.78 times as much mutation on each edge, and the results are show in Table 1(b). As we go up the tree, the consensus sequences may include errors caused by mistakes in previous alignment phases. We are interested in how our anchor choosing algorithm performs assuming correct input, so we ignore anchors chosen in regions with existing errors. Noting this, we observe that we made almost no errors in anchoring, though there are errors in the alignment.

Increasing mutation. Here we examine only the consensus sequences since as mutation increases in frequency it quickly becomes less likely to find anchors as we go down. Using the same trees as before, but with still increasing mutation rates we still found correct consensus anchors, though in smaller numbers. We continued to avoid incorrect consensus anchors. Table 2 shows the results and compares alignment level accuracies of our program with that of MLAGAN. Our algorithm produces more errors, largely due to the algnment heuristic.

Increasing the frequency of insertion. Our previous experiments had few insertions, and they were short. With point mutation levels comparable to our second experiment, we then drastically increased the length of the insertions and doubled and quadrupled their frequency. Here again our algorithm performed well at choosing anchors, the majority of its mistakes again coming from the alignment heuristic. Results are shown in Table 3.

Table 2. Experiments with more mutation. As mutations increase, the number of anchors found decreases, but we still set the threshold so that few bad anchors are added. Anchors from "bad regions" are false anchors chosen from regions where the multiple alignment is already incorrect.

Tree	Mutation amount times reference	Mean number of anchors chosen	bad	bad region	Columns correct our program	MLAGAN
A	1	2566	0	2.5	99.79%	99.95%
A	1.8	799	0.15	1.45	99.66%	99.93%
A	2.2	295	0.2	0.4	99.39%	99.84%
A	2.6	101	0.25	0.05	98.59%	99.60%
B	1	3172	0.05	3.1	99.83%	99.94%
B	1.8	1140	0.1	1.24	99.72%	99.92%
B	2.2	483	0.05	0.85	99.67%	99.86%

Table 3. Experiments with more insertion. Added insertions diminish the quality of our alignments, largely due to the heuristic that aligns alignments; anchors chosen continue to be numerous and typically correct.

Tree	Number of insertions times reference	Mean number of anchors chosen	bad	bad region	Columns correct our program	MLAGAN
A	2	787	0.05	4.65	96.84%	99.81%
A	4	779	0.25	11.25	93.58%	99.49%
B	2	1133	0.05	4.2	98.15%	99.83%
B	4	1127	0.05	8.15	96.79%	99.68%

More sequences. We tested the algorithm on increasing numbers of taxa. Here we present only the quality of the alignment and the run time needed to produce it. In this case we are again using a small number of short insertions, but a large amount of point mutation, on random trees of 4, 8, 14, and 28 taxa. The results of these experiments are in Table 4. We examined results when we increase the score threshold used in anchoring; this tests what happens if we require slightly more stringent requirements. The quality of alignment increases notably, with moderate cost in run time. Our performance is comparable on large numbers of taxa to MLAGAN, and our run times are faster on many sequences, due to the consensus approach.

7 Conclusions and Future Work

We have presented a mathematical framework for anchor-based large-scale multiple alignment. Our framework consists of picking anchors in sequences subject to a phylogenetic tree, both working from the root of the tree to the leaves, which is effective for closely related sequences, and working from the leaves of the tree to the root, which is effective for more distant taxa.

Our ideas focus on avoiding bad anchors. By using spaced seeds [17], and computing the weight of a seed needed to avoid picking bad anchors, we continually find good sets of anchors.

When working from the leaves of the tree to the root, we make a potentially ambiguous estimate of the ancestral sequences found at each node of the tree.

Table 4. Experiments with many sequences. Our algorithm performs efficiently and with few errors, even as the number of sequences grows substantially.

# of sequences	Aligner	Run time (sec)	Columns correct
4	Ours (threshold +0)	212	98.88%
4	Ours (threshold +1)	299	99.63%
4	Ours (threshold +2)	410	99.74%
4	MLAGAN	42	99.88%
8	Ours (threshold +0)	326	98.20%
8	Ours (threshold +1)	414	99.33%
8	Ours (threshold +2)	505	99.63%
8	MLAGAN	215	99.84%
14	Ours (threshold +0)	1165	95.03%
14	Ours (threshold +1)	1591	97.57%
14	Ours (threshold +2)	1887	98.83%
14	MLAGAN	2503	99.14%
28	Ours (threshold +0)	2644	89.17%
28	Ours (threshold +1)	3072	94.02%
28	Ours (threshold +2)	3995	96.37%
28	MLAGAN	18003	98.57%

We anchor alignments of alignments in the tree, using hits to a spaced seed, scored with an appropriate log-odds scoring scheme. Again, we pick a threshold for the seed to avoid picking false anchors in this phase as well.

Our strategy of choosing consensus anchors based on ambiguous ancestral sequences and using a spaced seed allows us to have more, and more evenly spaced, anchors in this alignment phase. This allows us to produce global multiple alignments of several species in efficient run times.

Our experiments on simulated sequences, show that our method, despite using a heuristic for alignment of alignments that often makes insertion errors, is competitive with existing programs for this problem, and, as the number of sequences grows, our method is faster by a theoretical factor of the number of sequences (and is also faster in practice).

Future work. We have presented a threshold choosing algorithm for simple models of evolution, which straightforwardly adapts to more complicated ones, such as for regions of low and high conservation. However, we have not incorporated these enhancements into our exisiting implementation.

Also, our scoring strategy for consensus sequences could be adapted to full alignments. Most of the errors we currently produce are done between anchors: the anchors found are correct, but the alignment is bad. If we only considered unambiguous consensus sequences, this would be easy: we could just align them directly. However, we are exploring how to align consensus sequences with ambiguities in them, and also if it is possible to do this in a way that approximately respects proper gap penalties. Alternatively, we could extend the strategy so it includes weights on the sequences being aligned. Finally, our anchors are based only on hits in the forward directions of the sequences, and we are exploring how to extend this to the context of inversion.

References

1. M. Blanchette, W. J. Kent, C. Riemer, et al. Aligning multiple genomic sequences with the threaded blockset aligner. *Genome Res.*, 14:708–715, 2004.
2. N. Bray and L. Pachter. MAVID: Constrained ancestral alignment of multiple sequences. *Genome Res.*, 14:693–699, 2004.
3. B. Brejova, D. Brown, and T. Vinar. Vector seeds: an extension to spaced seeds allows substantial improvements in sensitivity and specificity. In *Proceedings of WABI 2003*, pages 39–54, 2003.
4. B. Brejova, D. Brown, and T. Vinar. Optimal spaced seeds for homologous coding regions. *J. Bioinf. and Comp. Biol.*, 1:595–610, January 2004.
5. D. Brown. Multiple vector seeds for protein alignment. In these proceedings.
6. M. Brudno, M. Chapman, B. Gottgens, S. Batzoglou, and B. Morgenstern. Fast and sensitive multiple alignment of large genomic sequences. *BMC Bioinf.*, 4:66, 2003.
7. M. Brudno, C. Do, G. Cooper, M. Kim, et al. LAGAN and Multi-LAGAN: Efficient tools for large-scale multiple alignment of genomic DNA. *Genome Res.*, 13:721–731, 2003.
8. M. Brudno and B. Morgenstern. Fast and sensitive alignment of large genomic sequences. In *Proceedings of CSB 2002*, pages 138–147, 2002.
9. H. Carrillo and D. Lipman. The multiple sequence alignment problem in biology. *SIAM J. Appl. Math.*, 48:1073–1082, 1988.
10. D. Eppstein, R. Giancarlo, Z. Galil, and G.F. Italiano. Sparse dynamic programming. I: Linear cost functions; II: Convex and concave cost functions. *J. ACM*, 39, 1992.
11. W. Feller. *An Introduction to Probability Theory and Its Applications*. John Wiley & Sons, New York, 1957.
12. W.M. Fitch. Toward defining the course of evolution: minimum change for a specified tree topology. *Syst. Zool.*, 20:406–416, 1971.
13. M. Hohl, S. Kurtz, and E. Ohlebusch. Efficient multiple genome alignment. *Bioinf.*, 18:S312–S320, 2002.
14. J. D. Kececioglu and W. Zhang. Aligning alignments. In *Proceedings of CPM 1998*, pages 189–208, 1998.
15. U. Keich, M. Li, B. Ma, and J. Tromp. On spaced seeds for similarity search. *Discrete Appl. Math.*, 138:253–263, 2004.
16. M. Li, B. Ma, D. Kisman, and J. Tromp. PatternHunter II: Highly sensitive and fast homology search. *J. Bioinf. and Comp. Biol.*, 2004. To appear.
17. B. Ma, J. Tromp, and M. Li. PatternHunter: faster and more sensitive homology search. *Bioinf.*, 18:440–445, March 2002.
18. B. Ma, Z. Wang, and K. Zhang. Alignment between two multiple alignments. In *Proceedings of CPM 2003*, pages 254–265, 2003.
19. B. Morgenstern, A. Dress, and T. Werner. Multiple DNA and protein sequence alignment based on segment-to-segment comparison. *Proc. Natl. Acad. Sci. USA*, 93:12098–12103, 1996.
20. J. Thompson, D. Higgins, and T. Gibson. CLUSTAL W: improving the sensitivity of progressive multiple sequence alignment through sequence weighting, position-specific gap penalties and weight matrix choice. *Nucl. Acids Res.*, 22:4673–4680, 1994.
21. Y. Zhang and M. Waterman. An eulerian path approach to global multiple alignment for DNA sequences. *J. Comp. Biol.*, 10:803–819, 2003.

Chaining Algorithms for Alignment of Draft Sequence

Mukund Sundararajan[1], Michael Brudno[1], Kerrin Small[2],
Arend Sidow[2,3], and Serafim Batzoglou[1]

[1] Stanford University, Department of Computer Science, Stanford, California, 94305 USA
{mukunds,brudno,serafim}@CS.Stanford.edu
[2] Stanford University, Department of Genetics, Stanford, California, 94305 USA
{kerrin,sidow}@Stanford.edu
[3] Stanford University, Department of Pathology, Stanford, California, 94305 USA

Abstract. In this paper we propose a chaining method that can align a draft genomic sequence against a finished genome. We introduce the use of an overlap tree to enhance the state information available to the chaining procedure in the context of sparse dynamic programming, and demonstrate that the resulting procedure more accurately penalizes the various biological rearrangements. The algorithm is tested on a whole genome alignment of seven yeast species. We also demonstrate a variation on the algorithm that can be used for co-assembly of two genomes and show how it can improve the current assembly of the *Ciona savignyi* (sea squirt) genome.

1 Introduction

In bioinformatics, the development of novel biological techniques has created new computational challenges. This is perhaps best epitomized in the problem of sequence alignment, where the development of new sequencing techniques has led to the demand for alignment algorithms capable of dealing with mega-base long regions while maintaining high sensitivity. While in the past it has been sufficient to use two general types of alignment methods, global (Needleman and Wunsch, 1970) and local (Smith and Waterman, 1982), both approaches have been shown recently to be insufficient for alignment of long genomic sequences that have undergone rearrangements (Brudno et al, 2003c). Alignment of draft sequence, where one or both of the sequences being aligned is split into a set of unordered contigs creates similar problems. Global algorithms cannot handle draft sequence at all, while local algorithms just report all the similarity that they find and do not reproduce the syntenic regions that exist between genomes. For many of the genomes currently being sequenced, there are no plans to finish the sequence, and as the number of draft genomes grows, there will be an increasing need for algorithms that can effectively align such sequences.

Unfortunately, the problem of ordering a set of DNA sequences based on another set of (possibly unordered) DNA sequences has been shown to be MAX-SNP hard (Veeramachaneni et al 2002), even in the simpler case of ordering a set of contigs against a finished chromosome. To our knowledge Avid (Bray et al. 2003) and MUMmer 2.0 (Delcher et al. 2002) were the first programs to order a draft sequence based on a second, "finished" sequence based on effective heuristics.

I. Jonassen and J. Kim (Eds.): WABI 2004, LNBI 3240, pp. 326–337, 2004.
© Springer-Verlag Berlin Heidelberg 2004

Recently, sparse dynamic programming based chaining techniques have come to the forefront as a successful approach for fast sequence alignment. Typically, a local aligner produces sections of homology, called fragments. Chaining involves selecting a high scoring subset of these based on some objective criteria. Since the original demonstration that fragment chaining could be achieved in O(nlgn) time by Eppstein and colleagues (Eppstein et al. 1992) the method has been used to speed up global alignment (Delcher et al 1999). Recently Abouelhoda and Ohlebusch (2003) and Brudno et al. (2003c), have used different variations of the sparse dynamic programming algorithm to find rearrangements between genomes. Lippert and colleagues have used a variation on the Brudno et al. (2003c) algorithm to find the differences between two assemblies of a genome (Lippert et al. 2004).

In this work we use a sparse dynamic programming chaining algorithm to align and co-assemble draft genomes. In particular, we make two important modifications to the chaining with rearrangements algorithm (Brudno et al. 2003c, Lippert et al. 2004). An *unrelated* gap penalty is introduced, to chain fragments that are on different contigs of the draft sequence. This formulation allows setting a threshold for the minimal sequence similarity necessary to include a contig in the ordering. We also introduce a new model for sparse dynamic programming that allows us to efficiently make intelligent chaining decisions and enforce penalties based not only on the last local alignment in the chain, but also on previous fragments. Though the solution is no longer optimal with respect to the fore defined objective criteria, this technique improves the chains formed in practice.

We test our draft alignment algorithm on sequences of the seven yeast genomes (Kellis et al. 2003, Cliften et al. 2003). The finished *Saccharomyces cerevisiae* genome was compared to four other yeasts of the sensu-stricto group, one from sensu-lato, and one petite-negative yeast. While we are able to align accurately the four genomes of sensu-stricto, the alignment of the two more distant genomes is less reliable, both indicating the necessity for future work, and suggesting the extent to which draft genomic sequence can be used to shed light on a distant relative.

We have also applied our chaining technique to the problem of co-assembling a genome. In this problem one is given two different assemblies, in draft form, of the same genome, with each assembly organized into contigs (contiguous pieces of the DNA sequence), which are joined into scaffolds (orderings of contigs) by the assembly program based on the paired-read information. When one is given two assemblies of the same data, it becomes possible to correct the potential errors in each assembly, fill the gaps between contigs and label contigs that are potentially misassembled. Here we present a practical method for sequence co-assembly and test it on 19 regions constituting 10% of the *Ciona savignyi* genome. Our method was able to identify more than 60 potential misassemblies with only one false positive, and in all but two cases correctly ordered the contigs to build an assembly with larger contigs and scaffolds.

2 Draft Sequence Alignment

The goal of the draft sequence alignment problem is to map the sections of the finished sequence (Sequence 1) to sections of the draft contigs (Sequence 2). This synteny map is used to construct a reference alignment with respect to the finished sequence. We present an algorithm to solve this problem based on the Sparse Dynamic Programming (DP) technique (Eppstein et al 1992) that builds chains of fragments (local alignments).

2.1 Fragment Chaining

Representation of Fragments: For the purpose of this paper, each fragment is represented as a 7-tuple (Start1, End1, Start2, End2, Score, Strand, CName). The first four numbers represent substrings of Sequence 1 and Sequence 2 respectively. Score is assigned to the fragment by the local alignment algorithm that produced it. Strand is either +/- , representing either a hit of the reference on the forward or reverse strand of sequence 2. Finally CName is an identifier indicating which contig of Sequence 2 the fragment belongs to.

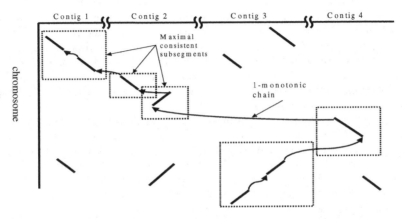

Fig. 1. Draft v/s Finished Alignment: Local alignments are chained together, series of fragments that belong to the same contig, and that are rearrangement free form a single syntenic block.

Scoring: Our scoring scheme is based on SLAGAN (Brudno et al 2003c). We find the highest scoring 1-monotonic chain, which is a subset of fragments that increase monotonically in Sequence 1 coordinates, but there is no requirement for monotonicity on the Sequence 2 axis. Relaxing this requirement allows us to capture rearrangements. We now describe a scoring scheme that scores chains, when the both sequences under consideration are finished sequences.

For a chain C, $\text{score}(C) = \Sigma_i \ (F_i.\text{score} - g_{case}(f_{i-1}, f_i))$.

Here g_{case} is one of 8 penalty functions. The 8 functions correspond to 8 cases based on (Strand$_{i-1}$, Strand$_i$, Transposition). Transposition is a boolean value that is TRUE if a transposition has occurred between the two fragments. Each of these functions is an affine expression on the L1 distance between the end of one fragment and the start of the next fragment, and the difference in the diagonals of the fragments. For a comprehensive description of the scoring functions and the algorithm to find the best such chain refer to (Brudno et al 2003c). The algorithm runs in $O(n\lg n)$, where n is the number of fragments.

Handling Contigs: We arrange the contigs in arbitrary order on the Sequence 2 axis. See Figure 1. We introduce a new scoring function $g_{unrelated}$, which is applied if the adjacent functions belong to different contigs. This function is a large constant penalty, though one could also use an affine function of the sequence 1 distance between the two fragments to penalize long stretches of sequence with no alignment.

We modify the original sparse DP algorithm to handle contigs on the Sequence 2 axis and the unrelated penalty without change in the asymptotic space or time requirements.

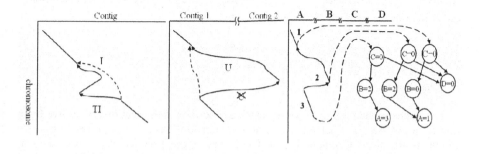

Fig. 2. 2a shows an inversion event being over penalized, 2b shows a contig splice event being penalized twice. 2c shows the structure of the overlap tree for three fragments (1,2,3) across two contigs (A,B).

2.2 Scoring Issues

In practice, the scoring scheme described above suffers from a few problems. Fragment chaining methods score transitions between adjacent fragments on a chain instead of scoring rearrangement events. Here are two cases where it affects the quality of the map produced.

CASE 1: We pay both an inversion penalty, and a translocated inversion penalty for a simple inversion (Figure 2a).

CASE 2: For local alignments from one contig spliced between local alignments from another contig, we pay the unrelated penalty twice (Figure 2b).

One possible model that does capture nested rearrangements explicitly is a stochastic pair-CFG. This is a generalization of a stochastic CFG (Eddy,Durbin 1994) that generates two sequences simultaneously. Though it is possible to write down a set of productions that capture all nested rearrangements, the task of finding the optimal parse is computationally expensive ($O(N^6)$ for two sequences of length N each.). To our knowledge, this formulation has not been proposed previously. One possible heuristic is to work with fragments produced by a local aligner, rather than individual basepairs. We introduce a further heuristic that adds information to the chaining procedure, to make better chaining decisions.

The chaining procedure works in a top down manner. The sparse DP works by maintaining a set of active fragments, to which other fragments yet to be considered can chain. For each such active fragment F, we maintain a $CINFO_F$ query structure. Such a structure maintains a reference to the last fragment on the chain ending at F, for each value for the tuple (CName, Strand), if such a fragment exists. Here CName $\in$ **ContigNames** (which is the set of identifiers of all the contigs), Strand $\in \{+, -\}$.

The scoring proceeds as follows: Fragment f_{Child} ($CName_{Child}$, $Strand_{Child}$) chains to f_{Parent} ($CName_{Parent}$, $Strand_{Parent}$), we apply a penalty corresponding to the minimum of the following quantities:

1. g_{case} (f_{Child}, $CINFO_{fParent}$ ($CName_{Child}$, +))
2. g_{case} (f_{Child}, $CINFO_{fParent}$ ($CName_{Child}$, -))
3. $g_{unrelated}$.

The issues of the previous scoring scheme described in the section are resolved: the transition in CASE 1 is penalized as a normal transition instead of an inverted translocation as the third fragment is penalized against the first instead of the second, though it is chained to the second. Similarly, the transition in the CASE 2 is penalized as a normal transition instead of a contig transition event.

2.3 Application of OvBSTs to the Context Fragment Chaining

The following observation allows us to implement the CINFO structure efficiently: If fragments f_{Child} chains to f_{Parent}, then the contents of $CINFO_{Child}$ differs from that of $CINFO_{Parent}$ only in the fragment corresponding to ($CName_{Child}$, $Strand_{Child}$). All other queries would return the same answer. We use a Overlapping Binary Search Tree (OvBST) (Burton and Huntbach 1985) to efficiently maintain the CINFO structure. An OvBST is a collection of binary search trees each with a distinct root, but share paths.

Let C be the number of contigs, n be the number of fragments.

DATA STRUCTURE: Each node of the OvBST has: key:= (CName, Strand); data:=(Reference to Fragment)

SETUP: Create statically a balanced binary search tree, over the key space (**ContigNameNames**, $\{+, -\}$). The time taken for this step is O(ClgC).

QUERY: A query to the CINFO structure of a fragment is a binary search starting at the root corresponding to the fragment. This operation is performed identically to a find key operation in a regular search tree. It takes time O(lgC).

CREATE: To create a CINFO structure for a fragment, given the CINFO structure for the parent. Let X be the node corresponding to the child's (CName, Strand) in the parent's CINFO structure. A copy is made of all the nodes on the path between X and the root of the parent. Note that all the pointers are copied as well. The data field in the node corresponding to the child is changed to refer to the child fragment. The operation takes time proportional to the height of the tree, which is O(lgC), and at most O(lgC) nodes are copied to make a CINFO structure for the child fragment.

The overlap tree modification allows the sparse DP algorithm to run in time O(nlgnlgC), and space O(nlgC) making it practical for genome-sized datasets.

3   Whole Genome Alignment and Results

Using the draft fragment chaining algorithm described above we have generated a whole genome alignment between *S. cerevisiae* and each of *S. paradoxus, S. mikatae, S. kudriavzevii, S. bayanus, S. castellii* and *S. kluyveri* by repeating the following steps for each of the 16 chromosomes of *S. cerevisiae* (the finished sequence) and each of the other genomes (the draft contigs):

1. Generation of local alignments between the sequences using CHAOS (Brudno and Morgenstern 2002). We used an exact matching 12-mer as a minimal seed, and each local alignment after extension had to have a score of at least 2500 (see the LAGAN toolkit manual available from http://lagan.stanford.edu for an explanation of these and all other parameters).

2. Generation of the 1-monotonic conservation map using the standard SLAGAN penalties. We used an unrelated penalty of 15000 for the five yeast genomes, and a lower 7000 unrelated penalty for *S. kluyveri*, which is not only the most distant to *S. cerevisiae* but also has the poorest assembly.

3. Extension and alignment of all consistent subsegments in the 1-monotonic conservation map with the LAGAN aligner (Brudno et al. 2003b).

4. Glueing together the global alignments from step 3 to form a single global alignment between the chromosome and the draft contigs. Note that the expansion step may cause an overlap in the first sequence. This overlap is resolved by clipping the generated alignments. The optimal clipping point is found by a linear pass over the overlapping region. See (Brudno et al 2003c) for further explanation.

The results of the alignment are summarized in Tables 1 and 2. In constructing the statistics for each species we considered only the exons alignable by a protein-based local aligner: those that were covered more that 90% of their length by TBLASTX (Altschul et al 1997) alignments with a protein-level identity > 50%.

Table 1. Nucleotide Coverage Statistics: Table number of times nucleotides of each species was aligned. The numbers are in percentages.

	0	1	2	3	4	5	>5-10	>10-20	>20
S-paradoxu	4.24	90.67	3.94	0.84	0.15	0.04	0.04	0.05	0.05
S-mikatae	5.83	90.02	2.97	0.58	0.17	0.12	0.18	0.09	0.05
S-kudriavz	5.41	90.94	2.54	0.40	0.23	0.08	0.18	0.16	0.05
S-bayanus	4.80	91.93	2.24	0.63	0.16	0.04	0.07	0.06	0.07
S-castellii	16.45	65.08	15.61	2.18	0.40	0.17	0.04	0.03	0.04
S-kluyveri	29.50	42.88	24.17	2.39	0.57	0.14	0.19	0.09	0.07

Table 2. Exon Conservation Statistics: Table shows the percentage of exons that have a particular percent conservation rate. The second column denotes the size of the TBLASTX-pruned set of exons for each species, as explained above.

Exon Conservation Stats for 14820 exons

	# Exons	<45	45-55	55-65	65-75	75-85	85-95	95-100
S-paradoxus	13674	0.44	0.16	0.23	1.50	9.22	77.63	10.82
S-mikatae	12620	1.16	0.46	0.96	5.19	50.73	33.66	7.83
S-kudriavzevii	11446	0.96	0.67	1.65	10.76	57.60	20.33	8.04
S-bayanus	11694	0.84	0.62	2.25	14.80	60.32	13.77	7.41
S-castellii	5004	11.22	3.90	20.51	31.30	13.88	9.57	9.61
S-kluyveri	4606	16.44	6.56	20.16	24.09	13.07	10.63	9.09

For all of the sensu-stricto sequences, more than 90% of the nucleotides of the assembly were mapped to a single location in the yeast genome. An additional 5% was unmapped, while no more than 1.2% of any genome was mapped more than twice (see below for a discussion of the nucleotides mapped twice). These statistics attest to the specificity of the algorithm when comparing the more closely related sequence. Additionally, none of these four genomes had more than 2% of exons conserved at less than 55%, indicating that the resulting algorithm is sensitive enough to align the important coding elements.

For the two more distant genomes our algorithm maintained a low false positive rate (~3% of the genome mapped to more than two places), but the exon conservation dropped significantly: 15% of the alignable *S. cerevisiae* exons were conserved less than 55% with *S. castellii* and 23% with *S. kluyveri*. These numbers reflect the lower biological conservation between *S. cerevisiae* and these genomes, and the difficulty in aligning a distant, highly fragmented genome (the *S. kluyveri* was the worst quality genome with the shortest contigs; *S castelli*, the third worst).

It has recently been reported that the yeast genome has undergone whole genome duplication, followed by extensive loss of up to 90% of all genes through short deletions (Kellis et al. 2004). This result correlates with the large number of nucleotides that we have found mapped twice from *S. kluyveri* and *S. castellii* to *S. cerevisiae*. Because the genes are lost over time, the most distant sequences are more likely to lose different genes, forcing the alignment of some contigs in two places, once for each of the two copies.

4 Draft Genome Co-assembly

In this section we describe an algorithm to co-assemble two assemblies of the same genome, based on the sparse DP-chaining technique from above. The genome of *Ciona savignyi* has been sequenced to draft quality by a standard whole genome shotgun (WGS) strategy (*http://www.broad.mit.edu/annotation/ciona/*). WGS sequencing and assembly entails randomly breaking DNA from a genome into fixed-size pieces (inserts), and sequencing a 'read' from both ends of each insert (Fleischmann, Adams et al. 1995). Reads from opposite ends of a single insert are termed paired reads, and the distance between them on the sequence can be estimated by the size of the insert. An assembler program rebuilds the genomic sequence by combining reads with sequence overlaps into contigs. Contigs are organized into scaffolds by linking contigs using paired read information.

Like most multi-cellular organisms, *C. savignyi* individuals carry exactly two copies of every chromosome in their genome, which are referred to as the two haplotypes. One copy of each chromosome in an individual is inherited from each parent, and it is not possible to separate out individual chromosomes prior to WGS sequencing. Differences between copies of the same chromosome, called polymorphisms, can take the form of individual base pair substitutions or insertions and deletions that can range from a few base pairs to several thousand base pairs in length. It is these polymorphisms that make every individual within a species unique.

Polymorphism rates vary between species, and current WGS assembly programs (Arachne (Batzoglou et al. 2002, Jaffee et al. 2003), Phusion (Mullikin et al. 2003)) are not designed for highly polymorphic genomes: the polymorphism rate in the human genome is estimated to be only 0.1%. *C. savignyi*, on the other hand, has a very high rate of polymorphism, estimated to be more than 7%. When dealing with such a genome, the assembler (Arachne) places reads from the same position in the genome but from different haplotypes into separate contigs. The resultant assembly of *C. savignyi* thus contains two distinct copies of the genome, each of which is fractured. According to our observations, contigs and scaffolds contain several misassemblies because of the assembler's handling of the varying rate of polymorphism across the genome.

It is desirable to build a joint assembly of the two haplotype assemblies to each other for two primary reasons: (1) to identify regions of disagreement which highlight potential errors in the assembly, and (2) to provide a global alignment from which a *C. savignyi* single reference sequence can be built. Such a sequence can then be used to more accurately predict genes and other functional elements in the genome.

4.1 Strategy for Draft – Draft Alignment

Construction of a Bipartition: Initially the scaffolds are separated into two sets, each set corresponds to a single haplotype. Each scaffold is represented as a node, and there is an edge between two contigs if there is significant local similarity between them. In the absence of large-scale duplications the resulting graph is bipartite, and all the nodes in each partition come from a single haplotype.

Genome Co-assembly: The underlying idea behind the co-assembly of two haplotypes is that each haplotype can be used to establish an ordering of the contigs and scaffolds in the other haplotype. Each haplotype is now a set of contigs (contiguous stretches of DNA sequence). The contigs are ordered into scaffolds by assembly links. These assembly links are based on paired reads, and are less reliable than the contigs that they join. To order haplotype X we use all contigs of haplotype Y as the "chromosomes" for the sparse DP-chaining algorithm described above. Because the two genomes being co-assembled are very similar (these genomes are from the same individual), we use high thresholds for homology.

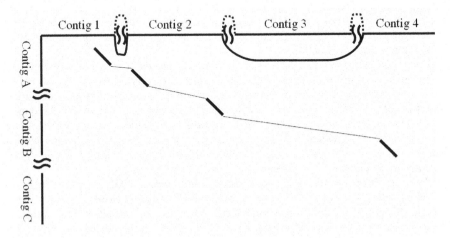

Fig. 3. Dotted connections between contigs are assembly links and solid connections are alignment links. Contigs 1 & 2 are joined by both as the alignment link (through Contig A) confirmed the assembly link. Contigs 2 & 4 were joined by an alignment link, causing the rejection of the assembly links from contig 3 to both contigs 2 and 4.

This step is summarized in Figure 3. Initially we set forward and backward links for each contig to its neighbors in the scaffold. Whenever two contigs 1 and 2 of X are aligned next to each other and against a single contig A of haplotype Y based on the chain from our sparse DP algorithm, 1 and 2 are said to be joined by an alignment link, and the forward and backward links of the two contigs are set to each other. Note that any contigs that lie between 1 and 2 can be separated out into a new scaffold: if the in-between contigs match any sequence, they will be aligned separately; if they do not match any sequence, we use the sequence from haplotype Y to fill the sequence gap (as between contigs 2 and 4 in Figure 3). If a contig has multiple forward or backward alignment links, it is labeled unreliable, as it could be a site of a misassembly on the contig level (or a biological rearrangement). All links to unreliable contigs are removed. All connected components of the link graph are now joined into a contiguous sequence and the process is repeated in order to get a relative ordering of the connected components. During this step, only the reliable "scaffolds" of haplotype Y are used as a basis for ordering all of the "scaffolds" of haplotype X.

4.2 Co-assembly Results

We applied the algorithm to nineteen genomic regions representing approximately 10% of the *C. savignyi* genome. The nineteen regions contain a total of 38.8 Mbp of sequence, and are comprised of 3,283 contigs arranged into 211 scaffolds. The algorithm above was used to order the contigs and scaffolds, and the results were analyzed manually for 1) incorrect ordering of scaffolds, 2) change of order between contigs within a scaffold, and 3) locations where one scaffold was inserted into another one.

For some regions the algorithm ordered the scaffolds in two or three groups and did not order the groups; however, the overall ordering was correct in all cases. In nineteen cases the ordering of the contigs within a scaffold that was reported by our algorithm differed from the ordering given by the assembler (3 regions had 3 rearrangements, 3 regions had 2, and 5 had 1). Of these 19 rearrangements, 18 were judged to be true positives, with one false positive. In 16 of the 18 true rearrangements the resulting sequence was correctly ordered with respect to the opposite haplotype, as determined by manual inspection of the resulting alignments. Additionally, in 42 instances an entire scaffold was inserted into the middle of another scaffold, and all 42 of these cases were detected and ordered correctly. A full analysis of the co-assembly of the genome of *C. savignyi* will be published separately (K. Small, A. Sidow, et al., manuscript in preparation).

5   Conclusions

One of the main requirements of sequence comparison algorithms is the ability to process megabase-long sequences in a reasonable amount of time. While there has been extensive work in developing effective local and global alignment algorithms, these methods have commonly been too inaccurate to align draft genomic sequences. Chaining is an attractive technique for alignment because it provides an $O(n \lg n)$ (n is the number of fragments) method for building larger alignments from short similarities. The technique is also applicable for multiple alignment (Abouelhoda and Ohlebusch 2003), as each additional sequence only requires an additional factor of $O(\lg n)$, while in most other alignment methods the running time is exponential in the number of sequences. Recent work has shown chaining to be one of the most effective ways of speeding up global alignment and reducing the false positive rate of local alignment. Additionally sparse DP has been successfully used to detect and classify rearrangements and to compare genome assemblies. In the current work we apply chaining techniques to alignment of draft (incomplete) genomic sequence, and show that it can be used both to align draft contigs against a finished chromosome and to co-assemble two sets of draft contigs in order to build a more accurate assembly of a genome based on two independent assemblies.

One shortcoming of chaining (and all other sparse DP) methods is that they are memoryless: the decision about the chaining of a particular fragment can only depend on the score of the chain leading up to it and not on the previous history. While it is possible to retrace the chain in order to find the relevant previous information (such as whether some piece of a sequence was already aligned) this would add an extra factor of $O(n)$ to the run time, making the chaining approach much less practical. In this paper we introduce a method to add state to the sparse chaining methods while suffering an O(lg C) runtime hit, where C is the number of "events" that is necessary to remember. This state information helps the fragment chaining algorithm to properly charge penalties for the various events that happen along the chain, and similar approaches can be useful in other algorithms that use sparse dynamic programming.

References

Abouelhoda, M.I., Ohlebusch, E. 2003. A Local Chaining Algorithm and Its Applications in Comparative Genomics. WABI, 1-16.

Altschul, SF, Madden TL, Schäffer AA, Zhang J, Zhang Z, Miller W, and Lipman DJ. 1997 Gapped BLAST and PSI-BLAST: a new generation of protein database search. Nucleic Acids Res 25(17):3389-3402

Batzoglou S, Jaffe D, Stanley K, Butler J, Gnerre S, Mauceli E, Berger B, Mesirov JP, Lander ES, 2002 ARACHNE: A whole genome shotgun assembler. *Genome Research* 12:177-189,

Bray, N., Dubchak, I., Pachter, L. 2003. AVID: A Global Alignment Program.Genome Research, 13:97-102.

Brudno, M., Chapman, M., Gottgens, B., Batzoglou, S., and Morgenstern, B. 2003a. Fast and sensitive multiple alignment of large genomic sequences. BMC Bioinformatics, 4(1):66.

Brudno, M., Do, CB, Cooper, GM, Kim, MF, Davydov, E, Green, ED, Sidow, A, and Batzoglou, S. 2003b. LAGAN and Multi-LAGAN: Efficient Tools for Large-Scale Multiple Alignment of Genomic DNA. Genome Research, 13(4): 721-731.

Brudno M., Malde S., Poliakov A., Do C.B., Couronne O., Dubchak I., Batzoglou S. 2003c. Glocal alignment: finding rearrangements during alignment. Bioinformatics. 19 Suppl 1:i54-62.

Brudno M, Morgenstern B. Fast and sensitive alignment of large genomic sequences. Proceedings of the IEEE Computer Society Bioinformatics Conference (CSB) 2002.

Burton, FW, Huntbach, MM. 1985. Multiple Generation Text Files Using Overlapping Tree. The Computer Journal, 28(4):414-416

Cliften P, Sudarsanam P, Desikan A, Fulton L, Fulton B, Majors J, Waterston R, Cohen BA., and Johnston M. 2003. Finding functional features in *Saccharomyces* Genomes by phylogenetic footprinting. Science, 301:71-76

Delcher AL, Kasif S, Fleischmann RD, Peterson J, White O, and Salzberg SL 1999. Alignment of Whole Genomes. Nucleic Acids Research, 27:11, 2369-2376

Delcher AL, Phillippy A, Carlton J, and Salzberg SL 2002. Fast Algorithms for Large-scale Genome Alignment and Comparision., Nucleic Acids Research , Vol. 30, No. 11 2478-2483.

Eddy SR and Durbin R. 1994 RNA sequence analysis using covariance models. Nucl Acids Res. 22:2079-2088,

Eppstein, D., Galil, R., Giancarlo, R., and Italiano, G.F. 1992. Sparse dynamic programming I: linear cost functions. J. ACM, 39:519-545.

Fleischmann, RD, Adams, MD, White, O, Clayton, RA, Kirkness, EF, Kerlavage, AR, Bult, CJ, Tomb, JF, Dougherty, BA, Merrick, JM, et al. 1995. Whole-genome random sequencing and assembly of Haemophilus influenzae. Science, 269(5223):496-512.

Jaffe DB, Butler J, Gnerre S, Mauceli E, Lindblad-Toh K, Mesirov JP, Zody MC, and Lander ES. 2003. Whole-genome sequence assembly for mammalian genomes: Arachne 2. Genome Res. 13(1):91-6.

Kellis, M., Birren, B., Lander, ES. 2004. Proof and evolutionary analysis of ancient genome duplication in the yeast Saccharomyces cerevisiae. Nature, 428:617-624.

Kellis, M., Patterson, N., Endrizzi, M., Birren, B., Lander, ES. 2003. Sequencing and comparison of yeast species to identify genes and regulatory elements. Nature, 423:241-54.

Lippert, R.A., Zhao, X., Florea, L., Mobarry, C., and Istrail, S. 2004. Finding Anchors for Genomic Sequence Comparison. Proceedings of ACM RECOMB 2004.

Mullikin, J.C., Ning, Z. 2003. The phusion assembler. Genome Res, 13(1):81-90.

Needleman, SB. and Wunsch, CD. 1970. A general method applicable to the search for similarities in the amino acid sequence of two proteins. J. Mol. Biol. 48, 443-453.

Smith, TF and Waterman, MS. 1981. Identification of common molecular subsequences. J. Mol. Biol. 147, 195-197.

Tzouramanis, T., Vassilakopoulos, M., Manolopoulos, Y. 2000. Multiversion Linear Quadtree for Spatio-Temporal Data. DASFAA.

Veeramachaneni, V., Berman, P., Miller, W. 2003. Aligning two fragmented sequences. Discrete Applied Mathematics, 127(1):119-143.

Translation Initiation Sites Prediction
with Mixture Gaussian Models

Guoliang Li[1], Tze-Yun Leong[1], and Louxin Zhang[2]

[1] Medical Computing Laboratory, School of Computing, National University of Singapore
3 Science Drive 2, Singapore, 117543
{ligl,leongty}@comp.nus.edu.sg
[2] Department of Mathematics, National University of Singapore
3 Science Drive 2, Singapore, 117543
matzlx@nus.edu.sg

Abstract. Translation initiation sites (TIS) are important signals in cDNA sequences. Many research efforts have tried to predict TIS in cDNA sequences. In this paper, we propose using mixture Gaussian models to predict TIS in cDNA sequences. Some new global measures are used to generate numerical features from cDNA sequences, such as the length of the open reading frame downstream from ATG, the number of other ATGs upstream and downstream from the current ATGs, etc. With these global features, the proposed method predicts TIS with sensitivity 98% and specificity 92%. The sensitivity is much better than that from other methods. We attribute the improvement in sensitivity to the nature of the global features and the mixture Gaussian models.

1 Introduction

Translation Initiation Sites (TIS) are the positions in cDNA sequences to start constructing proteins. The translation from cDNA to proteins, as we know, starts from TIS in a cDNA sequence and ends at the first in-frame stop codon downstream. It means if we know the TIS in one cDNA sequence, we will know the corresponding protein. Therefore, correct recognition of TIS can help us understand the gene structure, and its product.

Recognition of TIS in cDNA sequences is an important research topic that has been and is still being extensively examined [4,11-15]. In most cases, TIS is a trinucleotide ATG[1] (in DNA or cDNA) or AUG (in mRNA). However, there are numerous ATGs in cDNA sequences and only about one in thirty ATGs acts as TIS – this ATG is a functional ATG.

TIS is dependent on the position of ATG to the 5'-end of the cDNA sequences. As indicated in biological experiments, the first occurrence of codon ATG in a full-length, error-free cDNA sequence is a TIS in most of the known messenger RNA sequences. This inspired the scanning model hypothesis [4,12,15], which postulates that the small (40S) subunit of eukaryotic ribosomes initially binds at the 5'-end of messenger RNA, migrates linearly downstream of the sequence, stops at the first AUG codon [15], and then translation process starts. Moreover, TIS is dependent on

---

[1] There are rare cases that other codons, such as ACG and CUG, are served as translation initiation sites. These will not be considered in this paper.

I. Jonassen and J. Kim (Eds.): WABI 2004, LNBI 3240, pp. 338–349, 2004.
© Springer-Verlag Berlin Heidelberg 2004

the context of the AUGs. Kozak first derived the consensus motif GCCRCCatgG around the TIS with statistical method in [11]. Within this motif, the purine in position[2] -3 and G in position +4 are the most highly conserved.

Although the scanning model hypothesis and consensus motif apply to most of the known messenger RNA sequences well, there are some notable exceptions [7,13,14] – the first ATGs are not TIS due to: 1) leaky scanning – the ribosome bypasses the first ATG codon – the putative start site – due to the very weak context, and translation starts from a downstream ATG with more optimal context; 2) reinitiation – translation starts from an ATG near the 5'-end of the messenger RNA and a small open reading frame (ORF) will be translated, but the ribosome continues scanning until the authentic ATG is reached to construct the protein; 3) internal initiation – the ribosome binds near the real ATG codon directly without scanning, which is reported for several viral mRNAs.

Advancement in technology has enabled more and more TIS to be verified by biological experiments. However, biological experiments are expensive and time-consuming. Therefore, computational methods are needed to help predict TIS in a cDNA sequence accurately and efficiently.

The consensus motif GCCRCCatgG around the TIS [11] was possibly the first attempt to identify TIS with statistical meaning. Although it is often used in biological experiments as a preliminary step to identify TIS in cDNA sequences, the motif is very rough and can't predict TIS well, since many fragments in cDNA sequences can match this consensus motif. Different data mining methods have been tried on TIS prediction problem, such as neural network [19], linear discriminant analysis [22], and support vector machine [26]. The common approach to solving the TIS prediction problem is to generate the numerical data from the cDNA sequences first, and then apply some computational methods to predict TIS.

To date, however, most of relevant features used in the existing prediction algorithms are local information. Little attempt has been made to generate numerical global features to predict TIS.

In this paper, we propose some measures to generate global features, and apply mixture Gaussian models to predict functional ATGs (which act as TIS) from all the occurrences of ATGs in cDNA sequences. With the global features, the proposed method can predict TIS with 98% sensitivity and 92% specificity, which represents a significant improvement in performance with respect to sensitivity.

2 Related Works

Several past efforts focused on TIS prediction using different data mining methods. Stormo *et al* used neural network to identify TIS in E. coli [24], which is probably the first application of neural network technique to predict TIS. The other applications of neural network with comparable prediction performance can be found in [6,9,19]. The numerical features used in these works are direct coding – which is a general way to generate numerical features from cDNA sequences: each nucleotide is encoded by four bits, 0001 for A, 0010 for C, 0100 for G, 1000 for T and 0000 for others. Coding difference between the region before and after TIS was used in [9].

---

[2] Numbering begins with the A of ATG as position +1 and increases downstream. The position just before ATG is numbered as -1 and decreases upstream.

Salamov *et al* developed the system *ATGpr* to identify TIS with linear discriminative analysis [22]. Six characteristics around ATG, such as the positional triplet weight matrix around an ATG and the ORF hexanucleotide characteristics were used to generate numerical data. Zien *et al* [26] engineered support vector machine to recognize TIS, probably with the best prediction performance so far. The measures they used to generate numerical features are as follows: direct coding, positional conditional matrix and other measures. The first-order Markovian dependencies and a dynamic program have been applied to the TIS prediction problem in [20,23], and the generalized second-order profiles were used in [1]. In addition to statistical information, Nishikawa *et al* [18] took the similarity of the cDNA sequences with protein sequences into consideration. The recent work by Nadershahi *et al* [17] compared several available computational methods for identifying TIS in EST data, and concluded that *ATGpr* is the best in the examined methods.

In the above efforts, the features used mainly take local information into account. As observed in [22,26], the data encoding measures affect the performance to recognize TIS from cDNA sequences. The experiments conducted by Pedersen and Nielsen [19] showed that relevant global information could improve prediction significantly.

3   Methods

In this work, we first propose some global measures to generate numerical data from the cDNA sequences. Then we apply the mixture Gaussian models to predict TIS from all occurrences of ATGs.

3.1   Proposed Measures to Generate Numerical Data

Many different data encoding measures can be used to generate numerical data from genomic sequences [8], including special data encoding measures for recognition of TIS, such as the consensus motif GCCACCatgG [11], the positional triplet weight matrix around an ATG, and the ORF hexanucleotide characteristics [22]. All of these features are local features, which take only the local information into account.

In the literature, some simple global features are also used, such as whether the ATG is the first ATG in the cDNA sequence. The first ATG is a strong feature to predict TIS from all occurrences of ATGs, which is supported by the scanning model hypothesis.

After examining the sequences, we observe that the length of the ORF in the cDNA sequences follows different distributions conditioned on the starting ATG – whether it is a TIS or not. From the histograms of the proposed features in the later section, this property is very clear. In another work (under preparation), we have compared several local features and drawn some preliminary conclusions as follows: 1) the direct coding measure generates too many features which makes each feature less meaningful and 2) the higher-order position weight matrix and ORF hexanucleotide characteristics easily overfit the training data.

After careful consideration and comparison, the following measures are chosen for our experiments.

1) Length of upstream sequence from current ATG
2) Length of downstream sequence from current ATG
3) Log ratio of values in (2) / values in (1)
4) Number of upstream ATGs from current ATG
5) Number of downstream ATGs from current ATG
6) Log ratio of values in (5) / values in (4)
7) Number of inframe upstream ATGs from current ATG
8) Number of inframe downstream ATGs from current ATG
9) Log ratio of values in (8) / values in (7)
10) Number of upstream stop codons from current ATG
11) Number of downstream stop codons from current ATG
12) Log ratio of values in (11) / values in (10)
13) Number of inframe upstream stop codons from current ATG
14) Number of inframe downstream stop codons from current ATG
15) Log ratio of values in (14) / values in (13)
16) Length of open reading frame from current ATG

The numbers of ATGs and stop codons have been used in previous works, but the log ratio and the length of the open reading frame have not been used before. When we calculate the log ratio, we add pseudo count 1 to each value involved, since some values may be 0 under some cases. Note that some features in the list are functions of other features, e.g., feature 3 is the log ratio of feature 2 and feature 1. There are five such groups in total. We have done experiments to drop some features to avoid this type of functions. But, under all those cases, the performance of the system would deteriorate. A reasonable explanation is that the algorithm cannot learn the derived functions among the features properly.

3.2 Histograms of the Numerical Features

A histogram is simply a pictorial representation of a collection of observed data. It is particular useful in forming a clear image of the true character of the data from a representative sample of the population. Usually the range of the attribute is divided into 7~15 bins and the frequencies of observations in each bin are counted and displayed in a graph. The graph will show how the observations distribute among the range. If the data follows a Gaussian distribution, the graph tends to cluster around an average and then taper away from this average on each side.

The histograms of the positive and negative training data are shown in Figure 1 and Figure 2 separately. Comparing these two figures, we can see that the distributions of positive and negative data of each feature are quite different. In particular, the distributions of all the log ratio features are near Gaussian. The means for the log ratio features for positive data are near 1, and the means for the log ratio features for negative data are around 0. It implies that the distributions of the features depend on the class labels.

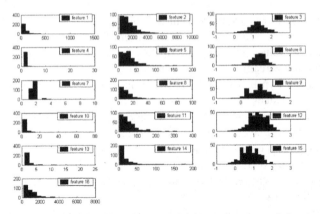

Fig. 1. Histograms of the positive training data – one for each feature

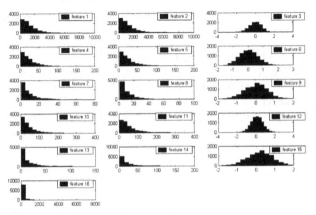

Fig. 2. Histograms of the negative training data – one for each feature

3.3 Mixture Gaussian Models

The task of predicting TIS from all occurrences of ATGs in cDNA sequences is carried out as follows. Considering that the values of each feature are not randomly distributed, we make use of the histogram of each feature to distinguish TIS from all occurrences of ATGs. Referred to Figure 1 and 2, the peaks in the histograms of the positive data and negative data of each feature are quite different. The distributions of the positive data and the negative data of each feature can be approximated by mixture Gaussian models, since mixture models can approximate any continuous density to arbitrary accuracy provided the model has sufficiently large number of components and the parameters of the model are chosen correctly [2].

Mixture Gaussian model is a type of density model, which comprises a number of component Gaussian functions. Suppose the number of the components in the mixture Gaussian model is M. The class conditional density function of a data point $\vec{x}$ belonging to class C is given by

$$p(\vec{x}\,|\,C) = \sum_{m=1}^{M} p(\vec{x}\,|\,m,C)p(m\,|\,C) \tag{1}$$

where $p(m\,|\,C)$ is the prior probability of the data point $\vec{x}$ to be generated from component m of the mixture with probability $p(\vec{x}\,|\,m,C)$, which is a Gaussian as

$$p(\vec{x}\,|\,m,C) = (2\pi)^{-d/2}(\det \Sigma_m^C)^{-1/2}\exp\left\{-\frac{1}{2}(\vec{x}-\mu_m^C)\Sigma_m^{C^{-1}}(\vec{x}-\mu_m^C)\right\}$$

where d is the dimension of the vector $\vec{x}$, μ_m^C is the mean vector of component m of class C and Σ_m^C is the covariance matrix of component m of class C. Here we assume that the covariance matrix of each Gaussian is some scalar multiple of the identity matrix, $\Sigma_m^C = (\delta_m^C)^2 I$.

In the TIS prediction problem, the model is a two-class, two-component mixture model ($M=2$) – later section shows that $M=2$ is enough to model the data. Class 1 represents TIS. It is modeled by two 16 dimensional Gaussians (means and covariances) with associated mixing parameters. Class 2 represents the non-functional ATGs. It has a similar model as Class 1, but with different parameter values.

The structure of the model is shown in Figure 3 as a graphical model. Note that the square nodes represent discrete values and the round nodes represent continuous values. Node *class1/2* means that there are two classes, node *component1/2* with arrows from node *class* means that there are two components for each class, and node *Gaussian* means that the parameters μ_m^C and Σ_m^C depend on both the class and the component. Given the model, the parameters μ_m^C and Σ_m^C of the Gaussian mixture can be determined by the EM algorithm [5] with the training data belonging to that class.

In the E-step, the probability of each feature vector under different Gaussian components is calculated based on the existing parameters of the model. The recurrent equation is as follows.

$$p^{new}(m\,|\,\vec{x},C) = \frac{p^{old}(m\,|\,C)p^{old}(\vec{x}\,|\,m,C)}{\sum_{m=1}^{M} p^{old}(m\,|\,C)p^{old}(\vec{x}\,|\,m,C)}$$

In the M-step, the parameters of the model are re-calculated as the sufficient statistics with the probabilities from the E-step.

$$p^{new}(m\,|\,C) = \frac{1}{N^C}\sum_{n=1}^{N^C} p^{new}(m\,|\,\vec{x}_n,C)$$

$$(\mu_m^C)^{new} = \frac{\sum_{n=1}^{N^C} p^{new}(m\,|\,\vec{x}_n,C)\vec{x}_n}{\sum_{n=1}^{N^C} p^{new}(m\,|\,\vec{x}_n,C)}$$

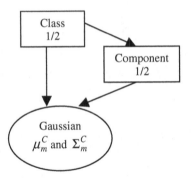

Fig. 3. The graphical representation of the mixture Gaussian model

$$((\sigma_m^C)^{new})^2 = \frac{1}{d} \frac{\sum_{n=1}^{N^C} p^{new}(m \mid \vec{x}_n, C) \left\| \vec{x}_n - (\mu_m^C)^{new} \right\|^2}{\sum_{n=1}^{N^C} p^{new}(m \mid \vec{x}_n, C)}$$

where N^C is the number of feature vectors belonging to class C.

With two Gaussian distributions for the representation of feature values of TIS and non-functional ATGs, we can get the probability of each feature vector generated by different classes with formula (1). Then the probability of each feature vector belonging to class C is

$$p(C \mid \vec{x}) = \frac{p(\vec{x} \mid C) p(C)}{\sum_C p(\vec{x} \mid C) p(C)}$$

where $p(C)$ is the probability that a random feature vector belongs to class C, which is simply estimated from the training data.

In our study, there are only two classes. An optimal threshold to classify a feature vector is determined as the probability to generate the feature vector under Class 1 model which balances the sensitivity and specificity on the training data.

4   Experiments

To illustrate our observation we applied the mixture Gaussian models to a validated sequences set and compared the result with three other data mining methods. We also list the results from literature for reference.

4.1   Data Set

Data is always a big issue in any data mining research. Here we choose the validated sequences set which has already been used successfully in [9] (personal communication with A. G. Hatzigeorgiou). The original sequence set is extracted from Swissprot. The steps are as follows: 1) collect human protein sequences whose N-terminal sites are sequenced at the amino acid level (sequences manually checked by Amos Bairoch); 2) retrieve the full-length mRNAs for these proteins whose TIS had been indirectly experimentally verified. 480 completely-sequenced and annotated human cDNAs were found; 3) divide the sequences set into training set and testing set before the experiments, 325 for training and 155 for testing.

After we got the original sequences, we generated numerical data with the proposed features. The feature vectors around TIS are positive data and those around the non-functional ATGs are negative data. There are 480 positives and 13628 negatives in total – the negatives are much more than the positives. This bias makes other classification methods difficult to recognize positives – many positives will be classified wrongly as negatives. This will be shown in the later section.

4.2 Evaluation Measures

Prediction accuracy is measured by sensitivity and specificity. Let RP be the number of the total real positive ATGs in the data set, TP be the number of the total real positive ATGs predicted as positive, RN be the number of the total real negative ATGs in the data set, and TN be the number of the total real negative ATGs predicted as negative. Sensitivity (*Se*) is defined as TP/RP, and specificity (*Sp*) is defined as TN/RN.

4.3 Experiment Design

Generating the numerical data is the first important step of the experiment. The sequences and the encoding measures have been mentioned above. The total sequences are split into two sets: 325 for training and 155 for testing. There are 325 positive ATGs and 9489 negative ATGs in training set, and there are 155 positive ATGs and 4139 negative ATGs.

The mixture Gaussian model is built in Matlab with support of the Bayes Net Toolbox (BNT) [16]. The BNT is a special software package for manipulating graphical models and Bayesian networks. It supports most of the important methods (inference, parameter learning, and structure learning) for graphical models. This makes training mixture Gaussian models easier.

In our model, it contains three nodes: one for different classes, one for different components and one for all the Gaussian vectors. The parameters are learned by the EM algorithm. The EM algorithm works by starting with randomly initialized parameters, and then iteratively refines the model parameters to produce a locally optimal maximum-likelihood fit or stops when the number of the maximum iterations reaches. The number of the maximum iterations is determined by experiments. It is set to 10 in our experiments.

After the trained model is available, it is used to classify the training samples to determine the threshold. We adopt such a strategy to choose a threshold: when such a threshold is chosen, the sensitivity and specificity on the training set should be balanced – the sensitivity is equal to or approximately equal to the specificity. The value of the threshold is model-dependent, since each model is trained with random initial parameters. However, the performance on the testing data is similar when the testing data is tested with the same model.

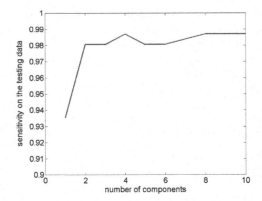

Fig. 4. The relationship of the number of components and the sensitivity of the system on the testing data

Another factor that affects the performance of the system is the number of components in the model. We have tested with 1, 2, 3, 4, 5, 6, 8, 10 components separately. The classification result on the testing data is shown in the following Figure 4.

From Figure 4, we know that there is an increase in sensitivity when the Gaussian model changes from one- to two-components. After that, the sensitivity remains around a stable value. Therefore we chose 2 as the number of components in our experiments.

4.4 Comparison with Other Methods

Three other data mining methods – decision tree, support vector machine, and logistic regression – are adopted to build classification model with the same data to build mixture Gaussian model for comparison purpose. Decision tree method [21] is a *de facto* classification method to evaluate other classification method. Support vector machine [3] is possibly the classification method which have the best prediction result up to date, although it sometimes suffers from noisy data. Logistic regression [10] is a non-linear transformation of the linear regression, which applies to the case when the dependent variable is discrete. A well-known machine learning package Weka [25] has these three methods implemented. We ran these three methods with the default parameters on the same training and test data used in mixture Gaussian model. The results are shown in Table 1.

From Table 1, we can see the mixture Gaussian model can predict TIS with very high sensitivity, and the system performs with balanced sensitivity and specificity. The other three methods are better in terms of specificity of the prediction results, but their sensitivities are quite lower. It means that many TIS are missed in the prediction – this makes the prediction less meaningful.

The best results in the literature to predict TIS are listed in Table 2. In these works, the sequences used to generate these results are ESTs, not full-length cDNA sequences. The global features defined in this work are not considered in these works. Hence, we cannot run our method on their sequences. Consequently, we also cannot at this point conclude that our method is better than all or most of the existing meth-

Table 1. The comparison of the results from 4 different methods – mixture Gaussian models, decision tree, support vector machine and logistic regression

Method	Sensitivity (%)	Specificity (%)
Mixture Gaussian model	98.06	92.41
Decision tree	80	99.68
Support vector machine	67.74	99.37
Logistic regression	76.77	99.46

Table 2. The results from literature. Note: The data for these three methods are from [26]. 1) The original work for neural network is in [19]. 2) The original work for Salzberg method is in [23]. 3) The original work for SVM is in [26]

Method	Sensitivity (%)	Specificity (%)
Neural network[1]	82.4	64.5
Salzberg method[2]	68.1	73.7
SVM[3]	78.4	76

ods, although in terms of sensitivity and specificity our numbers are "better". These results are listed here for reference only.

5 Conclusion

In our system, we have proposed new measures to generate global features and applied mixture Gaussian models for predicting TIS in cDNA sequences. The numbers of ATGs and stop codons around ATG have been used in literature before. Other features, such as the log ratio of the lengths of down stream sequences and upstream sequences from ATGs, the log ratio of the numbers of down stream ATGs and stop codons, the length of the open reading frame down stream the ATGs, are used for the first time to predict functional ATGs from non-functional ATGs. From the histograms of these features, we can observe that these features follow the Gaussian distribution or approximate Gaussian distribution. The mixture Gaussian models are natural and efficient to model these phenomena.

Our mixture Gaussian model is trained with the EM algorithm. When the trained model is applied on the TIS prediction problem, it performs much better than other methods in terms of sensitivity. This means that the proposed global features and mixture Gaussian models are good for the TIS prediction problem. Two specific features and their related features should be mentioned: One is the number of upstream ATGs, which contains the information of whether one ATG is the first ATG in the cDNA sequence and coincides with the scanning model hypothesis. The other is the number of downstream stop codons, which contains the information of whether there is a stop codon downstream – important information about the completeness of the ORF.

A possible problem in the proposed method is that it requires full-length cDNA sequences to generate global features. Since it is getting easier to get full-length cDNA sequences and more full-length cDNA sequences are available now, this problem will be alleviated in the future.

Acknowledgments

We are grateful to Artemis Hatzigeorgiou, who kindly provided the original cDNA sequences. This research is supported by Research Grant No. R-252-000-111-112/303 from the Biomedical Research Council (BMRC) of the Agency for Science, Technology, and Research (A*Star) and the Ministry of Education in Singapore. L. Zhang was supported by BMRC Research Grant BMRC01/1/21/19/140.

References

1. P.K. Agarwal, V. Bafna, Detecting non-adjoining correlations within signals in DNA, in: Proceeding of the 2nd Annual International Conference on Computational Molecular Biology RECOMB (1998) 2-8.
2. C.M. Bishop, Neural networks for pattern recognition (Clarendon Press, Oxford, 1995).
3. C.J.C. Burges, A Tutorial on Support Vector Machines for Pattern Recognition, Data Mining and Knowledge Discovery 2 (1998) 121-167.
4. A. Cigan, L. Feng, T. Donahue, tRNAi(met) functions in directing the scanning ribosome to the start site of translation, Science 242 (1988) 93-97.
5. A.P. Dempster, N.M. Laird, D.B. Rubin, Maximum Likelihood from Incomplete Data via The EM Algorithm, Journal of Royal Statistical Society 39 (1977) 1-38.
6. C. Derst, M. Reczko, A. Hatzigeorgiou, Prediction of human translational initiation sites using a multiple neural network approach, The International Journal of Computers, Systems and Signals 1 (2000) 169-179.
7. T.E. Dever, Gene-specific regulation by general translation factors, Cell 108 (2002) 545-556.
8. J.W. Fickett, The gene identification problem: an overview for developers, Computer & Chemistry 20 (1996) 103-108.
9. A.G. Hatzigeorgiou, Translation initiation start prediction in human cDNAs with high accuracy, Bioinformatics 18 (2002) 343-350.
10. D.W. Hosmer, S. Lemeshow, Applied logistic regression (John Wiley & Sons, New York, 2000).
11. M. Kozak, At least six nucleotides preceding the AUG initiator codon enhance translation in mammalian cells, Molecular Biology 196 (1987) 947-950.
12. M. Kozak, How do eucaryotic ribosomes select initiation regions in messenger RNA?, Cell 15 (1978) 1109-1123.
13. M. Kozak, Interpreting cDNA sequences: some insights from studies on translation, Mammalian Genome 7 (1996).
14. M. Kozak, Pushing the limits of the scanning mechanism for initiation of translation, Gene 299 (2002).
15. M. Kozak, The scanning model for translation: an update, Cell Biology 108 (1989) 229-241.
16. K. Murphy, Bayes Net Toolbox for Matlab, http://www.ai.mit.edu/~murphyk/Software/BNT/bnt.html, in: (2004).
17. A. Nadershahi, S.C. Fahrenkrug, L.B.M. Ellis, Comparison of computational methods for identifying translation initiation sites in EST data, BMC Bioinformatics 5 (2004).
18. T. Nishikawa, T. Ota, T. Isogai, Prediction whether a human cDNA sequence contains initiation codon by combining statistical information and similarity with protein sequences., Bioinformatics 16 (2000) 960-967.
19. A. Pedersen, H. Nielsen, Neural network prediction of translation initiation sites in eukaryotes: prespectives for EST and genome analysis, in: T. Gaasterland, P.D. Karp, K. Karplus, C.A. Ouzounis, C. Sander, A. Valencia (Eds.), Proceedings of the 5th International Conference on Intelligent Systems for Molecular Biology ISMB'97 (AAAI Press, Halkidiki, Greece, 1997) 226-233.

20. M. Pertea, S. Salzberg, A Method to Improve the Performance of Translation Start Site Detection and Its Application for Gene Finding, in: Proceeding of the 2nd Workshop on Algorithms in BioInformatics (WABI2002) (2002) 210-219.
21. J.R. Quinlan, C4.5: programs for machine learning (Morgan Kaufmann, San Mateo, Calif., 1993).
22. A. Salamov, T. Nishikawa, M.B. Swindells, Assessing protein coding region integrity in cDNA sequencing projects, Bioinformatics 14 (1998) 384-390.
23. S. Salzberg, A method for identifying splice sites and translational start sites in eukaryotic mRNA, Computer Applications in Biosciences (CABIOS) 13 (1997) 365-376.
24. G.D. Stormo, T.D. Schneider, L. Gold, A. Ehrenfeucht, Use of the 'Perceptron' algorithm to distinguish translational initiation sites in E. coli., Nucleic Acids Res 10 (1982).
25. I.H. Witten, E. Frank, Data mining: practical machine learning tools and techniques with Java implementations (Morgan Kaufmann, San Francisco, 1999).
26. A. Zien, G. Ratsch, S. Mika, B. Scholkopf, T. Lengauer, K.-R. Muller, Engineering support vector machine kernels that recognize translation initiation sites, Bioinformatics 16 (2000) 799-807.

Online Consensus and Agreement
of Phylogenetic Trees

Tanya Y. Berger-Wolf

Department of Computer Science, University of New Mexico
Albuquerque, NM 87131, USA
tanyabw@cs.unm.edu

Abstract. Computational heuristics are the primary methods for reconstruction of phylogenetic trees on large datasets. Most large-scale phylogenetic analyses produce numerous trees that are equivalent for some optimization criteria. Even using the best heuristics, it takes significant amount of time to obtain optimal trees in simulation experiments. When biological data are used, the score of the optimal tree is not known. As a result, the heuristics are either run for a fixed (long) period of time, or until some measure of a lack of improvement is achieved. It is unclear, though, what is a good criterion for measuring this lack of improvement. However, often it is useful to represent the collection of best trees so far in a compact way to allow scientists to monitor the reconstruction progress. Consensus and agreement trees are common such representations. Using existing static algorithms to produce these trees increases an already lengthy computational time substantially. In this paper we present efficient online algorithms for computing strict and majority consensi and the maximum agreement subtree.

1 Introduction

Reconstruction of the evolutionary history (phylogeny) of a set of organisms is one of the fundamental problems in biology. Computational heuristics are the primary methods of phylogeny reconstruction on large datasets (for example [3, 6, 7, 16, 22, 23, 26]). Most large-scale phylogenetic analyses (including Bayesian methods) produce numerous trees that are equivalent for some optimization criteria (such as maximum parsimony or maximum likelihood). Even using the best heuristics, it takes significant amount of time to obtain optimal trees in simulation experiments. As the number of taxa increases, the running time of various heuristics increases substantially. When biological data are used, the score of the optimal tree is not known. Therefore, at any given point in running the heuristic, we do not know whether the current best score can be improved if the program is run for longer time. As a result, the heuristics are either run for a fixed (long) period of time, or "long enough" until some measure of a lack of improvement is achieved (*e.g.* tree scores do not improve). It is unclear, though, what is a good criterion for measuring this lack of improvement (one such recently proposed criterion is the small topological difference between the

I. Jonassen and J. Kim (Eds.): WABI 2004, LNBI 3240, pp. 350–361, 2004.
© Springer-Verlag Berlin Heidelberg 2004

majority consensus of the best and the second best trees so far [28]). However, to allow scientists to monitor the reconstruction progress, often it is useful to represent the collection of the best trees so far in a compact way. Consensus and agreement trees are common such representations.

The existing static consensus and agreement methods take time polynomial in the number of input trees, multiplied by the number of taxa, to compute a single tree. Repeating this computation at every iteration of a phylogeny reconstruction heuristic, when a new tree is added to the set of best-scoring trees, is impractical. Such approach would significantly slow down an already lengthy computation. The only way to avoid this repetitive computational penalty is to update the consensus tree iteratively, using an on-line algorithm. This paper introduces on-line algorithms for computing the two most common types of consensus trees, strict and majority, and the maximum agreement subtree. The consensus and binary tree agreement algorithms are efficient, robust, and are simple to implement. To the best of our knowledge, this is the first paper explicitly addressing the issue of designing on-line algorithms for computing consensus and agreement of phylogenetic trees.

The rest of this paper is organized as follows. Section 2 provides basic definitions and a description of consensus and agreement techniques. The algorithms for on-line strict and majority consensus are given in Sections 3 and 4, respectively. The algorithm for online maximum agreement subtree is given in Section 5. Conclusions and directions of future work are discussed in Section 6.

2   Definitions

Taxon is the representation of the biological entity for which a phylogeny is desired. We denote the set of taxa by $S = \{s_1, ..., s_n\}$ and let n denote the number of taxa.

A *(rooted) phylogenetic or evolutionary tree* is a (rooted) tree with every internal (non-leaf and non-root) node of degree at least three and the leaves labeled by taxa. We denote a particular tree by T. Unless otherwise specified, a "tree" refers to a "phylogenetic tree". A tree is *binary* or *fully resolved* if every internal node has degree exactly three.

A *bipartition* is a pair of subsets of taxa defined uniquely by the deletion of an edge in a tree. We denote a bipartition by $A|B$ where $A, B \subseteq S, B = S - A$, and the set of all the bipartitions of a tree T by $C(T)$.

A collection of bipartitions is *compatible* if there exists a tree T such that the set of its bipartitions, $C(T)$, is exactly the given collection. A set of bipartitions is compatible if and only if it is pairwise compatible [10, 11]. A pair of bipartitions $A_1|B_1$ and $A_2|B_2$ is compatible if and only if at least one of the intersections $A_1 \cap A_2$, $A_1 \cap B_2$, $B_1 \cap A_2$, or $B_1 \cap B_2$ is empty [5,20]. A pair of clades A_1 and A_2 is compatible if and only if $A_1 \cap A_2 \in \{A_1, A_2, \emptyset\}$. Determining whether a collection of m bipartitions over a set of n taxa is compatible can be done in $O(mn)$ time [15, 27].

A *consensus method* is a technique that combines a collection of trees (called a profile) on the same set of taxa into a single tree representative of the profile for some criteria. We denote the number of trees in a profile by k.

Strict consensus [19] is the most conservative of the consensus methods and produces a tree with only those bipartitions that are common to all the trees in the profile. That is, given a profile $T_1, ..., T_k$ over a set of taxa S, the strict consensus tree $SC(T_1, ..., T_k)$ is the tree uniquely defined by the set of bipartitions $C(SC) = \cap_{i=1}^{k} C(T_i)$. The strict consensus tree of k trees can be computed in time $O(kn)$ [9].

A *majority rule* [2,18,19,25], consensus tree is defined by the set of bipartitions that appear in more than half of the trees in the profile. That is, given a profile $T_1, ..., T_k$ over a set of taxa S, the majority rule consensus tree $MRC(T_1, ..., T_k)$ is the tree uniquely defined by the set of bipartitions $C(MRC) = \{\pi, s.t. |\{\pi \in C(T_i), 1 \leq i \leq k\}| > k/2\}$. The majority consensus tree always exists and is unique. The majority consensus tree of k trees can be computed in time $O(kn)$ [21].

A subtree of T *induced* by a subset of taxa $R \subseteq S$ is a phylogenetic tree on the leaves labeled by R that contains only the paths in T between the leaves in R and the degree 2 nodes removed. We denote such a subtree $T|_R$.

Given a collection of k trees $T_1, T_2, ..., T_k$ with the leaves labeled by $S_1, S_2, ..., S_k$ respectively, an *agreement subtree* is a subtree induced by a set $L \subseteq S = \cap_{i=1}^{k} S_i$ such that $T_1|_L = T_2|_L = ... = T_k|_L$. *Maximum agreement subtree*, denoted $MAST(T_1, ..., T_k)$ is an agreement subtree with the maximum size of the set L [13]. There can be exponentially many (in the number of leaves) MAST for a given collection of trees [17].

A *rooted triple* is a binary subtree of a rooted phylogenetic tree induced by three leaves. If the leaves are labeled by a, b, c and a and b have a common ancestor which is not the root of the subtree, it is denoted by $ab|c$. The set of all the rooted triples of a tree T is denoted by $r(T)$. A subtree induced by three leaves in which all the leaves have the subtree root as their least common ancestor is called a *fan* and is denoted by (abc). The set of all the fans of a tree T is denoted by $f(T)$.

A *quartet* is a binary subtree of an unrooted phylogenetic tree induced by four leaves. We denote the quartet by its (unique) bipartition. The set of all the quartets of a tree T is denoted by $q(T)$. A subtree induced by four leaves with a single internal node is called a *star* and is denoted by $(abcd)$, The set of all the stars of a tree T is denoted by $s(T)$.

3   Online Strict Consensus

First, for the sake of completeness, we present the simple online algorithms for the strict consensus. The strict consensus tree contains bipartitions common to all the source trees. That is, if SC is the strict consensus tree of the set of source trees $T_1, T_2, ..., T_k$, then

$$C(SC) = \cap_{i=1}^{k} C(T_i).$$

We formulate the on-line strict consensus problem as follows.

Input: A set of evolutionary trees $T_1, T_2, ..., T_i, ..., T_k$ arriving online one at a time. All the trees are over the same set of leaves $S = \{s_1, ..., s_n\}$.

Output: At each step i we wish to maintain the strict consensus tree SC_i of the trees $T_1, ..., T_i$.

Solution: The strict consensus tree contains only those bipartitions that appear in all the source trees. Hence, given a strict consensus tree of the first $i - 1$ trees, SC_{i-1}, the strict consensus tree of the first i trees is the strict consensus of SC_{i-1} and T_i. That is,

$$C(SC_i) = \cap_{j=1}^{i} C(T_j) = \cap_{j=1}^{i-1} C(T_j) \cap C(T_i) = C(SC_{i-1}) \cap C(T_i).$$

The intersection of the sets of bipartitions of SC_{i-1} and T_i can be computed in $\Theta(n)$ time.

Since it takes $O(n)$ time to process the tree T_i, this solution is time-optimal. A tree is uniquely defined by a set of its (compatible) bipartitions and can be computed in linear time [15, 27]. Thus, there is no additional space requirements beyond storing the set of $O(n)$ bipartitions of the current consensus tree.

This algorithm is essentially Day's strict consensus algorithm [9] (assuming that the number of bits used to store each bipartition is $O(\log n)$. It can also be viewed as a repetitive application of Day's algorithm, which takes $O(kn)$ time to compute the strict consensus of k trees. Thus, we have shown that the following statement is true.

Proposition 1. *The time it takes to compute the strict consensus tree SC_i with the arrival of each new tree T_i is $O(n)$ and the total time to compute the strict consensus tree of k trees online is $O(kn)$ and is optimal.*

4   Online Majority Consensus

The majority rule tree contains those bipartitions that appear in more than half of the source trees. That is, if M is the majority consensus tree of the set of source trees $T_1, T_2, ..., T_k$, then

$$A|B \in C(M) \text{ if and only if } |\{C(T_i) \text{ s.t. } A|B \in C(T_i)\}| > \frac{k}{2}.$$

The solution for the on-line majority consensus is slightly more complicated than that for the strict consensus.

Input: A set of evolutionary trees $T_1, T_2, ..., T_i, ..., T_k$ arriving online one at a time. All the trees are over the same set of leaves $S = \{s_1, ..., s_n\}$.

Output: At each step i we wish to maintain the majority consensus tree M_i of the trees $T_1, ..., T_i$.

Solution: We maintain a set of bipartitions that have appeared in all the trees seen so far. For each bipartition we keep a count of the number of trees it has appeared in up to this point. The majority tree at any point, by definition, is the collection of bipartitions that have appeared in the majority of the trees. When a new tree T_i arrives, we update the count on all the bipartitions that appear in that tree. If any of these now make the majority, we add them to the majority tree M_i. We check the counts on the bipartitions that were in the previous majority tree M_{i-1}. If any of the bipartitions now drop below the majority, then we remove them from the majority tree. Below is the formal description of the algorithm.

Algorithm ONLINEMAJORITY

```
1      C = ∅
2      FOR each new tree Tᵢ DO
3              FOR each bipartition c ∈ C(Tᵢ) DO
4                      IF c ∉ C THEN
5                              C = C ∪ {c}
6                              count(c) = 0
7                      count(c) + +
8                      IF count(c) > i/2 THEN
9                              C(Mᵢ) = C(Mᵢ₋₁) ∪ {c}
10             FOR each bipartition c ∈ C(Mᵢ₋₁) DO
11                     IF count(c) ≤ i/2 THEN
12                             C(Mᵢ) = C(Mᵢ₋₁) − {c}
13             Build the tree Mᵢ from C(Mᵢ)
14     RETURN Mᵢ
```

Line 3 is a Depth First Search (DFS) traversal of the tree T_i. The FOR loop is executed in the order of the finish times of the DFS for the nodes associated with the bipartitions. The set of bipartitions is maintained using any efficient implementation of a set with a SEARCH operation (a dictionary). The correctness and the time complexity of the algorithm are discussed below.

4.1   Correctness of the ONLINE MAJORITY Algorithm

Proposition 2. *Given a majority tree M_{i-1} and a new tree T_i, only the bipartitions in M_{i-1} or in T_i can be in the new majority tree M_i:*

$$C(M_i) \subseteq C(M_{i-1}) \cup C(T_i)$$

Proof. The count of any bipartition which is not in M_{i-1} was at most $(i-1)/2 < i/2$. If this bipartition is not in T_i then the number of trees it appears in has not increased with the arrival of T_i. Thus, it cannot be in M_i.

With the arrival of a new tree $T - i$ the algorithm checks the count of every bipartition in $C(M_{i-1}) \cup C(T_i)$ and retains only those whose count is greater than $i/2$. Thus, at every step i the tree T contains only the bipartitions that appear in the majority of trees, therefore, by definition, T is the majority consensus tree of the trees so far.

4.2   Running Time of the ONLINE MAJORITY Algorithm

Lemma 1. *The time it takes to compute the majority consensus tree M_i with the arrival of each new tree T_i is $O(n \times f(n))$, where $f(n)$ is the time of the SEARCH operation of a dictionary data structure implementation.*

Proof. We use several data structures to store the information. The underlying dictionary data structure is discussed below. The bipartitions of the current majority tree in addition are stored in a linked list (although not duplicated) with a pointer to the root (or the head of the linked list) which is null if the bipartition is not in the current tree.

As we mention above, each new arriving tree T_i is traversed using DFS and the bipartitions are processed in the order of their completion times. When each new bipartition is processed, the time it takes to check whether it is in the current set of bipartitions C (line 4) depends on the implementation of the dictionary structure. Since the bipartitions in the majority tree have a pointer to the non-null root, it takes constant additional time to check whether that bipartition is in the current majority tree as well. If the bipartition is not in the current majority tree, we add it to the linked list (in constant time, after the head) and update its root pointer. After we finish processing the T_i tree, we traverse the linked list of the majority tree, discarding the bipartitions whose count is less than the majority. Again, we use a linear time algorithm [15, 27] to build the tree in line 13. Thus, the total time for lines 7–14 is $O(n)$. Therefore, the total time *for each new tree* is $O(n \times f(n))$, where $f(n)$ is the time of the SEARCH operation for a dictionary data structure and depends on the implementation.

Theorem 1. *The time it takes to compute the majority consensus tree M_i with the arrival of each new tree T_i is at most $O(n)$ and the total time to compute the majority consensus tree of k trees online is $O(kn)$ and is optimal.*

Proof. From Lemma 1 the time needed to compute the majority consensus with the arrival of each new tree is $O(n \times f(n))$, where $f(n)$ is the time of the SEARCH operation of a dictionary data structure. A standard implementation of a dictionary data structure is a *hash table*. Given a uniform universal hash function with h keys, the expected SEARCH time to access a table with s items is at most $O(1 + s/h)$. In our case, $s = |C|$ is the total number of bipartitions in k trees, which is at most kn. Making h a constant fraction of s gives a constant expected time SEARCH operation, albeit with a possibly high constant. Thus, the total running time for recomputing majority consensus using a hash table is $O(n)$ with $O(|C|)$ space requirements. The $O(n)$ running time is optimal since it takes $O(n)$ time process the input of a new tree and the $O(|C|)$ space is necessary. For a detailed implementation of a hashing table for bipartitions see Amenta *et al.* [21]. Notice that for k trees the running time of our algorithm and Amenta et al.'s algorithm is the same $O(kn)$. Thus we provide yet another optimal linear time algorithm for majority consensus, either online or off line.

5   Online Maximum Agreement Subtree

Maximum agreement subtree (MAST) represents yet another valuable piece of information about a collection of trees. It shows how much of the phylogenetic information is common to all the trees in the input. It looks at all the phylogenetic relationships, not only the bipartitions. We formulate the online version of the MAST problem as follows.

Input: A set of evolutionary trees $T_1, T_2, ..., T_i, ..., T_k$ arriving online one at a time. All the trees are over the same set of leaves $S = \{s_1, ..., s_n\}$.

Output: At each step i we wish to maintain a maximum agreement subtree $MAST_i$ of the trees $T_1, ..., T_i$.

MAST of two arbitrary trees is polynomial [14, 24] and for two binary trees Cole *et al.* [8] present an $O(n \lg n)$ algorithm. For an arbitrary collection of $k \geq 3$ trees the offline MAST problem is NP-hard [1]. However, it is fixed parameter tractable. When the degree of even one tree in the input is bounded by d, both Bryant [4] and Farach *et al.* [12] present an $O(kn^3 + n^d)$ offline algorithm.

5.1   Greedy Online MAST

The simplest online algorithm is, of course, a greedy one:

$$MAST_i = MAST(MAST_{i-1}, T_i).$$

This algorithm is polynomial and for k trees its running time is $O(kf(n))$, where $f(n)$ is the running time for the MAST of two trees. However, when $MAST_i$ is not unique we must make a decision which tree to retain. We cannot retain all the trees since there may be exponentially many of them. We show that retaining the wrong tree may have a dramatic effect on the size of subsequent agreement subtrees.

Proposition 3. *The greedy algorithm for the online MAST problem can produce trees with unbounded size ratio with respect to the optimal tree.*

Proof. Consider the following example described in Figure 1. The input trees T_1, T_2, T_3 arrive in order. There are three $MAST(T_1, T_2)$ trees, all with 3 leaves. However, if the algorithm chooses $MAST_2 = ((13)4)$ then $MAST_3$ has three leaves (and is optimal), while if $MAST_2 = ((12)4)$ or $MAST_2 = ((23)4)$ then $MAST_3$ is empty. Now consider the case when instead of the leaves there are subtrees of size t each. In this case, all the choices for $MAST_2$ are of size $3t$. However, one of them produces a $MAST_3$ of size $3t$, while the other two result in a $MAST_3$ of size $2t$. We can now recursively build the subtrees 1, 2, 3, and 4, each a copy of the structure of the larger tree. Figure 2 shows the first level of the recurrence. After j levels of the recursive structure being repeated, there are 3^{3^j+1} possible $MAST_2$ trees, each with $3^{j+1}t$ leaves. However, only one of them gives rise to the $MAST_3$ with $3^{j+1}t$ leaves, while others produce $MAST_3$ with only $2^{j+1}t$ leaves. Thus the wrong choice of a $MAST_i$ by the algorithm at an earlier stage can arbitrarily badly affect the size of $MAST_k$.

5.2   Online MAST Algorithm

As we have mentioned, we cannot retain all the MAST trees at every stage of the algorithm. Recomputing MAST in a straight forward way when every new tree T_i arrives is too expensive: it increases the computational time by a factor of k. Instead, we modify a MAST algorithm to maintain a structure that allows to

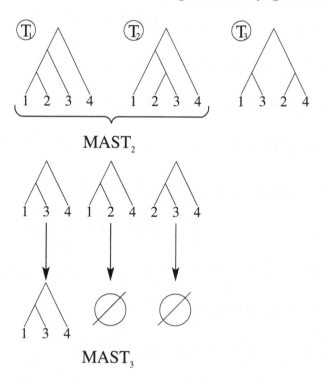

Fig. 1. An example of a choice of MAST in earlier stages of the greedy algorithm affecting the outcome of the later stages.

compute $MAST_i$ at every stage. Specifically, we will follow Bryant's algorithm [4, page 180] for computing MAST. The algorithm is a dynamic programming algorithm that relies on the following fact ([4], Lemma 6.6), which we restate here.

Lemma 2. *A tree T is an agreement subtree of $T_1, ..., T_i$ if and only if*

$$r(T) \subseteq R_i = \bigcap_{j=1}^{i} r(T_j) \text{ and } f(T) \subseteq F_i = \bigcap_{j=1}^{i} f(T_j).$$

Note that using the property of intersection for any collection of sets $A_1, ..., A_i$

$$\bigcap_{j=1}^{i} A_j = \bigcap_{j=1}^{i-1} A_j \cap A_i,$$

it is easy to maintain the sets R_i and F_i online greedily.

For the simplicity of the discussion we also restate Bryant's algorithm here.

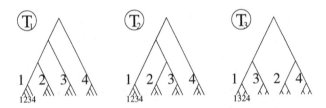

Fig. 2. The input trees T_1, T_2, T_3 with the structure of the tree recursively repeated at every leaf.

Algorithm BRYANTMAST(a, b)

1 **IF** $a = b$ **THEN RETURN** 1
2 Construct:
$$A \leftarrow \{x : ax|b \in R\} \cup \{a\}$$
$$B \leftarrow \{x : by|a \in R\} \cup \{b\}$$
$$C \leftarrow \{z : (azb) \in F\}$$
3 Choose $x^* \in A$ that maximizes $MAST(a, x^*)$
4 Choose $y^* \in B$ that maximizes $MAST(b, y^*)$
5 **FOR** (each $z \in C$) **DO**
   Choose $z^* \in \{z' \in C : zz'|a \in R\} \cup \{z\}$
   that maximizes $MAST(z, z^*)$
6 Construct a weighted graph $G = (V, E)$
   $V = C$, $(v, w) \in E \Leftrightarrow (avw) \in F$, $w(v) = MAST(v, v^*)$
7 Choose maximum weight clique Q in G
8 **RETURN** $MAST(a, x^*) + MAST(b, y^*) + \sum_{z \in Q} MAST(z, z^*)$

The algorithm fills in an $n \times n$ matrix of all leaf pairs with the sizes of the respective MASTs. The actual MAST can be reconstructed from this matrix using the standard dynamic programming trace of computation. The running time of Bryant's algorithm is $O(kn^3 + n^d)$, where all trees have the maximum degree at least d. When even one tree is binary (a common case when the input trees are a result of a tree reconstruction heuristic), then there is no set F and the algorithm reduces to steps 1, 2 (sets A and B only), 3, 4, and 8. The running time in this case is $O(kn^3)$, which is the time it takes to construct the set R.

We are now ready to state and analyze the online MAST algorithm.

Theorem 2. *Let the input trees $T_1, ..., T_i, ..., T_k$ have the corresponding maximum degrees $d_1, ..., d_i, ..., d_k$ and let $min_i = \min_{1 \le i \le k} d_i$. The MAST of these k trees can be maintained online in $O(kn^3 + \sum_{i=1}^{k} n^{min_i})$ time.*

Before we prove Theorem 2, we state an important corollary.

Corollary 1. *If the tth input tree is the first binary tree in the input then the MAST of these k trees can be maintained online in $O(kn^3 + \sum_{i=1}^{t} n^{min_i})$. Specifically, if the first tree of the input is binary, the MAST of k trees can be maintained online in $O(kn^3)$.*

Proof. To maintain $MAST_i$ online, we use Bryant's algorithm stated earlier. When a new tree T_i with the maximum degree d_i arrives we update the sets R_i and F_i of the common triples and fans using the intersection property:

$$R \leftarrow R \cap r(T_i), \ F \leftarrow F \cap f(T_i)$$

Note, that the vertices of T_i of degree greater than min_{i-1} do not contribute fans to the intersection and hence the set F is limited by the smallest d_i so far. We then update the MAST matrix using Bryant's algorithm. To maintain the structure needed to compute the MAST, we need to maintain the sets A, B, and C *for every pair of vertices.* Let us examine what can happen to those sets as a new tree T_i arrives. As we are computing the new R, some triples may be eliminated from the set. This would possibly eliminate elements from the sets A and B. If this happens and those elements were the elements x^* and y^* (lines 3, 4) that maximize the corresponding MASTs, new maxima need to be found. However, recalculating the sets A and B and choosing new maxima takes $O(n)$. Thus, redoing the procedure for each for each vertex pair would take $O(n^3)$ time and, hence, would not increase the (asymptotic) running time over k input trees.

When the minimum maximum degree in first i trees is $min_i > 2$, the most computationally expensive step of the offline algorithm is line 7. The lines 5 and 6 still take $O(n)$ time for each pair, $O(n^3)$ total for a new tree. However, any time a fan is deleted from the set F, both a vertex and an edge can potentially be removed from the graph G *for every vertex pair.* Moreover, the vertex weights need to be recomputed. Thus, we need to recompute the maximum weight clique S, possibly from scratch, for every pair, for every new tree. The maximum degree of the graph G is no greater than the maximum size of a fan in the set F, that is at most min_i at the time $MAST_i$ is calculated. Thus, the line 7 adds $O(n^{min_i-2})$ running time to the algorithm per vertex pair, $O(n^{min_i})$ total for each new tree T_i.

Thus, the total time to compute $MAST_i$ online using Bryant's algorithm is $O(n^3 + n^{min_i})$ adding up to the total of $O(kn^3 + \sum_{i=1}^{k} n^{min_i})$ for k trees.

The arrival of a binary tree T_t sets $F = \emptyset$, therefore all the calculations related to the set C from then on are eliminated and only the lines 1,2,3,4, and 8 are executed. Thus, the computational time from that point on for each new tree is $O(n^3)$. Therefore, the total computational time for k trees in this case is $O(kn^3 + \sum_{i=1}^{t} n^{min_i})$. When $t = 1$ this, of course, becomes $O(kn^3)$.

6 Conclusions

Phylogeny reconstruction heuristics on biological data cannot recognize when an optimal or a "true" tree is produced. Thus, they need to use some criteria for termination. Initial experiments show [28] that the lack of difference between the consensus of the trees with the top score and the trees with the second best score may be a good criterion. To use these criteria, we need to have algorithms that maintain consensus of a sequence of trees on-line, as the new trees are generated by a heuristic. Another or additional option is to rely on expert knowledge to determine when the reconstruction process has been run sufficiently long. For this, scientists need to monitor the progress of the reconstruction presented in a meaningful and compact way. Consensus and agreement trees are common such representations. We have proposed and analyzed algorithms for the on-line computation of strict and majority consensus trees and maximum agreement subtree. We have shown that the on-line strict and majority consensus algorithms are time and space-optimal. Thus these easy to implement algorithms can be used to maintain stopping criteria or monitor the reconstruction progress without substantially increasing the overall running time of the heuristic search. Finally, we have also proposed an online maximum agreement subtree algorithm which does not increase the overall computational time with respect to the best known offline algorithms.

Clearly, we need experimental results to verify that the algorithms are efficient in practice. The ultimate goal is to develop a suite of tools that allow compact and efficient online representation of the various aspects of the collections of trees arising in the phylogenetic reconstruction process.

Acknowledgments

This work is supported by the National Science Foundation Postdoctoral Fellowship grant EIA 02-03584. The author is deeply grateful to Tandy Warnow for suggesting the problem and for many insights.

References

1. A. Amir and D. Keselman. Maximum agreement subtree in a set of evolutionary trees - metrics and efficient algorithms. In *Proceedings of the 35th Annual Symposium on Foundations of Computer Science*, pages 758–769, 1994.
2. J. P. Barthélemy and F. R. McMorris. The median procedure for *n*-trees. *Journal of Classification*, 3:329–334, 1986.
3. M. L. Berbee. The phylogeny of plant and animal pathogens in the Ascomycota. *Physiological and Molecular Plant Pathology*, 2001.
4. D. Bryant. *Building trees, hunting for trees, and comparing trees: Theory and methods in phylogenetic analysis*. PhD thesis, University of Canterbury, 1997.
5. P. Buneman. The recovery of trees from measures of dissimilarity. In F.R.Hodson, D.G.Kendall, and P.Tautu, editors, *Mathematics in the Archeological and Historical Sciences*, pages 387–395. Edinburgh University Press, Edinburgh, 1971.
6. R. M. Bush, W. M. Fitch, C. A. Bender, and N. J. Co. Positive selection on the H3 hemagglutinin gene of human influenza virus A. *Molecular Biology and Evolution*, 16:1457–1465, 1999.
7. M. W. Chase, D. E. Soltis, R. G. Olmstead, D. Morgan, D. H. Les, B. D. Mishler, M. R. Duvall, R. A. Price, H. G. Hills, Y. L. Qiu, K. A. Kron, J. H. Rettig, E. Conti, J. D. Palmer, J. R. Manhart, K. J. Sytsma, H. J. Michaels, W. J. Kress, K. G. Karol, W. D. Clark, M. Hedren, B. S. Gaut, R. K. Jansen, K. J. Kim, C. F. Wimpee, J. F. Smith, G. R. Furnier, S. H. Strauss, Q. Y. Xiang, G. M. Plunkett, P. S. Soltis, S. M. Swensen, S. E. Williams, P. A. Gadek, C. J. Quinn, L. E. Eguiarte, E. Golenberg, G. H. Learn, Jr., S. W. Graham, S. C. H. Barrett, S. Dayanandan, and V. A. Albert. Phylogenetics of seed plants: an analysis of nucleotide sequences from the plastid gene rbcL. *Annals of the Missouri Botanical Garden*, 80:528–580, 1993.

8. R. Cole, M. Farach-Colton, R. Hariharan, T. M. Przytycka, and M. Thorup. An $O(n \log n)$ algorithm for the maximum agreement subtree problem for binary trees. *SIAM Journal of Computing*, 30(5):1385–1404, 2000.

9. W. H. E. Day. Optimal algorithms for comparing trees with labeled leaves. *Journal of Classification*, 2:7–28, 1985.

10. G. F. Estabrook, Jr. C. S. Johnson, and F. R. McMorris. An algebraic analysis of cladistic characters. *Discrete Mathematics*, 16:141–147, 1976.

11. G. F. Estabrook and F. R. McMorris. When is one estimate of evolutionary history a refinement of another? *Mathematical Biology*, 10:367–373, 1980.

12. M. Farach, T. Przytycka, and M. Thorup. On the agreement of many trees. *Information Processing Letters*, 55:297–301, 1995.

13. C. R Finden and A. D. Gordon. Obtaining common pruned trees. *Journal of Classification*, 2:255–276, 1985.

14. W Goddard, E. Kubicka, G. Kubicki, and F. R. McMorris. The agreement metric for labelled binary trees. *Mathematical Biosciences*, 123:215–226, 1994.

15. D. Gusfield. Efficient algorithms for inferring evolutionary trees. *Networks*, 21:12–28, 1991.

16. M. Källersjö, J. S. Farris, M. W. Chase, B. Bremer, M. F. Fay, C. J. Humphries, G. Pedersen, O. Seberg, and K. Bremer. Simultaneous parsimony jackknife analysis of 2538 rbcl DNA sequences reveals support for major clades of green plants, land plants, seed plants and flowering plants. *Plant Systematics and Evolution*, 213:259–287, 1998.

17. E. Kubicka, G. Kubicki, and F. R. McMorris. On agreement subtrees of two binary trees. *Congressus Numerantium*, 88:217–224, 1992.

18. T. Margush and F. R. McMorris. Consensus n-trees. *Bulletin of Mathematical Biology*, 43(2):239–244, 1981.

19. F. R. McMorris, D. B. Meronik, and D. A. Neumann. A view of some consensus methods for trees. In J. Felsenstein, editor, *Numerical Taxonomy*, pages 122–125. Springer-Verlag, 1983.

20. F. R. McMorris. On the compatibility of binary qualitive taxonomic characters. *Bulletin of Mathematical Biology*, 39:133–138, 1977.

21. Katherine St. John Nina Amenta, Frederick Clarke. A linear-time majority tree algorithm. In Gary Benson and Roderic D. M. Page, editors, *Algorithms in Bioinformatics, Third International Workshop, WABI 2003, Budapest, Hungary, September 15-20, 2003, Proceedings*, volume 2812 of *Lecture Notes in Computer Science*, pages 216–227. Springer, 2003.

22. V. Savolainen, M. W. Chase, S. B. Hoot, C. M. Morton, D. E. Soltis, C. Bayer, M. F. Fay, A. Y. De Bruijn, S. Sullivan, and Y. L. Qiu. Phylogenetics of flowering plants based on combined analysis of plastid atpB and rbcL gene sequences. *Systematic Biology*, 49:306–362, 2000.

23. P. S. Soltis, D. E. Soltis, and M. W. Chase. Angiosperm phylogeny inferred from multiple genes as a tool for comparative biology. *Nature*, 402:402–404, 1999.

24. M. Steel and T. Warnow. Kaikoura tree theorems: Computing the maximum agreement subtree. *Information Processing Letters*, 48(2):77–82, 1993.

25. D. L. Swofford. When are phylogeny estimates from molecular and morphological data incongruent? In M. M. Miyamoto and J. Cracraft, editors, *Phylogenetic Analysis of DNA Sequences*, pages 295–333. Oxford University Press, 1996.

26. Y. Van de Peer and R. De Wachter. Evolutionary relationships among the eukaryotic crown taxa taking into account site-to-site rate variation in 18S rRNA. *Journal of molecular evolution*, 45:619–630, 1997.

27. T. J. Warnow. Three compatibility and inferring evolutionary history. *Journal of Algorithms*, 16:388–407, 1991.

28. T. L. Williams, T. Y. Berger-Wolf, B. M. E. Moret, U. Roshan, and T. J. Warnow. The relationship between maximum parsimony scores and phylogenetic tree topologies. Technical Report TR-CS-2004-04, University of New Mexico, 2004.

Relation of Residues in the Variable Region of 16S rDNA Sequences and Their Relevance to Genus-Specificity

Maciej Liśkiewicz[1,*], Hemant J. Purohit[2], and Dhananjay V. Raje[2]

[1] Institut für Theoretische Informatik, Universität zu Lübeck
Ratzeburger Allee 160, 23538 Lübeck, Germany
liskiewi@informatik.mu-luebeck.de
[2] Environmental Modeling and Genomics Division
National Environmental Engineering Research Institute
Nehru Marg, Nagpur 440020, India
{hemantdrd,dv_raje}@hotmail.com

Abstract. It has been observed that the short nucleotide sequences in a variable region, representing species level diversity in a set of 16S rDNA sequences carries the genus specific signature. In this study our aim is to assess the relationship of residues at different positions and thereby obtain consensus patterns using different statistical tools. If such patterns are found genus-specific then it would facilitate in designing hybridization arrays to target even unexplored species of the same genus in complex samples such as environmental DNA.
For obtaining consensus pattern from a set of aligned sequences, four different methods were used on five bacterial genera. The patterns were tested for genus-specificity using BLAST. In two out of the five genera, the consensus pattern was highly genus-specific and was identified as a signature pattern representing the genera. In other genera, although the sample sub-sequences had the edge on the consensus pattern with respect to genus-specificity, there was not much difference between the consensus pattern and the signature pattern of these genera.

1 Introduction

Determining patterns unique to a particular organism has been the issue of growing interest since last decade. A promising target for use in microbial monitoring has been the small sub-unit 16S rRNA sequence of ribosome. Molecular characterization of 16S rRNA made it possible to unravel evolutionary relationships amongst bacteria for the first time in early 1980s (Fox *et al.*, 1980). Large number of 16S rDNA sequences representing essentially all known genera of bacteria are now available in databases like Ribosomal Database Project (RDP) (http://rdp.cme.msu.edu/html) or the ARB project site (http://www.mpi bremen.de/molecol/arb/main.html). These sites provide necessary software support to better understand the hierarchical database of sequences with associated information (Zhang *et al.*, 2002). Efforts are being made to use the available data to design rapid systems to detect targeted bacteria.

It has been observed that 16S rDNA sequences are widely conserved, but despite being conserved on evolutionary scale, they are still diverse enough to be distinguished

---

* On leave from Instytut Informatyki, Uniwersytet Wrocławski, Wrocław, Poland.

I. Jonassen and J. Kim (Eds.): WABI 2004, LNBI 3240, pp. 362–373, 2004.
© Springer-Verlag Berlin Heidelberg 2004

from each other (Aman *et al.*, 1995). Recently, we have shown that computationally derived signature from the variable region of 16r DNA sequences could be used to design genus-specific primers, which was successfully tested for genus *Pseudomonas* through wet experiments (Purohit *et al.*, 2003). Prior to this, Monstein *et al.* (1996) demonstrated that genus-specific hybridization probes for *Chlamydia, Helicobacter,* and *Mobiluncus* located within variable regions V3, V4 and V9 of 16S rDNA specifically bound to the corresponding polymerase chain reaction (PCR) products obtained from pure cultures of the three genera. Also, genus-specific oligonucleotides were designed, which allow rapid detection of members of the genera *Pseudonocardia* and *Saccharopolyspora* by measure of PCR specific amplification (Moron *et al.*, 1999). In another study by Chen *et al.* (2000), they identified a 20-bp oligonucleotide that is highly characteristic of *Xyl. Fastidiosa* using the 16S rDNA sequence data. A signature region was identified within 16S rDNA sequences of *Aeromonas popoffii*, which differentiates *A. popoffii* strains from all other members of the genus *Aeromonas* (Demarta *et al.*, 1999). There are some online reports mentioning about the presence of signature oligonucleotides to position any unknown bacterium in the established phylogenetic tree. A variable α region in 16S rDNA sequences of genus *Streptomyces* has been used to select short nucleotide sequences for detecting the species of this genus (Kataoka, 1997).

On similar lines, our main interest in this exercise is to select short signature pattern(s) from the variable region of 16S rDNA sequence that would represent the genus. Here, we envisaged two possibilities - first, the sub-sequence belonging to variable region itself can act as a signature to a genus and second, the consensus pattern/string derived from the variable region of a set of aligned sequences belonging to genus can also act as a signature to a genus. By consensus pattern we mean, a pattern developed from the aligned sequences with or without considering the correlations of bases in the columns of the alignment. To explore the second possibility, we used the existing method of finding consensus patterns based on weight matrix as well as developed two new methods that makes use of correlation of bases in the columns of aligned sequences. Also a new method based on minimization of Hamming's distance has been used to determine the consensus pattern. Further, we define fitness of a pattern in terms of its specificity to a genus using L_2-norm. The system and methods part of the manuscript describes the four methods and the fitness procedure, while the implementation part describes an experiment using 16S rDNA sequence data belonging to five closely related bacterial genera. The results and discussion part details the fitness of consensus patterns as well as the sample sub-sequences in each genus to arrive at the signature string for each bacterial genus.

2 System and Methods

2.1 Consensus Pattern/String

To find a signature pattern s for a given set of sequences $\mathcal{S} = \{s_1, .., s_n\}$ common approaches studied in the literature use *consensus pattern*: a sequence s minimizing $\sum_{i=1}^{n} d(s_i, s)$ and *closest string*: a sequence s minimizing δ such that for every $s_i \in \mathcal{S}$, $d(s_i, s) \leq \delta$. In both cases d can be considered as Hamming or as edit distance. However, for certain problems in DNA and protein sequence analysis, both consensus pattern and closest string deliver highly nonspecific signatures. Particularly in determining the genus specific signature in short 16S rDNA sequences the strings do not deliver satisfactory solutions.

Below, the methods for deriving consensus pattern for a given aligned sequence $s_1, .., s_n$, each of length m, used in our experiments are described. The sequences are typically over alphabet $\Sigma = \{A, C, G, T\}$ for un-gapped alignment and in case of gapped alignment $\Sigma = \{A, C, G, T, -\}$.

a) **Weight matrix (Hertz and Stormo, 1999)** We consider a simple matrix whose rows correspond to symbols $a \in \Sigma$, columns correspond to positions i in sequences, with $1 \leq i \leq m$, and elements of the matrix are the weights obtained by using following expression:

$$\ln \frac{(f_i(a) + p(a))/(m+1)}{p(a)},$$

where $f_i(a)$ is the number of times symbol a appears at i^{th} position and $p(a)$ is the a priori probability of a (0.25 for each letter) (Hertz and Stormo, 1999). The bases with maximum scores in each column are selected to obtain the overall score; and the bases with maximum scores in each column result into the consensus pattern. This method does not consider correlation of bases amongst the columns of alignment.

b) **Mutual information** This method provides the inter-relationship of two or more variables in a system, which has been successfully used for gene expression data by Steuer et al., (2002). The variables, which are not statistically independent, suggest the existence of some functional relation between them. There are several approaches to quantify the linear dependence between the variables; and mutual information based on information theory provides a general measure of dependencies between the variables (Shannon, 1948).

The Shannon entropy is a good measure to study the positional conservation of bases along the columns of alignment. It provides the average uncertainty of an outcome at different positions and for the ith position is given by the expression

$$H(i) = - \sum_{a \in \Sigma} p_i(a) \log_2 p_i(a),$$

where $p_i(a)$ is the probability that i^{th} column is in state a. The joint entropy $H(i, j)$ for two columns i and j of alignment is defined analogously as

$$H(i, j) = - \sum_{a \in \Sigma} \sum_{b \in \Sigma} p_{i,j}(a, b) \log_2 p_{i,j}(a, b),$$

where $p_{i,j}(a, b)$ is the joint probability that i^{th} column is in state a, and j^{th} column in state b. Accordingly, the mutual information (MI) between the ith and the jth columns is given by

$$M(i, j) = H(i) + H(j) - H(i, j).$$

Thus the alignment of length m results into $m \times m$ matrix of mutual information. Considering only upper-diagonal elements of the matrix M, the maximum of MI (*maxMI_i*) for each row i across the columns $j = i + 1, \ldots, m$ is determined. The *maxMI*'s obtained for each row are ranked from highest to lowest. By following the rank order, select the row i and column j corresponding to *maxMI$_i$* and determine the most frequent base pair combination for the columns i and j of the alignment. Accordingly these bases block the positions i and j in the consensus pattern. In the same way, the remaining positions in the consensus pattern are blocked by strictly following the rank order of *maxMI*'s.

c) Bivariate frequency distribution The correlation of bases in any two columns of the aligned sequences can also be studied by considering the bivariate frequency distribution of bases. Any two columns of an un-gapped alignment will have four states viz. A, C, G and T. The paired states in these columns yields a frequency value for each cell of the bivariate frequency table constituting a bivariate frequency distribution. In case of gapped alignment, inclusion of an additional 'gap' character yields a 5×5 table. The four symbols A, C, G and T can be assigned numeric scores $1, 2, 3$ and 4 respectively; while for gapped alignment score 5 is assigned to gap to quantify the correlation C (Gupta, 1986), given by

$$C(i,j) = \frac{\frac{1}{m} \sum_{a=1}^{k} \sum_{b=1}^{k} a\, b f_{i,j}(a,b) - \mu_i \mu_j}{\sigma_i \sigma_j}$$

for $i, j = 1, 2, \ldots, m$ (number of columns) with $i \neq j$ (in general m is the total number of observations in all the cells), k takes value 4 or 5 depending on the un-gapped or gapped alignment respectively, $f_{i,j}(a,b)$ is the number of times pairs of symbols corresponding to a and b appears simultaneously at i^{th} and j^{th} positions. The expected value μ_i is given by

$$\mu_i = \frac{1}{m} \sum_{a=1}^{k} a f_i(a)$$

and the standard deviation σ_i by

$$\sigma_i = \left(\frac{1}{m} \sum_{a=1}^{k} a^2 f_i(a) - \mu_i^2 \right)^{1/2} .$$

The values μ_j and σ_j are defined analogously.

Such correlation between the pairs of aligned columns can be obtained for all the possible combinations (i,j) to generate a correlation matrix. The maximum of absolute correlation along the rows of matrix is determined that gives the best correlating columns. The correlation coefficients so obtained can be ranked from highest to lowest and the most frequent combination of bases in the columns is picked following the rank order of correlation coefficients, resulting into a consensus pattern for a set of sequences.

d) Weighted consensus string The problem of finding a string s close to each string in set $\mathcal{S} = \{s_1, s_2, \ldots, s_n\}$ of sequences, each of length m, is well studied in computer science. In a basic version, one defines closest string as a sequence minimizing

δ such that $d(s_i, s) \leq \delta$ for all $s_i \in \mathcal{S}$. In the present context, the closest string s is termed as consensus pattern for a set $\mathcal{S}$ if it minimizes $\sum_{i=1}^{n} d(s_i, s)$. However, our experiments showed that this approach gives unsatisfactory solutions to our problem delivering highly non-specific consensus patterns. In this paper we define for a function $F_m : [0, m] \to \mathbb{R}$, a weighted consensus string as a median sequence s of length m minimizing

$$\sum_{i=1}^{n} F_m(d(s_i, s)),$$

where $d(s_i, s)$ denotes Hamming distance. The weighted consensus string generalizes in an obvious way the consensus pattern. We use in this case just weight function $F_m(x) = x$. On the other hand for $F_m(x) = m^x$, the weighted consensus string coincides with the closest string (we assume here, without loss of generality, that $m > n$). Another interesting function is $F_m(x) = -m^x$. In this case the weighted consensus string coincides with the farther string s defined as: a sequence maximizing δ such that for every sequence $s_i \in \mathcal{S}$, $d(s_i, s) \geq \delta$. The biological motivation and the computational complexity of finding the farther string has been described by Lanctot et al., (1999). It is well known that both problems of finding the closest string and the farther string are NP-complete even for any alphabet Σ with $|\Sigma| = 2$. In this paper we show a general NP-completeness result such that the result of (Frances and Litman, 1997) follows as corollary. Define formally **Weighted Consensus String** problem for a given family of functions $F_m : [0, m] \to \mathbb{R}$ as follows:

INSTANCE: Set $\mathcal{S}$ of strings of length m over an alphabet Σ and a number B.
QUESTION: Is there a string $s \in \Sigma^m$ such that $\sum_{v \in \mathcal{S}} F_m(d(v, s)) \leq B$?

Recall that a function is called strictly convex if its value at the midpoint of every interval in its domain is less than the average of its values at the ends of the interval. In other words, a function $f(x)$ is strictly convex on an interval $[a, b]$ if for any two points $x_1, x_2 \in [a, b]$, $f\left(\frac{x_1 + x_2}{2}\right) < \frac{f(x_1) + f(x_2)}{2}$. If the sign of the inequality is reversed, the function is called strictly concave.

We say that the family of functions $F_m : [0, m] \to \mathbb{R}$, with $m = 1, 2, 3, \ldots$, is p-computable if there exists an algorithm A and some polynomials p and q such that for every integers $m > 0$ and $x \in [0, m]$ A with input x computes $A(x)$ in time $p(m)$ such that

$$|A(x) - f(x)| \leq 2^{-q(m)}$$

and for all integers $x_1, x_2 \in [0, m]$ if $f(x_1) \neq f(x_2)$ then $|f(x_1) - f(x_2)| > 2^{-q(m)}$.

Theorem 1 *If $F_m : [0, m] \to \mathbb{R}$ is a p-computable family of functions such that every F_m is strictly convex in $[0, m]$ then the Weighted Consensus String problem for F_m is NP-complete even for any alphabet Σ with $|\Sigma| = 2$.*

Note that $F_m(x) = m^x$ is p-computable and strictly convex in $[0, m]$. Hence the completeness results for finding the closest strings of (Frances and Litman, 1997) follows. On the other hand if F_m is a linear function then, the problem can be solved in linear time.

In our experiments we have used the following weight functions: $\log_2(x+1)$, x, x^2, 2^x, m^x. The most specific signatures to the considered bacterial groups have been obtained for $F_m(x) = \log_2(x+1)$. Note that our first three methods discussed in the previous sections are computationally tractable. On the other hand, from the theorem above, finding weighted consensus string for $x^2, 2^x$, and m^x is hard. We conjecture that the problem remains NP-complete for p-computable strictly concave functions. In particular we conjecture that solving the problem for $F_m(x) = \log_2(x+1)$ is hard, too. However, our heuristics of finding the weighted consensus strings for these functions on the high performance computer (SunFire 15K at the University of Lübeck, Germany) work very well even for $n = 50$ and $m = 32$.

2.2 Fitness of Pattern Based on BLAST

The fitness of any pattern, in terms of its specificity to genus, can be tested based on BLAST results. One of the online servers with BLAST program is available at Swiss Bioinformatics Institute (SBI) web site. For a given query sequence s, the program performs search against the target database and displays results with sequence Id, bacterial name, and matching score between the query sequence and sequences in the target database. For a query sequence belonging to a bacterial genus, the BLAST program results into specific/non-specific or both the type of hits to the genus depending upon the query sequence. Here by non-specific hit, we mean the hit other than the parent bacteria to which the query sequence belongs. The fitness of a query sequence can be decided based on number of specific hits (n_sp), sum of scores of specific hits (sum_sp), number of non-specific hits (n_nsp) and the sum of non-specific hits (sum_nsp). Thus, we define ordered pairs (n_sp, sum_sp) and (n_nsp, sum_nsp) for a query sequence. L_2-norm for both these vectors can be obtained as $L_2\text{-}sp = (n\_sp^2 + sum\_sp^2)^{1/2}$ and $L_2\text{-}nsp = (n\_nsp^2 + sum\_nsp^2)^{1/2}$ respectively. These L_2-norm values can be used to represent a query sequence as ($L_2\text{-}sp, L_2\text{-}nsp$). A sequence/pattern with high $L_2\text{-}sp$ and low $L_2\text{-}nsp$ can be regarded as fittest and can be regarded as the signature pattern of the bacterial genus.

3   Implementation

We had shown that for *Pseudomonas*, the genus-specific signature could be developed from the variable region of 16S rDNA gene, which subsequently could be used to design primers for rapid detection of the genus in environmental samples. The identified variable region is enclosed between the repeating marker CAGCAG. A sub-sequence belonging to this variable region and representing one of the sequences from the training set yielded good specificity to genus *Pseudomonas* and was considered as signature pattern for the genus (Purohit *et al.* 2003). However, in this exercise, our interest is to derive consensus pattern from the training set comprising of all the sub sequences. The consensus pattern may not be from the sample set but still is specific to the genus and could be regarded as signature to the genus. Accordingly, the methods described above were used on sub-sequences from the same variable region of *Pseudomonas*. In addition four other bacterial families viz., *Acinetobacter, Alcaligenes, Burkholderia* and *Moraxella* were also considered in the case study.

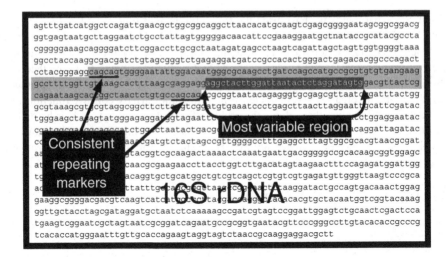

Fig. 1. The Most variable region enclosed between the two fixed repeating markers CAGCAG in 16S rDNA sequence.

The 16S rDNA sequences representing different species of each genus were retrieved from NCBI site. Only the complete sequences with size of around 1.5 kb were considered during sampling, while partial sequences were ignored. Since the selected five genera are taxonomically closely related, the sequences of respective families were searched for the presence of above repeating hexamer. A program 'Repeat Tuple Search' available at www.ebi.ac.uk/~lijnzaad/RepeatTupleSearch was used to identify repeats in sequences. The repeating pattern CAGCAG was dominantly observed in the sequences of all the genera with the same separating distance of 167 nucleotides, except for *Acinetobacter* in which one insertion was observed making the separating distance 168 in sequences of this group. To know whether the repeating markers possess a variable region like *Pseudomonas*, sub-sequences enclosed between the repeats of CAGCAG in the respective groups were aligned. The randomness of nucleotides across the columns of alignment was quantified using Shannon entropy. The most variable region was observed between these repeats in each genus and surprisingly the length and location of the region in each group were almost the same. Figure 1 shows the repeating pattern CAGCAG and the location of most variable region between the repeating markers in sequences. A contiguous stretch of 31 nucleotides (32 in case of gapped alignment), showing high variability was retrieved from each sequence of the respective genera. During extraction, base pair AA at the 5′-end and a base pair TG at the 3′-end were treated as fixed markers in *Alcaligenes, Burkholderia, Moraxella* and *Pseudomonas*. In *Acinetobacter*, base pair AG was considered at 5′-end and the same base pair TG at the 3′-end were treated as fixed markers constituting a stretch of 31 bases. Finally, the aligned sets of sub-sequences of length (31/32) of respective genera were analysed independently to obtain consensus pattern(s) using the methods described above.

Table 1. Consensus strings obtained for selected bacterial groups using four methods: Mutual information (the 1-st string in each group), Bivariate frequency distribution (2-nd string), Weight matrix (3-th string), and Weighted consensus string (4-th string).

Genus	Consensus String	Genus	Consensus String
Acinetobacter	AGGCTACTTTAGTTAATACCTAGGGATAGTG	*Moraxella*	AAAAGCTTATGGTTAATACCCATAAGCCCTG
	AGGCTACTTTAGTTAATACCTAGGGATAGTG		AAAAGCTTATGGTTAATACCCATAAGCCCTG
	AGGCTACTTTAGTTAATACCTAGGGATAGTG		AAAAGCTTATGGTTAATACCCATAAGCCCTG
	AGGCTACTTTAGTTAATACCTAGAGATAGTG		AAAAGCTAATGGTTAATACCCATGAGCCCTG
Alcaligenes	AAACGGCGCGGGCTAATACCCCG_TGCTACTG	*Pseudomonas*	AAGGGCAGTAACCTAATACGTTACT_GTTTTG
	AAACGGCACGGGCTAATACCCCG_TGGAACTG		AAGGGCATTAACCTAATACGTTAGT_GTTTTG
	AAACGGCGCGGGCTAATACCCCG_CGGAACTG		AAGGGCAGTAACCTAATACGTTGCT_GTTTTG
	AAACGTCGCGGGCTAATACCCCG_TGGAACTG		AAGGGCATTAACCTAATACGTTAGT_GTTTTG
Burkholderia	AAATCCTCGGGGCTAATACCGCCGGGGGATG		
	AAATCCTTGGCTCTAATATAGCCGGGGGATG		
	AAATCCTTGGTTCTAATATAGCCGGGGGATG		
	AAATCCTTGGTTCTAATATAGCCGGGGGATG		

Although, the same repeating marker was observed dominantly in the sequences of these five genera, it has been observed that *Alcaligenes* and *Acinetobacter* are not very close to the other three genera based on their dinucleotide frequencies (Raje *et al.* 2002). Hence these genera were considered to evaluate whether the variable region within the same repeating marker yield the signature pattern for the respective genus.

4   Result and Discussion

The application of methods resulted into consensus patterns in different genera, which are as shown in Table 1. In *Acinetobacter*, first three methods yielded the same consensus pattern, while the pattern generated by last method differed only at one base position compared to others. Similarly in *Moraxella*, the consensus pattern obtained by using first three methods was same, while the last method yielded a pattern, which differed at two base positions compared to others. In *Alcaligenes, Burkholderia* and *Pseudomonas*, the patterns obtained by different methods were different from each other. Another observation was that there are few positions in this stretch of 31 bases that are highly conserved across the bacterial genera. For instance, a 5-tuple TAATA starting at position 14 with reference to base $5'$ base A is conserved in all the sequences of the bacterial genera; while other base positions except base A at $5'$-end and TG at $3'$-end show variations across the genera.

4.1   Fitness of Patterns in Different Groups

The fitness of each pattern was determined in terms of L_2-norm by considering the specific and non-specific hits for each pattern, which resulted into scatter plots as shown in Figure 2(a-e). This analysis provides critical evaluation of the pattern with their non-specific hits since these consensus patterns finally would be used in the experiments where due to heterogeneity of the target DNA their specificity is important. In order to test the fitness of consensus patterns obtained using different methods, each pattern

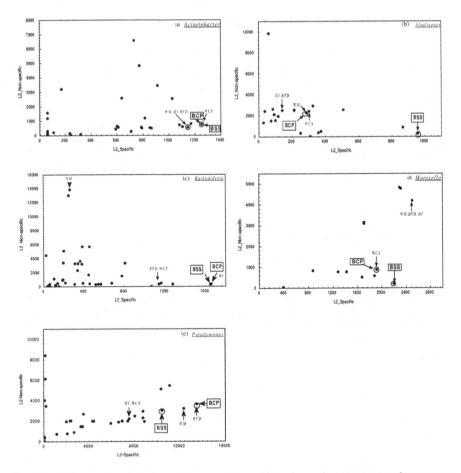

Fig. 2(a–e). The scatter plot of specifity vs. non-specifity of sample sequences and consensus patterns based on L2-norm. The points indicated by arrow are the consensus patterns obtained by using that method. The rest of the points indicate sample squences from the training set. BCS and BCP are the best sample sequence and the best consensus pattern respectively.

was tested for its specificity using BLAST against 16S rDNA database. Similarly, all the sample sequences from the training set were also tested for their specificity. Each pattern resulted into specific and non-specific hits along with a matching score corresponding to each hit. Each point on the $(L_2\text{-}sp, L_2\text{-}nsp)$ plane represents either a consensus or a sample sequence. The encircled points in each figure represent either the Best Consensus Pattern (BCP) or the Best Sample Sequence (BSS). We select the best pattern/sample sequence for each group as follows. Let $max\text{-}sp$ be the maximum value among all $L_2\text{-}sp$ values for the group. The best sample/pattern is the sequence, which minimises the Euclidean distance to $(max\text{-}sp, 0)$. Accordingly, the best pattern amongst the set of consensus patterns and the set of sample sequences were identified.

It has been observed that the consensus pattern is not necessarily the best representative pattern of the genus. For instance, in *Alcaligenes* and *Moraxella*, a sample sequence from the training set was found better than the best consensus pattern, while in *Acinetobacter*, the best consensus pattern coincides with the best sample sequence. However, in *Burkholderia* and *Pseudomonas*, the consensus pattern was found better than the respective sample sequences. The methods providing the best consensus patterns in different genera are shown in Table 2. The extent of similarity between the BCP and the BSS in each genus has been expressed in terms of edit distance as shown in Table 2. Except *Alcaligenes*, there is marginal difference between these two types of best patterns. In *Alcaligenes*, the BCP and BSS show mismatch at 12 positions, which is also reflected on the BLAST results in Figure 2b. The specificity of BCP is much lower than that of BSS unlike in other genera. In case of *Burkholderia*, the presence of base T at the 19th position results into a signature sequence of the genus. In *Moraxella*, the simultaneous occurrence of base A and G at positions 8 and 24, while in *Pseudomonas*, the simultaneous occurrence of C and G at positions 12 and 20 results into the respective signature patterns of the genera. In brief, Table 2 reveals that there is not much difference between the best consensus pattern and the best sample sequence in four out of five genera and there are few sensitive base positions and the bases that govern the specificity of patterns to the genus. The patterns indicated by asterisk can be regarded as signature for the respective bacterial genus that could be used in designing genus specific primers or even in designing hybridization arrays.

Earlier, Zhang *et al.* (2002) have established that highly specific patterns exist in the bacterial 16S rDNA sequence data set in large numbers using a measure of signature quality index. Although, the objective of our work matches with that of the work carried out by Zhang *et al.*, the approaches are different. They worked with oligonucleotides of specific length in each sequence of database starting from the 5'-end till the 3'-end of sequence. Contrary to this, we are targeting only the mis-matched region within the sequence and focussing on the sub-sequences in this region. Secondly, they defined signature quality index as a measure to decide the belongingness of the oligonucleotide to a particular phylogenetic cluster. We have used L_2-norm based on the number of specific and non-specific hits as well as the extent of matching of the query sequence with the target data set, which ultimately decides the fitness of query sequence as a signature of the bacterial family. Thus, with an additional level of information i.e. degree of matching, we have established the suitability of a sequence as a signature to a particular genus.

5   Conclusions

The genesis of this study was to derive consensus pattern for a bacterial genus and check its fitness/specificity to the parent genus to decide whether it could be regarded as a signature pattern. In this paper, we have demonstrated how consensus patterns could be derived from the most variable region of 16S rDNA using four different methods. One of patterns was identified as the best consensus pattern based on BLAST results. The case study with five different genera revealed that there is no single method that could provide the best consensus pattern.

Further, in *Alcaligenes* and *Moraxella*, the BCPs are not the signature patterns, rather, the sample sequence provided better specificity than the BCPs. The reasons

Table 2. Methods giving the best consensus string (BCS) and its edit distance from the best sample string (BSS).

Group	Method				Edit distance[#]
	WCS	MI	BFD	WM	
Acinetobacter	×				(0) AGGCTACTTTAGTTAATACCTAGAGATAGTG (BCS)⋆ AGGCTACTTTAGTTAATACCTAGAGATAGTG (BSS)⋆
Alcaligenes				×	(12) AAACGGCGCGGGCTAATACCCCG-TGCTACTG (BCS) | |||| | | ||| || AAAAGGTTTCGGATAATACCC-GGAACTGATG (BSS)⋆
Burkholderia		×			(1) AAATCCTTGGCGCTAATACAGCCGGGGATG (BCS) | AAATCCTTGGCTCTAATACAGCCGGGGATG (BSS)⋆
Moraxella	×				(2) AAAAGCTAATGGTTAATACCCATGAGCCCTG (BCS) | | AAAAGCTTATGGTTAATACCCATAAGCCCTG (BSS)⋆
Pseudomonas			×		(2) AAGGGCAGTAACCTAATACGTTGCT-GTTTTG (BCS)⋆ | | AAGGGCAGTAAGCTAATACCTTGCT-GTTTTG (BSS)

⋆: indicates signature pattern for the group being better than the other; #: numbers in the parenthesis; MI: Mutual information; BFD: Bivariate frequency distribution; WM: Weight matrix; WCS: Weighted consensus string.

could be the limited sequence data available on these genera, as regards species level diversity. Secondly, it is quite possible that the consensus pattern derived from some other variable region of 16S rDNA could be the signature pattern. Hence, it is necessary to target different variable regions in 16S rDNA sequences to generate consensus patterns, some of which could be the BCPs representing the genus. The combination of such BCPs could be used to design hybridization arrays for authentic identification of specific bacteria or even a group of bacteria.

Thus, a case study with few bacterial genera suggests that the relationship of bases in the most variable region of 16S rDNA, as established through consensus pattern, could have relevance to genus-specificity. However, to substantiate this conclusion, the approach needs to be tested in large number of bacteria that would lead to a library of genus-specific signatures. It is anticipated that such characteristic signatures could be used in rapid identification of unknown cultures through wet experimentation.

References

1. R.I. Amann, W. Ludwig, and K-H. Schleifer (1995), *Phylogenetic identification and in situ detection of individual microbial cells without cultivation*, Microbiol. Rev., 59, 143-169.
2. J. Chen, D. Banks, R.L. Jarret, C.J. Chang, B.J. Smith (2000), *Use of 16S rDNA sequences as signature characters to identify Xylella fastidiosa*, Curr Microbiol., 40, 29-33.

3. A. Demarta, A.-P. Tonolla, N. Caminada, R. Ruggeri,and R. Peduzzi (1999), *Signature region within 16S rDNA sequences of Aeromonas popofii*, FEMS Microbiol. Letts, 172, 239-246.

4. G.E. Fox, E. Stackebrandt, R.B. Hespel, J. Gibson, J. Maniloff, T.A. Dyer, R.S. Wolfe, W.E. Balch, R.S. Tanner, L.J. Magrum, L.B. Zablen, R. Blakemore, R. Gupta, L. Bonen, B.J. Lewis, D.A. Stahl, K.R. Luchrsen, K.N. Chen, and C.R. Woese (1980), *The phylogeny of prokaryotes*, Science, 209, 457-463.

5. M. Frances and A. Litman (1997), *On covering problems of codes*, Theory of Comput. Syst. 30(2), 113-119.

6. S. Gupta and V. Kapoor (1986), *Fundamentals of Mathematical Statistics*, Sultan Chand & Sons, New Delhi.

7. G. Hertz and G. Stormo (1999), *Identifying DNA and Protein patterns with statistically significant alignments of multiple sequences*, Bioinformatics, 15, 563-577.

8. M. Kataoka, K. Ueda, K. Takuji, S. Tatsuji, and Y. Toshiomi (1997), *Application of the variable region in 16S rDNA to create an index for rapid species identification in the genus Streptomyces* FEMS Microbiol. Letts. 151, 249-255.

9. J. Lanctot, M. Li, B. Ma, S. Wang, and L. Zhang (1999), *Distinguishing string selection problems*, Proc. 10th ACM-SIAM Symposium on Discrete Algorithms (SODA), 633-642, ACM Press.

10. H.J. Monstein, E. Kihlostrom, and A. Tiveljung (1996), *Detection and identification of bacteria using in-house broad range 16S rDNA PCR amplification and genus-specific hybridization probes, located within variable regions of 16S rRNA genes*, APMIS, 104(6), 451-458.

11. R. Moron, I. Gonzalez, O. Genilloud (1999), *New genus-specific primers for the PCR identification of members of the genera Pseudonocardia and Saccharopolyspora*, Int. J. Syst. Bacteriol. 49, 142-162.

12. H.J. Purohit, D.V. Raje, and A. Kapley (2003), *Identification of signature and primers specific to genus Pseudomonas using mismatched patterns of 16S rDNA sequences*, BMC Bioinformatics, 4, 19.

13. D.V. Raje, H.J. Purohit, and R.S. Singh (2002), *Distinguishing features of 16S rDNA gene for five dominating bacterial genus observed in bioremediation*, Journal of Computational Biology, 9(6), 819-829.

14. C. Shannon (1948), *A mathematical theory of communication*, The Bell System Techn. Journal, 27, 379-423.

15. R. Steuer, J. Kurths, C.O. Daub, J. Weise, and J. Selbig (2002), *The mutual information: Detecting and evaluating dependencies between variables*, Bioinformatics, 18, S231-S240.

16. Z. Zhang, R.C. Willson, and G. Fox, (2002), *Identification of characteristic oligonulceotides in the bacterial 16S ribosomal RNA sequence dataset*, Bioinformatics, 18, 244-250.

Topological Rearrangements and Local Search Method for Tandem Duplication Trees

Denis Bertrand and Olivier Gascuel

Equipe Méthodes et Algorithmes pour la Bioinformatique LIRMM-CNRS
161 rue Ada 34392 Montpellier Cedex 5 - France
{dbertran,gascuel}@lirmm.fr

Abstract. The problem of reconstructing the duplication history of a set of tandemly repeated sequences was first introduced by Fitch (1977). Many recent works deal with this problem, showing the validity of the unequal recombination model proposed by Fitch, describing numerous inference algorithms, and exploring the combinatorial properties of these new mathematical objects, which are duplication trees (DT). In this paper, we deal with the topological rearrangement of these trees. Classical rearrangements used in phylogeny (NNI, SPR, TBR, ...) cannot be applied directly on DT. We demonstrate that restricting the neighborhood defined by the SPR (Subtree Pruning and Re-grafting) rearrangement to valid duplication trees, allows exploring the whole space of DT. We use these restricted rearrangements in a local search method which improves an initial tree via successive rearrangements and optimizes the parsimony criterion. We show through simulations that this method improves all existing programs for both reconstructing the initial tree and recovering its duplication events.

1 Introduction

Tandemly repeated DNA sequences consist of two or more adjacent copies of a stretch of DNA. They arise from tandem duplication, in which a sequence of DNA (which may itself contain several copies) is transformed into two adjacent and identical sequences. Since copies are then free to evolve independently and likely to undergo additional mutation events (substitutions, insertions, deletions, etc.), they become approximate over time. Unequal recombination is widely viewed as the predominant biological mechanism responsible for the production of tandemly repeated sequences [6, 10, 19, 21, 26], at least when the basic repeated motif is large, *e.g.* when it contains a gene.

Gene duplication (in tandem or not) is one of the most important evolutionary mechanisms for producing genes with novel functionalities [21]. Reconstructing the history of duplicated sequences would be very beneficial for scientists studying their function and evolution. The problem of reconstructing the duplication history of tandemly repeated sequences was first considered by Fitch [10]. It has not received much attention until recently, probably due to the lack of available repeated sequence data, and also since there has been no dedicated algorithm.

I. Jonassen and J. Kim (Eds.): WABI 2004, LNBI 3240, pp. 374–387, 2004.
© Springer-Verlag Berlin Heidelberg 2004

However, several specific algorithms which take the ordered nature of tandemly repeated sequences into account were recently described in the literature. In [3,18,30] the authors provide reconstruction algorithms based on the parsimony principle, but these are limited either to the special case where only single copy duplications occurred (*i.e.* when the duplicated fragment always contains a single basic copy), or to the analysis of short tandem repeats (minisatellites). In [6] we present an exhaustive algorithm that computes the most parsimonious duplication tree; this algorithm (by its exhaustive approach) is limited to datasets of less than 15 repeats. Several distance based methods have also been described. The WINDOW method [30] uses an agglomeration scheme similar to UPGMA [27] and NJ [24], but the cost function used to judge potential duplication is based on the assumption that the sequences follow a molecular clock mode of evolution. The DTSCORE method [4] uses the same scheme but corrects this limitation using a score criterion [2], like ADDTREE [25]. This method can be used with sequences that do not follow the molecular clock mode of evolution which is, for example, essential when dealing with gene families containing pseudo-genes that evolve much faster than functional genes. An exact and polynomial distance based algorithm to reconstruct single copy tandem duplication trees is also described in [5]. Finally, GREEDY SEARCH [32] corresponds to a different approach divided into two steps: first a phylogeny is computed with a classical reconstruction method (NJ), then, with nearest neighbor interchange (NNI) rearrangements, a duplication tree close to this phylogeny is computed. This approach is noteworthy since it implements topological rearrangements which are highly useful in phylogenetics [29], but it works blindly and does not ensure that good duplication trees will be found (cf. section 5.2).

Topological rearrangements have an essential function in phylogenetic inference, where there are used to improve an initial phylogeny by subtree movement or exchange. Rearrangements are very useful for all common criteria (parsimony, distance, maximum likelihood) and are integrated into all classical programs like PAUP* [28] or PHYLIP [7]. Furthermore, they are used to define various distances between phylogenies and are the foundation of many mathematical works. Unfortunately, they cannot be directly used here, as shown by a simple example given later. Indeed, when applied to a duplication tree, they do not guarantee that another valid duplication tree will be produced.

In this paper, we describe a set of topological rearrangements to stay inside the DT space and explore the whole space from any of its elements, then we show the advantages of this approach for duplication tree inference from sequences. This paper is organized as follows. In section 2, we describe the duplication model introduced by [6,10,30], as well as an algorithm for recognition of duplication trees. Thanks to this algorithm, we restrict the neighborhood defined by NNI and SPR to valid duplication trees (Section 3). We demonstrate that, for NNI movement, this restricted neighborhood does not allow the exploration of the whole DT space. On the other hand, we demonstrate that the restricted neighborhood of SPR rearrangement allows the whole space of DT to be explored. In this way, we define a local search method, applied here to parsimony (section 4). In section 5, we compare (using simulations) this method to all the other approaches, and in section 6 we conclude by discussing the positive results obtained by our method, and indicate directions for further research.

2   Model

2.1   Duplication History and Duplication Tree

The tandem duplication model used in this article was first introduced by Fitch (1977) [10] then studied independently by [6, 30]. It is based on unequal recombination which is assumed to be the sole evolution mechanism (except punctual mutations) acting on sequences. Although being a completely different biological mechanism, slipped-strand mispairing leads to the same duplication model [1, 3].

Let $O = (1, 2, \ldots, n)$ be the ordered set of sequences representing a locus, as can now be observed. Initially containing a single copy, the locus grows through a series of consecutive duplications. As shown in Figure 1(a), a **duplication history** may contain simple duplication events. When the duplicated fragment contains 2, 3 or k repeats, we say that it involves a multiple duplication event. Under this duplication model, a duplication history is a rooted tree with n labelled and ordered leaves, in which internal nodes of degree 3 correspond to duplication events. In a real duplication history (Figure 1(a)), the time intervals between consecutive duplications are completely known, and the internal nodes are ordered from top to bottom according to the moment they occurred in the course of evolution. Any ordered segment set of the same height then represents an ancestral state of the locus. We call such a set a **floor**, and we say that two nodes i, j are **adjacent** ($i \prec j$) if there is a floor where i and j are consecutive.

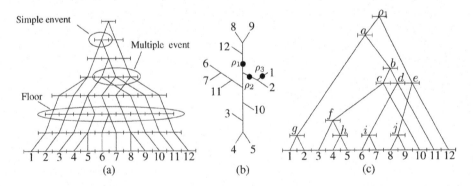

Fig. 1. (a) Duplication history; each segment represents a copy; extant segments are numbered. (b) Duplication tree (DT); the black points show the possible root locations. (c) Rooted duplication tree (RDT) corresponding to history (a) and root position ρ_1 on (b).

However, in the absence of a molecular clock mode of evolution (which is usually the case), it is impossible to recover the order between the duplication events of two different lineages from the sequences. In this case, we are only able to infer a **duplication tree (DT)** (Figure 1(b)) or a **rooted duplication tree** (Figure 1(c)).

A duplication tree is an unrooted phylogeny with ordered leaves, whose topology is compatible with at least one duplication history. Also, internal nodes of duplication trees are partitioned into **events** (or "blocks" following [30]), each containing one or more (ordered) nodes. We distinguish "simple" duplication events that contain a unique internal node (e.g. b and f in Figure 1(c)) and "multiple" duplication events which group a series of adjacent and simultaneous duplications (e.g. c, d and e in Figure 1(c)). Let $E = (s_i, s_{i+1}, \ldots, s_k)$ denote an event containing internal nodes s_i, s_{i+1}, ..., s_k in left to right order. We say that two consecutive nodes of the same event are **adjacent** ($s_j \prec s_{j+1}$) just like in histories, as any event belongs to a floor in all of the histories that are compatible with the DT being considered. The same notation will also be used for leaves to express the segment order in the extant locus. When the tree is rooted, every internal node u is unambiguously associated to one parent and two child nodes; moreover, one child of u is "left" and the other one is "right", which is denoted as l_u and r_u, respectively. In this case, for any duplication history that is compatible with this tree, son nodes of an event are organized as follows: $l_{s_i} \prec l_{s_{i+1}} \prec \ldots \prec l_{s_k} \prec r_{s_i} \prec r_{s_{i+1}} \prec \ldots \prec r_{s_k}$.

In [6, 13] it was shown that rooting a duplication tree is different from rooting a phylogeny: the root of a duplication tree necessarily lies on the tree path between the most distant repeats on the locus, *i.e.* 1 and n; moreover, the root is always located "above" all multiple duplications, e.g. Figure 1(b) shows that there are only three valid root positions, the root cannot be a direct ancestor of 12.

2.2   Recursive Definition of Rooted and Unrooted Duplication Trees

A duplication tree is compatible with at least one duplication history. This suggests a recursive definition, which progressively reconstructs a possible history, given a phylogeny T and a leaf ordering O. We define a **cherry** (l, u, r) as a pair of leaves (l and r) separated by a single node u in T and we call $C(T)$ the set of cherries of T. This recursive definition reverses evolution: it searches for a "visible duplication event", "agglomerates" this event and checks whether the "reduced" tree is a duplication tree. In case of rooted trees, we have:

(T, O) defines a duplication tree with root ρ if and only if:

(i) (T, O) only contains ρ, or
(ii) there is in $C(T)$ a series of cherries $(l_i, u_i, r_i), (l_{i+1}, u_{i+1}, r_{i+1}), \ldots, (l_k, u_k, r_k)$ with $k \geq i$ and $l_{s_i} \prec l_{s_{i+1}} \prec \ldots \prec l_{s_k} \prec r_{s_i} \prec r_{s_{i+1}} \prec \ldots \prec r_{s_k}$ in O,

such that (T', O') defines a duplication tree with root ρ, where T' is obtained from T by removing $l_i, l_{i+1}, \ldots, l_k, r_i, r_{i+1}, \ldots, r_k$, and O' is obtained by replacing ($l_i, l_{i+1}, \ldots, l_k, r_i, r_{i+1}, \ldots, r_k$) by $(u_i, u_{i+1}, \ldots, u_k)$ in O.

The definition for unrooted trees is quite similar:

(T, O) defines an unrooted duplication tree if and only if:

(i) (T, O) contains 1 segment, or
(ii) same as for rooted trees with (T', O') now defining an unrooted duplication tree. .

Those definitions provide a recursive algorithm, **PDT** (Possible Duplication Tree), to check whether any given phylogeny with ordered leaves is a duplication tree. In case of success, this algorithm can also be used to reconstruct duplication events: at each step the series of internal nodes above denoted as $(u_i, u_{i+1}, \ldots, u_k)$ is a duplication event. When the tree is rooted, l_j is the left child of u_j and r_j its right child, for every $j, i \leq j \leq k$. This algorithm can be implemented in $O(n)$ [13] where n is the number of leaves. Another linear algorithm is proposed by Zhang *et al.* [32] using a top down approach instead of a bottom-up one and applying only to RDT.

3   Topological Rearrangements for DT

This section shows how to explore the DT space thanks to SPR rearrangements. First, we describe some NNI and SPR rearrangement properties with standard phylogenies. But these rearrangements cannot be directly used to explore the DT space. Indeed, when applied to a duplication tree, they do not guarantee that another valid duplication tree will be produced. So we decided to restrict, using PDT, the neighborhood defined by those rearrangements to duplication trees. For the NNI movements the neighborhood is too restricted (as shown by a simple example), and do not allow the whole DT space to be explored. On the other hand, we can distinguish two types of SPR rearrangements which, when applied to a rooted duplication tree guarantee that another valid duplication tree will be produced. Thanks to this specific rearrangements, we demonstrate that restricting the neighborhood of SPR rearrangements, using PDT, allows the whole space of DT to be explored.

3.1   Topological Rearrangements for Phylogeny

There are many ways of carrying out topological rearrangements on phylogeny [29]. We only describe **NNI** (Nearest Neighbor Interchange) and **SPR** (Subtree Pruning Re-grafting) rearrangements.

The NNI movement is a simple rearrangement which exchanges a subtree adjacent to the same internal edge (Figures 2 and 3). There are two possible neighbors for each internal edge, so $2n - 6$ neighbors for one tree (n represents the number of leaves). This rearrangement allows the whole space of phylogeny to be explored, *i.e.* there is a succession of NNI movements making it possible to transform all phylogeny P_1 into any phylogeny P_2 [29].

The SPR movement consists in pruning a subtree and re-grafting it, by its root, to an edge of the resulting tree (Figure 5 and 6). We notice that the neighborhood defined by the NNI rearrangement is included into the neighborhood defined by SPR. This last rearrangement has a larger neighborhood of size $O(n^2)$.

3.2 NNI Rearrangements Do Not Stay in DT Space

The classical phylogenetic rearrangements (NNI, SPR, TBR, ...) do not stay in DT space. So, if we apply an NNI to a DT (Figure 2), the resulting tree is not always a valid DT. This property is also true for SPR and TBR rearrangements, since NNI rearrangements are included into this two rearrangement classes.

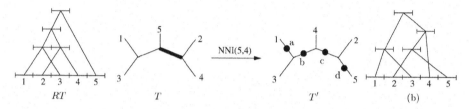

Fig. 2. Tree obtained by applying an NNI movement to DT is not always a valid DT: T whose RT is an unrooted version; T' is obtained by applying NNI(5,4) around the bold edge; none of the possible root positions of T' (a, b, c et d) leads to a valid RDT, *cf.* tree (b) which correspond to root b in T'.

3.3 Restricted NNI Does Not Allow the Whole Space of DT to Be Explored

To restrict the neighborhood defined by NNI rearrangements to duplication trees, each element of the neighborhood is filtered thanks to the PDT algorithm. But this restricted neighborhood does not allow the whole space of DT to be explored. Figure 3 gives an example of a duplication tree, T, the neighborhood of which does not contain any DT. So its restricted neighborhood is empty, and there is no succession of restricted NNI, allowing T to be transformed into any DT.

3.4 Restricted SPR Allows the Whole Space of DT to Be Explored

As before, we restrict, using PDT, the neighborhood defined by SPR rearrangements to duplication trees. We name **restricted SPR**, SPR movements that lead to duplication trees.

Main Theorem. *Let T_1 and T_2 be any given DT; T_1 can be transformed into T_2 via a succession of restricted SPR.*

Proof. To demonstrate the Main Theorem, we define two types of special SPR that ensure to stay within the space of rooted duplication trees (RDT). Thanks

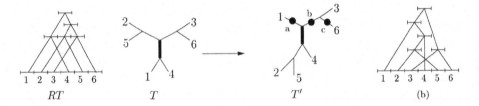

Fig. 3. Neighborhood computation of T an unrooted version of which is RT, with an NNI movement; T' is obtained by exchanging subtrees 1 and (2, 5); none of the possible root positions of T' (a, b et c) leads to a valid duplication tree, $cf.$ tree (b) which corresponds to root b in T'; and it is the same for each neighbor of T obtained by NNI.

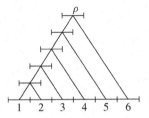

Fig. 4. A 6 leaf caterpillar.

to these two types of SPR, we demonstrate that it is possible to transform all RDT T into a **caterpillar**, $i.e.$ a rooted tree in which all internal nodes belong to the tree path between the leaf 1 and the tree root ρ, $cf.$ Figure 4.

This result demonstrates the theorem. Indeed, let T_1 and T_2 be two RDT. We can transform T_1 and T_2 into a caterpillar by a succession of restricted SPR. So it is possible to transform T_1 into T_2 by a succession of restricted SPR, with (eventually) a caterpillar as intermediate tree. This property holds since the reciprocal movement of an SPR is an SPR. As the two SPR types proposed ensure that we stay within the space of RDT, we have the desired result. Notice that we reasoned with RDT, instead of DT, to simplify the proof. But it is possible, with this property on RDT to infer the same property on DT. Indeed two DT can be arbitrarily rooted, transformed one on an other using restricted SPR, then unrooted.

The first special SPR allows multiple duplication events to be destroyed. Let $E = (s_i, s_{i+1}, \ldots, s_k)$ be a duplication event, r_{s_i} and l_{s_k} respectively right son of s_i and left son of s_k, and let p_{s_i} be the father of s_i. The **DELETE** rearrangement consists of pruning the subtree of root r_{s_i} and grafting this subtree on the edge (s_k, l_{s_k}), while l_{s_i} is renamed s_i and the edge (l_{s_i}, s_i) is deleted. Figure 5 represents this rearrangement.

Lemma 1. *DELETE preserves the RDT property.*

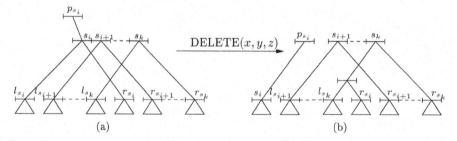

Fig. 5. DELETE rearrangement.

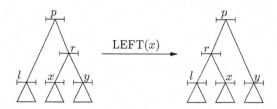

Fig. 6. LEFT rearrangement.

Proof. Let T be the initial tree (Figure 5(a)), $E = (s_i, s_{i+1}, \dots, s_k)$ be an event of T and T' be the resulting tree of the DELETE applying on T (Figure 5(b)). Sons of nodes s_j ($i \leq j \leq k$) are noted l_{s_j} and r_{s_j}. By definition, for any duplication history being compatible with T we have $l_{s_i} \prec l_{s_{i+1}} \prec \dots \prec l_{s_k} \prec r_{s_i} \prec r_{s_{i+1}} \prec \dots \prec r_{s_k}$. So there is a way to (partly) agglomerate T (using PDT) such as these nodes becomes leaves. The same agglomeration can be applied to T'. Now, if we agglomerate the event E of T, and if we reduce T' by agglomerating the cherry (l_{s_k}, g_{s_i}) then the event $(s_{i+1}, \dots, s_k)$, follows two identical trees. ∎

By successively applying DELETE to any duplication tree, we remove all multiple duplication events. The following SPR rearrangement allows duplications to be moved, on **"simple" RDT**, *i.e.* an RDT containing only simple duplications. Let p be a node of a simple RDT T, l its left son, r its right son, and x the left son of r. This rearrangement consists of pruning the subtree of root x and grafting it on the edge (l, p) (Figure 6). This rearrangement is an SPR (even an NNI); we name it **LEFT** since it moves the subtree root towards the left. It is obvious that the tree obtained by applying such a rearrangement to a simple RDT, is a simple RDT. We now establish the following lemma which shows that any simple tree can be transformed into a caterpillar.

Lemma 2. Let T be a simple RDT, T can be transformed into a caterpillar by a LEFT rearrangement succession.

Proof. In a caterpillar all internal nodes belong to the tree path between the leaf 1 and the tree root ρ, and are ancestors of 1. If T is not a caterpillar, there is an internal node d that does not belong to the path between 1 and ρ. So d

is not an ancestor of 1. If r is the right son of its father, we can apply LEFT to the left son of r. If r is the left son of its father, we consider its father: it cannot be the ancestor of 1 since its sons are r and a node on the right of r. And r is not on the path between 1 and ρ. So we can apply the same argument: either it is adequate for performing the movement, or we consider its father again. In this way, we necessarily obtain a node for which the rearrangement is possible. T can be transformed into a caterpillar by successively applying the LEFT rearrangement to all nodes which are not on the path between 1 and ρ, or to one of their ancestors that is necessarily adequate. This concludes the proof of the lemma 2 and therefore of the Main Theorem of the article. ∎

4   Local Search Method

Given a set of aligned and ordered sequences (DNA or proteins), the aim is to find the duplication tree that best explains these sequences, according to usual criteria in phylogenetics, *e.g.* parsimony or minimum evolution. Few studies have focused on the computational hardness of this problem. Jaitly *et al.* [18] shows that finding the optimal simple event duplication tree with parsimony is NP-Hard and that this problem has a PTAS (Polynomial Time Approximation Scheme). Another closely related PTAS is given by Tang *et al.* [30] for the same problem. Moreover, a polynomial distance-based algorithm of time complexity $O(n^3)$ for reconstructing optimal single copy tandem duplication trees is described in [5]. But it is commonly believed, as in phylogeny, that most duplication tree (especially multiple) inference problems are NP-Hard. This explains the development of heuristic approaches [3, 6, 30, 32], and the same line is followed here.

4.1   The LSDT Method

The topological rearrangements described in the previous section naturally yield a local search method for DT. Our method, **LSDT**, follows a classical local search procedure in which, at each step, we try to strictly improve the current tree. This approach can be used to optimize various criteria, like least-squares, parsimony or maximum likelihood. In this study, we adopted a parsimony approach, as parsimony is commonly acknowledged [29] as being a good criterion when dealing with little divergent sequences, which is the case with tandem duplicated genes [6].

Algorithm 1 summarizes our local search procedure. The neighborhood of the current DT, $T_{current}$, is computed using SPR. As we explained before, we use the PDT procedure to restrict the neighborhood to the valid DT. When a tree is a valid DT, its parsimony is computed. That way, we select the best neighbor of $T_{current}$, for this criterion. If this DT improves the best parsimony value obtained so far (the T_{best} value), the local search restarts with this new topology. If no neighbor of $T_{current}$ improves T_{best}, the local search is stopped.

Data: An initial DT, T_{best}, its parsimony P_{best}, and a set S of aligned and ordered sequences.

Result: A locally optimal DT for the parsimony criterion.

while $T_{best} \neq T_{current}$ **do**

$\quad$ $T_{current} \leftarrow T_{best}$

$\quad$ **foreach** DT T_{new} obtained by applying an SPR to $T_{current}$ **do**

$\quad\quad$ **if** PDT(T_{new}) = True **then**

$\quad\quad\quad$ **if** parsimony(T_{new}, S) < P_{best} **then**

$\quad\quad\quad\quad$ replace T_{best} by T_{new} and P_{best} by parsimony(T_{new}, S)

return T_{best}

Algorithm 1. Basic local search algorithm for DT.

4.2 Computing Parsimony

The algorithm that computes the parsimony value of a given tree topology [9, 17] has a relatively high computational complexity: $O(kn)$ where k represents the site number of the alignment. The PDT algorithm, has a linear complexity in n. Hence, the time complexity required by a rough implementation is $O(n^3 + kn^3)$ per local search step; as the neighborhood of any given DT is in $O(n^2)$: n^3 corresponds to the computation of the restricted neighborhood and kn^3 to the parsimony computation of the neighborhood elements.

To speed up this process, we adapted techniques commonly used in phylogeny for fast calculation of parsimony. Our implementation uses a data structure implemented (among others) in DNAPARS [7] and described in [11, 15], and it computes parsimony incrementally, as described in [14]. Thanks to these improvements computing the neighbor with minimum parsimony of any given duplication tree is achieved in $O(n^3 + kn^2)$.

5 Results

5.1 Simulation Protocol

We applied our method, as well as other existing methods, to simulated datasets obtained using the procedure described in [4]. We uniformly randomly generated tandem duplication trees (see [13]) with 12, 24 and 48 leaves and assigned lengths to the edges of these trees using the coalescent model [20]. We then obtained molecular clock trees, which might be unrealistic in numerous cases, e.g. when the sequences being studied contain pseudo-genes which evolve much faster than functional genes. Therefore, we generated non-molecular clock trees from the previous ones by independently multiplying every edge length by $1+0.8X$, where X was drawn from an exponential distribution with parameter 1.0. The trees obtained have a maximum leaf-to-leaf divergence in the range $[0.1, 0.35]$, and the ratio between the longest and shortest root-to-leaf lineages (a measure of departure from molecular clock) is about 3.0 on average. Both values are in accordance with real data, e.g. gene families[6] or repeated protein domains [30].

SEQGEN [23] was used to produce a 1000 bp-long nucleotide multiple alignment from each of the generated trees using Kimura two parameter model of substitution [8], and a distance matrix was computed by DNADIST [7] from this alignment using the same substitution model. 1000 trees (and then 1000 sequence sets and 1000 distance matrices) were generated per tree size. These data sets were used to compare the ability of the various methods to recover the original trees from the sequences or from the distance matrices, depending on the method being tested. We measured the percentage of trees (out of 1000) being correctly reconstructed (%tr). Due to the random process used for generating these trees and datasets, some short branches might not have undergone any substitution (as during Evolution), and thus are unobtainable, except by chance. When n and thus the branch number is high, it becomes hard or impossible to find the entire tree. So we also measured the percentage of duplication events in the true tree being recovered by the inferred tree (%ev). A duplication event involves one or more internal nodes and is the lowest common ancestor of a set of leaves; we say it "covers" its descendent leaves. However, the leaves covered by a simple duplication can change when the root position changes. As regards the true tree, the root is known and each event is defined by the set of leaves which it covers. But the inferred tree is unrooted. To avoid ambiguity, we then tested all possible root positions and chose the one which gave the highest proximity in number of events detected between the true tree and the inferred tree, where two events are identical if they cover the same leaves. Finally, we kept the average parsimony value of each method ($pars$).

5.2   Performance and Comparison

Using this protocol, we compared: NJ [24], GREEDY-SEARCH (GS) [32] when starting from the NJ tree, a modified version of GREEDY TRHIST RESTRICTED (GTR) [3] to infer multiple duplication trees, WINDOWS [30], DTSCORE [4], and 4 versions of our local search method corresponding to different starting duplication trees: GS+LSDT, GTR+LSDT, WINDOW+LSDT and DTSCORE+LSDT.

The results are given in Table 1. First, we observe that with $n = 48$ the true tree is almost never entirely found, for the reasons explained earlier. On the other hand, the best methods recover 80 to 95% of the duplication events, indicating that the tested datasets are relatively easy. NJ performs relatively well, but it often outputs trees that are not duplication trees, which is unsatisfactory (for 12 leaves 63% of the inferred trees are DT, only 0.7% for 48 leaves). The GS approach is noteworthy, since it modifies the trees inferred by NJ to transform them into duplication trees. However, GS is only slightly better than NJ regarding the proportion of correctly reconstructed trees, but considerably degrades the number of recovered duplication events, which could be explained by the blind search it performs to transform NJ trees into duplication trees. WINDOW also obtains relatively poor results, since the molecular clock is not respected. When it is respected, its performance is similar to that of NJ (results not shown), with the advantage that the inferred trees are duplication trees.

Table 1. Performance comparison using simulations.

	12			24			48		
	% tr	% ev	pars	% tr	% ev	pars	% tr	% ev	pars
NJ	44.6	90.8	–	9.0	86.0	–	0.0	79.7	–
GS	46.8	82.8	464	9.6	65.3	833	0.0	46.0	1493
GS+LSDT	61.9	93.5	−23	26.6	89.4	−110	1.9	82.2	−329
GTR	25.7	82.9	449	2.2	71.9	749	0.0	61.8	1223
GTR+LSDT	65.0	94.9	−9	28.5	91.2	−31	2.1	85.7	−77
WINDOW	26.4	85.5	445	3.3	75.8	736	0.0	67.2	1186
WINDOW+LSDT	65.1	95.0	−6	25.4	91.5	−19	1.7	86.1	−42
DTSCORE	55.2	92.5	442	18.8	89.0	722	0.7	83.6	1155
DTSCORE+LSDT	62.6	94.5	−2	28.9	92.3	−6	2.5	88.5	−15

GTR also obtains relatively poor results. Finally, DTSCORE obtains the best performance among the 5 existing methods, whatever the topological criterion considered.

Applying our method to starting trees produced by GS, GTR, WINDOW, and DTSCORE reveals the advantages of the local search approach. The trees produced by GS, GTR and WINDOW are very clearly improved, and, for most, are better than those obtained by DTSCORE. DTSCORE trees are also improved, even though this improvement is not very high from a topological point of view. This could be explained by the fact that DTSCORE is already an accurate method with respect to the datasets used. When we consider the parsimony criterion, the gain achieved by LSDT is very appreciable for each start method. This could be expected for GS, WINDOW and DTSCORE which do not optimize this criterion. With $n = 48$, the gain for GS is about 329, thus confirming that this method is clearly suboptimal; the gains for WINDOW and DTSCORE are about 42 and 15, which are lower but still significant. The GTR results, which optimizes parsimony, are more surprising since the gain (always with $n = 48$) is about 77 on average, which is very high. It should be stressed that these gains are obtained at low computational cost as dealing with any of the 48-taxon datasets only requires about twenty seconds on a standard PC-Pentium 4.

6   Conclusion and Prospects

We demonstrated that restricting the neighborhood defined by the SPR rearrangement to valid duplication trees, allows the whole space of DT to be explored. Thanks to these rearrangements, we defined a general local search method which we used to optimize the parsimony criterion. We thus improved the topological accuracy of all the tested methods.

Several research directions are possible. Finding the set of combinatorial configurations for the SPR movement which necessarily produce a duplication tree, could allow the neighborhood computation to be accelerated (for $n = 48$ only 5% of trees which belong to the neighborhood defined by the SPR rearrangement are duplication trees) and, furthermore, gain more insight into duplication trees which are just starting to be investigated mathematically [12, 13, 31]. Our local search method could be improved with the help of different stochastic approaches (taboo, noising, ...) in order to avoid local minima. Moreover, it would be relevant to test this local search method with other criteria like maximum likelihood. Finally, combining the tandem duplication events with speciation events, as described in [22] and [16] for non-tandem duplications, would be relevant for real applications where we have repeated sequences from several genomes.

References

1. B. Alberts, D. Bray, J. Lewis, M. Raff, K. Koberts, and J.D. Waston. *Molecular biology of the cell. 3rd ed.* Garland Publishing Inc., New York, USA, 1995.
2. J.P. Barthélemy and A. Guénoche. *Trees and proximity representations.* Wiley and Sons, Chichester, UK, 1991.
3. G. Benson and L. Dong. Reconstructing the duplication history of a tandem repeat. In *Proceedings of Intelligent Systems in Molecular Biology (ISMB1999)*, pages 44–53. AAAI, 1999.
4. O. Elemento and O. Gascuel. A fast and accurate distance-based algorithm to reconstruct tandem duplicatin trees. *Bioinformatics*, 18:92–99, 2002. Proceedings of European Conference on Computational Biology (ECCB2002).
5. O. Elemento and O. Gascuel. An exact and polynomial distance-based algorithm to reconstruct single copy tandem duplication trees. In *Proceedings of Combinatorial Pattern Matching*, Lecture Notes in Computer Science, pages 96–108. Springer, 2003.
6. O. Elemento, O. Gascuel, and M-P. Lefranc. Reconstructing the duplication history of tandemly repeated genes. *Molecular Biology and Evolution*, 19:278–288, 2002.
7. J. Felsenstein. PHYLIP - PHYLogeny Inference Package. *Cladistics*, 5:164–166, 1989.
8. J. Felsenstein and G.A. Churchill. A hidden markov model approach to variation among sites in rate of evolution. *Molecular Biology and Evolution*, 13:93–104, 1996.
9. W.M. Fitch. Toward defining the course of evolution: minimum change for a specified tree topology. *Systematic Zoology*, 20:406–416, 1971.
10. W.M. Fitch. Phylogenies constrained by cross-over process as illustrated by human hemoglobins in a thirteen-cycle, eleven amino-acid repeat in human apolipoprotein A-I. *Genetics*, 86:623–644, 1977.
11. G. Ganapathy, V. Ramachandran, and T. Warnow. Better hill-climbing searches for parsimony. In *Proceedings of the 3nd International Workshop on Algorithms in Bioinformatics*, Lecture Notes in Computer Science, pages 245–258. Springer, 2003.
12. O. Gascuel, D. Bertrand, and O. Elemento. Reconstructing the duplication history of tandemly repeated sequences. In O. Gascuel, editor, *Mathematics of Evolution and Phylogeny*. Oxford University Press, 2004. In press.
13. O. Gascuel, M. Hendy, A. Jean-Marie, and S. McLachlan. The combinatorics of tandem duplication trees. *Systematic Biology*, 52:110–118, 2003.
14. D. S. Gladstein. Efficient incremental character optimization. *Cladistics*, 13:21–26, 1997.
15. P. A. Goloboff. Methods for faster parsimony analysis. *Cladistics*, 12:199–220, 1996.
16. M. Hallett, J. Lagergren, and A. Tofigh. Simultaneous Identification of Duplications and Lateral Transfers. In *RECOMB*, 2004. in press.

17. J.A Hartigan. Minimum mutation fits to a given tree. *Biometrics*, 29:53–65, 1973.
18. D. Jaitly, P. Kearney, G. Lin, and B. Ma. Methods for reconstructing the history of tandem repeats and their application to the human genome. *J. of Computer and System Sciences*, 65:494–507, 2002.
19. A.J. Jeffreys and S. Harris. Processes of gene duplication. *Nature*, 296:9–10, 1981.
20. M.K. Kuhner and J. Felsenstein. A simulation comparison of phylogeny algorithms under equal and unequal evolutionary rates. *Molecular Biology and Evolution*, 11:459–468, 1994.
21. S. Ohno. *Evolution by gene duplication*. Springer Verlag, New York, 1970.
22. D.M. Page and M.A. Charleston. From gene to organisal phylogeny: Reconciled trees and the gene tree/species tree problem. *Melecular Phylogenetics and Evolution*, 7:231–240, 1997.
23. A. Rambault and N.C. Grassly. Seq-Gen: An application for the Monte Carlo simulation of DNA sequence evolution along phylogenetic trees. *Computer Applied Biosciences*, 13:235–238, 1997.
24. N. Saitou and M. Nei. The neighbor-joining method: a new method for reconstructing phylogenetic trees. *Molecular Biology and Evolution*, 4:406–425, 1987.
25. S. Sattath and A. Tversky. Additive similarity trees. *Psychometrika*, 42:319–345, 1977.
26. G.P. Smith. Evolution of repeated DNA sequences by unequal crossover. *Science*, 191:528–535, 1976.
27. P. Sneath and R. Sokal. *Numerical Taxonomy*, pages 230–234. W.H. Freeman and Company, 1973.
28. D.L. Swofford. *PAUP*. Phylogenetic Analysis Using Parsimony (*and Other Methods). Version 4.* Sinauer Associates, Sunderland, Massachusetts, 1999.
29. D.L. Swofford, P.J. Olsen, P.J. Waddell, and D.M. Hillis. *Molecular Systematics*, chapter Phylogenetic Inference, pages 407–514. Sinauer Associates, Sunderland, Massachusetts, 1996.
30. M. Tang, M.S. Waterman, and S. Yooseph. Zinc finger gene clusters and tandem gene duplication. *Journal of Computational Biology*, 9:429–446, 2002.
31. Y. Yang and L. Zhang. On counting tandem duplication trees. *Molecular Biology and Evolution, 2004* (in press).
32. L. Zhang, B. Ma, L. Wang, and Y. Xu. Greedy method for inferring tandem duplication history. *Bioinformatics*, 19:1497–1504, 2003.

Phylogenetic Super-networks from Partial Trees

Daniel H. Huson[1], Tobias Dezulian[1], Tobias Klöpper[1], and Mike A. Steel[2]

[1] Center for Bioinformatics (ZBIT), Tübingen University
Sand 14, 72076 Tübingen, Germany
{huson,dezulian,kloepper}@informatik.uni-tuebingen.de
[2] Biomathematics Research Centre, University of Canterbury
Christchurch, New Zealand
m.steel@math.canterbury.ac.nz

Abstract. In practice, one is often faced with incomplete phylogenetic data, such as a collection of partial trees or partial splits. This paper poses the problem of inferring a phylogenetic super-network from such data and provides an efficient algorithm for doing so, called the Z-closure method. Application to a set of five published partial gene trees relating different fungal species illustrates the usefulness of the method and an experimental study confirms its potential. The method is implemented as a plug-in for the program SplitsTree4.

Keywords: Molecular evolution, phylogeny, partial trees, networks, closure operations.

1 Introduction

Traditionally, in molecular phylogenetics, 16S rRNA has been used as the phylogenetic marker of choice to infer the evolutionary history of a collection of different species [18], which we will refer to as the *species tree.*

A *gene* tree is generated by considering some specific gene that is present in all given species. Given the current and growing abundance of whole genome sequences (see, e.g. [11]), for an increasing number of species it is now possible to compute gene trees for many different genes. Comparison of individual gene trees with the corresponding putative species tree may be useful e.g. when trying to determine whether a given gene may have been involved in *horizontal gene transfer.* Also, one may attempt to obtain a more reliable species tree by forming a consensus tree from a set of gene trees.

A more recent approach to the gene trees vs. species tree problem is to compute a consensus network [6, 8] that attempts to represent *all* phylogenetic signals present in the given set of gene trees, simultaneously, up to a given level of complexity. In the resulting network, regions of the evolutionary history that are undisputed within the set of gene trees appear tree-like, whereas regions containing conflicts are shown as "incompatibility boxes", whose "dimensionality" reflects the number of conflicting signals.

In practice, for a given set of taxa of interest, it is often the case that some of the genes under consideration are not present in all genomes, or, although

I. Jonassen and J. Kim (Eds.): WABI 2004, LNBI 3240, pp. 388–399, 2004.

present, their sequence is unavailable. This problem is usually addressed by removing those taxa from the analysis for which one of the genes is missing.

Therefore, it would be desirable to have a method that takes as input a
collection of *partial* trees defined on subsets of the full taxa set, and produces as
output a phylogenetic network representing all phylogenetic signals present in
the input partial trees.

In this paper we describe a first such *super-network method*, which we call
the *Z-closure* construction, and demonstrate its utility both by an experimental
study and also by application to a typical biological data set. In this approach,
partial splits are repeatedly extended using the *Z-rule*, a simple binary modification rule, until a closure under this rule has been obtained. The number of
computed splits is at most equal to the number of input splits and such a closure
can be computed efficiently.

In Section 2 we briefly introduce the basic underlying concepts of splits and
splits graphs. Then, in Section 3, we define the *complete* Z-closure and provide
an efficient algorithm for computing a (fixed order) Z-closure. We then define
two important properties and prove that one can be used to assign weights to
the edges in the network, and the other can often be used to compute a form of
strict consensus tree embedded in the network, in Section 4. This is followed, in
Section 5 by a discussion of how to compute the weights of the network edges.
The results of a first experimental study of the method is presented in Section 6.
In Section 7, we apply the method to five different partial gene trees showing
the phylogenetic relationships amongst different species of fungi, published in
[12, 13]. Finally, we discuss a number of variants of our approach and some other
possible application scenarios in Section 8.

We have implemented the Z-closure in the program SplitsTree4 [8].

2   Trees, Splits and Networks

Suppose we are given a set of taxa $X = \{x_1, \ldots, x_n\}$. A *(phylogenetic) X-
tree* $T = (V, E, \nu, \omega)$ is a connected, acyclic graph with node set V and edge
set $E \subseteq 2^V$, together with a *node labeling* $\nu : X \to V$ and *edge weighting*
$\omega : E \to \mathbb{R}^{\geq 0}$, such that every node of degree 1 receives a label. An X'-tree T is
called a *partial X-tree*, if $X' \subseteq X$, and we define $X(T) = X'$.

Suppose we are given an X-tree T. Every edge $e \in E$ partitions the tree T
into two components and thus defines a bipartitioning of X into two non-empty
and disjoint sets A and B, with $A \cup B = X$. Any such bipartitioning of X is
called an *X-split*, written as $S = \frac{A}{B}$ (or, equivalently, $S = \frac{B}{A}$) [1]. For an X-split
S obtained from a tree in this way, we define the *weight* of S to be $\omega(S) = \omega(e)$.
A split S is called *trivial*, if $|A| = 1$ or $|B| = 1$.

An X'-split S is called a *partial X-split*, if $X' \subseteq X$, and sometimes we will
call S a *full X-split*, if $X' = X$. For an X-split $S = \frac{A}{B}$ and taxon set $X' \subseteq X$,
we define the split *induced* on X' as $S|_{X'} = \frac{A \cap X'}{B \cap X'}$. Note that this may yield an
improper split $\frac{\emptyset}{X'}$. We say that a split $S = \frac{A}{B}$ in an extension of a second split
$S' = \frac{A'}{B'}$, if $A' \subseteq A$ and $B' \subseteq B$, where at least one of the inclusions is proper.

Suppose we are given two taxa $x, y \in X$. We say that a split $S = \frac{A}{B}$ *separates* x and y, if $x \in A$ and $y \in B$, or vice versa, and we use $\Sigma(x, y)$ to denote the set of all splits $S \in \Sigma$ that separate x and y.

Let $\Sigma(T)$ denote the *split encoding of* T, that is, the set of all X-splits defined by edges in T. Two X-splits $S_1 = \frac{A_1}{B_1}$ and $S_2 = \frac{A_2}{B_2}$ are called *compatible*, if one of the four following intersections is empty: $A_1 \cap A_2$, $A_1 \cap B_2$, $B_1 \cap A_2$ or $B_1 \cap B_2$ – otherwise they are said to be *incompatible*.

We have the following well-known result [2]: Suppose we are given an arbitrary set Σ of X-splits. Then Σ is the split encoding of some X-tree T, if and only if Σ is *compatible*, that is, if all pairs of splits in Σ are compatible.

The split encoding of trees plays an important computational role in phylogenetics. For example, given a set of X-trees $T_1, \ldots, T_k$, we obtain the split encoding of the *strict consensus tree* T_{strict} as the set of all X-splits that occurs in every input set $\Sigma(T_i)$. Similarly, we obtain the *majority consensus tree* $T_{majority}$ via the set of splits that occur in more than 50% of all input sets. Finally, we obtain a *(d-dimensional) consensus network* as the set of all splits that occur in a proportion of more than $\frac{1}{d+1}$ of all input sets [6].

Suppose we are given an arbitrary set Σ of X-splits, not necessarily compatible. *Every* such set of splits can be represented by a *splits graph* $G = (V, E, \nu, \omega, \sigma)$, which consists of a connected graph G with vertex set V and edge set $E \subseteq 2^V$, together with a node labeling ν, edge weighting ω and a surjective edge coloring $\sigma : E \to \Sigma$. Additionally, we require that the coloring σ is *isometric*, that is, for each pair of nodes $v, w \in V$, every shortest path from v to w uses the same set $\Sigma(v, w) \subseteq \Sigma$ of edge colors, and each such color is used precisely once. Moreover, we require that $\Sigma(\nu(x), \nu(y)) = \Sigma(x, y)$ for all pairs of taxa $x, y \in X$. Finally, we assume that $\omega(e) = \omega(\sigma(e))$ for all edges $e \in E$. For details, see [4].

Such a splits graph has the property that if one deletes all edges colored by a given split $S = \frac{A}{B} \in \Sigma$, then one obtains precisely two components, one containing $\nu(A)$ and the other containing $\nu(B)$. Thus, splits graphs contain *phylogenetic trees* as a special case and generalize them to a specific type of *phylogenetic network*. We have developed and implemented algorithms for constructing and visualizing splits graphs, see [4, 7, 8].

3 The Z-Closure Network

Suppose we are given a set of *partial X-splits* Σ. Our goal is to modify splits in Σ so as to obtain a collection of *full X-splits*. We propose to achieve this by repeatedly applying the following simple transformation, which we call the *Z-rule*:

For any two splits $S_1 = \frac{A_1}{B_1} \in \Sigma$ and $S_2 = \frac{A_2}{B_2} \in \Sigma$:
if $A_1 \cap A_2 \neq \emptyset$, $A_2 \cap B_1 \neq \emptyset$, $B_1 \cap B_2 \neq \emptyset$ and $A_1 \cap B_2 = \emptyset$, then
replace S_1 and S_2 by $S_1' = \frac{A_1}{B_1 \cup B_2}$ and $S_2' = \frac{A_1 \cup A_2}{B_2}$.

In short-hand, we write $\frac{A_1}{B_1} \, Z \, \frac{A_2}{B_2} \longrightarrow \frac{A_1}{B_1 \cup B_2}, \frac{A_1 \cup A_2}{B_2}$, where the three lines arranged in a "Z" connect those pairs of split parts that are required to have a non-empty intersection, hence the name "Z"-rule. This rule was introduced by C. Meacham in the context of inferring phylogenies from multi-state characters [3, 10, 15]. Note that application of the Z-rule will sometimes simply reproduce the two input splits S_1 and S_2, in which case we say that the Z-rule *does not apply*.

We obtain a *(fixed order) Z-closure* $\bar{\Sigma}$ by repeatedly applying the Z-operation to all splits originally contained in, or derived from, the input set Σ, in some fixed order. We define the *complete Z-closure* as the set of all splits that occur in at least one (fixed order) Z-closure. To avoid excessive notation, we will also use $\bar{\Sigma}$ to denote the complete Z-closure, but will always distinguish sharply between *a* Z-closure and *the complete* Z-closure. In both cases, we are particularly interested in the set $\bar{\Sigma}^*$ consisting of all *full* X-splits contained in $\bar{\Sigma}$, together with all trivial X-splits, which we will also refer to as a or the Z-closure, respectively. We define a *Z-closure network* $Z(T_1, \ldots, T_k)$ for $T_1, \ldots, T_k$ to be a splits graph representing $\bar{\Sigma}^*$.

The following algorithm computes a (fixed order) Z-closure, maintaining all partial splits in an array **data** and using three sets of indices, **old**, **active** and **new**, indicating which splits in the array were produced in an earlier, the previous or the current iteration of the algorithm, respectively:

Algorithm 1 (Z-Closure)

Input: A set of partial trees $\mathcal{T} = \{T_1, \ldots, T_k\}$
Output: A Z-closure $\bar{\Sigma}^*$
Initialization:
  Let **data** *be an array initialized to the set of all non-trivial splits in* $\bigcup_i \Sigma(T_i)$.
  Let **old** *be a set of indices, initially empty*
  Let **active** *be a set of indices, initially empty*
  Let **new** *be a set of indices, initialized to the index set of* **data**
while new $\neq \emptyset$ **do***:*
    Append **active** *to* **old**, *set* **active** = **new** *and set* **new** = $\emptyset$
    for each $i \in$ old $\cup$ active **do***:*
      for each $j \in$ active **do***:*
        Let $S_1 = \frac{A_1}{B_1} =$ data$[i]$ *and $S_2 = \frac{A_2}{B_2} =$* data$[j]$
        if $A_1 \cap A_2 \neq \emptyset$, $A_2 \cap B_1 \neq \emptyset$, $B_1 \cap B_2 \neq \emptyset$ *and* $A_1 \cap B_2 = \emptyset$ **then***:*
          Define $S_1' = \frac{A_1}{B_1 \cup B_2}$ and $S_2' = \frac{A_1 \cup A_2}{B_2}$
          if $S_1' \neq S_1$ **then** *set* data$[i] = S_1'$ *and add i to* new
          if $S_2' \neq S_2$ **then** *set* data$[j] = S_2'$ *and add j to* new
Return the set of all X-splits in old $\cup$ active $\cup$ *all trivial X-splits.*

We claim:

Theorem 1. *Let $n = |X|$ and $m = |\Sigma|$. Algorithm 1 computes a Z-closure $\bar{\Sigma}^*$ in at most $O(nm^3)$ steps. The space requirement is $O(nm)$ and the resulting number of splits is at most $n + m$.*

Proof. The algorithm operates as follows: Originally, all splits are considered "new". In any iteration, the set of all splits deemed "new" in the previous iteration are considered "active" and are compared with themselves and with all "old" splits. The resulting "new" splits become the "active" splits of the next iteration, repeatedly, until no "new" splits are generated. In the worst case, each iteration of the algorithm will extend only one split by one taxon. There are m splits, each requires $O(n)$ iterations to extend to size n, and each such iteration requires $O(m^2)$ comparisons, thus leading to a naive bound of $O(nm^3)$. As the algorithm operates "in place", the space requirement is simply the size of the input set, $O(nm)$. Moreover, the number of output splits is at most the number of input splits, m, plus the number of trivial splits on X, which is n.   □

In practice, we can expect each split to be extended at least by one taxon in every iteration, leading to a runtime bound of $O(nm^2)$.

Algorithm 1 computes a (fixed order) Z-closure efficiently. It would be desirable to have an efficient algorithm for computing the complete Z-closure, but it is an open problem whether such an algorithm exists. The results described below indicate that a (fixed order) Z-closure is a very good approximation to the complete Z-closure. Moreover, any order-dependence of the algorithm can be addressed by running the Z-closure algorithm a number of times using random input orders and retaining all splits computed, and our implementation of the method supports this feature. Moreover, the following result adds justification to the use of a (fixed order) Z-closure, see [3] for details:

Lemma 1. *If all input splits are compatible with each other, then any (fixed order) Z-closure is equal to the complete Z-closure.*

4   The Weak- and Strong Induction Properties

Suppose we are given an input set of partial trees $T = \{T_1, \ldots, T_k\}$. In the following, let X_i, Σ_i and ω_i denote the taxa set, split encoding and split weights for tree T_i, respectively, for all $i = 1, \ldots, k$. Let $\bar{\Sigma}^*$ denote the complete Z-closure.

We say that a split $S \in \bar{\Sigma}^*$ has the *weak-*, or *strong-induction property*, if there *exists* a tree $T_i \in T$, such that $S|_{X_i} \in \Sigma_i$, or, if for *every* tree $T_i \in T$ such that $S|_{X_i}$ is a proper X_i-split, we have $S|_{X_i} \in \Sigma_i$, respectively. We say that $\bar{\Sigma}^*$ has the *weak-* or *strong-induction property*, if every split $S \in \bar{\Sigma}^*$ has the weak- or strong-induction property, respectively.

The following result shows that the complete Z-closure does not contain any superfluous splits. Moreover, it is used in Section 5 to define the weights of the splits in $\bar{\Sigma}^*$:

Theorem 2. *The complete Z-closure of any set of partial X-trees $T = \{T_1, \ldots, T_k\}$ has the weak induction property.*

Proof. We will show by induction that for any split $S \in \bar{\Sigma}$, there exists an input tree T_i such that $S|_{X_i} \in \Sigma_i$. Let Σ^p denote the set of all splits obtained by p

applications of the Z-rule (in some order). Induction start: Consider $S \in \Sigma^0 = \Sigma$. By definition of Σ, there exists a tree T_i with $S = S|_{X_i} \in \Sigma_i$. Induction step: Consider a split $S \in \Sigma^{p+1}$. If $S \in \Sigma^p$, then S has the desired property. If $S \in \Sigma^{p+1} \setminus \Sigma^p$, then S was obtained by extension of some split $S' \in \Sigma^p$ using the Z-rule. By the induction hypothesis, there exists a tree T_i with $S'|_{X_i} \in \Sigma_i$. As S is an extension of S', we have $S|_{X_i} = S'|_{X_i}$, and thus $S|_{X_i} \in \Sigma_i$.   □

Suppose we are given a set of partial trees $\mathcal{T} = \{T_1, \ldots, T_k\}$. Let $\bar{\Sigma}^*_{SIP}$ be the set of all splits in $\bar{\Sigma}^*$ that have the strong induction property.

Often, although not always, $\bar{\Sigma}^*$ will be compatible and thus can provide a method for extracting a kind of *strict consensus tree* from the complete Z-closure $\bar{\Sigma}^*$.

5   Computing Weights

In Section 3, we described how to compute the set of splits $\bar{\Sigma}^*$ for the super-network $Z(T_1, \ldots, T_k)$. We now address the question of how to assign weights to the splits in $\bar{\Sigma}^*$, and thus, to the edges of the network $Z(T_1, \ldots, T_k)$.

Suppose we are given a set of input trees $\mathcal{T} = \{T_1, \ldots, T_k\}$. Let X_i, Σ_i and ω_i denote the taxa set, split encoding and split weights for tree T_i, respectively, for all $i = 1, \ldots, k$.

Consider an X-split $S = \frac{A}{B} \in \bar{\Sigma}^*$. Let $I(S)$ denote the set of all $i \in \{1, \ldots, k\}$ such that the induced split $S|_{X_i}$ is contained in Σ_i. By Theorem 2 we have $I(S) \neq \emptyset$. Thus, we propose to give S the smallest weight provided, that is, to set $\omega(S) = \min\{\omega_i(S|_{X_i}) \mid i \in I(S)\}$.

Alternatively, let us say that an input set $\mathcal{T}$ has the *all pairs* property, if for every pair of $x, y \in X$ there exists an input tree $T_i \in \mathcal{T}$ that contains both x and y. If this property holds, then we can define a pair-wise distance between any two taxa $x, y \in X$, e.g. as their minimum-, mean- or median distance in the input set. We then can obtain split weights for $\bar{\Sigma}^*$ using a least squares fit of the distance matrix [17].

6   Experimental Study

The performance of a phylogenetic inference method is sometimes evaluated in a simulation study, see e.g. [9]. This involves repeating the following three steps a sufficiently large number of times: First, input data is generated according to some model of evolution, guided by a specific *model* phylogeny. Second, this data is fed to the phylogenetic method as input. Third, the resulting tree is compared with the model tree to assess the accuracy of the method.

In this section we report on the results of a simulation study that we have performed. Our experiments are run as follows. Suppose we are given a set of taxa $X = \{x_1, x_2, \ldots, x_n\}$. We first choose a single model tree M, which we call the *model species tree*. From this species tree, we obtain a set of h different *model gene trees* $M_1, \ldots, M_h$, by performing a number of SPR operations on each (Subtree Pruning and Regrafting as defined in e.g. [16]), with the goal of producing model gene phylogenies that are related to, but different from the model species tree.

To obtain a collection of partial trees, for each model gene tree M_i we randomly select a subset of $X_i \subseteq X$ and let T_i denote the resulting induced *model partial gene tree*.

We then compute a Z-closure $\bar{\Sigma}^*$ for the set of all such partial trees $T = \{T_1, \ldots, T_h\}$.

The accuracy of $\bar{\Sigma}^*$ is evaluated as follows. Any split $S \in \bar{\Sigma}^*$ that is not present in some input tree T_i could be considered a *false positive* partial split. However, by Theorem 2, the number of such false positives is always zero.

As the Z-closure is not a phylogenetic inference method but rather a method for summarizing a collection of partial trees within a consistent super-network, the main question is, how successful the method is at representing the partial splits in the input set. A *false negative* is any split $S \in \Sigma(T_i)$ that is not represented in the Z-closure, that is, for which no split $S' \in \bar{\Sigma}^*$ exists with $S'|_{X_i} = S$.

The false negative rate will depend primarily on how large the partial trees are and how well they overlap, as well as how similar the trees are. In our simulations we apply a varied number of SPR operations, between 0.1 and 3.2 on average, and thus obtain gene trees of varying degrees of similarity. Additionally, we assume that the partial trees cover a large proportion of the taxon set X, and we choose the average size of the partial taxon sets to lie in the range $40 - 95\%$, with a standard deviation of 10%. Our study considers 10 gene trees.

These choices are motivated by the following use-case: a phylogeny of a group of species is to be studied based on a small number of important genes. In practice, it is usually the case that some gene sequences are missing for some of the taxa.

The results of our study are summarized in Figure 1. As expected, the simulations confirm that the number of false negatives depends strongly on the average coverage of the partial trees, and also on the similarity of the underlying gene trees, represented here by the number of SPR moves performed. Surprisingly, when the average coverage is larger than 50%, then the performance appears to be practically independent of the level of similarity. Moreover, the rate of false negatives drops well below 10%. In summary, these results suggest that the method should work well on data sets with an average coverage of more than 50%, say, regardless of how similar the input trees are.

7   Application to Real Data

To illustrate the application of this method, we obtained five gene trees relating different fungal species from TreeBASE [14], that were published in [12,13]. The three trees obtained from the first paper are based on the nuclear *internal transcribed spacer* (ITS), on the mitochondrial small subunit (SSU) ribosomal DNA (rDNA), and on a portion of the glyceraldehyde-3-phosphate gene (gpd), and are shown in Figure 2 as (a), (b) and (c), respectively. The two trees taken

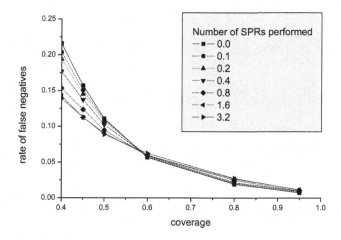

Fig. 1. Here we plot the proportion of non-trivial partial splits present in the input that are not represented in the Z-closure-network, as a function of the average proportion of taxa contained in the input trees and as a function of the number of SPR moves applied in the modification of the model gene trees. Every data-point represents the average score for 1000 repetitions, each using a different 60-taxon model tree, randomly generated under the Yule-Harding model.

from the second paper are also based on the ITS and SSU sequences, and are shown in Figure 2(d) and (e). Unfortunately, edge lengths were not available for the trees. In our experience, edge lengths greatly enhance the readability of the resulting network.

Calculation of a Z-closure network of the five trees took less than 20 seconds. The resulting graph is shown in Figure 3. We also considered a sixth tree based on 18S rDNA that contains a large number of taxa not present in the other five. As to be expected, many of these taxa remained unresolved in the resulting network (not shown here).

The network depicted in Figure 3 is based on a (fixed order) Z-closure. To determine how order-dependent the resulting network is, we re-ran Algorithm 1 a total of 1000 times, using different random input orders. In Figure 4, we see clearly that the input order has very little effect on the computed network. Indeed, in every single case the derived Z-closure contains at least 97% of the union of all non-trivial splits obtained within the 1000 runs. In this example, re-running the algorithm a small number of times suffices to produce all 71 splits. Our naive implementation of the Z-closure algorithm took 25 minutes on a $1.2GHz$ laptop to complete all 1000 runs.

8 Discussion

We believe that the concept of a super-network will prove to be very useful, especially as applied to partial gene trees, where there is reason to believe that

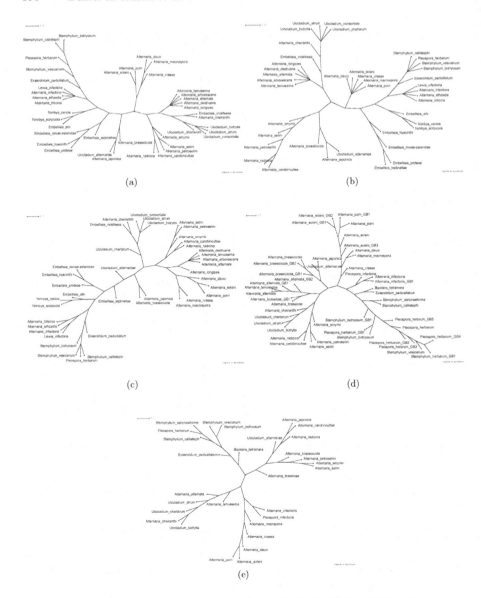

Fig. 2. The gene trees show here are based on (a) ITS, (b) SSU and (c) *gpd* sequences, as published in [12], and on (d) ITS and (e) SSU sequences, as published in [13]. These trees have varying numbers of taxa.

the underlying trees are incongruent and thus should not be "forced" into any particular super-tree. The Z-closure approach provides a simple and efficient method for computing super-networks, with some nice mathematical properties.

The Z-closure method has many potential applications. For example, in a first application it has provided useful insights into the evolution of close rela-

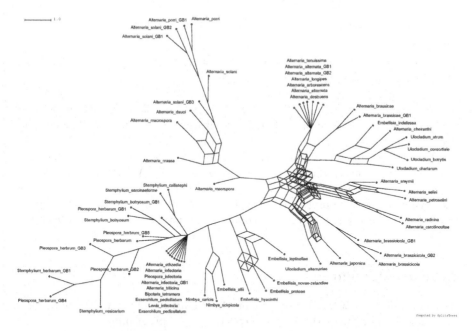

Fig. 3. A phylogenetic super-network on 63 taxa, obtained by applying the Z-closure method to the five partial trees depicted in the previous figure. This graph clearly shows on which parts of the phylogeny all partial gene trees agree, and where there exist contradicting signals. Note that this type of graph is *not* a model of evolution but rather a graphical summary of multiple phylogenies.

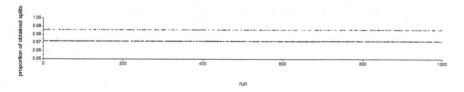

Fig. 4. For each of 1000 different random input orders, we plot the number of splits obtained by the Z-closure algorithm, as a proportion of the union of all splits obtained in all 1000 runs. This is based on the input trees depicted in Figure 2.

tives of Arabidopsis Thaliana and these will be reported in a forthcoming paper (Lockhart *et al.*, in preparation).

The approach formulated here can be extended in a number of different ways. First of all, note that the input can be an arbitrary collection of split systems, and is not restricted to split systems coming from trees. In this case, Theorem 2 still holds.

Secondly, one can consider other rules of the type defined in [3, 10]. Our main motivation for focusing on the Z-rule is that this rule can be applied "in place" and does not increase the number of splits, leading to a particularly efficient algorithm. For example, another rule that one might additionally consider is the following:

Suppose we are given $S_1 = \frac{A_1}{B_1}$ and $S_2 = \frac{A_2}{B_2}$ with $A_1 = A_2$ and $B_1 \cap B_2 \neq \emptyset$. Then replace S_1 and S_2 by $S' = \frac{A_1}{B_1 \cup B_2}$.

In short-hand, $\frac{A_1}{B_1} = \frac{A_2}{B_2} \longrightarrow \frac{A_1}{B_1 \cup B_2}$. One draw-back of this *transitive* rule is that the condition $A_1 = A_2$ implies that the resulting set of splits will depend very strongly on the order of application, in contrast to the Z-rule.

One additional application of the Z-closure that we are currently studying is as the merge step in the "disk-covering method" [9]. A further potential application is to the problem of haplotype assignment: given a set of n partial haplotypes, assign a complete haplotype to each [5]. Each partial haplotype is considered to be a partial split and complete haplotypes are obtained from the Z-closure network.

Acknowledgments

We thank the New Zealand Institute for Mathematics and its Applications (NZ-IMA) for support under the *Phylogenetic Genomics* programme.

References

1. H.-J. Bandelt and A. W. M. Dress. A canonical decomposition theory for metrics on a finite set. *Advances in Mathematics*, 92:47–105, 1992.
2. P. Buneman. The recovery of trees from measures of dissimilarity. In F. R. Hodson, D. G. Kendall, and P. Tautu, editors, *Mathematics in the Archaeological and Historical Sciences*, pages 387–395. Edinburgh University Press, 1971.
3. T. Dezulian and M. Steel. Phylogenetic closure operations and homoplasy-free evolution. In *Proceedings of the 2004 International Federation of Classification Societies*. Springer-Verlag, 2004. In press.
4. A.W.M. Dress and D.H. Huson. Algorithms for drawing phylogenetic networks. In preparation, 2004.
5. E. Halperin and R. M. Karp. Perfect phylogeny and haplotype assignment. In *Proceedings of the Eighth Annual Internation Conference on Research in Computational Molecular Biology*, pages 10–19. ACM Press, 2004.
6. B. Holland and V. Moulton. Consensus networks: A method for visualizing incompatibilities in collections of trees. In G. Benson and R. Page, editors, *Proceedings of "Workshop on Algorithms in Bioinformatics"*, volume 2812 of *LNBI*, pages 165–176. Springer, 2003.
7. D. H. Huson. SplitsTree: A program for analyzing and visualizing evolutionary data. *Bioinformatics*, 14(10):68–73, 1998.
8. D. H. Huson and D. Bryant. SplitsTree4 - a framework for building phylogenetic trees and networks. Manuscript in preparation, software available from http://www-ab.informatik.uni-tuebingen.de/software/splits, 2004.
9. D. H. Huson, S. Nettles, L. Parida, T. Warnow, and S. Yooseph. The disk-covering method for tree reconstruction. In *Proceedings of "Algorithms and Experiments" (ALEX98), Trento*, pages 62–75, 1998.

10. C.A. Meacham. Theoretical and computational considerations of the compatibility of qualitative taxonomic characters. In J. Felsenstein, editor, *Numerical Taxonomy*, volume G1 of *NATO ASI Series*. Springer, Berlin, 1983.

11. NCBI. Microbial complete genomes taxonomy. http://www.ncbi.nlm.nih.gov/PMGifs/Genomes/new_micr.html, 2003.

12. B.M. Pryor and D. M. Bigelow. Molecular characterization of Embellisia and Nimbya species and their relationship to Alternaria, Ulocladium and Stemphylium. *Mycologia*, 95(6):1141–1154, 2003.

13. B.M. Pryor and R.L. Gilbertson. Phylogenetic relationships among Alternaria and related fungi based upon analysis of nuclear internal transcribed sequences and mitochondrial small subunit ribosomal DNA sequences. *Mycological Research*, 104(11):1312–1321, 2000.

14. M.J. Sanderson, M.J. Donoghue, W. Piel, and T. Eriksson. Treebase: a prototype database of phylogenetic analyses and an interactive tool for browsing the phylogeny of life. *Amer. Jour. Bot.*, 81(6):183, 1994.

15. C. Semple and M.A. Steel. Tree recontruction via a closure operation on partial splits. In *Computational Biology (proceedings of JOBIM 2000), LNCS 2066*. Springer-Verlag, 2001.

16. D. L. Swofford, G. J. Olsen, P. J. Waddell, and D. M. Hillis. Chapter 11: Phylogenetic inference. In D. M. Hillis, C. Moritz, and B. K. Mable, editors, *Molecular Systematics*, pages 407–514. Sinauer Associates, Inc., 2nd edition, 1996.

17. R.C. Winkworth, D. Bryant, P. Lockhart, D. Havell, and V. Moulton. Biogeographic interpretation of splits graphs: least squares optimization of branch lengths. *Systematic Biology*, 2004. In press.

18. C. R. Woese. Bacterial evolution. *Microbiol. Rev.*, 51:221–272, 1987.

Genome Identification and Classification
by Short Oligo Arrays

Stanislav Angelov[1,*], Boulos Harb[1,**], Sampath Kannan[1,***],
Sanjeev Khanna[1,†], Junhyong Kim[2,‡], and Li-San Wang[2,§]

[1] Department of Computer and Information Science
School of Engineering and Applied Sciences
University of Pennsylvania, Philadelphia, PA 19104, USA
{angelov,boulos,kannan,sanjeev}@cis.upenn.edu
[2] Department of Biology, School of Arts and Sciences,
University of Pennsylvania, Philadelphia, PA 19104, USA
junhyong@sas.upenn.edu, lswang@mail.med.upenn.edu

Abstract. We explore the problem of designing oligonucleotides that
help locate organisms along a known phylogenetic tree. We develop a
suffix-tree based algorithm to find such short sequences efficiently. Our
algorithm requires $O(Nm)$ time and $O(N)$ space in the worst case where
m is the number of the genomes classified by the phylogeny and N is their
total length. We implemented our algorithm and used it to find these discriminating sequences in both small and large phylogenies. We believe
our algorithm will have wide applications including: high-throughput
classification and identification, oligo array design optimally differentiating genes in gene families, and markers for closely related strains and
populations. It will also have scientific significance as a new way to assess
the confidence in a given classification.

1   Introduction

Short sequence tags, typically oligonucleotides of 20–100 base pairs (bp), have
been widely used in genomic technologies, e.g. the Affymetrix GeneChip; SAGE
[1], EST [2], and STS [3]. The idea is that a sequence tag or a set of tags will be
a unique subsequence of a longer target sequence and, therefore, the tag(s) will
serve to identify the target sequence. In the limit, the tags may represent the
whole genome or a set of genomes. More recently, short sequence tags have been

* Supported in part by NIGMS 1-P20-GM-6912-1 and NSF ITR 0205456.
** Supported in part by NIH Training Grant T32HG0046.
*** Supported in part by ARO grant DAAD 19-01-1-0473 and NSF grant CCR-
0105337.
† Supported in part by Alfred P. Sloan Research Fellowship, NSF Career Award
CCR-0093117, and NSF ITR 0205456.
‡ Supported in part by NIGMS 1-P20-GM-6912-1.
§ Supported in part by NIH Training Grant in Cancer and Immunopathobiology
(1 T32 CA101968).

I. Jonassen and J. Kim (Eds.): WABI 2004, LNBI 3240, pp. 400–411, 2004.
© Springer-Verlag Berlin Heidelberg 2004

proposed as IDs for whole organisms [4], with the related idea that some reasonably long stretch of homologous sequence can be used as sufficient information for taxonomy. In this application, the tags are assayed not only for presence and absence but information about their string difference is also utilized in the taxonomic procedures. With shorter oligo tags, the string differences between different tags carry little information. In addition, string difference cannot be easily assayed by high-throughput hybridization based technologies. Therefore, with short oligo array technique the information is contained in the configuration of presence and absence of the suite of oligonucleotides assayed (called *probes*) by the hybridization array. Hypothetically, if the set of probes is very large and exhaustively assays a sufficiently large address space, say, the set of all 20 base pair strings, we can use the presence/absence information to construct or recover detailed taxonomic or phylogenetic information. However, practical devices are limited to assaying on the order of 10^4 to 10^5 different probes. Fortunately, genomes are not random collection of strings but they are related to each other by phylogenetic relationships, thus an efficient set of probes can be designed to recover genomic information from a limited but a judicious selection of the address space.

Databases have been set up for organism specific oligonucleotides and primer design (see for example [5], and [6]). Heuristic construction of short sequence tags has been used for identification and classification (reviewed in [7]), but no algorithm has been presented for tag design. Previous computational oligo-sequence design work has mostly concentrated on optimizing hybridization kinetics or assembly into longer fragments (for example, [8], [9], [10], and [11]) but has not considered the problem of optimal design for classifications.

We explore the problem of designing short sequence tags (simply called tags from here on) that help locate organisms along a known phylogenetic or a classification tree with binary splits. (For expedience, we will only refer to phylogenetic trees from here on, however many biologists make distinctions between phylogenetic trees and other classification trees.) Specifically, given complete genomes for a set of organisms and the binary phylogenetic tree that describes their evolution, we develop an efficient algorithm for detecting all *discriminating* tags. We will say that a tag t is discriminating at some node u of the phylogeny if all genomes under one branch of u contain t while none of the genomes under any other branch contain t. Thus, a set of discriminating tags for all the nodes of a phylogeny allows us to place a genome in the phylogeny by a series of binary decisions starting from the root. This procedure can be implemented experimentally as a microarray hybridization assay, enabling a rapid determination of an unidentified organism in a predetermined phylogeny or classification.

Using suffix trees [12], we implement an $O(Nm)$ time algorithm that computes all discriminating tags at all nodes of the phylogeny, where N is the total length of the genomic sequences under consideration and m is their number. The suffix tree (as well as its generalization for multiple sequences) is a compact tree representation of all suffixes of a given input text. The algorithm enumerates all substrings of the input via a bottom-up walk of the computed generalized suffix tree, maintaining the set of input sequences that contain each substring. This information is sufficient to determine for which nodes of the phylogeny the current substring is discriminating. We also analyze special cases of the problem that can be solved in $O(N \log m)$ time and $O(N)$ space. Finally, we demonstrate the existence of discriminating tags by running our algorithm on biological data sets whose estimated phylogenies we obtained from [13] and [14].

2   Finding Discriminating Tags

Recall that a discriminating tag for a node u in a given phylogeny is a substring that is present in every genome under one branch of u and is not present in any genome under all other branches of u. In this section, we outline our basic algorithm for finding discriminating tags, and we present an enhancement to reduce the algorithm's space requirement. Both the algorithm, and its enhancement build on the solution to the multiple common substring problem described in [12].

2.1   Preliminaries

We state the problem more precisely as follows. Let $S = \{s_1, \cdots, s_m\}$ be a set of m strings, also called *species*, and let P be a phylogeny represented by a rooted binary tree with m leaves labeled by S. For each internal node u of P, we wish to find the set of discriminating tags for u. Without loss of generality, we will focus on finding discriminating tags present in the left subtree of every node. We assume that the strings are drawn from a bounded-size alphabet Σ, and that $\sum_{i=1}^{m} |s_i| = N$ where $|s_i|$ denotes the number of characters in the string s_i.

Suffix Trees. Suffix trees, first introduced by [15], play a central role in our algorithms. A suffix tree is a rooted directed tree that compactly represents all suffixes of a string. Each edge in a suffix tree for a string s is labeled with a non-empty substring of s. We adopt the following definition from [12]. The *path-label* of a node v in a suffix tree T for s is the string formed by following the path from the root of T to v. We will denote it by $path(v)$. The path-labels of the $|s|$ leaves of T spell out the suffixes of s, and the path-labels of internal nodes spell out substrings of s. Furthermore, the suffix tree ensures that there is a unique path from the root, not necessarily ending at an internal node, that represents each substring of s (see Fig. 1(a)). The time and space requirements of building a suffix tree are linear in the size of the string it represents [15–17].

Our algorithms, more specifically, are based on *generalized suffix trees*. A generalized suffix tree extends the idea of a suffix tree for a single string to a suffix tree for a set of strings. As described in [12] such a tree can conceptually be constructed by appending a unique terminating marker not in Σ to each string in the given set, concatenating the strings, and building a suffix tree for the resulting string. The tree is post-processed so that each path-label of a leaf in the tree spells a suffix of one of the strings in the set and, hence, is terminated

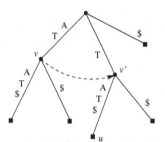

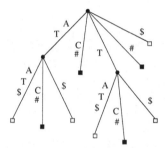

(a) Suffix tree for the string ATAT.   (b) Generalized suffix tree for
{ATAT, ABC}.

Fig. 1. Examples of a suffix tree and a generalized suffix tree. Note for instance (in the left figure) that the path-label of node v is AT, and the substring TA is a prefix of $path(u)$ and ends at the edge (v', u).

with that string's unique marker. Figure 1(b) gives an example of a generalized suffix tree for the set {ATAT, ATC} where the empty boxes represent leaves that have been trimmed after post-processing.

2.2 Basic Approach

We first build a generalized suffix tree T for our set of strings $S = \{s_1, \cdots, s_m\}$ in $O(N)$ time. We then preprocess T as in the $O(Nm)$ time algorithm for the multiple common substring problem [12, pp. 127-129]. If t is a substring of any string in S, then the preprocessing computes for each unique t the subset of strings in S that contain t. Finally, we use this information to find the set of nodes in the given phylogeny P for which each t is a discriminating tag. The details of the algorithm follow. We start with the preprocessing phase for clarity.

1. Preprocess the generalized suffix tree T.

   For each node v of T, we wish to compute the set

   $$B(v) = \{s \in S \; : \; \text{the path-label } path(v) \text{ is a substring of } s\} \; .$$

   That is, $B(v)$ is the set of species that share the substring $path(v)$. Note that,

   (a) The path-label of a leaf v is a substring only of $s_i \in S$, whose unique marker terminates the path; therefore, $B(v) = \{s_i\}$.
   (b) The path-label of an internal node v is a substring of the path-labels of its children; therefore,

   $$B(v) = \bigcup_{u \in child(v)} B(u) \; .$$

Each $B(v)$ is represented by a m-bit vector b_v where $b_v[i] = 1$ iff $s_i \in S$. We compute all the bit vectors by performing a post-order walk on T in $O(N)$ time. Since each bit-vector participates in one union operation, the running time of this phase is $O(Nm)$. Clearly, $O(Nm)$ space is required to store all the bit vectors.

2. Compute the discriminating tags for each u of P.

Observe that the set $B(v)$ for $v \in T$ contains sufficient information to determine if $path(v)$ is a discriminating tag for any $u \in P$. We determine all the nodes of P for which $path(v)$ is a discriminating tag efficiently as follows: Assume $w.l.o.g.$ that the leaves of P are labeled with $s_1, s_2, \cdots, s_m$ from left to right. We first label each internal node of P with the index i of the rightmost string s_i in it's left subtree. This labeling is unique since we can assume that each internal node has exactly 2 children. We then compute the number of leaves in both the left and right subtrees of each node. Denote these two quantities for $i \in P$ by $left\text{-}size(i)$ and $right\text{-}size(i)$ respectively. Now, $path(v)$ is a discriminating tag for i if the last run of 1's in b_v up to and including i is at least $left\text{-}size(i)$ and $b_v[i]$ is followed by at least $right\text{-}size(i)$ 0's. Note that if $path(v)$ is discriminating for $i \in P$, then the prefixes of $path(v)$ that are not prefixes of v's parent are also discriminating for i. Call these tags the discriminating tags *induced* by v, and note that they are determined by the label of the edge $(parent(v), v)$ in T.

We can compute the left-to-right running count of 1's, and the right-to-left running count of 0's for each b_v in $O(m)$ time. Since there are N bit vectors, the process requires $O(Nm)$ time.

Remark. If all input strings have length $O(n)$ such that $N = O(nm)$, then we need only store $O(n)$ bit vectors at a time, reducing the space requirement to $O(N)$.

2.3   Improving the Time and Space Requirement

We can improve the running time and space requirement of the algorithm above by slightly modifying the linear time algorithm used to solve the multiple common substring problem introduced in [18]. The algorithm computes $C_S(v) = |B(v)|$, the number of species in S containing $path(v)$, for all v of T in $O(N)$ time and space. It achieves this running time by utilizing suffix trees and constant-time *lowest common ancestor* (*lca*) queries [19, 20]. The *lca* of two nodes v and v' in a rooted tree is defined to be the deepest node in the tree that is an ancestor of both v and v'. Observe that $C_S(v)$ is equal to the number of *distinct* string markers terminating the path-labels of the leaves under v. The total number of string markers in v's subtree can be computed via a post-order walk of T. However, this number will be an over-count of $C_S(v)$ if there are duplicate markers under v. Suppose l and l' are two leaves such that l' is the first leaf with the same marker as l that the walk encounters after l. We then know that if v is an ancestor of $lca(l, l')$, the walk will over-count $C_S(v)$ by one due to l and l'. Now, for each such pair l and l', the algorithm first increments a counter maintained at $lca(l, l')$. Then, using the counters, it determines the number of duplicate markers under each node in a bottom-up fashion.

Our modification is straightforward, and we demonstrate it for the root of our phylogeny P. Suppose that the root is labeled r. Then, a tag is discriminating for r if it is a substring of all the strings of $L = \{s_1, \cdots, s_r\}$ and none of the strings of $R = \{s_{r+1}, \cdots, s_m\}$. Instead of just maintaining one counter $C_S(v)$ for each $v \in T$, we maintain $C_L(v)$ and $C_R(v)$, the number of strings of L and R containing $path(v)$ respectively. Now, $path(v)$ is discriminating for r if $C_L(v) = |L|$ and $C_R(v) = 0$.

The running time of this procedure is optimal for computing the discriminating tags for one node of P: It is linear in the total length of the strings in the subtree rooted at the node. Now, for $u \in P$, let $S_u \subseteq S$ denote those strings in u's subtree. The running time for computing the discriminating tags for all the nodes of P is then:

$$\sum_{u \in P} \sum_{s \in S_u} |s| \;=\; \sum_{s \in S} \sum_{u \in P : s \in S_u} |s| \;=\; O(N \cdot depth(P)) \;,$$

which is $O(N \log m)$ for a balanced phylogeny and $O(Nm)$ in the worst case. However, the space requirement is $O(N)$.

2.4   Minimal and Maximal Tags

We now consider the problem of finding *minimal* and *maximal* discriminating tags. A discriminating tag is said to be minimal if no substring of it is also discriminating. Correspondingly, a discriminating tag is said to be maximal if it is not a substring of another discriminating tag. We are interested in finding minimal and maximal tags for mainly two reasons. First, as implied by the claim below, the set of discriminating tags can potentially be enormous; hence, finding the minimal and maximal tags can be viewed as a data reduction step. More importantly, however, are the limitations on the lengths of tags that can be used in hybridization arrays. For example, tags used in the Affymetrix oligonucleotide microarrays are 25 bp long [21, pp. 3-4], whereas tags used in spotting arrays are between 100 and 5000 bp long [22, p. 11]. Consequently, the length distribution of minimal and maximal tags indicates the feasibility of executing our discrimination scheme in actual hybridization assays.

We start with a claim that will help us efficiently identify the subsets of minimal and maximal tags for a node in the phylogeny after computing the discriminating tags for the node.

Claim. Let t and t' be discriminating tags for some node $u \in P$, where t' is a substring of t. Then, any string t'' such that t' is a substring of t'' and t'' is a substring of t is also a discriminating tag for u.

Proof. Since t'' is a substring of t, it is present in all the sequences in the left subtree of u. Further, since t' is a substring of t'' and it is not present in any sequence in the right subtree of u, t'' will not be present in any of these sequences.

Corollary 1. *A discriminating tag is minimal if and only if its longest proper prefix and suffix are not discriminating.*

Corollary 2. *A discriminating tag is not maximal if and only if it is the longest proper prefix or suffix of a discriminating tag.*

Recall that in a suffix tree, the path-label of a node v, $path(v)$ is a prefix of $path(u)$: $u \in child(v)$ and that $path(parent(v))$ is a prefix of $path(v)$. We can also quickly determine the corresponding suffix relationships for $path(v)$ via the *suffix links* in the tree. The suffix link of an internal node v in the suffix tree is a directed non-tree edge from v to the longest proper suffix of $path(v)$. For example, if $path(v) = a\beta$ where $a \in \Sigma$ and $\beta \in \Sigma^*$, then (v, v') : $path(v') = \beta$ will be the suffix link of v. By construction, every internal node of the suffix tree has a suffix link [17]. As an example, the non-tree edge (v, v') in Fig. 1(a) is a suffix link.

Our task is then simple. Suppose t is a discriminating tag induced by v. In order for t to be minimal, it should be the shortest discriminating tag induced by v; $path(parent(v))$ should not be discriminating since it is the longest prefix of t; and, $path(u)$: (v, u) is a suffix link should not be discriminating since otherwise the longest proper suffix of t would be discriminating. In order for t to be maximal, on the other hand, t must equal $path(v)$, the path labels of the children of v should not be discriminating; and, $path(u)$ should not be discriminating if (u, v) is a suffix link.

Conceptually, we can ensure that all tags are induced by internal nodes for which suffix links are available by duplicating each input string and assigning each copy a distinct terminating character. Finally, the outlined computation can be readily incorporated in the algorithms above with no effect on the asymptotic running time and space requirements.

3   Experimental Results

As proof of concept, we first ran our algorithm on the estimated phylogeny for the 13 species: human, baboon, chimpanzee, rat, mouse, cat, dog, cow, pig, chicken, fugu, tetraodon, and zebrafish [13]. The phylogeny (Fig. 2(a)) is deduced from the genomic region orthologous to a segment of about 1.8 megabases (Mb) on human chromosome 7 containing the Cystic Fibrosis Transmembrane Conductance Regulator (CFTR) gene and 9 other genes. We will refer to this as the *CFTR data set*.

We also produced discriminating tags for some of the phylogenetic trees and the corresponding small subunit (SSU) prokaryotic ribosomal RNA sequences obtained from the Ribosomal Database Project (RDP) [14].

3.1   The CFTR Data Set

The CFTR data set we used to perform our first set of experiments consists of 13 sequences[1] of total length greater than 14 Mb. The estimated phylogeny, as well as an alternative grouping of the mammals, are shown in Fig. 2 [13].

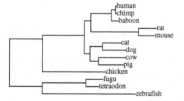

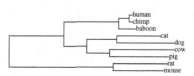

(a) Topology grouping rodents with primates. The root of the tree separating the fish from the other species was chosen arbitrarily.

(b) Alternative topology placing the rodents as an outgroup with respect to the other mammals.

Fig. 2. Maximum-liklihood phylogenies for the CFTR data set.

Our algorithm produced discriminating tags for each node in the phylogeny of the CFTR data set. Figure 3 plots the distribution of discriminating tag lengths for two nodes. We also produced the discriminating tags for the phylogeny shown in Fig. 2(b). The results for the root are shown in Fig. 4. Note that figures 3(b) and 4 show discriminating tags for the two alternative high-level partitions of the mammals.

Random Leaf Permutation. As a demonstration that the discriminating tags found are in fact dependent on the classification, we compared the distribution of the tags induced by the estimated phylogeny with the distribution of the tags induced by a random permutation of the sequences at the leaf nodes. The latter essentially represents a null hypothesis. Our results show that the number of discriminating tags generated in the latter case significantly decreases. For example, at the root node of the phylogeny for our 13 species, we only found two discriminating tags for the right branch as opposed to nearly 55, 000 tags we discovered when using the original phylogeny.

Random Tags from Conserved Regions. We examine the selectivity of our procedure by testing whether random substrings drawn from sequences' exon regions form discriminating tags or not. We obtained the annotations for the CFTR data set from the NISC Comparative Sequence Program website[1], and performed this experiment for the root of the phylogeny in Fig. 2(a). We observed that random substrings of such highly conserved regions of length more than 15, drawn from random sequences under one branch of the root, are rarely contained

---

[1] http://www.nisc.nih.gov/data

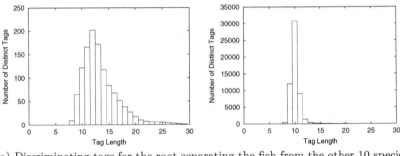

(a) Discriminating tags for the root separating the fish from the other 10 species.

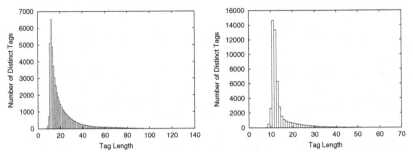

(b) Discriminating tags for the node separating the artiodactyls and carnivores from the rodents and primates.

Fig. 3. Length distribution of discriminating tags for two nodes of the CFTR phylogeny in Fig. 2(a). The left (right) panel displays discriminating tags present in the left (right) subtree of the corresponding node.

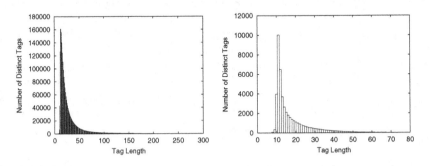

Fig. 4. Length distribution of discriminating tags for the root of the alternative grouping of the mammals shown in Fig. 2(b).

in any of the sequences under the other branch; however, each species from the same branch contained only a small fraction of the substrings, implying that those random substrings are not discriminating.

3.2   Large Phylogenies

In order to test the existence of tags in large phylogenies, we ran our algorithm on some of the phylogenies in the RDP-II database [14]. As an example, Fig. 5 shows the phylogeny for the Crenarch tree. The tree classifies 85 organisms with average sequence length equal to 1448 bp. We found discriminating tags for all nodes except the root of the tree. In general, for all the trees we considered (87–218 organisms), we found discriminating tags for most nodes except those that are close to the root of each tree, which is not surprising given the short sequence lengths of the SSU rRNA.

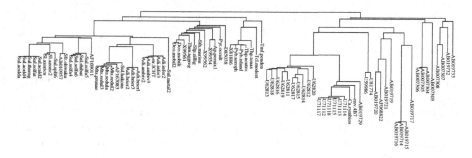

Fig. 5. The Crenarch 85-organism tree.

4   Discussion and Conclusion

In this paper, we proposed an efficient algorithm to generate short oligonucleotide tags for the purpose of identifying genomes within a preexisting phylogenetic (or classification) tree. We demonstrated tag feasibility using actual biological datasets and showed that typically a large number of useful tags can be obtained for each vertex in the phylogenetic tree. We also showed that the frequency and the lengths of tags depend on the tree structure – i.e., the tree structure that closely reflects the generative process of the targeted sequences yielded the largest collection of tags. There are other aspects of the tags such as the length of the minimal and maximal tags that depend on the stochastic process of evolution and the phylogenetic tree. Future work will be required to understand the statistical properties of discriminating tags vis-à-vis the phylogenies.

There are now multiple efforts to use modern molecular techniques for understanding the diversity of life (http://www.nsf.gov/bio/progdes/bioatol.htm) and cataloging organisms [4]. It is clear that these efforts will not succeed without a significant solution to scaling up both the data collection and the computational analysis. The current algorithm can be directly applied to the problem of obtaining efficient molecular markers (so-called bar-coding) to identify field samples against existing phylogenies or classification trees. However, it differs from some of the bar-coding efforts in that individual nucleotide identity is not required; assaying sequence identity is currently less amenable to high-throughput assays. More generally, we would like to obtain a high-throughput data collection method that will place novel organisms within an existing tree or *de novo* build phylogenetic trees for a large collection of organisms where prior information is available for only a small subset of taxa.

The current algorithm represents the first step in approaching the above goal. There are multiple remaining problems to convert this into a practical solution. First, given a set of discriminating tags, we need additional procedures to select optimal tags in terms of their biochemical characteristics including hybridization kinetics, potential for mismatch vis-à-vis other probes, optimal length, and so on. Second, the total genomic sequence may not be available for many organisms and if we generate tags from partial sequence information (such as the CFTR region above) we would like the tags to be valid for the whole genome. Third, our current procedure concentrates on placing known organisms on known trees, but ideally we would like a procedure that will place novel organisms in their proper locations by phylogenetic or classification criteria. That is, we would like the tags to generalize to novel information both in the tree and in the genome. These problems lead to the fourth problem of simultaneously optimizing the tags in terms of their potential instrument error, their generalization probability, and discriminating sensitivity. Finally, existing phylogenetic methods or classification methods are not based on models of presence/absence of genome-scale oligo tags; additional studies will need to be carried out for the tree reconstruction or recovery methods.

Rapid detection of unknown biological agents or large-scale construction of genomic phylogenies or classifications all demand the development of high-throughput analysis procedures. Many of the existing biochemical techniques are adequate to meet this challenge. The remaining problems are computational, both for efficient instrument design and for subsequent analysis. We believe the class of algorithms studied here and solutions to the problems proposed here will be critical toward designing high-throughput data collection strategies, maximizing efficiency by leveraging prior phylogenetic or taxonomic information.

References

1. Velculescu, V., Zhang, L., Vogelstein, B., Kinzler, K.: Serial analysis of gene expression. Science **270** (1995) 484–487
2. Adams, M., Kelley, J., Gocayne, J., Dubnick, M., Polymeropoulos, M., Xiao, H., C.R. Merril, e.a.: Complementary DNA sequencing: expressed sequence tags and human genome project. Science **252** (1991) 1651–1656
3. Olson, M., Hood, L., Cantor, C., Botstein, D.: A common language for physical mapping of the human genome. Science **245** (1989) 1434–1435
4. Hebert, P., Cywinska, A., Ball, S., deWaard, J.: Biological identifications through DNA barcodes. Proc. of the Royal Society of London **270** (2003) 313–321
5. Onodera, K., Melcher, U.: Viroligo: a database of virus-specific oligonucleotides. Nucl. Acids. Res. **30** (2002) 203–204
6. Ashelford, K.E., Weightman, A.J., Fry, J.C.: Primrose: a computer program for generating and estimating the phylogenetic range of 16S rRNA oligonucleotide probes and primers in conjunction with the rdp-ii database. Nucl. Acids. Res. **30** (2002) 3481–3489

7. Amann, R., Ludwig, W.: Ribosomal rna-targeted nucleic acid probes for studies in microbial ecology. FEMS Microbiology Reviews **24** (2000) 555–565

8. Matveeva, O.V., Shabalina, S.A., Nemtsov, V.A., Tsodikov, A.D., Gesteland, R.F., Atkins, J.F.: Thermodynamic calculations and statistical correlations for oligo-probes design. Nucl. Acids. Res. **31** (2003) 4211–4217

9. Kaderali, L., Schliep, A.: Selecting signature oligonucleotides to identify organisms using DNA arrays. Bioinformatics **18** (2002) 1340–1349

10. Frieze, A.M., Halldorsson, B.V.: Optimal sequencing by hybridization in rounds. Journal of Computational Biology **9** (2002) 355–369

11. Mitsuhashi, M., Cooper, A., Ogura, M., Shinagawa, T., Yano, K., Hosokawa, T.: Oligonucleotide probe design - a new approach. Nature **367** (1994) 759–761

12. Gusfield, D.: Algorithms on Strings, Trees, and Sequences. Cambridge University Press, New York (1997)

13. Thomas, J., et al.: Comparative analyses of multi-species sequences from targeted genomic regions. Nature **424** (2003) 788–793

14. Maidak, B.L., Cole, J.R., Lilburn, T.G., Parker, Charles T., J., Sax man, P.R., Farris, R.J., Garrity, G.M., Olsen, G.J., Schmidt, T.M., Tie dje, J.M.: The rdp-ii (ribosomal database project). Nucl. Acids. Res. **29** (2001) 173–174

15. Weiner, P.: Linear pattern matching algorithms. In: Proc. of the 14th IEEE Symposium on Switching and Automata Theory. (1973) 1–11

16. McCreight, E.M.: A space-economical suffix tree construction algorithm. Journal of the ACM (JACM) **23** (1976) 262–272

17. Ukkonen, E.: On-line construction of suffix-trees. Algorithmica **14** (1995) 249–260

18. Hui, L.: Color set size problem with applications to string matching. In: 3rd Symposium on Combinatorial Pattern Matching. Volume 644 of Lecture Notes in Computer Science., Springer (1992) 227–240

19. Harel, D., Tarjan, R.E.: Fast algorithms for finding nearest common ancestors. SIAM Journal of Computing **13** (1984) 338–355

20. Schieber, B., Vishkin, U.: On finding lowest common ancestors: Simplifications and parallelization. SIAM Journal of Computing **17** (1988) 1253–1262

21. Knudsen, S.: A Biologist's Guide to Analysis of DNA Microarray Data. Wiley Pub. (2002)

22. Baldi, P., Hatfield, G.W.: DNA Microarrays and Gene Expression. Cambridge University Press (2002)

Novel Tree Edit Operations
for RNA Secondary Structure Comparison

Julien Allali[1] and Marie-France Sagot[2,3]

[1] Institut Gaspard-Monge, Université de Marne-la-Vallée, Cité Descartes
Champs-sur-Marne, 77454 Marne-la-Vallée Cedex 2, France
allali@univ-mlv.fr
[2] Inria Rhône-Alpes, Université Claude Bernard, Lyon I
43 Bd du 11 Novembre 1918, 69622 Villeurbanne cedex, France
Marie-France.Sagot@inria.fr
[3] King's College, London, UK

Abstract. We describe an algorithm for comparing two RNA secondary
structures coded in the form of trees that introduces two novel operations, called *node fusion* and *edge fusion*, besides the tree edit operations
of deletion, insertion and relabelling classically used in the literature.
This allows us to address some serious limitations of the more traditional tree edit operations when the trees represent RNAs and what is
searched for is a common structural core of two RNAs. Although the
algorithm complexity has an exponential term, this term depends only
on the number of successive fusions that may be applied to a same node,
not on the total number of fusions. The algorithm remains therefore efficient in practice and is used for illustrative purposes on ribosomal as
well as on other types of RNAs.

Keywords: tree comparison, edit operation, distance, RNA, secondary
structure

1 Introduction

RNAs are one of the fundamental elements of a cell. Their role in regulation has
been shown recently to be far more prominent than initially believed (20 Decem-
ber 2002 issue of *Science*, which designated small RNAs with regulatory function
as the scientific breakthrough of the year). It is now known, for instance, that
there is massive transcription of non-coding RNAs. Yet current mathematical
and computer tools remain mostly inadequate to identify, analyse and compare
RNAs.

An RNA may be seen as a string over the alphabet of nucleotides (also
called bases), {A, C, G, T}. Inside a cell, RNAs do not retain a linear form but
instead fold in space. The fold is given by the set of nucleotide bases that pair.
The main type of pairing, called canonical, corresponds to bonds of the type
$A - U$ and $G - C$. Other rarer types of bonds may be observed, most frequent
among them is $G - U$, also called the wobble pair. Figure 1 shows the sequence
of a folded RNA. Each box represents a consecutive sequence of bonded pairs,

I. Jonassen and J. Kim (Eds.): WABI 2004, LNBI 3240, pp. 412–425, 2004.

corresponding to a helix in 3D space. The secondary structure of an RNA is the set of helices (or the list of paired bases) making up the RNA. Pseudo-knots, which may be described as a pair of interleaved helices, are in general excluded from the secondary structure of an RNA. RNA secondary structures can thus be represented as planar graphs. An RNA primary structure is its sequence of nucleotides while its tertiary structure corresponds to the geometric form the RNA adopts in space.

Apart from helices, the other main structural elements in an RNA are: 1. hairpin loops which are sequences of unpaired bases closing a helix; 2. internal loops which are sequences of unpaired bases linking two different helices; 3. bulges which are internal loops with unpaired bases on one side only of a helix; 4. multi-loops which are unpaired bases linking at least three helices. Stems are successions of one or more among helices, internal loops and/or bulges.

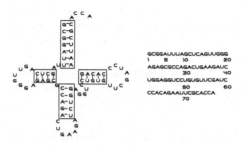

Fig. 1. Primary and secondary structures of a transfer RNA.

RNA secondary structure comparison is one of the main basic computational problems raised by the study of RNAs. It is the problem we address in this paper. The motivations are many. RNA structure comparison has been used in at least one approach to RNA structure prediction that takes as initial data a set of unaligned sequences supposed to have a common structural core [1]. For each sequence, a set of structural predictions are made (for instance, all suboptimal structures predicted by an algorithm like Zucker's MFOLD [9], or all suboptimal sets of compatible helices or stems). The common structure is then found by comparing all the structures obtained from the initial set of sequences, and identifying a substructure common to all, or to some of the sequences. RNA structure comparison is also an essential element in the discovery of RNA structural motifs, or profiles, or of more general models that may then be used to search for other RNAs of the same type in newly sequenced genomes. For instance, general models for tRNAs and introns of group I have been derived by hand [3] [5]. It is an open question whether models at least as accurate as these, or perhaps even more accurate, could have been derived in an automatic way. The identification of smaller structural motifs is an equally important topic that requires comparing structures.

As we saw, the comparison of RNA structures may concern *known* RNA structures (that is, structures that were experimentally determined) or *predicted* structures. The objective in both cases is the same: to find the common parts of such structures.

In [7], Shapiro suggested to mathematically model RNA secondary structures without pseudo-knots by means of trees. The trees are rooted and ordered, which means that the order among the children of a node matters. This order corresponds to the 5'-3' orientation of an RNA sequence. One way to compare two RNA secondary structures is then to apply a number of tree edit operations in one or both of the trees representing the RNAs until isomorphic trees are obtained. The tree edit operations considered are derived from the operations classically applied to sequences: substitution, deletion and insertion. In 1989, Zhang and Shasha proposed [8] a dynamic programming algorithm for comparing two trees. Shapiro and Zhang then showed [6] how to use tree editing to compare RNAs. The latter also proposed various tree models that could be used for representing RNA secondary structures. Each suggested tree offers a more or less detailed view of an RNA structure. Figure 2 (b) to (e) presents a few examples of such possible views for the RNA given in Figure 2 (a). In Figure 2, the nodes of the tree in (b) represent either unpaired bases (leaves) or paired bases (internal nodes). Each node is labelled with, respectively, a base or a pair of bases. A node of the tree in (c) represents a set of successive unpaired bases or of stacked paired ones. The label of a node is an integer indicating, respectively, the number of unpaired bases or the height of the stack of paired ones. The nodes of the tree in (d) represent elements of secondary structure: hairpin loop (H), bulge (B), internal loop (I) or multi-loop (M). The edges correspond to helices. Finally, the tree in (e) contains only the information concerning the skeleton of multi-loops of an RNA. The last representation, though giving a highly simplified view of an RNA, is important nevertheless as it is generally accepted that it is this skeleton which is usually the most constrained part of an RNA. The last two models may be enriched with information concerning, for instance, the number of (unpaired) bases in a loop (hairpin, internal, multi) or bulge, and the number of paired bases in a helix. The first label the nodes of the tree, the second its edges. Other types of information may be added (such as overall composition of the elements of secondary structure). In fact, one could consider working with various representations simultaneously or in an interlocked, multi-level fashion. This goes beyond the scope of this paper which is concerned with comparing RNA secondary structures using one among the many tree representations possible. We shall however comment further this multi-level approach later on.

Concerning the objectives of this paper, they are twofold. The first is to give some indications on why the classical edit operations that have been considered so far in the literature for comparing trees present some limitations when the trees stand for RNA structures. Three cases of such limitations will be illustrated through examples in Section 3. In Section 4, we then introduce two novel operations, so-called *node-fusion* and *edge-fusion*, that enable us to address some of these limitations and then give a dynamic programming algorithm for comparing two RNA structures with these two additional operations. Implementation

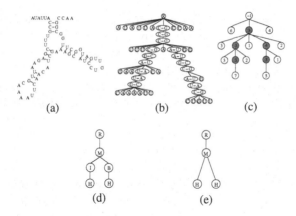

Fig. 2. Example of different tree representations of a same RNA.

issues and initial results are presented in Section 5. Before that, we start by introducing some notation and by recalling in the next section the basics about classical tree edit operations and tree mapping.

2   Tree Editing and Mapping

Let T be an ordered rooted tree, that is, a tree where the order among the children of a node matters. We define three kinds of operations on T: deletion, insertion and relabelling (corresponding to a substitution in sequence comparison). The operations are shown in Figure 3. The deletion (3 (b)) of a node u removes u from the tree. The children of u become the children of u's father. An insertion (3 (c)) is the symmetric of a deletion. Given a node u, we remove a consecutive (in relation to the order among the children) set $u_1, \ldots, u_p$ of its children, create a new node v, make v a child of u by attaching it at the place where the set was, and, finally, make the set $u_1, \ldots, u_p$ (in the same order) the children of v. The relabelling of a node (3 (d)) consists simply in changing its label.

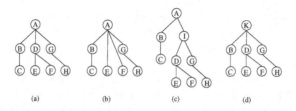

Fig. 3. Edit operations: (a) the original tree T, (b) deletion of node labelled D, (c) insertion of the node labelled I and (d) relabelling of a node in T (the label A of the root is changed into K).

Given two trees T and T', we define $S = \{s_1 \ldots s_e\}$ to be a series of edit operations such that if we apply successively the operations in S to the tree T, we obtain T' (*i.e.*, T and T' become isomorphic). A series of operations like S *realizes the editing* of T into T' and is denoted by $T \xrightarrow{S} T'$.

We define a function *cost* from the set of possible edit operations (deletion, insertion, relabelling) to the integers (or the reals) such that $cost_s$ is the score of the edit operation s. If S is a series of edit operations, we define by extension that $cost_S$ is $\sum_{s \in S} cost_s$. We can define the edit distance between two trees as the series of operations that performs the editing of T into T' and such that its cost is minimal: $distance(T, T') = \{\min(cost_S) | T \xrightarrow{S} T'\}$.

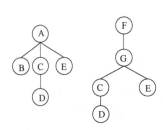

Let an insertion or a deletion cost one and the relabelling of a node cost zero if the label is the same and one otherwise. For the two trees of the figure on the left, the series $relabel(A \to F).delete(B).insert(G)$ realizes the editing of the left tree into the right one and costs 3. Another possibility is the series $delete(B).relabel(A \to G).insert(F)$ which also costs 3. The distance between these two trees is 3.

Given a series of operations S, let us consider the nodes of T that are not deleted (in the initial tree or after some relabelling). Such nodes are associated with nodes of T'. The *mapping M_S* relative to S is the set of couples (u, u') with $u \in T$ and $u' \in T'$ such that u is associated with u' by S.

The operations described above are the "classical tree edit operations" that have been commonly used in the literature for RNA secondary structure comparisons. We now present a few results obtained using such classical operations that will allow us to illustrate a few limitations they may present when used for comparing RNA structures.

3   Limitations of Classical Tree Edit Operations for RNA Comparison

As suggested in [6], the tree edit operations recalled in the previous section can be used on any type of tree coding of an RNA secondary structure.

Figure 4 shows two RNAsePs extracted from the database [2] (they are found, respectively, in *Thermotoga maritima* and *Streptococcus gordonii*). For the example we discuss now, we code the RNAs using the tree representation indicated in Figure 2 (b) where a node represents a base pair and a leaf an unpaired base. After applying a few edit operations to the trees, we obtain the result indicated in Figure 2, with deleted/inserted bases in grey. We have surrounded a few regions that match in the two trees. Bases in the box at the bottom of the RNA on the left are thus associated with bases in the rightmost box of the RNA on the right. Such matches illustrate one of the main problems with the classical tree edit operations: bases in one RNA may be mapped to identically labelled bases in the other RNA to minimise the total cost, while such bases should not be associated in terms of the elements of secondary structure to which they belong. In fact, such elements are often distant from one another along the common RNA structure. We call this problem the "scattering effect". It is related to the definition of tree edit operations. In the case of this example and of the representation adopted, the problem might have been avoided if structural information had been used. Indeed, the problem appears also because the structural location of an unpaired base is not taken into account. It is therefore possible to match, for instance, an unpaired base from a hairpin loop with an unpaired base from a multi-loop. Using another type of representation, as we shall do, would, however, not be enough to solve all problems as we see next.

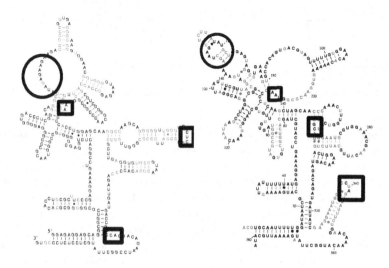

Fig. 4. Result of the matching of the two RNAs in Figure 4 using the model (b) given in Figure 2.

Indeed, to compare the same two RNAs, we can also use a more abstract tree representation such as the one given in Figure 2 (d). In this case, the internal nodes represent a multi-loop, internal-loop or bulge, the leaves code for hairpin loops and edges for helices. The result of the edition of T into T' for some cost function is presented in Figure 5 (we shall come back later to the cost functions used in the case of such more abstract RNA representations; for the sake of this example, we may assume an arbitrary one is used).

The problem we wish to illustrate in this case is shown by the boxes in the figure. Consider the boxes at the bottom. In the left RNA, we have a helix made up of 13 base pairs. In the right RNA, the helix is formed by 7 base pairs followed by an internal loop and another helix of size 5. By definition (see Section 2), the algorithm can only associate one element in the first tree to one element in the second tree. In this case, we would like to associate the helix of the left tree to the two helices of the second tree since it seems clear that the internal loop

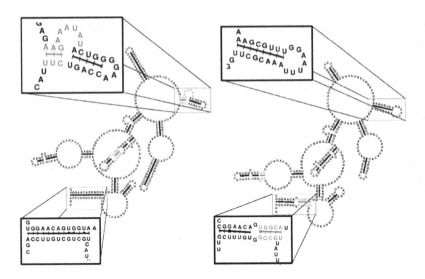

Fig. 5. Result of the matching of the two RNAs using the model (d) given in Figure 2.

represents either an inserted element in the second RNA, or the unbonding of one base pair. This, however, is not possible with classical edit operations.

A third type of problem one can meet when using only the three classical edit operations to compare trees standing for RNAs is similar to the previous one but concerns this time a node instead of edges in the same tree representation. Often, an RNA may present a very small helix between two elements (multi-loop, internal-loop, bulge or hairpin-loop) while such helix is absent in the other RNA. In this case, we would therefore have liked to be able to associate one node in a tree representing an RNA with two or more nodes in the tree for the other RNA. Once again, this is not possible with any of the classical tree edit operations. An illustration of this problem is shown in Figure 11 (see Section 5).

We shall use RNA representations that take the elements of the structure of an RNA into account to avoid some of this scattering effect. Furthermore, in addition to considering information of a structural nature, labels are attached, in general to both nodes and edges of the tree representing an RNA. Such labels are numerical values (integers or reals). They represent in most cases the size of the corresponding element, but may also further indicate its composition etc. Such additional information is then incorporated into the cost functions for all three edit operations.

It remains now to deal with the last two problems that are a consequence of the one-to-one associations between nodes and edges enforced by the classical tree edit operations. To that purpose, we introduce two novel tree edit operations, called the *edge fusion* and the *node fusion*.

4   Introducing Novel Tree Edit Operations

4.1   Edge Fusion and Node Fusion

In order to address some of the limitations of the classical tree edit operations that were illustrated in the previous section, we need to introduce two novel operations. These are the *edge fusion* and the *node fusion*. They may be applied to any of the tree representations given in Figure 2(c) to (e).

An example of edge fusion is shown in Figure 6. Let e_u be an edge leading to a node u, c_i a child of u and e_{c_i} the edge between u and c_i. The edge fusion of e_u and e_{c_i} consists in replacing e_{c_i} and e_u with a *new* single edge e. The edge e links the father of u to c_i. Its label then becomes a function of the (numerical) labels of e_u, u and e_{c_i}. For instance, if such labels indicated the size of each element (*e.g.* for a helix, the number of its stacked pairs, and for a loop, the min, max or the average of its unpaired bases on each side of the loop), the label of e could be the sum of the sizes of e_u, u and e_{c_i}. Observe that merging two edges implies deleting all subtrees rooted at the children c_j of u for j different from i. The cost of such deletions is added to the cost of the edge fusion.

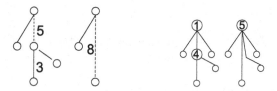

Fig. 6. On the left, an example of edge fusion. On the right, an example of node fusion.

An example of node fusion is given in Figure 6. Let u be a node and c_i one of its children. Performing a node fusion of u and c_i consists in making u the father of all children of c_i and in relabelling u with a value that is a function of the values of the labels of u, c_i and of the edge between them.

Observe that a node fusion may be simulated using the classical edit operations by a deletion followed by a relabelling. However, the difference between a node fusion and a deletion/relabelling is in the cost associated with both operations. We shall come back to this point later.

Obviously, like insertions or deletions, edge fusions and node fusions have of course symmetric conuterparts, which are the *edge split* and the *node split*.

We now present an algorithm to compute the tree edit distance between two trees using the classical tree edit operations plus the two operations just introduced.

4.2   Algorithm

The method we introduce is a dynamic programming algorithm based on the one proposed by Zhang and Shasha. Their algorithm is divided in two parts:

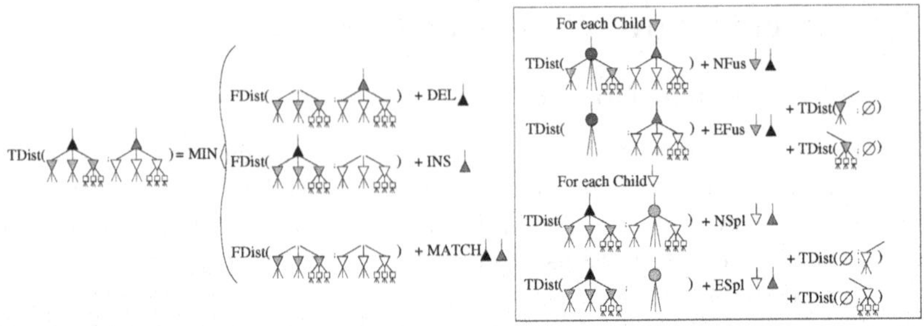

Fig. 7. Zhang and Sasha's dynamic programming algorithm: the tree distance part. The right box corresponds to the additional operations added to take fusion into account.

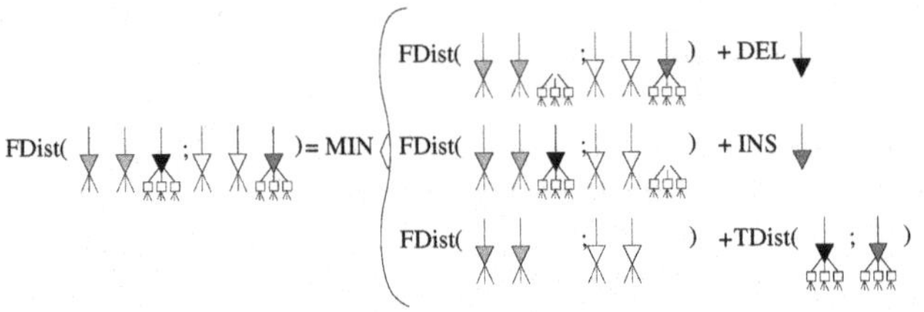

Fig. 8. Zhang and Sasha's dynamic programming algorithm: the forest distance part.

they first compute the edit distance between two trees (this part is denoted by $TDist$) and then the distance between two forests (this part is denoted by $FDist$). Figure 7 illustrates in pictorial form the part $TDist$ and Figure 8 the $FDist$ part of the computation.

In order to take our two new operations into account, we need to compute a few more things in the $TDist$ part. Indeed, we must add the possibility for each tree to have a node fusion (inversely, *node split*) between the root and one of its children, or to have an edge fusion (inversely *edge split*) between the root and one of its children. These additional operations are indicated in the right box of Figure 7.

We present now a formal description of the algorithm. Let T be an ordered rooted tree with $|T|$ nodes. We denote by t_i the ith node in a postfix order. For each node t_i, $l(i)$ is the index of the leftmost child of the subtree rooted at t_i. Let $T(i \ldots j)$ denote the forest composed by the nodes $t_i \ldots t_j$ ($T \equiv T(0 \ldots |T|)$). To simplify notation, from now on, when there is no ambiguity, i will refer to the node t_i. In this case, $distance(i_1 \ldots i_2, j_1 \ldots j_2)$ will be equivalent to $distance(T(i_1 \ldots i_2), T'(j_1 \ldots j_2))$.

The algorithm of Zhang and Sasha is fully described by the following recurrence formula:

$$distance(i_1 \ldots i_2, j_1 \ldots j_2) =$$

if $((i_1 == l(i_2))$ and $(j_1 == l(j_2)))$

$$MIN \begin{cases} distance(\ i_1 \ldots i_2 - 1\ , j_1 \ldots j_2 \quad\ ) + cost_{del}(i_2) \\ distance(\ i_1 \ldots i_2 \quad\ , j_1 \ldots j_2 - 1\ ) + cost_{ins}(j_2) \\ distance(\ i_1 \ldots i_2 - 1\ , j_1 \ldots j_2 - 1\ ) + cost_{match}(i_2, j_2) \end{cases} \quad (1)$$

else

$$MIN \begin{cases} distance(\qquad i_1 \ldots i_2 - 1 \quad , j_1 \ldots j_2) \qquad\ ) + cost_{del}(i_2) \\ distance(\qquad i_1 \ldots i_2) \qquad , j_1 \ldots j_2 - 1\ ) + cost_{ins}(j_2) \\ distance(\qquad i_1 \ldots l(i_2) - 1\ , j_1 \ldots l(j_2) - 1\ ) \\ \qquad + distance(\ l(i_2) \ldots i_2 \quad , l(j_2) \ldots j_2 \quad) \end{cases} \quad (2)$$

Part (1) of the formula corresponds to Figure 7 while part (2) corresponds to Figure 8. In practice, the algorithm stores in a matrix the score between each subtree of T and T'. The space complexity is therefore $O(|T| * |T'|)$. To reach this complexity, the computation must be done in a certain order (see [8] for further details). The time complexity of the algorithm is $O(|T| * min(leaf(T), height(T)) * |T'| * min(leaf(T'), height(T')))$ where $leaf(T)$ and $height(T)$ represent, respectively, the number of leaves and the height of a tree T.

Follows now the formula to compute the edit score allowing for both node and edge fusions.

$$distance(\{i_1, \ldots, i_k\}, path, \{j_1, \ldots, j_{k'}\}, path') =$$

if $((i_1 \geq l(i_k))$ and $(j_1 \geq l(j_{k'})))$

$$MIN \begin{cases} distance(\ \{i_1 \ldots i_{k-1}\}\ , \emptyset \quad\ , \{j_1 \ldots j_{k'}\} \quad\ , path'\ ) + cost_{del}(i_k) \\ distance(\ \{i_1 \ldots i_k\} \quad , path\ , \{j_1 \ldots j_{k'-1}\}\ , \emptyset \quad\ ) + cost_{ins}(j_{k'}) \\ distance(\ \{i_1 \ldots i_{k-1}\}\ , \emptyset \quad\ , \{j_1 \ldots j_{k'-1}\}\ , \emptyset \quad\ ) + cost_{match}(i_k, j_{k'}) \\ for\ each\ child\ i_c\ of\ i_k\ in\ \{i_1, \ldots, i_k\},\ set\ i_l = l(i_c) \\ \quad distance(\{i_1 \ldots i_{c-1}, i_{c+1} \ldots i_k\}, path.(u, i_c), \{j_1 \ldots j_{k'}\}, path') \\ \quad + cost_{node\_fusion}(i_c, i_k)\ \textbf{note:}\ i_k\ data\ are\ changed \\ \quad distance(\{i_l \ldots i_{c-1}, i_k\}, path.(e, i_c), \{j_1 \ldots j_{k'}\}, path') \\ \quad + cost_{edge\_fusion}(i_c, i_k) + distance(\{i_1 \ldots i_{l-1}\}, \emptyset, \emptyset, \emptyset) \\ \quad + distance(\{i_{c+1} \ldots i_k - 1, \emptyset, \emptyset, \emptyset)\ \textbf{note:}\ i_k\ data\ are\ changed \\ for\ each\ child\ j_{c'}\ of\ j_{k'}\ in\ \{j_1, \ldots, j_{k'}\},\ set\ j_{l'} = l(j_{c'}) \\ \quad distance(\{i_1 \ldots i_k\}, path, \{j_1 \ldots j_{c'-1}, j_{c'+1} \ldots j_{k'}\}, path'.(u, j_{c'})) \\ \quad + cost_{node\_split}(j_{c'}, j_{k'})\ \textbf{note:}\ j_{k'}\ data\ are\ changed \\ \quad distance(\{i_1 \ldots i_k\}, path, \{j_{l'} \ldots j_{c'}, j_{k'}, path'.(e, j_{c'})\}) \\ \quad + cost_{edge\_split}(j_{c'}, j_{k'}) + distance(\emptyset, \emptyset, \{j_1 \ldots j_{l'-1}\}, \emptyset) \\ \quad + distance(\emptyset, \emptyset, j_{c'+1} \ldots j_{k'-1}, \emptyset)\ \textbf{note:}\ j_{k'}\ data\ are\ changed \end{cases}$$

$$(3)$$

else *set* $i_l = l(i_k)$ *and* $j_{l'} = l(j_{k'})$

$$MIN \begin{cases} distance(& \{i_1 \ldots i_{k-1}\} , \emptyset & , \{j_1 \ldots j_{k'}\} & , path') + del(i_k) \\ distance(& \{i_1 \ldots i_k\} & , path , \{j_1 \ldots j_{k'-1}\} , \emptyset &) + ins(j_{k'}) \\ distance(& \{i_1 \ldots i_{l-1}\} , \emptyset & , \{j_1 \ldots j_{l'-1}\} , \emptyset &) \\ & +distance(\{i_l \ldots i_k\} & , path , \{j_{l'} \ldots j_{k'}\} & , path') \end{cases}$$

$$(4)$$

Given two nodes u and v such that v is a child of u, $node\_fusion(u, v)$ is the fusion of node v with u and $edge\_fusion(u, v)$ is the edge fusion between the edges leading to, respectively, nodes u and v. The symmetric operations are denoted by, respectively, $node\_split(u, v)$ and $edge\_split(u, v)$.

The distance computation takes two new parameters $path$ and $path'$. These are sets of pairs (e or u, v) which indicate, for node i_k (resp. j_k), the series of fusions that were done. Thus a pair (e, v) indicates that an edge fusion has been perfomed between i_k and v, while for (u, v) a node v has been merged with node i_k. The notation $path.(e, v)$ indicates that the operation (e, v) has been performed in relation to node i_k and the information is thus concatenated to the set $path$ of pairs currently linked with i_k.

Observe that the computation of the forest distance does not change in relation to the original algorithm. We therefore need a matrix to store all edit scores between ($i, path$) and ($j, path'$).

Let d be the maximum degree of the two trees. Then each node has $O((2d)^l)$ different ways of participating in l **consecutive** fusions (that is, in fusions with successive nodes along a same branch). If we limit the number of consecutive fusions to l, the total memory complexity of our algorithm is $O(n(2d)^l)$. By using the same arguments as in [8], we prove that the time complexity of the new algorithm to be $O((2d)^l * |T| * min(leaf(T), height(T)) * |T'| * min(leaf(T'), height(T')))$.

This complexity suggests that the fusion operations may be used only for reasonable trees (typically, less than 100 nodes) and small values of l (typically, less then 4). It is however important to observe that the overall number of fusions one may perform can be much greater than l without affecting the worst-case complexity of the algorithm. Indeed, any number of fusions can be made while still retaining the bound of $O((2d)^l * |T| * min(leaf(T), height(T)) * |T'| * min(leaf(T'), height(T')))$ so long as one does not realize more than l consecutive fusions for each node.

In general also, most interesting tree representations of an RNA are of small enough size as will be shown next, together with some initial results obtained in practice.

5   Implementation and Results

The algorithm presented in the previous section has been coded using C++. An online version is available at http://www-igm.univ-mlv.fr/~allali/migal/.

We recall that RNAs are relatively small molecules with sizes limited to a few kilobases. For instance, the small ribosomal subunit of *Sulfolobus acidocaldarius*

(D14876) is made up of 1147 bases. Using the representation shown in Figure 2 (b), the tree obtained contains 440 internal nodes and 567 leaves, that is 1007 nodes overall. Using the representation 2 (d), the tree is composed of 78 nodes. Finally, the tree obtained using the representation given in Figure 2 (e) contains only 48 nodes. We therefore see that even for large RNAs, any of the known abstract tree-representations (that is, representations which take the elements of the secondary structure of an RNA into account) that we can use leads to a tree of manageable size for our algorithm. In fact, for small values of l (2 or 3), the tree comparison takes reasonable time (a few minutes) and memory (less than 1Gb).

As we already mentioned, a fusion (resp. split) can be viewed as an alternative to a deletion (resp. insertion) followed by a relabelling. Therefore, the cost function for a fusion must be chosen carefully.

Let us assume that the cost function returns a real value between zero and one. If we want to compute the cost of a fusion between two nodes u and v, the aim is to give to such fusion a cost slightly greater than the cost of deleting v and relabelling u; that is, we wish to have $cost_{node\_fusion}(u, v) = min(del(v) + t, 1)$. The parameter t is a *tuning parameter* for the fusion. Suppose that the new node w resulting from the fusion of u and v matches with another node z. The cost of this match is $cost_{match}(w, z)$. If we do not allow for node fusions, the algorithm will first match u with z, then will delete v. If we compare the two possibilities, on one hand we have a total cost of $cost_{node\_fusion}(u, v) + cost_{match}(w, z)$ for the fusion, that is $del(v) + t + cost_{match}(w, z)$, on the other hand a cost of $del(v) + cost_{match}(u, z)$. Thus t represents the gain that must be obtained by $cost_{match}(w, z)$ with regard to $cost_{match}(u, z)$, that is by a match without fusion. This is illustrated in Figure 9.

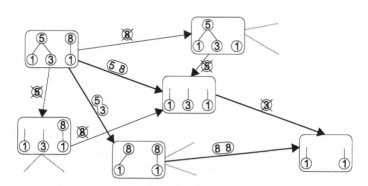

Fig. 9. Illustration of the gain that must be obtained using a fusion instead of a deletion/relabelling.

The tuning parameter t is thus an important parameter that allows us to control fusions. Always considering a cost function that produces real values between 0 and 1, if t is equal to 0.1, a fusion will be performed only if it improves the score by 0.1. In practice, we use values of t between 0 and 0.2.

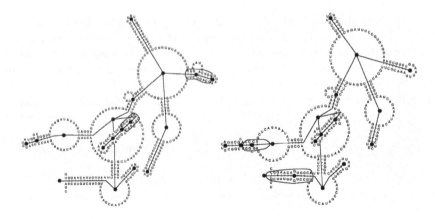

Fig. 10. Result of the editing between the two RNAs shown in the Figure 4 allowing for node and edge fusions. The boxes represent edge fusions.

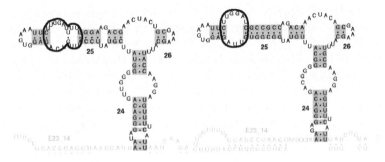

Fig. 11. Part of a mapping between two rRNA small sub-units. The node fusion is circled.

We have applied the new algorithm on the two RNAs shown in Figure 5 (these are eukaryotic nuclear P RNAs from *Saccharomyces uvarum* and *Saccharomyces kluveri*) and coded using the same type of representation as in Figure 2 (d). We have limited the number of consecutive fusions to one ($l = 1$). The computation of the edit distance between the two trees taking node and edge fusions into account besides deletions, insertions and relabelling has taken less than a second. The total cost allowing for fusions is 6.18 with $t = 0.05$ against 7.42 without fusions. As indicated in Figure 10, the last two problems discussed in Section 3 disappear thanks to some edge fusions (represented by the boxes).

An example of node fusions required when comparing two "real" RNAs is given in Figure 11. The RNAs are coded using the same type of representation as in Figure 2 (d). The figure shows part of the mapping obtained between the small sub-units of two ribosomal RNAs retrieved from [4] (from *Bacillaria paxillifer* and *Calicophoron calicophorum*). The node fusion has been circled.

6   Further Work and Conclusion

We have proposed an algorithm that addresses two main limitations of the classical tree edit operations for comparing RNA secondary structures. Its complexity is high in theory if many fusions are applied in succession to any given (the same) node, but the total number of fusions that may be performed is not limited. In practice, the algorithm is fast enough for most situations one can meet in practice.

To provide a more complete solution to the problem of the scattering effect, we propose the following scheme. Given an RNA, we build four tree representations for it instead of just one. Each tree corresponds to a different level of abstraction of the secondary structure of the RNA. The higher level represents the multi-loop skeleton while the lower level represents the sequence of paired and unpaired bases. For each tree except the one at the lowest level, there is a one-to-many relation between nodes, that is, a node of the tree at a given level is related with at least one node in the tree at the levels below. Two elements at the lowest level of abstraction, that is, two bases or base pairs can be matched only if they are part of a same structural element, the correspondence between such elements having being identified at a previous step by using the algorithm for comparing two structural tree representations introduced in this paper. Such approach thus allows us to address the problem of the scattering effect in an efficient way. A full description and practical consequences of such 4-level comparison method will be the subject of another paper.

References

1. D. Bouthinon and H. Soldano. A new method to predict the consensus secondary structure of a set of unaligned RNA sequences. *Bioinformatics*, 15(10):785–798, 1999.
2. James W. Brown. The ribonuclease p database. *Nucleic Acids Research*, 24(1):314, 1999.
3. N. el Mabrouk and F. Lisacek. Very fast identification of RNA motifs in genomic DNA. application to tRNA search in the yeast genome. *J. Mol. Biol.*, 264(1):46–55, 1996.
4. T. Winkelmans J. Wuyts, Y. Van de Peer and R. De Wachter. The european database on small subunit ribosomal rna. *Nucleic Acids Research*, 30(1):183–185, 2002.
5. F. Lisacek, Y. Diaz, and F. Michel. Automatic identification of group I intron cores in genomic DNA sequences. *J. Mol. Biol.*, 235(4):1206–1217, 1994.
6. B. A. Shapiro and K. Zhang. Comparing multiple RNA secondary structures using tree comparisons. *Comput. Appl. Biosci.*, 6(4):309–318, 1990.
7. B. Shapiro. An algorithm for multiple rna secondary structures. *Comput. Appl. Biosci.*, 4(3):387–393, 1988.
8. K. Zhang and D. Shasha. Simple fast algorithms for the editing distance between trees and related problems. *SIAM J. Comput.*, 18(6):1245–1262, 1989.
9. M. Zuker. Mfold web server for nucleic acid folding and hybridization prediction. *Nucleic Acids Res.*, 31(13):3406–3415, 2003.

The Most Probable Labeling Problem in HMMs and Its Application to Bioinformatics

Broňa Brejová, Daniel G. Brown*, and Tomáš Vinař**

School of Computer Science, University of Waterloo, Waterloo ON N2L 3G1 Canada
{bbrejova,browndg,tvinar}@uwaterloo.ca

Abstract. Hidden Markov models (HMMs) are often used for biological sequence annotation. Each sequence element is represented by states with the same label. A sequence should be annotated with the labeling of highest probability. Computing this most probable labeling was shown NP-hard by Lyngsø and Pedersen [9]. We improve this result by proving the problem NP-hard for a fixed HMM. High probability labelings are often found by heuristics, such as taking the labeling corresponding to the most probable state path. We introduce an efficient algorithm that computes the most probable labeling for a wide class of HMMs, including models previously used for transmembrane protein topology prediction and coding region detection.

1 Introduction

We present several contributions towards understanding the most probable labeling problem in hidden Markov models (HMMs) and its impact on bioinformatics applications. We prove the problem NP-hard even for a fixed HMM; previous NP-hardness proofs constructed HMM topologies whose size depended on an input instance, which is not appropriate in the context of bioinformatics applications. We also characterize a class of HMMs where the problem can be solved in polynomial time, and finally we demonstrate the usefulness of our findings, using examples from bioinformatics literature and simple experiments.

HMMs are often used in bioinformatics for sequence annotation tasks, such as gene finding (*e.g.*, [1]), protein secondary structure prediction (*e.g.*, [8]), and transmembrane protein topology (*e.g.*, [7]). An HMM is a generative probabilistic model composed of states and transitions. In each step of the generative process, the current state randomly generates one character of the DNA or protein sequence according to an emission probability table associated with the state. An outgoing transition is randomly chosen and followed to a new state, according to the transition probability table at that state.

In each application, biological knowledge is reflected in the model's topology. Parameters are set through automatic training so that the model generates sequences like those in nature. States represent distinct biological features, such

---

* Supported by NSERC and the Human Frontier Science Program.
** BB and TV supported by NSERC grant RGPIN46506-01 and by CITO.

I. Jonassen and J. Kim (Eds.): WABI 2004, LNBI 3240, pp. 426–437, 2004.

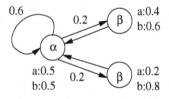

Fig. 1. An HMM with the multiple path problem. The most probable path for the string "ababa" is "$\alpha\alpha\alpha\alpha\alpha$," with probability 0.004, while the most probable labeling is "$\alpha\beta\alpha\beta\alpha$," with probability 0.01. The highest probability path with the same labeling has probability only 0.003.

as introns, exons, and splicing signals in gene finding, or transmembrane helices, cytoplasmic and non-cytoplasmic loops in transmembrane topology prediction.

For sequence annotation, we label each state with the feature it represents. For a given sequence, we examine paths through the model (or *state paths*) generating it. The labelings of such paths represent potential annotations. The HMM defines a probability distribution over all possible labelings: a given labeling's probability is the sum of the probabilities of the paths that generate it. Finding high probability labelings of a sequence in an HMM is *HMM decoding*.

The most widely used decoding method is the *Viterbi algorithm* [11]. It uses dynamic programming to find the most probable state path and reports the labeling associated with it. However, one labeling may have multiple paths, with its probability being the sum of probabilities of these paths. We would prefer to find the *most probable labeling* instead of the *most probable state path*.

Consider the example in Fig.1. The most probable labeling and most probable state path yield different annotations, which is surely a source for concern. We say that such HMM exhibits the *multiple path problem*. Moreover, as the sequence $(ab)^n$ grows in our example, the number of paths forming the most probable labeling increases exponentially. The probability of each single path is very low compared to the probability of the most probable path.

This problem has been recognized before [1, 6] and various heuristics have been suggested. A common (but rarely implemented) idea is to compute the k most probable paths, providing k candidate labelings [3]. These paths can be found efficiently with an algorithm for finding the k shortest paths in a directed acyclic graph [4]. However, this approach may fail, since the probability of each path in the most probable labeling may be small.

A different heuristic, called the 1-best algorithm, was suggested by Krogh [6]. The algorithm, similar to Viterbi, maintains a pool of several candidates for the labeling. The algorithm guarantees only that the probability of the resulting labeling is at least as high as the probability of the most probable state path.

Finally, one can apply *a posteriori* decoding – using the forward-backward algorithm [3] to compute the most probable label at each sequence position. However, no state path may correspond to this annotation, so we cannot guarantee that it is consistent with biological constraints of the model. To complete the heuristic, a second step is required to modify such a labeling to obtain a plausible annotation (see, *e.g.*, [10]).

In general, it is unlikely that there exists an efficient algorithm for finding the most probable labeling: the problem is NP-hard [9] (see also earlier work on a related model [2]). However, in these NP-hardness proofs, the size of the

HMM obtained by the reduction is polynomial in the size of the input instance. This is not appropriate, since in our applications the model is fixed (constant size) and the input sequence can be very long. This leads to the question of whether there is an algorithm for this problem whose runtime is polynomial in the sequence length but exponential in the model size. In Section 2 we show that this is unlikely: the problem is NP-hard even for a *fixed HMM of constant size*.

Ideally, one would like to distinguish between HMMs for which the most probable labeling problem is NP-hard and HMMs for which it is possible to solve the problem in polynomial time. As a first step in this direction, we present an algorithm that can find the most probable labeling for a wide class of HMMs in polynomial time, and we give a sufficient condition characterizing this class. The class includes topologies commonly used for various bioinformatics applications. Finally, we provide simple experiments that show that using the most probable labeling instead of the most probable paths may increase accuracy.

2   NP-Completeness Proof

Here, we give a new NP-completeness proof for the most probable labeling problem. We show a construction of a *specific* HMM, for which if we could compute the most probable labeling, we could solve instances of SAT. We first present a path counting problem on directed acyclic "layered" graphs, which we show is NP-complete. Then, we show a reduction from this problem to the HMM problem, and demonstrate that for SAT instances, we will have a specific HMM of constant, though large, size for which decoding is hard.

Layered Digraphs. A *colored proper layered digraph* is a directed graph with vertices arranged in layers $L_1, L_2, \ldots, L_w$. Each edge connects a vertex in some layer L_i to a vertex in layer L_{i+1}. Each vertex is colored white or black.

A *layer coloring* is an assignment of a color (white or black) to each layer. A directed path from layer L_1 to layer L_w is *consistent with a layer coloring* if the colors of the vertices on the path match the layer coloring.

Definition 1 (BEST-LAYER-COLORING). *Given a colored proper layered digraph G and a threshold T, is there a layer coloring which has at least T paths consistent with it?*

Theorem 1. BEST-LAYER-COLORING *is NP-complete, even if each layer has at most a constant number of vertices.*

Proof. BEST-LAYER-COLORING is in NP: for a given layer coloring, the number of consistent paths can be computed by simple dynamic programming.

To prove NP-hardness, we reduce SAT to BEST-LAYER-COLORING. Consider an instance of SAT with n variables $u_1, u_2, \ldots, u_n$ and m clauses $c_1, c_2, \ldots, c_m$. We give an overview of the construction in Fig.2.

Goal of the Construction. The graph consists of $m + 1$ blocks $0, 1, 2, \ldots, m$, each with $4n$ layers. The layer coloring of each block represents a truth assignment of variables $u_1, \ldots, u_n$. The truth assignment of each variable is encoded by four

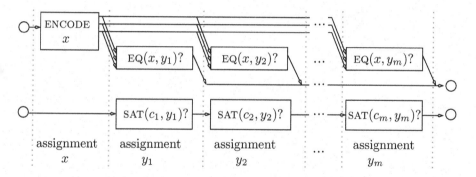

Fig. 2. Overview of the graph construction. The boxes represent components of the graph construction. Lines connecting the components represent layered subgraphs that propagate the number of paths from left to right regardless of the layer coloring (this is achieved by using one black and one white vertex in each layer).

consecutive layer colors: $\circ \circ \circ \circ$ (false) or $\circ \circ \bullet \circ$ (true). Layer colorings that are not of this form have no corresponding paths, so we do not consider them. Thus we use the terms "layer coloring of a block" and "truth assignment of a block" interchangeably. Let x be the truth assignment of block 0 and let $y_1, \ldots, y_m$ be the truth assignments of blocks $1, \ldots, m$.

In a "yes" instance of SAT, we want the truth assignments $x, y_1, \ldots, y_m$ to be the *same* satisfying truth assignment.

Details of Construction. We will decompose the structure of the graph into several components, each of them having several inputs and outputs. An input of a component is the number of consistent paths ending in a designated vertex on the left-most layer of the component. Similarly, an output of a component is the number of consistent paths ending in a designated vertex on the right-most layer of the component. Let $A \xrightarrow{x} B$ denote a component transforming a vector of inputs A to a vector of outputs B when corresponding layers have coloring x.

The component ENCODE(x) in Fig.2 encodes the truth assignment x as a vector of three integers $v(x)$ on its output. In each of the blocks $1, 2, \ldots, m$, we enforce the truth assignment to be the same as x with component EQ(x, y_i). The input of this component is the vector $v(x)$, and the output is the number $2K(n)$, where $K(n) = 4^n - 2^{n+1} + 1$, if the truth assignments x and y_i are the same, or a number smaller than $2K(n)$ otherwise. In particular, the two components have the following specification: ENCODE(x) : $1 \xrightarrow{x} (1, K(n) - b(x)^2, 2b(x))$ and EQ(x, y, i) : $(1, \alpha, \beta) \xrightarrow{y} K(n) - b(y)^2 + \beta \cdot b(y) + \alpha$, where $b(x)$ is the number whose binary representation encodes the truth assignment x of variables $u_1, \ldots, u_n$ with u_1 as the highest-order bit and u_n as the lowest-order bit.

Finally, component SAT(c_i, y_i) outputs its input, if truth assignment y_i satisfies clause c_i, or 0 otherwise. The input to SAT(c_1, y_1) is 1. Details are omitted.

There is an additional layer before the first block and after the last block to ensure the proper start and end of each consistent path.

The threshold T is $2m \cdot K(n) + 1$. For the number of consistent paths to reach this threshold, all of the block colorings must represent the same truth

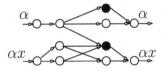

Fig. 3. One section of component
MULT$(x) : \alpha \xrightarrow{x} \alpha b(x)$.

assignment and the assignment must satisfy all clauses (otherwise the number
of consistent paths will be smaller).

Components ENCODE *and* EQ. These can be built from subcomponents MULT(x) :
$\alpha \xrightarrow{x} \alpha b(x)$ and SQUARE(x) : $1 \xrightarrow{x} K(n) - b(x)^2$. They consist of identical 4-layer sections, each processing one bit of $b(x)$. Section k of component
SQUARE(x) has four inputs and outputs $(1, B(k-1), C(y, k-1), D(y, k-1)) \xrightarrow{t}$
$(1, B(k), C(z, k), D(z, k))$, where y is the binary representation of truth assignment of the first $k - 1$ variables, t is the truth assignment of the k-th variable,
$z = 2y + t$, and B, C, D are defined as follows:

$$B(k) = 2^{k+2} - 4 = 2B(k-1) + 4 \tag{1}$$

$$C(z,k) = B(k) - 4z = \begin{cases} 2C(y, k-1) + 4, & \text{if } t = 0 \\ 2C(y, k-1), & \text{if } t = 1 \end{cases} \tag{2}$$

$$D(z,k) = 4^k - 2^{k+1} + 1 - z^2 = \begin{cases} 4D(y, k-1) + B(k-1) + 1, & \text{if } t = 0 \\ 4D(y, k-1) + C(y, k-1), & \text{if } t = 1 \end{cases} \tag{3}$$

The D output of the last section has value $D(b(x), n) = K(n) - b(x)^2$, as
desired. Fig.3 shows the construction of the simpler subcomponent for multiplication, MULT(x). Other drawings are omitted due to space.

Summary. The total number of vertices in each layer is at most 29. Thus, an
instance of SAT can be reduced to BEST-LAYER-COLORING with a constant
number of nodes per layer.   □

Connection to HMMs. Now, with the hardness of our problem shown, we connect
it to HMMs by reducing it to the most probable labeling problem.

Theorem 2. *For a given constant k and a colored proper layered digraph G with
at most k vertices in each layer, there exists an HMM M and a binary string S
such that the most probable labeling of string S in M represents the best layer
coloring of G. Moreover, the topology, emission, and transition probabilities of
M depend only on the constant k, not on the size of G, and the size of S is
polynomial in the number of layers of G.*

Proof. (Sketch.) We can construct an HMM and an input string so that paths in
the HMM will correspond to paths in G, and all paths have the same probability.
The states represent possible configurations of edges in G outgoing from each
vertex of one layer (pairs (i, V'), where i is a vertex in a layer, and V' is a subset
of vertices in the next layer). Since there is only a constant number of vertices
in each layer, this HMM is fixed and does not depend on the structure of G.
The structure of G is encoded in the input string: each symbol represents the
configuration of edges of one layer in G. The alphabet size is constant and can
be further reduced to binary by replacing every symbol by a binary string.   □

Corollary 1. *There exists a specific HMM for which it is NP-hard to find the most probable labeling of an input binary string.*

Proof. Theorem 1 shows there exists a constant k such that finding the best layer coloring is NP-hard, even if each layer has size at most k. Theorem 2 shows this problem can be reduced to the most probable labeling of a binary string in a fixed HMM. □

The HMM obtained in the proof of Corollary 1 is very large. It is possible to use ideas from the proof of Theorem 1 to reduce SAT to the most probable labeling problem directly, obtaining a much smaller HMM.

3 Computing the Most Probable Labeling

We have shown that in general it is NP-hard to compute the most probable labeling for a given HMM. However, we can characterize special classes of HMMs for which the most probable labeling can be computed efficiently. For example, consider an HMM where any two state paths Π_1 and Π_2 have different labelings. The most probable state path can be computed with the Viterbi algorithm, and thus so can be the most probable labeling. The runtime is $O(nm\Delta)$ time, where n is the length of the sequence, m is the number of states in HMM, and Δ is the maximum in-degree of a state.

Here, we introduce *extended labelings* and give an algorithm that computes the *most probable extended labeling* in polynomial time. We characterize a class of HMMs for which an extended labeling uniquely determines a single state labeling. For this class, our algorithm finds the most probable labeling. Even if the input HMM does not belong to the class, our algorithm returns a labeling with probability at least as high as the probability of the most probable state path. We give several practical examples found in this class of HMMs.

3.1 Most Probable Extended Labeling

Recall that a hidden Markov model is a generative probabilistic model, consisting of *states* and *transitions*. The HMM starts in a designated initial state s. In each step, a symbol is generated according to the emission probabilities of the current state. Unless the HMM has reached a designated final state f, it follows one of the transitions to another state. Let $e_u(x)$ be the emission probability of generating character x in state u, $a_{u,v}$ the probability of a transition from u to v, $\ell(u)$ the label of state u, and $\text{in}(u)$ the set of states having a transition to state u.

Definition 2 (Extended labeling). *A critical edge is an edge between states of different label. The extended labeling of a state path $\pi_1\pi_2\ldots\pi_n$ is the pair (L, C), where $L = \lambda_1, \lambda_2, \ldots, \lambda_n$ is the sequence of labels of each state in the path and $C = c_1, c_2, \ldots, c_k$ is the sequence of critical edges followed on the path.*

Theorem 3 (The most probable extended labeling). *For a given sequence* $S = x_1 \ldots x_n$ *and an HMM with m states, it is possible to compute the most probable extended labeling in* $O(n^2 m L_{\max} \Delta)$, *where* $L_{\max}$ *is the maximum number of states with the same label, and* Δ *is the largest in-degree of a state.*

Proof. We modify the Viterbi algorithm for computing the most probable state path. Let $V[u, i] = \max \Pr(x_1 \ldots x_i, \pi_1 \ldots \pi_i)$, where the maximum is taken over all state paths $\pi_1 \ldots \pi_i$ starting in state s and ending in state u. The Viterbi algorithm computes the values $V[u, i]$ by dynamic programming with the following recurrence, examining all possible options for the second to last state:

$$V[u, i] = \max_{v \in \text{in}(u)} V[v, i - 1] \cdot a_{v,u} \cdot e_u(x_i). \tag{4}$$

We modify the Viterbi algorithm as follows. Let $L[u, i] = \max \Pr(x_1 \ldots x_i, (L, C), \pi_i = u)$, where the maximum is taken over all extended labelings (L, C) and the generating process ends in state u. Instead of considering possible options for the second last state, we will examine all possible durations of the last segment with the same label and instead of the most probable path in such segment, we will compute the sum of all possible state paths in this segment. If the segment starts at position $j \leq i$ of the sequence, such sum $P[v, u, j, i]$ is the probability of generating the sequence $x_j \ldots x_i$ starting in state v and ending in state u, using only states with label $\lambda(u)$. We get the following recurrence:

$$L[u, i] = \max_{j \leq i} \max_{\substack{v:\lambda(v)=\lambda(u)}} \max_{\substack{w \in \text{in}(v): \\ \lambda(w) \neq \lambda(v)}} L[w, j - 1] \cdot a_{w,v} \cdot P[v, u, j, i] \tag{5}$$

We compute values of L in order of increasing i. For each i, we compute all relevant values of $P(v, u, j, i)$ in order of decreasing j by an algorithm similar to backward algorithm (see, *e.g.*, [3]), using the following recurrence:

$$P[v, u, j, i] = \sum_{w:v \in \text{in}(w), \lambda(v)=\lambda(w)} e_v(x_j) \cdot a_{v,w} \cdot P[w, u, j + 1, i] \tag{6}$$

When the computation of L is finished, the most probable extended labeling can be reconstructed by tracing back labels and critical edges used to obtain the value of $L[f, n]$, as in the Viterbi algorithm. $\qquad\square$

Note that the probability of the extended labeling returned by the algorithm is always at least as high as the probability of the most probable path Π found by Viterbi algorithm. This is because the probability of the extended labeling corresponding to Π must be at least as high as the probability of Π itself.

3.2   The Sufficient Condition

The algorithm defined above is guaranteed to compute the most probable labeling for a much wider class of HMMs than the Viterbi algorithm. Here is a sufficient condition for this class:

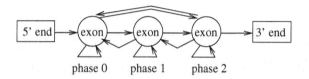

phase 0   phase 1   phase 2

Fig. 4. Simplified model of ESTScan. ESTScan uses an HMM for predicting coding part of an EST. Compared to a typical coding region predictor, ESTScan needs to handle insertions and deletions within the coding sequence caused by the low quality of EST sequencing. The exact place of a sequencing error cannot be easily identified, so to simply distinguish coding part of ESTs from non-coding, we assign the same label to all states corresponding to coding sequence. The resulting HMM has the multiple path problem, with each path corresponding to some combination of insertions and deletions. The actual model used in ESTScan has a more complicated topology, ensuring for example that only one insertion or deletion can occur within the same codon.

Definition 3. *An HMM satisfies the critical edge condition for an input sequence S if any two paths with the same labeling have the same sequence of critical edges.*

Corollary 2. *If an HMM satisfies the critical edge condition for a sequence S, then the above algorithm computes the most probable labeling of sequence S.*

Proof. We call a labeling Λ (extended labeling, state path) *possible* with respect to sequence S, if $\Pr(\Lambda \mid S) > 0$.

The algorithm above computes the most probable extended labeling. Therefore for the statement to be false, the most probable labeling and the most probable extended labeling must be different.

This happens only if two different extended labelings Λ_1 and Λ_2 correspond to the most probable labeling. Let Π_1 be a state path corresponding to Λ_1, and Π_2 be a state path corresponding to Λ_2. Since Λ_1 and Λ_2 are different, the paths Π_1 and Π_2 must differ in at least one critical edge; yet both Π_1 and Π_2 produce the same labeling. Therefore the HMM cannot satisfy the critical edge condition. □

Examples. The simplified model of ESTScan [5] shown in Fig.4 has the multiple path problem. Therefore the Viterbi algorithm is not appropriate for decoding it. The model satisfies the critical edge condition, so our algorithm finds the most probable labeling. The condition is satisfied because the states labeled "exon" are grouped in a subgraph with only one incoming and outgoing edge.

A more complicated example is the simple model of exon/intron structure of eukaryotic genes in Fig.5. Multiple copies of the same intron model preserve three-periodicity of coding regions. The multiple path problem is caused by ambiguity of transition between the two "intron" states. The model does not violate our condition, since the length of the exonic sequence uniquely determines which critical edge will be used.

Testing the Critical Edge Condition. We can test algorithmically whether a given HMM topology (not considering emission probabilities) satisfies the critical edge

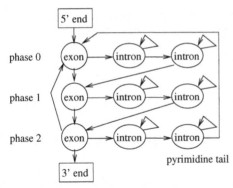

Fig. 5. Simple model of exon/intron structure. Intronic sequence in DNA contains a pyrimidine-rich tail close to the acceptor site. Its composition is very different from the rest of the intron, providing strong support for a neighboring acceptor site. The tail has variable length and does not have a clear boundary. This creates the multiple path problem because there are always several high-probability alternatives for the transfer from "intron" state to the "tail" state.

condition for every input sequence. We first use depth-first search to build a set S_s of all pairs of states that are reachable from the start state by the same labeling. We start from pair $(s, s) \in S_s$ and in each iteration we add a new pair (u, v) if $\lambda(u) = \lambda(v)$, and there exists $(u', v') \in S_s$ such that $u' \in \text{in}(u)$ and $v' \in \text{in}(v)$. Similarly, we also build a set S_f of all pairs of states, from which the final state can be reached by the same labeling. For the critical condition to be violated, there must exist a pair $(u, v) \in S_s$ and $(u', v') \in S_f$ such that $\lambda(u) \neq \lambda(u')$, and (u, u') and (v, v') are two different transitions. The algorithm takes $O(m^4)$ time.

It is possible to modify this verification algorithm to verify the critical edge condition in $O(m^4 |\Sigma|^2)$ time, if emission probabilities are given. Note that this test may yield a different result, since some states may not produce some of the alphabet symbols, making it impossible for two different paths with the same extended labeling to generate the same string; hence, this extended algorithm may find even more HMMs that satisfy the condition.

And finally, we can also verify the condition for a given HMM and input string in $O(nm^4)$ time. In that case, we will build a set of state pairs that can be reached by the same labeling for each position in the sequence.

3.3   Introducing Silent States

Silent states do not emit symbols. They are sometimes used in HMMs as a convenient modeling extension (see [3, Section 3.4]). Both the Viterbi algorithm and our algorithm can be easily extended to HMMs with silent states [3, Section 3.4]. The example in Fig.6 shows that some HMMs can be transformed to equivalent HMMs that satisfy the critical edge condition by addition of silent states. Thus, in our case, the silent states are a crucial modeling tool.

Example. Silent states are useful when one wants to provide several models for a particular sequence element. An example of such a model is TMHMM [7], shown in Fig.7. The two different models of non-cytoplasmic loops create the multiple path problem, potentially decreasing prediction accuracy if the Viterbi algorithm is used. We introduce silent states to ensure that the critical edge condition is satisfied.

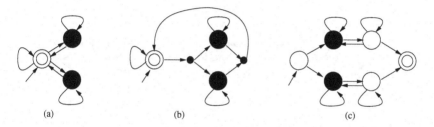

Fig. 6. Example of usefulness of silent states. The color of each state represents its label. Silent states are shown as smaller circles. The HMM (a) violates the critical edge condition and cannot be decoded by our algorithm. There is no equivalent topology without silent states satisfying the condition. Using silent states, we can construct an equivalent HMM (b) that satisfies the critical edge condition. However, the technique is not universal: HMM (c) cannot be so transformed to comply with the condition.

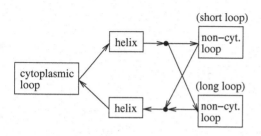

Fig. 7. TMHMM: prediction of topology of transmembrane proteins. The task is to predict positions of transmembrane helices, cytoplasmic, and non-cytoplasmic loops. Boxes in the figure represent groups of states with the same label.

4 Experiments

We have designed a simple experiment to test if decoding with most probable labeling increases accuracy. We used the HMM in Fig.8 to generate 5000 sequences of mean length about 500 for various combinations of the parameters p_1 and p_2.

We then used three decoding algorithms: standard Viterbi, our algorithm for most probable labeling, and Viterbi on a simplified model. In the simplified model, we replaced the two "gray" states with self-loops with one state and set its parameters to maximize likelihood (probability of the self-loop: 0.97, probability of emission of 1: $(2p_2 + p_1)/3$). This new HMM does not have the multiple path problem and therefore the Viterbi algorithm yields the most probable labeling.

We evaluated the error rate (percentage of the positions that were mislabeled compared to the labels on the state path that generated each sequence) for each algorithm. Fig.9 shows our results.

We have observed two trends in the data. First, using the most probable labeling does increases the accuracy. Second, the Viterbi algorithm on a simplified model, which does not have the multiple path problem, often performs better than the Viterbi algorithm on the full model. This behavior is paradoxical: using a model that is further from reality, we have obtained better results.

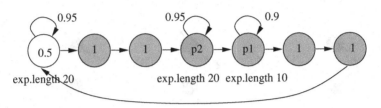

Fig. 8. HMM used in experiments. The HMM over alphabet $\{0, 1\}$ is inspired by intron model with composition changing towards the end of "gray" region. Colors (white or gray) represent the labels. The numbers inside states represent emission probability of symbol 1 (p_1 and p_2 are parameters).

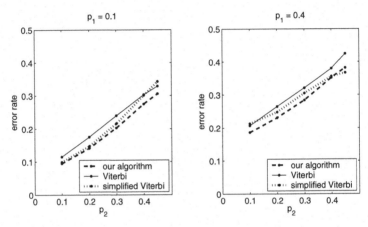

Fig. 9. Results of experiments. Error rate using three decoding algorithms: Viterbi, our algorithm for the most probable labeling, and Viterbi in simplified model.

5   Conclusions and Future Work

We have investigated the most probable labeling problem in HMMs. We showed that the problem is NP-hard, even for a fixed HMM constructed in the proof, in contrast to previous NP-hardness proofs, where the HMM constructed depended on the input instance (in most biological applications, the HMM is fixed).

Even though the problem is NP-hard in general, it is possible to compute the most probable labeling for some HMMs. We provided an $O(n^2)$ time decoding algorithm and characterized a wide class of HMMs that can be decoded by our algorithm. This run time may cause problems in applications with long input sequences, such as gene finding. Still, it is acceptable in other cases, such as analysis of protein sequences or ESTs. In practice, the running time can be further decreased by application of biological constraints (such as location of open reading frames) and various stopping conditions.

The model topologies that can be decoded by our algorithm include those for transmembrane protein topology prediction (TMHMM), distinguishing coding regions in ESTs (ESTScan), or intron model in gene finding. We also noted that the use of the Viterbi algorithm instead of the most probable labeling may lead to paradoxical behavior, where a more accurate model will yield worse results.

Several problems remain open. First, we do not know at present any polynomial algorithm for finding the most probable labeling for model shown in Fig.6c. Is decoding of this simple model NP-hard? Similar topologies are useful in various applications for providing alternative models for multi-label structures (such as different types of genes). More generally, can we provide a complete characterization of the models that are NP-hard to decode? Second, are there HMM topologies (other than ones without the multiple path problem) that can be decoded in subquadratic time? Such models may be useful in applications where the input sequence is long. Finally, we would like to test the most probable labeling algorithm in the applications mentioned above.

Acknowledgments

The authors would like to thank Ming Li and Prabhakar Ragde for useful suggestions.

References

1. C. Burge and S. Karlin. Prediction of complete gene structures in human genomic DNA. *Journal of Molecular Biology*, 268(1):78–94, 1997.
2. F. Casacuberta and C. de la Higuera. Computational Complexity of Problems on Probabilistic Grammars and Transducers. In *Grammatical Inference: Algorithms and Applications (ICGI)*, volume 1891 of *LNCS*, pages 15–24. Springer, 2000.
3. R. Durbin, S. Eddy, A. Krogh, and G. Mitchison. *Biological sequence analysis*. Cambridge University Press, 1998.
4. D. Eppstein. Finding the k shortest paths. *SIAM Journal on Computing*, 28(2):652–673, 1998.
5. C. Iseli, C. V. Jongeneel, and P. Bucher. ESTScan: a program for detecting, evaluating, and reconstructing potential coding regions in EST sequences. In *Seventh International Conference on Intelligent Systems for Molecular Biology (ISMB)*, pages 138–148, 1999.
6. A. Krogh. Two methods for improving performance of an HMM and their application for gene finding. In *Fifth International Conference on Intelligent Systems for Molecular Biology (ISMB)*, pages 179–186. AAAI Press, 1997.
7. A. Krogh, B. Larsson, G. von Heijne, and E. L. Sonnhammer. Predicting transmembrane protein topology with a hidden Markov model: application to complete genomes. *Journal of Molecular Biology*, 305(3):567–570, 2001.
8. P. Lio, N. Goldman, J. L. Thorne, and D. T. Jones. PASSML: combining evolutionary inference and protein secondary structure prediction. *Bioinformatics*, 14(8):726–733, 1998.
9. R. B. Lyngsø and C. N. S. Pedersen. The consensus string problem and the complexity of comparing hidden Markov models. *Journal of Computer and System Sciences*, 65(3):545–569, 2002.
10. P. L. Martelli, P. Fariselli, A. Krogh, and R. Casadio. A sequence-profile-based HMM for predicting and discriminating beta barrel membrane proteins. *Bioinformatics*, 18(Suppl 1):S46–53, 2002.
11. L. R. Rabiner. A tutorial on hidden Markov models and selected applications in speech recognition. *Proceedings of the IEEE*, 77(2):257–285, 1989.

Integrating Sample-Driven and Pattern-Driven Approaches in Motif Finding[*]

Sing-Hoi Sze[1,2], Songjian Lu[1], and Jianer Chen[1]

[1] Department of Computer Science
Texas A&M University, College Station, TX 77843, USA
[2] Department of Biochemistry and Biophysics
Texas A&M University, College Station, TX 77843, USA

Abstract. The problem of finding conserved motifs given a set of DNA sequences is among the most fundamental problems in computational biology, with important applications in locating regulatory sites from co-expressed genes. Traditionally, two classes of approaches are used to address the problem: sample-driven approaches focus on finding the locations of the motif instances directly, while pattern-driven approaches focus on finding a consensus string or a profile directly to model the motif. We propose an integrated approach by formulating the motif finding problem as the problem of finding large cliques in k-partite graphs, with the additional requirement that there exists a string s (which may not appear in the given sample) that is close to every motif instance included in such a clique. In this formulation, each clique represents the locations of the motif instances, while the existence of string s ensures that these instances are derived from a common motif pattern. The combined approach provides a better formulation to model motifs than using cliques alone, and the use of k-partite graphs allows the development of a fast and exact divide-and-conquer approach to handle the cases when almost every sequence contains a motif instance. Computational experiments show that this approach is feasible even on the most difficult motif finding problems of moderate size. When many sequences do not contain a motif instance, we complement the above approach by an optimized branch-and-bound algorithm that is much faster than standard clique finding approaches. We will discuss how to further generalize the formulation to better model biological reality.

1 Introduction

The motif finding problem is among the most well-studied problems in computational biology, with important applications in the computational identification of regulatory sites given a set of genes believed to be co-regulated. Most early motif finding approaches used conventional statistical optimization techniques to identify the most over-represented patterns in a sample of sequences (Stormo and Hartzell 1989; Lukashin et al. 1992; Lawrence et al. 1993; Bailey and Elkan

---

[*] This work was supported by NSF grant CCR-0311590.

I. Jonassen and J. Kim (Eds.): WABI 2004, LNBI 3240, pp. 438–449, 2004.

1994). Pevzner and Sze (2000) proposed a combinatorial graph-theoretic formulation and suggested algorithms WINNOWER and SP-STAR demonstrating that, at least for artificial samples with uniform background distribution, these previous approaches have not reached the limit of prediction yet: they are not able to find implanted patterns in the cases when these patterns should be found. More recently, a steady stream of combinatorial approaches were proposed which investigate the theoretical prediction limit of motif finding algorithms and further improve both prediction performance and computational speed (Marsan and Sagot 2000; Buhler and Tompa 2002; Eskin and Pevzner 2002; Keich and Pevzner 2002; Price et al. 2003). Instead of looking for the motif instances directly as in early sample-driven approaches, most of these later approaches assume the existence of a single string which is close to each motif instance and try to locate this string directly by a pattern-driven technique. These approaches fall roughly into two categories: either a fast heuristic is used which does not guarantee that the best solution is found, or an exact algorithm is employed which is often quite slow in practice.

In this paper, we propose an integrated approach utilizing both sample-driven and pattern-driven techniques which allows us to develop a fast exact approach for moderate size problems. We employ the clique formulation from Pevzner and Sze (2000) to represent each candidate motif as a clique containing a set of motif instances as in the sample-driven approach, while further requiring the existence of a string that is close to every motif instance in the clique as in the pattern-driven approach. This reduces the motif finding problem to finding large cliques which satisfy the close string constraint. Computational experiments show that the new technique is able to solve some of the most difficult motif finding problems which were unsolvable before using the clique formulation alone.

One possible approach to solve the problem is to employ a branch-and-bound technique to repeatedly expand a growing clique (Bomze et al. 1999), while pruning those branches which will not possibly lead to a clique that satisfies the constraint. However, a recent result (Chen et al. 2004) showed that unless unlikely consequences occur in parameterized complexity theory, the clique finding problem (and many other NP-hard problems) cannot be solved in $n^{o(k)}$ time, where n is the size of the graph (number of vertices) and k is the size of the maximum clique. Thus this kind of clique-based approach is not likely to be efficient enough for large samples if we do not take advantage of special structures inherent in the given graphs.

The key observation is that we can restrict our attention to k-partite graphs, which are graphs consisting of k parts with no edges between vertices within the same part. Although the clique finding problem in k-partite graphs is still an intrinsically difficult computational problem, we will show that a divide-and-conquer approach can be used to circumvent the difficulties in the cases when every sequence contains a motif instance, by subdividing the original graph into smaller subgraphs, using a branch-and-bound algorithm to solve each subproblem independently, and combining the results. To cope with imperfect biological samples, we will demonstrate how to generalize this approach to handle the cases when almost all sequences contain a motif instance. When many sequences do not contain a motif instance, we will show an improved branch-and-bound approach to handle realistically sized samples. When there can be more than one motif instance in a sequence, we will show how to reduce the more general problem of having at most p instances per sequence, for a small p, to the original problem so that we still have an efficient algorithm.

One very important requirement of the proposed algorithm is to be able to quickly decide if a close string exists when given a clique containing a set of strings of the same length. Unfortunately, this problem has been proven to be NP-hard (Lanctot et al. 2003). Although branch-and-bound techniques for this problem have been proposed (Gramm et al. 2001), they are too slow for our purpose if used directly. One way to solve this problem is not to check for the existence of a close string every time when we expand a clique, but check only when a clique potentially larger than before is found. To reduce the number of intermediate cliques, weaker necessary conditions are used instead during every clique expansion to prune impossible branches.

A further concern is that while this model is adequate in many cases, it is not a sufficient model when some position within the motif has more than one dominant letter, since we are forced to choose only one letter in each position of the close string. We will discuss various possibilities to address this problem. Recently, Price et al. (2003) and Eskin (2004) tried to address this problem by using a profile instead of a single string to represent a motif.

2   Problem Formulation

We use the combinatorial (l, d)-motif model proposed in Pevzner and Sze (2000): start from a set of random sequences $\{s_1, \ldots, s_k\}$, each of a fixed length, implant an (l, d)-motif by fixing a pattern P of length l and putting a randomly mutated pattern with exactly d substitutions from P at a randomly chosen position in each sequence. Without knowing P and where the implantations are, the motif finding problem is to recover the locations of the implanted patterns. Notice that making exactly d substitutions gives a more difficult model than another alternative of making at most d substitutions since distances between the motif instances will be greater.

A graph-theoretic approach was proposed in Pevzner and Sze (2000): for each position j in sequence s_i, construct a vertex s_{ij} representing the substring of length l starting at position j in s_i. Connect two vertices s_{ij} and s_{pq} by an edge if $i \neq p$ and the distance (number of substitutions) between them does not exceed $2d$. In this formulation, an (l, d)-motif is modeled by a clique of size k. In light of substantial computational difficulty, Pevzner and Sze (2000) developed the WINNOWER algorithm to find large clique-like structures instead of cliques. Liang (2003) made a few refinements to WINNOWER to make it more sensitive.

The typical size of a motif finding problem consists of 10 to 50 sequences with sequence lengths ranging from a few hundred to about 2000, and thus there can be from a few thousand to tens of thousands of vertices in the constructed graph. Despite the large graph size, an important observation is that the size of the maximum clique is small. Since edges are constructed only between vertices in different sequences in the (l, d)-motif model, the resulting graph is k-partite, with all the vertices of a sequence residing in the same part. This restricts the size of the maximum clique to be at most k. In fact, in the (l, d)-motif model, each sequence contains exactly one motif instance and the maximum clique size is k. We will take advantage of these two facts to develop a divide-and-conquer approach for the problem. For simplicity of analysis, we assume that the length of each sequence is the same, and thus, the number of vertices in each part of the graph is the same. We formulate the unconstrained clique finding problem as follows.

Definition 1 ((CLIQUE$_{kP}$)). *Given a k-partite graph G with n vertices in each part, find a maximum clique in G.*

Unfortunately, the problem is still hard to solve.

Proposition 2. *CLIQUE$_{kP}$ is NP-hard.*

We introduce the close string constraint as follows.

Definition 3 ((CLIQUE$_{kP}(l, d)$)). *Given a k-partite graph G with n vertices in each part, where each vertex represents a string of length l and an edge is connected between two vertices if they are in different parts and their distance (number of substitutions) is at most $2d$, find a maximum clique in G such that there exists a string s of length l with distance at most d to every string in the clique.*

Computational experiments show that with this constrained formulation, we are able to solve very difficult motif finding problems, such as samples containing 20 sequences each of length 600 with (16,5)- or (18,6)-motifs, which were unsolvable before due to an exceedingly large number of unconstrained maximum cliques ($> 10^6$) that do not represent the implanted motif (these results were obtained by running our clique finding program without the close string constraint). In fact, if we think of a solution as the close string s instead of the corresponding clique, this constrained formulation has exactly the same solutions as a pure pattern-driven approach. The major advantage of including cliques in the formulation is that it allows the use of a fast divide-and-conquer approach when the size of the maximum clique is very close to k.

With the above formulation, there is no guarantee that the letter at each position of a close string s is among the most frequent letters at that position in the motif instances, and thus s may not characterize the motif very well. When it is desirable to address this requirement, define the consensus pattern of a set of strings to consist of the most frequent letter at each position (in case of ties, include all the tied letters in the consensus pattern at that position), and define a consensus string to be any string obtained by picking one letter from each position of the consensus pattern. We revise the model as follows.

Definition 4 ((CLIQUE$'_{kP}(l, d)$)). *The problem is CLIQUE$_{kP}(l, d)$, with the additional requirement that s is among one of the consensus strings.*

3   Finding Cliques of Size k in k-Partite Graph

Instead of addressing the full problem directly, we first consider the simpler problem of finding cliques of size k only. This corresponds to the case when every sequence contains a motif instance, which can be applied to more confident samples involving confirmed co-regulated genes. Later we will generalize the approach to find maximum cliques not necessarily of size k. We first consider the basic problem of finding k-cliques without imposing the close string constraint.

Definition 5 ((k-CLIQUE$_{kP}$)). *Given a k-partite graph G with n vertices in each part, find a k-clique in G.*

We will use a divide-and-conquer approach to handle the problem. The idea is to subdivide the given k-partite graph into several k_0-partite subgraphs with $k_0 < k$ and solve each smaller subproblem independently using a branch-and-bound approach, as long as the number of cliques of size k_0 in each subproblem is not too high. This approach reduces the waste due to repeated computations within the same subgraphs in conventional branch-and-bound techniques when the search is applied to the entire graph. These cliques can then be used as vertices in a second graph in which cliques from the same subproblem reside in the same part and two vertices are connected by an edge if the corresponding cliques together form a larger clique. There is an one-to-one correspondence between cliques in this graph and cliques in the original graph, and a recursive procedure can be used to find them.

Fig. 1 shows the divide-and-conquer algorithm utilizing a recursive procedure. One detail is that k_0 may not divide k evenly. In this case, the last subgraph will have overlapping parts with other subgraphs, requiring a slight modification to the algorithm.

Algorithm find-clique($G = (\{V^{(1)}, \ldots, V^{(k)}\}, E)$) {
$\quad$ for $i \leftarrow 1$ to k/k_0 do
$\quad\quad$ $C^{(i)} \leftarrow$ set of all k_0-cliques in the induced subgraph $G[V^{((i-1)k_0+1)}, \ldots, V^{(ik_0)}]$;
$\quad$ if ($k_0 < k$) {
$\quad\quad$ construct $G' = (\{C^{(1)}, \ldots, C^{(k/k_0)}\}, E')$, where for $c_p \in C^{(p)}$ and $c_q \in C^{(q)}$
$\quad\quad\quad$ with $p \neq q$, $(c_p, c_q) \in E'$ if and only if $c_p \cup c_q$ is a clique in G;
$\quad\quad$ $C^{(1)} \leftarrow$ set of all cliques from find-clique(G') transformed to cliques in G;
$\quad$ }
$\quad$ return $C^{(1)}$;
}

Fig. 1. Algorithm find-clique(G) for finding k-cliques in a k-partite graph.

One difficulty with the above approach is that it is hard to determine what k_0 to use. One could use a random k-partite graph model and estimate appropriate values of k_0 based on average case analysis, but significant correlations between adjacent vertices and severe non-randomness around the implanted patterns make such an estimate very unreliable. This problem can be corrected by an automatic approach which subdivides the original graph dynamically.

Given a threshold t denoting the maximum number of cliques allowed in a subgraph, the idea is to find the smallest induced subgraph $G[V^{(1)}, \ldots, V^{(i)}]$ such that the number of cliques of size i is at most t by trying each i incrementally. To utilize intermediate results as much as possible, we follow a combined breadth-first depth-first approach based on the branch-and-bound technique. Consider each target clique size i in turn (starting from 2), and keep an ordered list L of cliques representing all partial results computed so far, initialized to a single entry containing an empty clique with no vertices. Initialize a pointer P pointing to this entry. Repeat the following: suppose that the clique c pointed by P is of size j with one vertex from each of $V^{(1)}, \ldots, V^{(j)}$. If $j < i$, replace c by a list of all cliques of size $j + 1$ that form a clique with c by adding one vertex from $V^{(j+1)}$; otherwise $j = i$ and we advance P to the next entry in L. If the number of cliques in L before P is more than t, the current subgraph has too many cliques: increment i by 1, reset P to point to the first entry in L and continue to look into the next part $V^{(i)}$. If there is no more entry in L, we are done finding all cliques in the subgraph $G[V^{(1)}, \ldots, V^{(i)}]$, repeat this procedure starting with the next part $V^{(i+1)}$ to determine the next subgraph.

By choosing $t = O(n)$, we can make sure that the second clique finding problem is not more complicated than the original. Since the procedure does not waste any partial results, its computational performance should be similar to the previous approach. In terms of memory requirement, with appropriate data structures we only need space for at most t cliques (entries before P) in addition to the normal space requirement for depth-first search (entries after P). To further decrease the number of cliques that need to be considered, when expanding a clique of size j to a clique of size $j + 1$, vertices that do not have a neighbor in all the remaining parts are discarded.

We now incorporate the close string constraint in the model. As discussed before, instead of imposing it every time when a clique is expanded, we check only when the clique size becomes i. During every clique expansion, the following necessary conditions are used to prune impossible branches (proofs omitted).

Proposition 6. *Given three strings s_1, s_2 and s_3 each of length l, assume that the maximum pairwise distance (number of substitutions) is between s_1 and s_2 with $d(s_1, s_2) = d'$, where $d < d' \leq 2d$. Without loss of generality, assume that all these differences are in the last d' positions of s_1 and s_2, and cut each string s_i into two parts s_i' and s_i'' with $|s_i''| = d'$ (in other words, we have $s_1' = s_2'$ and $d(s_1'', s_2'') = d'$). Let $x = d(s_1', s_3') = d(s_2', s_3')$, $y = d(s_1'', s_3'')$, and $z = d(s_2'', s_3'')$. Then a string s of length l with distance at most d to each of the three strings exists if and only if $x + y + z \leq 3d$.*

Proposition 7. *Given a set S of strings all of length l, if there exists a string s of length l with distance (number of substitutions) at most d to every string in S, then (i) for every pair of strings s_1 and s_2 in S with distance $d(s_1, s_2) = 2d$, the jth letter of s must either be the jth letter of s_1 or the jth letter of s_2; (ii)* *for every pair of strings s_1 and s_2 in S with distance $d(s_1, s_2) = 2d - 1$, when the jth letter of s_1 and the jth letter of s_2 are the same, the jth letter of s must be this letter.*

In the first condition, if the maximum pairwise distance between three strings is at most d, a close string s always exists. Otherwise, it is in the form of an inequality which can be easily checked given any sub-clique of size 3. To use the second condition, whenever a clique c is expanded to include a vertex v, v is checked against each vertex in c and also against the old constraints on c. It is very useful since the vast majority of edges in the graph have a distance of either $2d$ (typically over 50%) or $2d - 1$ (typically over 20%). If any inconsistencies are found, the branch is pruned. Computational experiments show that these conditions help to reduce the subdivision size k_0 in many cases and make the problem much easier to solve (as opposed to not using them). When a new clique of size i is found, we use the branch-and-bound algorithm from Gramm et al. (2001) to search for a close string s, modifying it to avoid trying letters disallowed from the above constraints to further reduce the search space.

To address the revised model where the close string s is required to be among one of the consensus strings, when a clique reaches size i, we compute its consensus pattern (which may consist of more than one letter at some positions in case of frequency ties) and use it to further restrict the letters that can appear in each position of s. When i is large enough, most positions of the consensus pattern consist of only one letter and thus this requirement eliminates almost all the need to search for s. However, one must be careful not to use the consensus patterns to prune the search branch since it is still possible that the clique will expand into a solution.

Pevzner and Sze (2000) used samples containing 20 random sequences each of length 600 as test samples (which are typical sizes of realistic biological samples) and found that finding (15,4)-motifs is very challenging. Buhler and Tompa (2002) found that the (15,4)-motif problem is not the hardest, but (14,4)-, (16,5)- and (18,6)-motif problems are considerably more challenging. They showed, by a probabilistic analysis of the (l, d)-motif model, that these problems are already at the prediction limit of motif finding algorithms, which include any sample-driven or pattern-driven approaches. Table 1 shows the performance of the divide-and-conquer approach for two classes of limiting motif finding problems. The correct motif was found in all cases.

These results showed that in almost all cases we were able to subdivide the original problem containing 20 sequences into at least two subproblems. Note that for easier samples (data not shown), it is almost always the case that $k_0 = 2$ and thus the algorithm is very fast. We found that even for the most difficult motif finding problems, the graphs become so sparse in subsequent recursion levels that it is almost always the case that $k_0 = 2$ in these deeper levels and thus the running time becomes negligible when compared to the outermost level. For most samples in Table 1, the total number of 20-cliques with a close string was more than one, sometimes even in the tens or hundreds, suggesting that these samples become so difficult that there are random patterns which can

Table 1. Typical values of k_0 and typical computation times of the divide-and-conquer approach after the graph construction phase. In each case, the values are consistently obtained over a few runs with t set to 600 on random samples with 20 sequences each of length 600 containing an (l, d)-motif.

(l,d)	(11,2)	(13,3)	(15,4)	(17,5)	(19,6)	(10,2)	(12,3)	(14,4)	(16,5)	(18,6)
k_0	2	3	4	6	8	3	5	8	10	13
speed	secs	secs	secs	mins	mins	secs	secs	mins	mins	hrs

serve as motif variances. However, for all these cliques the close string is unique and is always the correct one, and thus at least for these simulated samples, it is sufficient to find only one solution. Price et al. (2003, Table 1) compared the performance of several algorithms on the (15,4)-motif problem. From this table, we found that on this problem while the exact divide-and-conquer approach is not as fast as the fastest heuristics with 99.7% success rate (in a few seconds), it is much faster than other exact algorithms with 100% success rate (in minutes).

4   Finding Maximum Cliques in k-Partite Graph

A similar divide-and-conquer technique can be used to find maximum cliques when their size k' is less than k, by subdividing the original k-partite graph into r subgraphs G_i, $1 \leq i \leq r$, such that G_i is k_i-partite and $\sum_{i=1}^{r} k_i = k$, and finding cliques within the subgraphs independently. In each subgraph G_i, we need to look for cliques of size $k_i' \leq k_i$ such that $\sum_{i=1}^{r} k_i' = k'$ over all possible combinations of $k_1', \ldots, k_r'$. Individual clique finding in subgraphs can be accomplished by branch-and-bound approaches. To find maximum cliques of unknown size, one can start the search with $k' = k$, and iteratively reducing k' until maximum cliques are found.

One drawback with the above approach is that it is difficult to determine the subdivision sizes k_i so that there are not too many cliques within each subgraph. Alternatively, a straightforward direct reduction is possible so that the original divide-and-conquer approach can be used. For each choice of k' parts out of k parts to form an induced k'-partite subgraph, run the original k'-partite clique algorithm (with automatic determination of subdivision sizes) on each subgraph to find k'-cliques. The problem is that there are a total of $\binom{k}{k'}$ possible choices and this number grows very quickly as $k - k'$ increases.

To avoid this combinatorial explosion, we employ a reduction at the graph level to construct a single k'-partite graph $G' = (\{W^{(1)}, \ldots, W^{(k')}\}, E')$ to represent all k'-cliques in the original k-partite graph $G = (\{V^{(1)}, \ldots, V^{(k)}\}, E)$, where $W^{(i)}$ $(= W^{(i)}(k, k'))$ is a set of new vertices representing the vertices in $V^{(i)} \cup \ldots \cup V^{(i+k-k')}$, and for $i < j$, a vertex v in $W^{(i)}$ is connected to a vertex w in $W^{(j)}$ if $v \in V^{(i+p)}$ and $w \in V^{(j+q)}$ are connected in G with $p \leq q$. It is evident that there is an one-to-one correspondence between k'-cliques in G' and k'-cliques in G, and thus only one application of the divide-and-conquer algorithm is needed to find them. One problem with this reduction is that the size of G' grows very quickly as $k - k'$ increases: if G contains nk vertices, G' contains $n(k - k' + 1)k$ vertices. Although G' is less dense, given threshold t the number of parts in the subgraphs after subdivision and the running time increase substantially as $k - k'$ increases.

The above two reductions stand at two extremes: the first reduction employs a large number of small k'-partite graphs while the second reduction uses a single k'-partite graph G' with many vertices. One approach that lies between these two extremes is to decompose G' into an intermediate number of k'-partite graphs with an intermediate number of vertices. Recall that in the first attempt we subdivide G into r subgraphs such that the ith subgraph G_i is k_i-partite and $\sum_{i=1}^{r} k_i = k$. Let s_i $(= \sum_{j=1}^{i-1} k_j + 1)$ be the index of the first part of G_i within G. By applying the previous reduction to each k_i-partite subgraph G_i, $W_i = \{W^{(s_i)}(k_i, k_i'), \ldots, W^{(s_i + k_i' - 1)}(k_i, k_i')\}$ includes all vertices for constructing a single k_i'-partite graph to represent all k_i'-cliques in G_i. For each fixed combination of $k_1', \ldots, k_r'$ such that $\sum_{i=1}^{r} k_i' = k'$, the induced k'-partite subgraph $G'[\cup_{i=1}^{r} W_i]$ represents all k'-cliques to which the dynamic subdivision algorithm can be applied.

In effect, we have decomposed G' into smaller induced subgraphs, one for each combination of $k_1', \ldots, k_r'$, which together specify exactly the same set of k'-cliques as G' and can be searched separately. For each value of r and for each choice of values of k_i, we get one decomposition and the number of decomposed subgraphs increases as r increases, which is a tradeoff against the smaller sizes of these subgraphs. We found that $r = 2$ and $k_1 = k_2$ work well, with a linear increase in the number of decomposed subgraphs as $k - k'$ increases. When almost all sequences contain a motif instance, the above technique reduces the motif finding problem to a manageable number of subproblems to which the divide-and-conquer approach is applied. We have successfully applied it (with reasonable computation times) to find motifs in very difficult motif finding problems such as samples containing (14,4)- or (16,5)-motifs where 20 out of 21, 22 or 23 sequences each of length 600 contain a motif instance.

To handle the more general case when there can be at most p motif instances per sequence, for a small p, we reduce this problem to the original problem as follows: instead of constructing only one part for each sequence, construct p parts each representing the same sequence, resulting in a kp-partite graph. As before, connect vertices from different parts by an edge if the distance between them does not exceed $2d$, except that vertices representing overlapping positions within the same sequence are not connected.

5   Branch-and-Bound Algorithm

The above divide-and-conquer approach is very fast when every sequence contains a motif instance (i.e., $k' = k$), but its running time requirement increases rapidly as $k - k'$ increases. To handle cases when many sequences do not contain a motif instance, we have to resort back to traditional branch-and-bound approaches. Standard branch-and-bound techniques employ the following recursive algorithm find-clique(G) to find a maximum clique in $G = (V, E)$: let C_v be the set of cliques returned from applying find-clique on the graph induced by all neighbors of vertex v, output the largest set $\{v\} \cup C_v$ over all $v \in V$ (Bomze et al. 1999). Although these approaches were shown to be feasible only on unrestricted graphs with at most a few hundred vertices, we will show that with appropriate optimizations these approaches can still be applied to motif finding.

If the goal is to find at most u maximum cliques, once a maximal clique of size m is found, there is no need to look for maximal cliques of size less than m. Also, once more than u maximal cliques of size m are found, there is no need to look for maximal cliques of size m. These observations help us to prune the search tree branches when it is no longer possible to obtain a large enough clique. With the above optimization, it is advantageous to consider the vertices in non-increasing degree order (denoted by $\succ$) so that it is more likely to get large cliques during the earlier branches. To make each maximum clique appear uniquely while simultaneously trying to further reduce the running time, when a vertex v is considered in the search tree, only the vertices w satisfying $v \succ w$ and $(v, w) \in E$ are included in the subgraph at the next recursion level to build larger cliques. During every clique expansion, we use the necessary conditions for the existence of a close string to prune impossible branches. Only when a new potential maximal clique is found, we look for a close string.

Computational experiments show that except in the cases when almost every sequence contains a motif instance (when the divide-and-conquer approach can be many times faster), this enhanced branch-and-bound algorithm is faster than the divide-and-conquer approach. By also allowing edges between vertices in the same sequence, the branch-and-bound approach can be further generalized to handle the cases when there can be at most p motif instances per sequence, for a small p, or even to the case when there can be at most p motif instances in total, without restrictions on the number of motif instances in any particular sequence.

6 Biological Samples

Since Price et al. (2003) reported considerable difficulties in obtaining experimentally confirmed samples with biological motifs that are very difficult to find, our goal here is simply to show that our model is appropriate and our new algorithms are effective by testing on a few published data sets. For these data sets, most standard sample-driven or pattern-driven approaches are also able to find the motifs. In each case, we assume that l is known while d is unknown and the general strategy is to start with $d = 0$ and try successively larger d. Since the motif finding problem is usually much easier when d is small, this approach does not add much to the computation time.

The sample in Stormo and Hartzell (1989) contains 18 sequences each of length 105 with experimentally confirmed $E.coli$ CRP binding sites. Using $l = 16$, the most accurate solution returned correct sites in 16 out of 18 sequences when $d = 6$. There are three samples in Sze et al. (2002) (also from $E.coli$), all containing sequences of length 200. The first (ARG) sample contains 9 upstream sequences from genes regulated by the arginine repressor ArgR, with a two-part site in each sequence, where each part is of length 18. The patterns in the two parts are very similar and they are separated by 3 positions in 8 out of 9 sequences. When the patterns were treated as one long motif of length 39, all the 8 two-part sites were found when $d = 14$. With such a large d, this is the only case when the consensus string requirement is required in addition to the close string constraint to limit the search space so as to get reasonable computation time (in the other cases, this is optional). When the patterns were treated as independent motifs of length 18 and we looked for at most two motif instances per sequence, most of the sites were found when $d = 6$. The second (PUR) sample contains 19 upstream sequences from genes regulated by the purine repressor PurR with sites of length 16, and most sequences contain one site. Almost all of these sites were found when $d = 5$. Many of the above cases reduce to either the (16,5)- or (18,6)-motif problems and the divide-and-conquer approach was able to finish the computation within seconds or minutes. The third (CRP) sample is a much more difficult variant of the sample in Stormo and Hartzell (1989) containing 33 sequences with variable number of weak sites in each sequence and many sequences contain no sites. Unfortunately, the divide-and-conquer approach was not applicable, and with $l = 16$, the branch-and-bound approach was not able to terminate in a reasonable amount of time when $d = 5$. Partial results revealed that more than 10 correct sites and an approximately correct close string were found when $d = 4$.

7   Discussion

By employing a constrained clique formulation to model motifs, we provide two exact algorithms: a very fast divide-and-conquer approach to handle the cases when almost all sequences contain a motif instance and an optimized branch-and-bound algorithm for the other cases when many sequences do not contain a motif instance. There are a few additional complications that need to be addressed before the new algorithms can be applied in all situations: most biological samples have a non-uniform nucleotide composition, and both l and d are unknown. A good strategy is to use weighted distance values according to the background distribution and try different values of d over a range of values of l, while employing heuristics to avoid wasting time on unrealistically large d for a given l. To compare motifs of different lengths and different number of instances, for each motif a score can be computed based on similarity between predicted motif instances.

To further improve the formulation to better model biological reality, one possibility is to allow degenerate letters in the close string. Although this allows motifs to have more than one dominant letter in some positions, the search space becomes much larger. In fact, one can go a step further to use a profile instead of a string to model a motif, which takes into consideration the relative frequency of all letters within a position. However, it is unclear whether a graph-theoretic approach will be able to help to limit the search space.

References

Bailey, T. L., Elkan, C. P.: Fitting a mixture model by expectation maximization to discover motifs in biopolymers. Proc. of the 2nd Int. Conf. on Intelligent Systems for Mol. Biol. (ISMB'1994) 28–36

Bomze, I., Budinich, M., Pardalos, P., Pelillo, M.: The maximum clique problem. Handbook of Combinatorial Optimization **4** D.-Z. Du and P. M. Pardalos ed. Kluwer Academic (1999)

Buhler, J., Tompa, M.: Finding motifs using random projections. J. Comp. Biol. **9** (2002) 225–242

Chen, J., Huang, X., Kanj, I., Xia, G.: Linear FPT reductions and computational lower bounds. Proc. of the 34th ACM Symp. on Theory of Computing (STOC'2004) 212–221

Eskin, E.: From profiles to patterns and back again: a branch and bound algorithm for finding near optimal motif profiles. Proc. of the 8th Ann. Int. Conf. on Comp. Mol. Biol. (RECOMB'2004) 115–124

Eskin, E., Pevzner, P. A.: Finding composite regulatory patterns in DNA sequences. Bioinformatics **18** (2002) S354–363

Gramm, J., Niedermeier, R., Rossmanith, P.: Exact solutions for closest string and related problems. Lecture Notes in Computer Science **2223** (ISAAC'2001) 441–453

Keich, U., Pevzner, P. A.: Finding motifs in the twilight zone. Proc. of the 6th Ann. Int. Conf. on Comp. Mol. Biol. (RECOMB'2002) 195–204

Lanctot, J. K., Li, M., Ma, B., Wang, S., Zhang, L.: Distinguishing string selection problems. Information and Computation **185** (2003) 41–55

Lawrence, C. E., Altschul, S. F., Boguski, M. S., Liu, J. S., Neuwald, A. F., Wootton, J. C.: Detecting subtle sequence signals: a Gibbs sampling strategy for multiple alignment. Science **262** (1993) 208–214

Liang, S.: cWINNOWER algorithm for finding fuzzy DNA motifs. Proc. of the 2nd IEEE Computer Society Bioinformatics Conf. (CSB'2003) 260–265

Lukashin, A. V., Engelbrecht, J., Brunak, S.: Multiple alignment using simulated annealing: branch point definition in human mRNA splicing. Nucleic Acids Res. **20** (1992) 2511–2516

Marsan, L., Sagot, M.-F.: Algorithms for extracting structured motifs using a suffix tree with an application to promoter and regulatory site consensus identification. J. Comp. Biol. **7** (2000) 345–362

Pevzner, P. A., Sze, S.-H.: Combinatorial approaches to finding subtle signals in DNA sequences. Proc. of the 8th Int. Conf. on Intelligent Systems for Mol. Biol. (ISMB'2000) 269–278

Price, A., Ramabhadran, S., Pevzner, P. A.: Finding subtle motifs by branching from sample strings. Bioinformatics (2003) SII149–155

Stormo, G. D., Hartzell, G. W.: Identifying protein-binding sites from unaligned DNA fragments. Proc. Natl. Acad. Sci. USA **86** (1989) 1183–1187

Sze, S.-H., Gelfand, M. S., Pevzner, P. A.: Finding weak motifs in DNA sequences. Pac. Symp. Biocomput. (PSB'2002) 235–246

Finding Optimal Pairs of Patterns

Hideo Bannai[1], Heikki Hyyrö[2], Ayumi Shinohara[2,3], Masayuki Takeda[3],
Kenta Nakai[1], and Satoru Miyano[1]

[1] Human Genome Center, Institute of Medical Science
The University of Tokyo, Tokyo 108-8639, Japan
{bannai,knakai,miyano}@ims.u-tokyo.ac.jp
[2] PRESTO, Japan Science and Technology Agency (JST)
helmu@cs.uta.fi
[3] Department of Informatics, Kyushu University 33, Fukuoka 812-8581, Japan
{ayumi,takeda}@i.kyushu-u.ac.jp

Abstract. We consider the problem of finding the *optimal pair of string patterns* for discriminating between two sets of strings, i.e. finding the pair of patterns that is best with respect to some appropriate scoring function that gives higher scores to pattern pairs which occur more in the strings of one set, but less in the other. We present an $O(N^2)$ time algorithm for finding the optimal pair of *substring patterns*, where N is the total length of the strings. The algorithm looks for all possible Boolean combination of the patterns, e.g. patterns of the form $p \land \neg q$, which indicates that the pattern pair is considered to match a given string s, if p occurs in s, AND q does NOT occur in s. The same algorithm can be applied to a variant of the problem where we are given a single set of sequences along with a numeric attribute assigned to each sequence, and the problem is to find the optimal pattern pair whose occurrence in the sequences is *correlated* with this numeric attribute. An efficient implementation based on suffix arrays is presented, and the algorithm is applied to several nucleotide sequence datasets of moderate size, combined with microarray gene expression data, aiming to find regulatory elements that *cooperate, complement, or compete with* each other in enhancing and/or silencing certain genomic functions.

1 Introduction

Pattern discovery from biosequences is an important topic in Bioinformatics, which has been, and is being, studied heavily with numerous variations and applications (see [1] for a survey on earlier work). Although finding the single, most significant pattern conserved across multiple sequences has important and obvious applications, it is known that in many, if not most, actual cases, more than one sequence element is responsible for the biological role of the sequences. There are several methods which address this observation, focussing on finding *composite* patterns. In [2], they develop a suffix tree based approach for discovering *structured motifs*, which are two or more patterns separated by a certain

I. Jonassen and J. Kim (Eds.): WABI 2004, LNBI 3240, pp. 450–462, 2004.
© Springer-Verlag Berlin Heidelberg 2004

distance, similar to text associative patterns [3]. MITRA is another method that looks for composite patterns [4] using *mismatch trees*. Bioprospector [5] applies the Gibbs sampling strategy to find gapped motifs.

The main contribution of this paper is to present an efficient $O(N^2)$ algorithm (where N is the total length of the input strings) and implementation based on suffix arrays, for finding the *optimal pair of substring patterns* combined with any Boolean function. Note that the methods mentioned above for finding composite patterns can be viewed as being limited to using only the $\wedge$ (AND) operation. The use of any Boolean function allows the use of the $\neg$ (NOT) operation, therefore making it possible to find not only sequence elements that *cooperate* with each other, but those of the form $p \wedge \neg q$, which can be interpreted as two sequence elements with *competing* functions (e.g. positive and negative elements). The pattern pairs discovered by our algorithm are optimal in that they are guaranteed to be the highest scoring pair of patterns with respect to a given scoring function, and also, a limit on the lengths of the patterns in the pair is not assumed. Our algorithm can be adjusted to handle several common problem formulations of pattern discovery, for example, pattern discovery from positive and negative sequence sets [6–9], as well as the discovery of patterns that *correlate* with a given numeric attribute (e.g. gene expression level) assigned to the sequences [10–14]. The significance of the algorithm in this paper lies in the fact that since there are indeed $O(N^2)$ possible substring pattern combinations, the information needed to calculate the score for each pattern pair can be gathered, effectively, in constant time.

The algorithm is presented conceptually as using a generalized suffix tree [15], which is an indispensable data structure for efficient processing of substring information. Moreover, the algorithm using the suffix tree can, with the same asymptotic complexity, be simulated very efficiently using suffix arrays, and is thus implemented. We apply our algorithm to 3'UTR (untranslated region) of yeast and human mRNA, together with data obtained from microarray experiments which measure the decay rate of each mRNA [16, 17]. We were successful in obtaining several interesting pattern pairs where some correspond to known mRNA destabilizing elements.

2   Preliminaries

2.1   Notation

Let Σ be a finite alphabet. An element of Σ^* is called a *string*. Strings x, y, and z are said to be a *prefix*, *substring*, and *suffix* of string $w = xyz$, respectively. The length of a string w is denoted by $length(w)$. The empty string is denoted by ε, that is, $length(\varepsilon) = 0$. The i-th character of a string w is denoted by $w[i]$ for $1 \leq i \leq length(w)$, and the substring of a string w that begins at position i and ends at position j is denoted by $w[i:j]$ for $1 \leq i \leq j \leq length(w)$. For convenience, let $w[i:j] = \varepsilon$ for $j < i$. For any set S, let $|S|$ denote the cardinality of the set.

Let $\psi(p, s)$ be a Boolean matching function that has the value **true** if the *pattern* string p is a substring of the string s, and **false** otherwise. We de-

Table 1. Summary of candidate Boolean operations on pattern pair $\langle p, F, q \rangle$.

	input				
$\psi(p,s)$	true	true	false	false	
$\psi(q,s)$	true	false	true	false	
	output $F(\psi(p,s), \psi(q,s))$				representation
F_0	false	false	false	false	false
F_1	false	false	false	true	$(\neg p) \wedge (\neg q)$
F_2	false	false	true	false	$(\neg p) \wedge q$
F_3	false	false	true	true	$(\neg p)$
F_4	false	true	false	false	$p \wedge (\neg q)$
F_5	false	true	false	true	$(\neg q)$
F_6	false	true	true	false	$(p \wedge (\neg q)) \vee ((\neg p) \wedge q)$
F_7	false	true	true	true	$(\neg p) \vee (\neg q)$
F_8	true	false	false	false	$p \wedge q$
F_9	true	false	false	true	$(p \wedge q) \vee ((\neg p) \wedge (\neg q))$
F_{10}	true	false	true	false	q
F_{11}	true	false	true	true	$(\neg p) \vee q$
F_{12}	true	true	false	false	p
F_{13}	true	true	false	true	$p \vee (\neg q)$
F_{14}	true	true	true	false	$p \vee q$
F_{15}	true	true	true	true	true

fine $\langle p, F, q \rangle$ as a *Boolean pattern pair* (or simply *pattern pair*), which consists of two patterns p and q and a Boolean function $F : \{\text{true}, \text{false}\} \times \{\text{true}, \text{false}\} \rightarrow \{\text{true}, \text{false}\}$. The matching function value $\psi(\langle p, F, q \rangle, s)$ is defined as $F(\psi(p,s), \psi(q,s))$. Table 1 lists all 16 possible Boolean functions of two Boolean variables, that is, all possible choices for F. We say that a pattern or Boolean pattern pair π *matches* string s if and only if $\psi(\pi, s) = \text{true}$.

For a given set of strings $S = \{s_1, \ldots, s_m\}$, let $M(\pi, S)$ denote the subset of strings in S that π matches, that is, $M(\pi, S) = \{s_i \in S \mid \psi(\pi, s_i) = \text{true}\}$. Now suppose that for each $s_i \in S$, we are given an associated numeric attribute value r_i. Let $\sum_{M(\pi,S)} r_i$ denote the sum of r_i over all s_i such that $\psi(\pi, s_i) = \text{true}$, that is, $\sum_{M(\pi,S)} r_i = \sum (r_i \mid \psi(\pi, s_i) = \text{true})$. For brevity, we shall omit S where possible and let $M(\pi)$ and $\sum_{M(\pi)} r_i$, be a shorthand for $M(\pi, S)$ and $\sum_{M(\pi,S)} r_i$, respectively.

2.2 Problem Definition

In general, the problem of finding a good pattern from a given set of strings S refers to finding a pattern π that maximizes some suitable scoring function *score* with respect to the strings in S. We concentrate on scoring functions whose values for a pattern π depend on values cumulated over the strings in S that match π. We also assume that the score value computation itself can be done in constant time if the required parameter values are known. More specifically, we concentrate on *score* that takes parameters of type $|M(\pi)|$ and $\sum_{M(\pi)} r_i$. The specific choice of the scoring function depends highly on the particular application. A variety of problems fall into the category represented by the following problem definition:

Problem 1 (Optimal pair of substring patterns). Given a set $S = \{s_1, \ldots, s_m\}$ of strings, where each string s_i is assigned a numeric attribute value r_i, and a scoring function $score : \mathbf{R} \times \mathbf{R} \rightarrow \mathbf{R}$, find the Boolean pattern pair $\pi \in \{\langle p, F, q \rangle \mid p, q \in \Sigma^*, F \in \{F_0, \ldots, F_{15}\}\}$ that maximizes $score(|M(\pi)|, \sum_{M(\pi)} r_i)$.

Below, we give examples of choices for $score$ and r_i.

Positive/Negative Sequence Set Discrimination. We are given two disjoint sets of sequences S_1 and S_2, where sequences in S_1 (the positive set) are known to have some biological function, while the sequences in S_2 (the negative set) are known not to. The objective is to find pattern pairs which match more sequences in one set, and less in the other.

We create an instance of the optimal pair of substring patterns problem as follows: Let $S = S_1 \cup S_2 = \{s_1, \ldots, s_m\}$, and let $r_i = 1$ if $s_i \in S_1$ and $r_i = 0$ if $s_i \in S_2$. Then, for each pattern pair π, the scoring function will receive $|M(\pi, S)|$ and $\sum_{(\pi, S)} r_i = |M(\pi, S_1)|$. Notice that $|M(\pi, S_2)| = |M(\pi, S)| - |M(\pi, S_1)|$. Common scoring functions that are used in this situation include the entropy information gain, the Gini index, and the chi-square statistic, which all are essentially functions of $|M(\pi, S_1)|$, $|M(\pi, S_2)|$, $|S_1|$ and $|S_2|$.

Correlated Patterns. We are given a set S of sequences, with a numeric attribute value r_i associated with each sequence $s_i \in S$, and the task is to find pattern pairs whose occurrences in the sequences correlate with their numeric attributes. For example, r_i could be the expression level ratio of a gene with upstream sequence s_i. The scoring function used in [11,13] is the inter class variance, which can be maximized by maximizing the scoring function $score(x, y) = y^2/x + (y - \sum_{i=1}^{|m|} r_i)^2/(m - x)$, where $x = |M(\pi)|$ and $y = \sum_{M(\pi)} r_i$. We will later describe how to construct a nonparametric scoring function based on the normal approximation of the Wilcoxon rank sum test, which can also be used in our framework.

2.3   Basic Data Structures

A *suffix tree* [15] for a given string s is a rooted tree whose edges are labeled with substrings of s, satisfying the following characteristics. For any node v in the suffix tree, let $l(v)$ denote the string spelled out by concatenating the edge labels on the path from the root to v. For each leaf node v, $l(v)$ is a distinct suffix of s, and for each suffix in s, there exists such a leaf v. Furthermore, each node has at least two children, and the first character of the labels on the edges to its children are distinct. A generalized suffix tree (GST) for a set of m strings $S = \{s_1, \ldots, s_m\}$ is basically a suffix tree for the string $s_1\$_1 \cdots s_m\$_m$, where each $\$_i$ ($1 \le i \le m$) is a distinct character which does not appear in any of the strings in the set. However, all paths are ended at the first appearance of any $\$_i$, and each leaf is labeled with id_i. It is well known that suffix trees (and generalized suffix trees) can be represented in linear space and constructed in linear time [15] with respect to the length of the string (total length of the strings for GST).

A *suffix array* [18] A_s for a given string s of length n, is a permutation of the integers $1, \ldots, n$ representing the lexicographic ordering of the suffixes of s. The value $A_s[i] = j$ in the array indicates that $s[j:n]$ is the ith suffix in the lexicographic ordering. The *lcp array* for a given string s is an array of integers representing the longest common prefix lengths of adjacent suffixes in the suffix array, that is $lcp_s[i] = \max\{k \mid s[A_s[i-1]:A_s[i-1]+k-1] = s[A_s[i]:A_s[i]+k-1]\}$. Recently, three methods for constructing the suffix array directly from a string in linear time have been developed [19–21]. The *lcp* array can be constructed from the suffix array also in linear time [22]. It has been shown that several algorithms (and potentially many more) which utilize the suffix tree can be implemented very efficiently using the suffix array together with its *lcp* array [22, 23] (the combination termed in [23] as the *enhanced suffix array*). This paper presents yet another example for efficient implementation of an algorithm based conceptually on suffix trees, but uses the suffix and *lcp* arrays.

The *lowest common ancestor* $lca(x, y)$ of any two nodes x and y in a tree is the deepest node which is common to the paths from the root to each of the nodes. The tree can be pre-processed in linear time to answer the lowest common ancestor (*lca-query*) for any given pair of nodes in constant time [24]. In terms of the suffix array, the *lca*-query is almost equivalent to a *range minimum query* (*rm-query*) on the *lcp* array, which, given a pair of positions i and j, $rmq(i, j)$ returns the position of the minimum element in the sub-array $lcp[i:j]$. The *lcp* array can also be pre-processed in linear time to answer the *rm*-query in constant time [24, 25].

The linear time bounds mentioned above for the construction of suffix trees and arrays, as well as the preprocessing for *lca*- and *rm*-queries are actually not required for the $O(N^2)$ overall time bound for finding optimal pattern pairs, since they need only be done once, and a naïve algorithm costs $O(N^2)$. However, they are very important for an efficient implementation of our algorithm.

3   Algorithm

Now we present algorithms to solve the optimal pair of substring patterns problem, given the set of strings $S = \{s_1, \ldots, s_m\}$, an associated attribute r_i for each string s_i, and a scoring function *score*. Also, let $N = \sum_{i=1}^{m} length(s_i)$. We first show that a naïve algorithm requires $O(N^3)$ time, and then describe the $O(N^2)$ algorithm. The algorithms calculate scores for all possible combinations of pattern pairs, from which finding the optimal pair is a trivial task.

3.1   An $O(N^3)$ Algorithm

We know that there are only $O(N)$ candidates for a single pattern, since the candidates can be confined to patterns of form $l(v)$, where v is a node in the generalized suffix tree over the set S. This is because for any pattern corresponding to a path that ends in the middle of an edge of the suffix tree, the pattern which corresponds to the path extended to the next node will match the same set of strings, and hence the score would be the same. Therefore, there are $O(N^2)$ possible candidate pattern pairs for which we must calculate the scoring function value. For a given pattern pair candidate $\pi = \langle l(v_1), F, l(v_2) \rangle$, where v_1, v_2 are nodes of the GST, the values $|M(\pi)|$ and $\sum_{M(\pi)} r_i$ can be computed in $O(N)$ time, by using any of the linear time substring matching algorithms. Then each corresponding scoring function value can be computed in constant time. Therefore, the total time required is $O(N^3)$.

3.2   An $O(N^2)$ Algorithm

Our algorithm is derived from the technique for solving the *color set size problem* [26], which calculates the values $|M(l(v))|$ in $O(N)$ time for all nodes v of a GST over the string set S. Let us first describe a slight generalization of this algorithm. The following algorithm computes the values $\sum_{M(l(v))} r_i$ for all v. Note that if we give each attribute r_i the value 1, then $\sum_{M(l(v))} r_i = |M(l(v))|$. Thus we do not need to consider separately how to compute $|M(l(v))|$.

First we introduce some auxiliary notation. Let $LF(v)$ denote the set of all leaf nodes in the subtree rooted by the node v, and let $c_i(v)$ denote the number of leaves in $LF(v)$ that have the label id_i. Let us also define the sum of leaf attributes for a node v as $\sum_{LF(v)} r_i = \sum(c_i(v)r_i \mid \psi(l(v), s_i) = \text{true})$.

For any node v in the GST over the string set S, the matching value $\psi(l(v), s_i)$ is true for at least one string s_i. Thus the equality $\sum_{M(l(v))} r_i = \sum(r_i \mid \psi(l(v), s_i) = \text{true}) = \sum_{LF(v)} r_i - \sum((c_i(v) - 1)r_i \mid \psi(l(v), s_i) = \text{true})$ holds. Let us define the preceding subtracted sum to be a *correction factor*, which we denote by $corr(l(v), S) = \sum((c_i(v) - 1)r_i \mid \psi(l(v), s_i) = \text{true})$.

Since the recurrence $\sum_{LF(v)} r_i = \sum(\sum_{LF(v')} r_i \mid v'$ is a child node of $v)$ clearly holds, the values $\sum_{LF(v)} r_i$ can be easily calculated for all v during a linear time bottom-up traversal of the GST.

The next step is to remove the redundancies, represented by the values $corr(l(v), S)$, from the values $\sum_{LF(v)} r_i$. Let $I(id_i)$ be the list of all leaves with the label id_i in the order they appear in a depth-first traversal of the tree. Clearly the lists $I(id_i)$ can be constructed in linear time for all labels id_i. We note the following four simple but useful properties: (1) The leaves in $LF(v)$ with the label id_i form a continuous interval of length $c_i(v)$ in the list $I(id_i)$. (2) If $c_i(v) > 0$, a length-$c_i(v)$ interval in $I(id_i)$ contains $c_i(v) - 1$ adjacent (overlapping) leaf pairs. (3) If $x, y \in LF(v)$, the node $lca(x, y)$ belongs to the subtree rooted by v. (4) For any $s_i \in S$, $\psi(l(v), s_i) = \text{true}$ if and only if there is a leaf $x \in LF(v)$ with the label id_i.

Assume that each node v has a correction value that has been initialized to 0. Consider now what happens if we go through all adjacent leaf pairs x, y in the list $I(id_i)$, and add for each pair the value r_i into the correction value of the node $lca(x, y)$. It follows from properties (1) - (3), that now for each node v in the tree, the sum of the correction values in the nodes of the subtree rooted by v

equals $(c_i(v)-1)r_i$. Moreover, if we repeat the process for each of the lists $I(id_i)$, then, due to property (4), the preceding total sum of the correction values in the subtree rooted by v becomes $\sum((c_i(v)-1)r_i \mid \psi(l(v), s_i) = \texttt{true}) = corr(l(v), S)$. Hence at this point a single linear time bottom-up traversal of the tree enables us to cumulate the correction values $corr(l(v), S)$ from the subtrees into each node v, and at the same time we may record the final values $\sum_{M(l(v))} r_i$.

The preceding process involves a constant number of linear time traversals of the tree, as well as a linear number of lca-queries. Since each lca-query can be done in constant time after a linear time preprocessing, the total time for computing the values $\sum_{M(l(v))} r_i$ is linear.

The above described algorithm permits us to compute the values $\sum_{M(l(v))} r_i$ and $|M(l(v))|$ in linear time, which in turn leads into a linear time solution for the problem of finding a good pattern when the pattern is a *single* substring: The scoring function can now be computed for each possible pattern candidate $l(v)$. Also the case of a Boolean pattern pair will be solved in this manner. That is, we will concentrate on how to compute the values $\sum_{M(\pi)} r_i$ (and $|M(\pi)|$) for all possible $O(N^2)$ pattern pair candidates, where $\pi = \langle l(v_1), F, l(v_2) \rangle$ and v_1, v_2 are any two nodes in the GST over S. If we manage to do this in $O(N^2)$ time, then the whole problem will be solved in $O(N^2)$ under the assumption that the scoring function can be computed in constant time for each candidate.

Naïve use of the information gathered by the single substring pattern algorithm is not sufficient for solving the problem for *pairs* of patterns in $O(N^2)$ time, since computing the needed values $\psi(\langle l(v_1), F, l(v_2) \rangle, s_1)$ requires us to somehow conduct an intrinsic set operation between the string subsets that match / do not match $l(v_1)$ and $l(v_2)$. However, an $O(N^2)$ algorithm for pattern pairs is fairly simple to derive from the linear time algorithm for the single pattern.

We go over the $O(N)$ choices for the first pattern, $l(v_1)$. For each such fixed $l(v_1)$, we use a modified version of the linear time algorithm in order to process the $O(N)$ choices for the second pattern, $l(v_2)$, in $O(N)$ time. Given a fixed $l(v_1)$, we additionally label each string $s_i \in S$, and the corresponding leaves in the GST, with the Boolean value $\psi(l(v_1), s_i)$. This can be done in $O(N)$ time. Now the trick is to cumulate the sums and correction factors *separately* for different values of the additional label. The end result is that we will have the values $\sum_{M(l(v)), b} r_i = \sum(r_i \mid \psi(l(v), s_i) = \texttt{true} \wedge \psi(l(v_1), s_i) = b)$ for all nodes in linear time. We note that $\sum(r_i \mid \psi(l(v), s_i) = \texttt{false} \wedge \psi(l(v_1), s_i) = b) = \sum(r_i \mid \psi(l(v_1), s_i) = b) - \sum_{M(l(v)), b} r_i$, where the value $\sum(r_i \mid \psi(l(v_1), s_i) = b)$ can easily be computed in linear time. Thus *all* cumulative values of form $\sum(r_i \mid \psi(l(v), s_i) = b_1 \wedge \psi(l(v_1), s_i) = b_2)$, where $b_1, b_2 \in \{\texttt{true}, \texttt{false}\}$, can be computed in linear time. And from these it is straightforward to compute the values $\sum_{M(\langle l(v_1), F, l(v_2) \rangle)} r_i = \sum(r_i \mid F(\psi(l(v_1), s_i), \psi(l(v_2), s_i)) = \texttt{true})$, as well as the corresponding scoring function values, in linear time. Thus, given a fixed $l(v_1)$, we can compute the scores for all pattern pair candidates of form $\langle l(v_1), F, l(v_2) \rangle$ in $O(N)$ time. Since there are only $O(N)$ candidates for $l(v_1)$, we have an $O(N^2)$ algorithm for evaluating all possible pattern pair candidates.

It is not difficult to see that the space complexity of the algorithm is $O(N)$. The algorithm is also highly parallelizable, since the $O(N)$ time calculations for each fixed $l(v_1)$ are independent of each other.

4 Implementation Using Suffix Arrays

The algorithm on the suffix tree can be simulated efficiently by a suffix array. We modify the algorithm of [22, 27] that simulates a bottom-up traversal of a suffix tree, using a suffix array. Notice that since each suffix of the string corresponds to a leaf in the suffix tree, each position in the suffix array corresponds to a leaf in the suffix tree. A subtlety in the modification lies in determining where to store the correction factors after calculating the lca, since the simulation via suffix arrays does not explicitly recreate the internal nodes of the suffix tree. For storing the correction factors, we construct another array C of the same length as the suffix array, which represents the internal nodes of the suffix tree. An element $C[i]$ in the array corresponds to the lca of suffix $A_S[i-1]$ and suffix $A_S[i]$. When adding up the correction factor for two leaves corresponding to $A_S[i]$ and $A_S[j]$ ($i < j$), the correction factor is added into $C[rmq(i+1,j)]$. Although it is the case that different positions in C can correspond to the same internal node when the node has more than two children, it is possible to correctly sum the values required for the score calculations, since all positions of the array are visited in the traversal simulation, and the addition operation on numeric values is commutative as well as associative.

5 Computational Experiments

The degradation of mRNA, in addition to transcription, is one of several important mechanisms which control the expression level of a gene (see [28] for survey). The half lives of mRNA are very diverse: some mRNAs can degrade 100 times faster than others, which allows their expression level to be adjusted more quickly. The degradation of mRNA is controlled by many factors, for example, it is known that some proteins bind to the UTR of the mRNA to promote its decay, while others inhibit it. Recently, the comprehensive decay rates of many genes have been measured using microarray technology [16, 17]. We consider the problem of finding substring pattern pairs related to the rate of mRNA decay to find possible binding sites of the proteins, in order to further understand this complex mechanism.

The algorithm was implemented using the C language, and uses POSIX threads to execute parallel computations. To give an idea of the efficiency of the algorithm, it can be run on a data set with $720,673$ candidates for a single pattern, meaning $720,673^2 = 519,369,572,929$ possible pattern pairs, requiring about half a day on a Sun Fire 15K with 96 CPUs (UltraSPARC III Cu 1.2GHz). To ease the interpretation process in the following experiments, we limit the search to Boolean combinations: $p' \wedge q'$ and $p' \vee q'$, where p' is either p or $\neg p$, and q' is either q or $\neg q$.

Table 2. Top 5 scoring pattern pairs found from yeast 3'UTR sequences.

rank	$\|M(\pi, S_f)\|$	$\|M(\pi, S_s)\|$	χ^2 (p-val)	pattern pair
1	55/393	7/379	38.5 ($< 10^{-9}$)	UAAAAAUA $\vee$ UGUAUAA
2	63/393	13/379	34.5 ($< 10^{-8}$)	UAUGUAA $\vee$ UGUAUAA
3	240/393	152/379	33.9 ($< 10^{-8}$)	($\neg$AUCC) $\wedge$ UGUA
4	262/393	174/379	33.8 ($< 10^{-8}$)	($\neg$UAGCU) $\wedge$ UGUA
5	223/393	136/379	33.7 ($< 10^{-8}$)	($\neg$GUUG) $\wedge$ UGUA

5.1 Positive/Negative Set Discrimination of Yeast Sequences

For our first experiment, we used the two sets of predicted 3'UTR processing site sequences provided in [29], which are constructed based on the microarray experiments in [16] that measure the degradation rate of yeast mRNA. One set S_f consists of 393 sequences which have a fast degradation rate ($t_{1/2} < 10$ minutes), while the other set S_s consists of 379 predicted 3'UTR processing site sequences which have a slow degradation rate ($t_{1/2} > 50$ minutes). Each sequence is 100 nt long, and the total length of the sequences is $77,200$ nt. The traversal on the suffix array on this dataset shows that there are $46,554$ candidates for a single pattern (i.e. the number of internal nodes in the suffix tree. Patterns corresponding to leaf nodes were ignored, since they are not "commonly occurring" patterns), meaning that there are $46,554^2 = 2,167,274,916$ possible pattern pairs. For the scoring function, we used the chi-squared statistic, calculated by $(|S_f| + |S_s|)(\mathtt{tp} * \mathtt{tn} - \mathtt{fp} * \mathtt{fn})^2/(\mathtt{tp} + \mathtt{fn})(\mathtt{tp} + \mathtt{fp})(\mathtt{tn} + \mathtt{fp})(\mathtt{tn} + \mathtt{fn})$ where $\mathtt{tp} = |M(\pi, S_f)|$, $\mathtt{fp} = |S_f| - \mathtt{tp}$, $\mathtt{tn} = |S_s| - \mathtt{fn}$, and $\mathtt{fn} = |M(\pi, S_s)|$.

The time required for computation was around $3 \sim 4$ minutes on the above mentioned computer. The 5 top scoring pattern pairs found are shown in Table 2. Several interesting patterns can be found in these pattern pairs. For all the patterns in the pairs that match more in the faster decaying set, the substring UGUA is contained. This sequence is actually known as a core consensus for the binding site of the PUF protein family that plays important roles in mRNA regulation [30], and has also been found in the previous analysis [29] to be significantly over-represented in the fast degrading set.

On the other hand, patterns which are combined with $\neg$ can be considered as sequence elements which *compete* with UGUA, and interfere with mRNA decay. The patterns AUCC and GUUG were in fact found to be substrings of a less studied mRNA *stabilizer* element, experimentally shown to be within a region of 65nt in the TEF1/2 transcripts [31]. We cannot say directly that the two substrings represent components of this stabilizer element, since it was reported that this stabilizer element should be in the translated region in order to function. However, the mechanisms of stabilizers are not yet well understood, and further investigation may uncover relationships between these sequences.

5.2 Finding Correlated Patterns from Human Sequences

For our second experiment, we used the decay rate measurements of the human hepatocellular carcinoma cell line HepG2 made available as Supplementary

Table 3. Top 5 scoring pattern pairs found from human 3'UTR sequences.

rank	$|M(\pi, S)|$	rank sum	avg rank	z (p-val)	pattern pair
1	1338/2306	1.7101×10^6	1278.1	10.56 ($< 10^{-25}$)	UUAUUU $\vee$ UGUAUA
2	904/2306	1.2072×10^6	1335.4	10.53 ($< 10^{-25}$)	UUUUAUUU $\vee$ UGUAUA
3	1410/2306	1.7900×10^6	1269.5	10.49 ($< 10^{-25}$)	UUUAAA $\vee$ UUUAUA
4	711/2306	9.7370×10^5	1369.5	10.40 ($< 10^{-24}$)	UAUUUAU $\vee$ UGUAUAU
5	535/2306	7.5645×10^5	1413.9	10.32 ($< 10^{-24}$)	UGUAAAUA $\vee$ UGUAUAU

Table 9 of [17]. 3'UTR sequences for each mRNA was retrieved using the EN-SMART [32] interface. We were able to obtain 2306 pairs of 3'UTR sequences and their decay rates, with the average length of the sequences being 925.54 nt, and the total length was $2,134,294$ nt.

Since the distribution of the turnover rates seemed to have a heavier tail than the normal distribution, we used a nonparametric scoring function that fits into our $O(N^2)$ total time bound: the normal approximation of the Wilcoxon rank sum test statistic. The set of sequences S is first sorted in increasing order according to its decay rate, and each sequence s_i is assigned its rank for r_i. For a pattern pair π, the rank sum statistic $\sum_{M(\pi)} r_i$ approximately depends on the normal distribution when the sample size is large. Therefore, we use the z-score defined by: $z(x,y) = \dfrac{(y - x(|S| + 1)/2)}{\sqrt{x(|S| - x)(|S| + 1)/12}}$ where $x = |M(\pi)|$ and $y = \sum_{M(\pi)} r_i$, with appropriate corrections for ranks and variance when there are ties in the decay rate values. The score function can be calculated in constant time for each x and y, provided $O(m \log m)$ time preprocessing for sorting of the data and assigning the ranks.

The 5 top scoring patterns are presented in Table 3. All pairs are of the form $p \vee q$, common to sequences with higher ranks, that is, sequences with higher decay rates. Notice that most of the highest scoring patterns contain UGUAUA, which was contained also in the results for yeast, which may indicate a possibility that these degradation mechanisms are evolutionarily conserved between eukaryotes. The other pattern in the pairs consist of A and U, and apparently captures the A+U rich elements (AREs) [28] which are known to promote rapid mRNA decay dependent on deadenylation. The form $p \vee q$ of the pattern pairs also indicates that the two elements may have *complementary* roles in the degradation of mRNA.

6 Discussion

We have presented an efficient $O(N^2)$ algorithm for finding the optimal Boolean pattern pair with respect to a suitable scoring function, from a set of strings and a numeric attribute value assigned to each string. The algorithm was applied to moderately sized biological sequence data and was successful in finding pattern pairs that captured known destabilizing elements, as well as possible stabilizing elements, from 3'UTR of yeast and human mRNA sequences, where each mRNA sequence is labeled with values depending on its decay rate.

Frequently in biological applications, motif models which consider ambiguity in the matching are preferred, rather than the "exact" substring patterns used in this paper. Nevertheless, the selection of the motif model for a particular application is still a very difficult problem, and substring patterns can be effective as shown in this paper and others [10]. As well as being efficient, simpler models also have the advantage of being easier to interpret, and can be used as a quick, initial scanning for the task.

Acknowledgments

This work was supported in part by Grant-in-Aid for Encouragement of Young Scientists (B), and Grant-in-Aid for Scientific Research on Priority Areas (C) "Genome Biology" from the Ministry of Education, Culture, Sports, Science and Technology of Japan. The authors would like to acknowledge Dr. Seiya Imoto (Human Genome Center, Institute of Medical Science, University of Tokyo) for helpful comments concerning the scoring functions.

References

1. Brazma, A., Jonassen, I., Eidhammer, I., Gilbert, D.: Approaches to the automatic discovery of patterns in biosequences. J. Comput. Biol. **5** (1998) 279–305
2. Marsan, L., Sagot, M.F.: Algorithms for extracting structured motifs using a suffix tree with an application to promoter and regulatory site consensus identification. J. Comput. Biol. **7** (2000) 345–360
3. Arimura, H., Wataki, A., Fujino, R., Arikawa, S.: A fast algorithm for discovering optimal string patterns in large text databases. In: International Workshop on Algorithmic Learning Theory (ALT'98). Volume 1501 of LNAI., Springer-Verlag (1998) 247–261
4. Eskin, E., Pevzner, P.A.: Finding composite regulatory patterns in DNA sequences. Bioinformatics **18** (2002) S354–S363
5. Liu, X., Brutlag, D., Liu, J.: BioProspector: discovering conserved DNA motifs in upstream regulatory regions of co-expressed genes. In: Pac. Symp. Biocomput. (2001) 127–138
6. Shimozono, S., Shinohara, A., Shinohara, T., Miyano, S., Kuhara, S., Arikawa, S.: Knowledge acquisition from amino acid sequences by machine learning system BONSAI. Transactions of Information Processing Society of Japan **35** (1994) 2009–2018
7. Shinohara, A., Takeda, M., Arikawa, S., Hirao, M., Hoshino, H., Inenaga, S.: Finding best patterns practically. In: Progress in Discovery Science. Volume 2281 of LNAI., Springer-Verlag (2002) 307–317
8. Takeda, M., Inenaga, S., Bannai, H., Shinohara, A., Arikawa, S.: Discovering most classificatory patterns for very expressive pattern classes. In: 6th International Conference on Discovery Science (DS 2003). Volume 2843 of LNCS., Springer-Verlag (2003) 486–493
9. Shinozaki, D., Akutsu, T., Maruyama, O.: Finding optimal degenerate patterns in DNA sequences. Bioinformatics **19** (2003) 206ii–214ii
10. Bussemaker, H.J., Li, H., Siggia, E.D.: Regulatory element detection using correlation with expression. Nature Genetics **27** (2001) 167–171

11. Bannai, H., Inenaga, S., Shinohara, A., Takeda, M., Miyano, S.: A string pattern regression algorithm and its application to pattern discovery in long introns. Genome Informatics **13** (2002) 3–11
12. Conlon, E.M., Liu, X.S., Lieb, J.D., Liu, J.S.: Integrating regulatory motif discovery and genome-wide expression analysis. Proc. Natl. Acad. Sci. **100** (2003) 3339–3344
13. Bannai, H., Inenaga, S., Shinohara, A., Takeda, M., Miyano, S.: Efficiently finding regulatory elements using correlation with gene expression. Journal of Bioinformatics and Computational Biology (2004) in press.
14. Zilberstein, C.B.Z., Eskin, E., Yakhini, Z.: Using expression data to discover RNA and DNA regulatory sequence motifs. In: The First Annual RECOMB Satellite Workshop on Regulatory Genomics. (2004)
15. Gusfield, D.: Algorithms on Strings, Trees, and Sequences. Cambridge University Press (1997)
16. Wang, Y., Liu, C., Storey, J., Tibshirani, R., Herschlag, D., Brown, P.: Precision and functional specificity in mRNA decay. Proc. Natl. Acad. Sci. **99** (2002) 5860–5865
17. Yang, E., van Nimwegen, E., Zavolan, M., Rajewsky, N., Schroeder, M., Magnasco, M., Jr., J.D.: Decay rates of Human mRNAs: correlation with functional characteristics and sequence attributes. Genome Res. **13** (2003) 1863–1872
18. Manber, U., Myers, G.: Suffix arrays: a new method for on-line string searches. SIAM Journal on Computing **22** (1993) 935–948
19. Kim, D.K., Sim, J.S., Park, H., Park, K.: Linear-time construction of suffix arrays. In: 14th Annual Symposium on Combinatorial Pattern Matching (CPM 2003). Volume 2676 of LNCS., Springer-Verlag (2003) 186–199
20. Ko, P., Aluru, S.: Space efficient linear time construction of suffix arrays. In: 14th Annual Symposium on Combinatorial Pattern Matching (CPM 2003). Volume 2676 of LNCS., Springer-Verlag (2003) 200–210
21. Kärkkäinen, J., Sanders, P.: Simple linear work suffix array construction. In: 30th International Colloquium on Automata, Languages and Programming (ICALP 2003). Volume 2719 of LNCS., Springer-Verlag (2003) 943–955
22. Kasai, T., Arimura, H., Arikawa, S.: Efficient substring traversal with suffix arrays. Technical Report 185, Department of Informatics, Kyushu University (2001)
23. Abouelhoda, M.I., Kurtz, S., Ohlebusch, E.: The enhanced suffix array and its applications to genome analysis. In: Second International Workshop on Algorithms in Bioinformatics (WABI 2002). Volume 2452 of LNCS., Springer-Verlag (2002) 449–463
24. Bender, M.A., Farach-Colton, M.: The LCA problem revisited. In: Proc. Latin American Theoretical Informatics (LATIN). Volume 1776 of LNCS., Springer-Verlag (2000) 88–94
25. Alstrup, S., Gavoille, C., Kaplan, H., Rauhe, T.: Nearest common ancestors: a survey and a new distributed algorithm. In: 14th annual ACM symposium on Parallel algorithms and architectures. (2002) 258–264
26. Hui, L.: Color set size problem with applications to string matching. In: Proceedings of the Third Annual Symposium on Combinatorial Pattern Matching (CPM 92). Volume 644 of LNCS., Springer-Verlag (1992) 230–243
27. Kasai, T., Lee, G., Arimura, H., Arikawa, S., Park, K.: Linear-time longest-common-prefix computation in suffix arrays and its applications. In: 12th Annual Symposium on Combinatorial Pattern Matching (CPM 2001). Volume 2089 of LNCS., Springer-Verlag (2001) 181–192
28. Wilusz, C.J., Wormington, M., Peltz, S.W.: The cap-to-tail guide to mRNA turnover. Nat. Rev. Mol. Cell Biol. **2** (2001) 237–246

29. Graber, J.: Variations in yeast 3'-processing cis-elements correlate with transcript stability. Trends Genet. **19** (2003) 473–476
    http://harlequin.jax.org/yeast/turnover/
30. Wickens, M., Bernstein, D.S., Kimble, J., Parker, R.: A PUF family portrait: 3'UTR regulation as a way of life. Trends Genet. **18** (2002) 150–157
31. Ruiz-Echevarria, M.J., Munshi, R., Tomback, J., Kinzy, T.G., Peltz, S.W.: Characterization of a general stabilizer element that blocks deadenylation-dependent mRNA decay. J. Biol. Chem. **276** (2001) 30995–31003
32. Kasprzyk, A., Keefe, D., Smedley, D., London, D., Spooner, W., Melsopp, C., Hammond, M., Rocca-Serra, P., Cox, T., Birney, E.: EnsMart: A generic system for fast and flexible access to biological data. Genome Research **14** (2004) 160–169

Finding Missing Patterns

Shunsuke Inenaga, Teemu Kivioja, and Veli Mäkinen

Department of Computer Science, P.O. Box 26 (Teollisuuskatu 23)
FIN-00014 University of Helsinki, Finland
{inenaga,kivioja,vmakinen}@cs.helsinki.fi

Abstract. Consider the following problem: Find the shortest *pattern* that does not occur in a given *text*. To make the problem non-trivial, the pattern is required to consist only of characters that occur in the text. This problem can be solved easily in linear time using the suffix tree of the text. In this paper, we study an extension of this problem, namely the *missing patterns problem*: Find the shortest *pair of patterns* that do not occur close to each other in a given text, i.e., the distance between their occurrences is always greater than a given threshold α. We show that the missing patterns problem can be solved in $O(\min(\alpha n \log n, n^2))$ time, where n is the size of the text. For the special case where both pairs are required to have the same length, we give an algorithm with time complexity $O(\alpha n \log \log n)$. The problem is motivated by optimization of multiplexed nested-PCR.

1 Introduction

For a decade, *pattern discovery* has played a central role in bioinformatics [15]. Especially extracting surprising and useful patterns is a core of knowledge discovery from textual data [4, 13]. One extreme example of surprising patterns is *missing patterns*, namely, patterns that do *not* appear in a given text T are sought. Amir et al. [1] introduced a generalized version of the missing pattern problem in such a way that pattern P may 'approximately' occur in T. They call this problem the *inverse pattern matching* problem. Some improvements for this inverse problem appeared in [5]. Another related work is the *farthest substring problem* by Lanctot et al. [10], where a set of text strings is considered as input.

In this paper, we explore another type of extension of the missing pattern problem: given a text T and threshold value α, find the *shortest pair of patterns* such that the distance between their occurrences in T is *always greater than* α (see Fig. 1). We show that this problem is solvable in $O(\min(\alpha n \log n, n^2))$ time, where n is the length of T. For the case that the lengths of the two patterns have to be the same, we present a simpler and slightly more efficient solution when α is small; we achieve time complexity $O(\alpha n \log \log n)$. Not only is our missing patterns problem interesting in theory, but it is also well-motivated in practice. Indeed, we will show how missing patterns can be used to optimize the sensitivity of PCR.

The rest of the paper is organized as follows: In Section 2 we give definitions, and introduce our biological motivations and data structures. Section 3 presents

I. Jonassen and J. Kim (Eds.): WABI 2004, LNBI 3240, pp. 463–474, 2004.

our $O(\min(\alpha n \log n, n^2))$-time algorithm for finding a missing patterns pair in general cases, and in Section 4 we develop specialized algorithms for cases where the two patterns are required to be of the same length. In section 5 we make some preliminary experimental observations.

2   Preliminaries

2.1   Definitions

A *string* $T = t_1 t_2 \cdots t_n$ is a sequence of *characters* from an ordered *alphabet* Σ of size σ. A *substring* of T is any string $T_{i...j} = t_i t_{i+1} \cdots t_j$, where $1 \leq i \leq j \leq n$. A substring of length k is called k-*mer*. A *suffix* of T is any substring $T_{i...n}$, where $1 \leq i \leq n$. A *prefix* of T is any substring $T_{1...j}$, where $1 \leq j \leq n$. Suffixes and prefixes can be identified by their starting and ending positions, respectively. A *pattern* is a short string over the alphabet Σ. We say that pattern $P = p_1 p_2 \cdots p_k$ *occurs* at position j of *text* string T iff $p_1 = t_j, p_2 = t_{j+1}, \ldots, p_k = t_{j+k-1}$. Such positions j are called the occurrence positions of P in T.

A *missing pattern* P (with respect to text T) is such that P is not a substring of T, i.e., P does not occur at any position j of T. Let $\alpha > 0$ be a threshold parameter. A *missing pattern pair* (A, B) is such that if A (resp. B) occurs at position j of text T, then B (A) does not occur at any position j' of T, such that $j - \alpha \leq j' \leq j + \alpha$. If (A, B) is a missing pair, we say that A and B do not occur α-*close* in T. These notions are illustrated in Fig. 1.

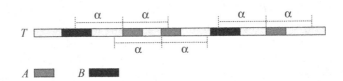

Fig. 1. Missing pattern pair (A, B).

We study the following problem:

Problem 1 *(Missing Patterns Problem) Given a text T and threshold α, find patterns A and B of minimum total length, such that (A, B) is a missing pattern pair with respect to T, i.e., A and B do not occur α-close in T.*

2.2   Biological Motivation

The missing patterns problem is biologically motivated by finding good adapters for primers used in a polymerase chain reaction (PCR). PCR is a standard technique for producing many copies of a region of DNA [3]. It is routinely used for example in medicine to detect infections and in forensic science to identify individuals even from tiny samples. In PCR a pair of short fragments of DNA called primers is specifically designed for the amplified region so that each of them is complementary to the 3' end of one of two strands of the region (see also Fig. 2). Given single-stranded DNA molecules, the primers hybridize to their binding sites flanking the target region. An enzyme called DNA polymerase adds nucleotides after each primer using the other strand as template and thus builds a copy of the strand started by the primer. The molecules are made single-stranded again by heating and the process is repeated many times (20-30) resulting in an exponential blow-up in the number of copies of the target region.

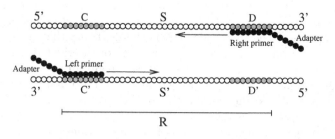

Fig. 2. PCR.

PCR is a sensitive method: in theory one can detect a couple of DNA molecules in a complex mixture of different molecules by using 20-30 amplification cycles but in practice the achieved sensitivity is lower. In particular, it is difficult to combine specificity of amplification with sensitivity. The sample can include homologous sequences that need to be separated from each other, such as sequences from several bacterial species, and the specific primers can thus bind to other sites than their intended target sites and start amplification of an incorrect region of DNA. The problem can be alleviated by using stringent conditions in PCR reaction but that can also decrease the amplification efficiency of the correct target leading to lower sensitivity. In order to achieve ultrasensitive detection, repeated PCR with nested primers, so-called nested PCR, is used.

Today, detection tests are performed preferably in a multiplexed fashion. In PCR, it means that several regions are amplified simultaneously in the same assay. Selecting specific primer pairs for multiplex PCR is a hard computational problem [12] but outside the scope of this paper.

In this work we describe adapters for specific PCR primers that facilitate the set-up of multiplexed nested-PCR assays. An adapter-primer pair is designed to further amplify any fragment amplified by the specific primers. Adapter primers work by binding to short sequences called adapters attached to the ends of all specific primers (see Fig. 2). Further amplification by adapter primers increases sensitivity of detection. On the other hand, using the same pair of primers for all fragments in the latter step of nested PCR simplifies multiplexing.

A crucial requirement for a good adapter primer pair is that it should only bind to the adapters and should not amplify anything directly from the sample. However, since the adapter sequences can be chosen freely, we have a huge number of options available. Motivated by the demand for ultrasensitive and multiplexed detection, we aim for optimal selection of adapter primers.

Assume that we want to select such a pair of primers that they do not amplify a certain region R ($|R| < \alpha$) and let that region be flanked by patterns C and D in the sequence S (see Fig. 2). The sequence S denotes all the sequences that can be present in a sample, such as the human genome and the genomes of the bacteria or viruses that possibly are the cause of an infection. Let us denote the reverse complement of a sequence S by S' (the choice of one strand as S and the other as S' is arbitrary). The region R can be amplified by a primer pair only if the left primer binds to C' and the right primer binds to D. Thus, if we have found a missing pair (A, B) for the strand S, then (A', B) is such a pair of patterns that there is no such region of length less than α that A' is the binding site of the left primer and B is the binding site of the right primer. However, (A', B) can also work the other way round, i.e. there can be such a region that A' is a binding site of the right primer and B is the binding site of the left primer. But in that case the pair (A, B) would occur α-close in the strand S'. In conclusion, if (A, B) is a missing pair for both strands of the sample sequence, i.e. for the string SS', then (A', B) is not a pair of primer binding sites for any region of length less than α. A safe threshold α can be determined based on the speed of the polymerase reaction and duration of the PCR cycle.

Until now we have ignored the fact that a primer can bind and initiate the polymerase reaction even if the primer and the binding site are not exact complements. Especially, if the 3' end of the primer binds to a site, it can initiate the polymerase reaction even if the 5' end of the primer dangles freely. That is why we search for the shortest missing pair (A_{min}, B_{min}), where $|A_{min}| = |B_{min}| = k$. Then, we can take the pair (A'_{min}, B_{min}) as the 5' ends of the adapters and (A_{min}, B'_{min}) as the 3' primer ends of the adapter primers. Such a selection guarantees that if the first k nucleotides of the left primer match a binding site in the sample sequence, the right primer has at least one mismatch in the first k nucleotides and vice versa. In addition, no pair of primers satisfies this condition for $k-1$. Therefore, using the shortest missing pair in the construction of adapter primers minimizes the risk of any unwanted amplification. Notice that Problem 1 is a generalization of the objective described above; all of our algorithms can be adopted for this special case.

The adapter primers also have to satisfy other requirements. For example, the melting temperature of both primers should be within a specified range. The primers should not form stable hairpin loops or bind to each other. These additional requirements can be satisfied because there most probably are many missing pairs of the same length and the missing pairs are short compared to the length of the primer which typically is around 17..25 bases. We show in Section 5 that these assumptions are valid for the yeast genome. Thus, the 3' ends of the adapter primers can be chosen from several alternatives and the 5' ends can be chosen freely to satisfy the other requirements.

2.3   Data Structures

We use some well-known string data structures in our algorithms, such as *keyword tries* and *suffix trees*. Let us briefly recall these structures.

Definition 2 *(Adopted from [6]) The* keyword trie *for set $\mathcal{P}$ of patterns is a rooted directed tree $\mathcal{K}$ satisfying three conditions: (1) Each edge is labeled with exactly one character; (2) any two edges out of the same node have distinct labels; (3) every pattern P of $\mathcal{P}$ maps to some node v of $\mathcal{K}$ such that the characters on the path from the root of $\mathcal{K}$ to v spell out P, and every leaf of $\mathcal{K}$ is mapped to by some pattern in $\mathcal{P}$.*

Definition 3 *The* suffix trie *of text T is a keyword trie for set $\mathcal{S}$, where $\mathcal{S}$ is the set of all suffixes of T.*

Definition 4 *The* suffix tree *of text T is the* path-compressed *suffix trie of T, i.e., a tree that is obtained by representing each maximal non-branching path of the suffix trie as a single edge labeled by the catenation of the labels in the corresponding edges of the suffix trie. The labels of the edges of suffix tree correspond to substrings of T; each edge can be represented as a pair (l, r), such that $T_{l...r}$ gives the label.*

Definition 5 *The* sparse suffix tree *of text T is a suffix tree built on a subset $\mathcal{S}'$ of suffixes of T, i.e., a path-compressed keyword trie for set $\mathcal{S}'$.*

The suffix tree of text $T = t_1 t_2 \cdots t_n$ takes $O(n)$ space, and it can be constructed in $O(n)$ time [16, 11, 14] (omitting the alphabet factors). A sparse suffix tree can be pruned from the (full) suffix tree in linear time [8, 2]. (Direct constructions of sparse suffix trees were also considered in [8, 2], but $O(n)$ time is enough for our purposes.)

The nodes of all the above-defined trees can be partitioned into two classes: (1) A node is *complete* if it has an edge $e(c)$ for each $c \in \Sigma$ such that the label of edge $e(c)$ starts with character c; (2) otherwise the node is *incomplete*. Let us denote by $label(v, s)$ the catenation of labels between two nodes v and s. With the *depth* of a node v we mean $|label(root, v)|$.

We sometimes refer to *implicit nodes* of the suffix tree, meaning, in addition to all (explicit) nodes of the suffix tree, also the positions on the edge labels of the suffix tree, as they all correspond to nodes of the corresponding suffix trie.

3   General Algorithm Using Sparse Suffix Trees

Let us first describe how the one pattern case (as mentioned in the abstract) can be solved. That is, we wish to find a pattern of minimum length that does not occur in a text of length n. The solution is as follows. Build the suffix trie of the text. Among all incomplete nodes of the trie, select the one that has the minimum depth. Let that node be v and let the character that makes the node incomplete be c. Then the answer to the question is $label(root, v)c$. The size of the suffix trie can be $O(n^2)$, which makes this algorithm inefficient. The same algorithm can be simulated using the suffix tree, which reduces the running time to $O(n)$; instead of scanning through all implicit nodes of the suffix tree, we can check the explicit nodes for incompleteness and for each edge whose label is longer than 1, we know that the implicit node corresponding to the first letter on the label is incomplete.

3.1 Basic Properties

The topic of this paper, the two-pattern case defined in Problem 1, is more challenging. However, some aspects of the one-pattern solution can be exploited, as summarized in the following observation.

Observation 6 *(Monotony property) Let v be a node of the suffix tree of text T, and let e be an edge out of v labeled $L = l_1 l_2 \cdot l_p$. Then, string $label(root, v)l_1$ occurs in T exactly at the same positions as any string $label(root, v)L_{1...i}$, where $1 < i \leq p$.*

Before using the monotony property, we also mention the following simple but important lemma.

Lemma 7 *(Substring property) It holds that either (i) there is a solution to the missing patterns problem, say pair (A, B), such that both A and B are substrings of the text; or (ii) the solution is a single pattern.*

Proof. Let (A, B) be a solution to the missing pairs problem such that A is not a substring of the text. Then (A, ϵ) is also a missing pair. Since $|A| + |\epsilon| = |A| \leq |A| + |B|$, pair (A, B) can not be the shortest missing pair, unless B is an empty string, in which case A is a single pattern solution. $\square$

The above lemma states that we can restrict to selecting both A and B as non-empty substrings of T. The case where one pattern is enough was considered at the beginning of this section.

3.2 Basic Algorithm

Let V be the set of all nodes of the suffix tree of text T, and let $\mathcal{P}$ be the set of strings obtained by adding to each $label(root, v)$, $v \in V$, all starting characters of labels on the out edges of v. It is easy to see that $|\mathcal{P}| \leq 2n - 1$; the size of $\mathcal{P}$ is bounded by the number of internal nodes in the tree. Finally, let $Occ(P)$ be the list of occurrences of pattern $P \in \mathcal{P}$ in T; it can be obtained in time $O(|Occ(P)|)$ from the suffix tree.

Recall that we are interested in finding a missing pair (A, B). Let us choose as A a string from $\mathcal{P}$. Our goal is to choose B so that (A, B) will be a missing pair. As A is now fixed, we try to choose B of minimum length. Let us, for now, assume that we have found pattern B of minimum length such that (A, B) is a missing pair. The crucial observation is that if we repeat this process for all

$A \in \mathcal{P}$, we can choose among all the missing pairs found so far, the one where the sum $|A| + |B|$ is minimized. The correctness of this procedure follows directly from Observation 6 and Lemma 7.

What is left is to explain how to choose B of minimum length so that (A, B) will be a missing pair. This is done as follows. Let us define a set $Zone(A, \alpha)$:

$$Zone(A, \alpha) = \cup_{j \in Occ(A)} [j - \alpha, j + \alpha].$$

We have the following observation:

Observation 8 *If and only if B is a prefix of any suffix $T_{j'...n}$ such that $j' \in Zone(A, \alpha)$, then pair (A, B) occurs α-close in T.*

Now, building the sparse suffix tree over suffixes $T_{j'...n}$, $j' \in Zone(A, \alpha)$, we can choose B exactly as in the algorithm sketched at the beginning of this section: Among all incomplete implicit nodes of the sparse suffix tree, select the one that has the minimum depth. Let that node be u and let the character that makes the node incomplete be d. Then $B = label(root, u)d$. The algorithm is illustrated in Fig. 3.

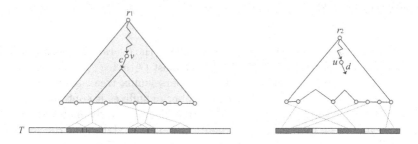

Fig. 3. Illustration of the algorithm to find a missing pair (A, B), where $A = label(r_1, v)c$, $B = label(r_2, u)d$, r_1 is the root of the full suffix tree, and r_2 is the root of the sparse suffix tree corresponding to A.

Theorem 9 *The missing patterns problem on text of length n can be solved in time $O(n^2)$ and space $O(n)$.*

Proof. The correctness of the algorithm should be clear from the above discussion. The time complexity follows from the facts that the size of $\mathcal{P}$ is at most $2n - 1$, and for each $A \in \mathcal{P}$ we use $O(n)$ time for constructing the set $Zone(A, \alpha)$ and the corresponding sparse suffix tree; To construct $Zone(A, \alpha)$ in linear time, one should first mark in a bit-vector of length n all suffixes in $Occ(A)$. Then for each marked suffix j, one should mark in some other bit-vectors the starting point $j - \alpha$ and the end point $j + \alpha$ of the influence region. Finally, scanning from left to right one can maintain a counter to know at each text position j' whether it is inside some influence region or not, i.e., whether it should be included in the sparse suffix tree or not. As mentioned earlier, the sparse suffix tree can be obtained from the full suffix tree in $O(n)$ time. Overall, we have $O(n \times n) = O(n^2)$ time. At each phase of the algorithm, we use $O(n)$ space.   □

3.3   Improved Algorithm

We will now improve Theorem 9 in the case where α is small. First, we observe that we can select pattern A near the root of the suffix tree because of the following lemma.

Lemma 10 *If $\sigma^k > n$, then there must be a missing pattern of length k.*

Proof. There are at most $n - k + 1$ different k-mers in T. Since σ^k is larger than this, there must be some $X \in \Sigma^k$ that is not a k-mer.   □

Hence, we can restrict to the case $k \leq \log_\sigma n$, as there must otherwise be a single pattern solution, which can be found in linear time as explained at the beginning of this section.

Let $\mathcal{P}^{\leq q}$ be a subset of $\mathcal{P}$ such that all strings in $\mathcal{P}^{\leq q}$ are at most of length q. Now, we make the following observation:

Observation 11 *For each suffix j, there are at most $q = \log_\sigma n$ strings $A \in \mathcal{P}^{\leq q}$ such that $j \in Occ(A)$.*

A direct consequence of Observation 11 is that the overall size of sparse suffix trees corresponding to strings $A \in \mathcal{P}^{\leq q}$ is at most $O(\alpha n \log_\sigma n)$; each suffix can belong to at most $(2\alpha + 1) \log_\sigma n$ different sparse suffix trees, and the size of a sparse suffix tree is proportional to the number of suffixes it contains.

Now, we can build the sparse suffix trees incrementally in linear time in their overall size as follows: make a depth-first search (DFS) on the full suffix tree limited to depth $\log_\sigma n$. Let SST_v be the sparse suffix tree corresponding to an internal node v; more formally, SST_v is the sparse suffix tree of suffixes $j \in Zone(A, \alpha)$, where $A = label(root, u)c$, u is the parent of v, and c is the first letter of the edge label from u to v. Let g be the child node of v to which we are proceeding in the DFS search. We make the observation that the sparse suffix tree SST_g corresponding to node g will contain a subset of suffixes represented by SST_v; we can prune SST_v to construct SST_g. To manage the incremental computation efficiently, we show in the next lemma that SST_g can be constructed from SST_v in linear time in the size of SST_v. To make this possible, we need to attach some additional information to the sparse suffix trees: We use *threaded* sparse suffix trees, where the leaves (suffixes) of the tree are linked together in a double linked list in increasing order of the suffix positions, and each leaf has a pointer to the corresponding leaf of the full suffix tree.

Lemma 12 *Let SST_v be the threaded sparse suffix tree corresponding to a node v of the full suffix tree (in the sense defined above). Then, the threaded sparse suffix tree SST_g corresponding to the child g of v can be constructed in linear time in the size of SST_v.*

Proof. The algorithm is as follows. We make a copy of SST_v and prune it (i.e. delete extra leaves) to construct SST_g. Let us simply use SST_v to denote the copy of it. The construction has three phases; (i) we mark all leaves (suffixes) of SST_v that are contained in the subtree of g in the full suffix tree, (ii) we mark all leaves of SST_v whose suffix positions are within α distance from the ones marked at phase (i), and (iii) we delete all unmarked leaves of SST_v to construct SST_g.

Phase (iii) is trivial; as a leaf is deleted (making some constant time local updates to the tree) we redirect the links between suffix positions to retain the threaded structure. In phase (ii) we extend the effect of the suffixes marked in phase (i) by scanning through the double linked list once from first to last and once from last to first. For phase (i) recall that the leaves of SST_v have pointers to the corresponding leaves of the full suffix tree. We reverse these pointers, so that we have pointers from some leaves of the full suffix tree to SST_v. Then we go through the leaves in the subtree of g, and follow the pointers from these leaves marking the corresponding leaves of SST_v. This concludes phase (i).

It is clear that after steps (i),(ii), and (iii), the remaining tree corresponds to SST_g, and the construction time is linear in the size of the tree SST_v.   □

After noticing that the threaded version of the full suffix tree is easy to obtain in linear time in its size, we get by induction using Lemma 12 the following result.

Theorem 13 *The missing patterns problem on text of length n can be solved in time $O(\alpha n \log n)$ and space $O(n \log n)$ on a constant alphabet.*

Proof. Lemma 12 states that we use linear time in the size of the parent sparse suffix tree to construct the child sparse suffix tree. Each node of the full suffix tree can have at most σ children, and hence we can use time at most σ times the size of each sparse suffix tree. This gives the claimed time bound on a constant alphabet. The space usage follows from the fact that we need to store at most $\log_\sigma n$ different sparse suffix trees at the same time during the DFS to manage the incremental computation.   □

The constant multiplicative factor σ occurring in the proof of the above theorem can be reduced to $\log \sigma$ by organizing the edges of each node of the full suffix tree in a balanced tree; we can build temporary sparse suffix trees for the nodes of each balanced tree. The overall size of the trees grows to $O(\log \sigma \alpha n \log_\sigma n)$, but each tree is scanned through only a constant number of times.

4   Algorithms for Patterns of Same Length

We now concentrate on the special case of the missing patterns problem, mentioned in Section 2.2 with a biological motivation, where the patterns are required to be of the same length. That is, we search for a missing pattern pair (A, B) such that $|A| = |B| = k$. For this special case we give a slightly faster algorithm than the algorithm of the previous section based on the sparse suffix trees, when α is not too large; we obtain time complexity $O(\alpha n \log \log n)$ on a constant alphabet.

In the sequel, we assume that the alphabet of the strings is $\Sigma = \{0, 1, \ldots, \sigma - 1\}$. This makes the exposition easier, and is not a crucial assumption, since it takes $O(n \log \sigma)$ time to map any other (ordered) alphabet into Σ. This is negligible to the time required by the algorithms for the missing patterns problem.

Let us start with a trivial algorithm, and then proceed with improvements that result into an improved bound. We will first consider checking if there is a missing pair for a fixed length k.

4.1   Trivial Algorithm

For a fixed k, search the text for each possible pattern pair and check whether any occurrences are too close. There are σ^{2k} pattern pairs of correct length. It takes $O(k + n)$ time and $O(k)$ space to check one pair: Run e.g. two Knuth-Morris-Pratt algorithms [9] in parallel. So, the total time requirement to report a possible missing pair for fixed k is $O(\sigma^{2k}(k + n))$. The algorithm only needs $O(k)$ space.

4.2   Simple Algorithm

For each k-mer pair (C, D) of T such that C and D are α-close in T, insert the concatenated string CD into a keyword trie and search the tree for missing pairs. The size of the keyword trie is $O(k\alpha n)$ and the time requirement for inserting the concatenated strings is $O(k\alpha n \log \sigma)$. Checking whether there exists a string P of length $2k$ which is not in the keyword trie, takes $O(k\alpha n)$ time. Such a string $P = AB$ defines a missing pair (A, B), $|A| = |B| = k$.

Alternatively, one can use a bit-table of size σ^{2k}, as there is a bijective mapping from the strings CD to integers $0, 1, \ldots, \sigma^{2k} - 1$. For each string CD its entry in the table can be computed in constant time using the well-known technique of computing the entry of string Yb knowing the entry of string aY (see e.g. [7])[1]. Then, marking the entries corresponding to the α-close k-mer pairs (C, D) takes $O(\sigma^{2k} + n\alpha)$ time. An unmarked entry corresponds to a missing pair.

Analysis. Notice that $\sum_{i=0}^{k-1} \sigma^i < \sigma^k$, since $\sigma \geq 2$. The term σ^k is a multiplicative factor in the previous algorithms, and hence if we run those algorithms for each value $k = 1, 2, \ldots$, until we find a missing pair (A, B) with some length $|A| = |B| = k$, the total time complexity will be at most two times the complexity of the last step.

Before running any of the algorithms, we can first check in $O(n)$ time whether there is a single missing pattern using the suffix tree approach. If such pattern

---

[1] The entry of aY is $v = a\sigma^{|Y|} + y_1\sigma^{|Y|-1} + y_2\sigma^{|Y|-2} + \cdots + y_{|Y|}$. The entry of Yb can be computed in constant time, as it is $\sigma(v - a\sigma^{|Y|-1}) + b$.

is not found, we know that $k \leq \log_\sigma n$ (due to the Lemma 10). Equivalently $\sigma^k \leq n$, which simplifies the bounds.

We notice also that we can bound σ^k with a function of n and α: If k is the smallest value such that $\sigma^{2k} > n\alpha$, then there must be a missing pair (A, B) such that $|A| = |B| = k$. Then equation $\sigma^{2k-2} \leq n\alpha$ gives an estimate $\sigma^{2k} \leq \sigma^2 n\alpha$.

Plugging these bounds to the complexity of the bit-table algorithm we get $O(\sigma^{2k} + n\alpha) = O(\sigma^2 n\alpha + n\alpha)$. This bound is for fixed k. We can search for the correct value of k using binary search among $1, 2, \ldots, \frac{\log_\sigma(n\alpha)}{2} + 1$. The overall work becomes $O(\sigma^2 n\alpha + n\alpha \log(\frac{\log_\sigma(n\alpha)}{2} + 1)) = O(\sigma^2 n\alpha + n\alpha \log\log_\sigma n)$, which gives the following result.

Theorem 14 *The missing patterns problem on a text of length n for patterns of the same length can be solved in $O(\alpha n \log\log n)$ time and $O(n\alpha)$ space on a constant alphabet.*

Notice also that the algorithms for the general case can be used to solve this restricted case. In fact, the result of Theorem 13 already gives an $O(\alpha n \log n)$ time solution with better constant factors on the alphabet size. Moreover, that algorithm uses considerably less space than the bit-table algorithm.

5   Experiments

We have run some preliminary tests with the baker's yeast (*Saccharomyces cerevisiae*) genome using the bit-table version of the simple algorithm described in Section 4.2. We set the distance α to a realistic value 5000 and searched for shortest missing pattern pairs of the same length k. There were solutions for $k = 8$ (i.e. both patterns of the pair are of length 8), in fact there were over 16 million such pairs.

Ultimately our aim is to find short missing pairs of patterns for the human genome in order to construct good adapter primers for biomedical applications. The test results with the yeast genome suggest that there most likely are missing pattern pairs for the human genome that are short enough to make the approach attractive from the biochemical point of view. The human genome is about 250 times larger than the baker's yeast genome, the size which size is about 12Mb, but on the other hand increasing k by 2 increases the number of pattern pairs 256 times. In addition, the human genome has a lot of repetitive elements. Processing a text as large as the human genome will be challenging but, in our opinion, feasible. Moreover, the time complexity of the simple algorithm depends on σ^k, which is in practice (at least in our experiment with yeast) much smaller than the theoretical bounds we derived for it.

Acknowledgements

The authors would like to thank Mr. K. Kataja and Dr. R. Satokari from VTT Biotechnology. Mr. Kataja was the first one to bring the adapter primer selection problem to our attention and Dr. Satokari has advised us on the biotechnological issues. The discussions with Juha Kärkkäinen from the University of Helsinki and Jens Stoye, Sven Rahmann, and Sebastian Böcker from Bielefeld University led to better understanding of the problem. Especially we wish to thank Matthias Steinrücken from Bielefeld University, as he found a fundamental fault in one of the algorithms we had in an earlier version of this paper.

References

1. A. Amir, A. Apostolico, and M. Lewenstein. Inverse Pattern Matching. *J. Algorithms*, 24(2):325–339,1997.
2. A. Andersson, N. J. Larsson, and K. Swanson. Suffix trees on words. *Algorithmica*, 23(3):246–260, 1999.
3. B. Alberts, A. Johnson, J. Lewis, M. Raff, K. Roberts, and P. Walter. *Molecular Biology of the Cell, fourth edition*. Garland Science, 2002.
4. A. Apostolico. Pattern discovery and the algorithmics of surprise. *Artificial Intelligence and Heuristic Methods for Bioinformatics*, pp. 111–127, 2003.
5. L. Gąsieniec, P. Indyk, P. Krysta. External Inverse Pattern Matching. In *Proc. Combinatorial Pattern Matching 97 (CPM'97)*, Springer-Verlag LNCS 1264, pp. 90–101, 1997.
6. D. Gusfield. *Algorithms on strings, trees and sequences: Computer science and computational biology*. Cambridge University Press, 1997.
7. R. Karp and M. Rabin. Efficient randomized pattern-matching algorithms. *IBM Journal of Research and Development*, 31:249–260, 1987.
8. J. Kärkkäinen and E. Ukkonen. Sparse suffix trees. In *Proc. Second Annual International Computing and Combinatorics Conference (COCOON '96)*, Springer-Verlag LNCS 1090, pp. 219–230, 1996.
9. D. Knuth, J. Morris, and V. Pratt. Fast pattern matching in strings. *SIAM Journal on Computing*, 6(2):323–350, 1977.
10. J. Lanctot, M. Li, B. Ma, S. Wang, and L. Zhang. Distinguishing string selection problems. *Information and Computation*, 185(1):41–55, 2003.
11. E. M. McCreight. A space economical suffix tree construction algorithm. *Journal of the ACM*, 23, pp. 262–272, 1976.
12. P. Nicodème and J.-M. Steyaert. Selecting optimal oligonucleotide primers for multiplex PCR. In *Proc. of the 5th International Conference on Intelligent Systems for Molecular Biology (ISMB'97)*, pp. 210–213, 1997.
13. A. Shinohara, M. Takeda, S. Arikawa, M. Hirao, H. Hoshino, and S. Inenaga. Finding Best Patterns Practically. In *Progress in Discovery Science (Final Report of the Japanese Discovery Science)*, Springer-Verlag LNAI 2281, pp. 307–317, 2002.
14. E. Ukkonen. On-line construction of suffix trees. *Algorithmica*, 14:249–260, 1995.
15. J. Wang, B. Shapiro, and D. Shasha. *Pattern Discovery in Biomolecular Data.*, Oxford University Press, 1999.
16. P. Weiner. Linear pattern matching algorithms. In *Proc. IEEE 14th Annual Symposium on Switching and Automata Theory*, pp. 1–11, 1973.

Author Index

Lecture Notes in Bioinformatics